T0281569

An Invitation to
Abstract Algebra

Textbooks in Mathematics

Series editors:
Al Boggess, Kenneth H. Rosen

Functional Linear Algebra
Hannah Robbins

Introduction to Financial Mathematics
With Computer Applications
Donald R. Chambers, Qin Lu

Linear Algebra
An Inquiry-based Approach
Jeff Suzuki

The Geometry of Special Relativity
Tevian Dray

Mathematical Modeling in the Age of the Pandemic
William P. Fox

Games, Gambling, and Probability
An Introduction to Mathematics
David G. Taylor

Linear Algebra and Its Applications with R
Ruriko Yoshida

Maple™ Projects of Differential Equations
Robert P. Gilbert, George C. Hsiao, Robert J. Ronkese

Practical Linear Algebra
A Geometry Toolbox, Fourth Edition
Gerald Farin, Dianne Hansford

An Introduction to Analysis, Third Edition
James R. Kirkwood

Student Solutions Manual for Gallian's Contemporary Abstract Algebra, Tenth Edition
Joseph A. Gallian

Elementary Number Theory
Gove Effinger, Gary L. Mullen

Philosophy of Mathematics
Classic and Contemporary Studies
Ahmet Cevik

An Introduction to Complex Analysis and the Laplace Transform
Vladimir Eiderman

An Invitation to Abstract Algebra
Steven J. Rosenberg

https://www.routledge.com/Textbooks-in-Mathematics/book-series/CANDHTEX-BOOMTH

An Invitation to Abstract Algebra

Steven J. Rosenberg

CRC Press
Taylor & Francis Group
Boca Raton London New York

CRC Press is an imprint of the
Taylor & Francis Group, an **informa** business

A CHAPMAN & HALL BOOK

Cover: Evariste Galois viewed through a kaleidoscope. Original Galois portrait © Emily Koch; kaleidoscope software by Steve J. Rosenberg.

First edition published 2022
by CRC Press
6000 Broken Sound Parkway NW, Suite 300, Boca Raton, FL 33487-2742

and by CRC Press
2 Park Square, Milton Park, Abingdon, Oxon, OX14 4RN

CRC Press is an imprint of Taylor & Francis Group, LLC

Library of Congress Cataloging-in-Publication Data

ISBN: 9780367748616 (hbk)
ISBN: 9781032171784 (pbk)
ISBN: 9781003252139 (ebk)

DOI: 10.1201/9781003252139

Publisher's note: This book has been prepared from camera-ready copy provided by the authors

To my wife Eleni,
without whom I would have been eaten by wolves long
ago.

Contents

Preface

This book arose to fill what I perceived as a hole in the available resources for introductory abstract algebra courses. Having taught such a course several times, I wanted to provide an experience which did not assume that abstract algebra was a bitter pill to be watered down as much as possible, and at the other extreme did not assume that everyone in the class had already decided to become a mathematics professor themselves. Instead I wanted to bring students with me on a shared adventure of mutual awe-inspiring discovery, hoping to instill the desire to understand and discover more for themselves: to appreciate the subject for its own sake. For abstract algebra is a beautiful, profound, and useful subject which is part of the shared language of many areas both within and outside of mathematics.

This text assumes a background in high-school algebra, with some trigonometry and analytic geometry (e.g., Cartesian coordinates in the plane and the ability to calculate the distance between two points). Some familiarity with the properties of the ordinary integers, including divisibility, prime numbers, and congruences, would be helpful. In addition, a first course in calculus would be useful in understanding the material presented in Chapter 16 and beyond. Finally, some experience with mathematical reasoning would be beneficial, as this text takes a fairly rigorous approach to its subject, and expects the student to understand and create proofs as well as examples throughout. Some of these topics are presented briefly in Chapter 1, mainly for reference, although a reader with limited background eager to delve into abstract algebra may be able to use this chapter as a starting point. For the reader desiring a more thorough treatment of sets, functions, and proofs at an introductory level, the first 12 chapters of [8] are recommended (please note that numbers within square brackets refer to entries in the Bibliography; for example, [8] refers to *Book of Proof* by Hammack).

The book follows a single story arc, starting from humble beginnings with arithmetic and high-school algebra, gradually introducing abstract structures and concepts, and culminating with Niels Henrik Abel and Evariste Galois' great achievement in understanding how we can—and cannot—represent the roots of polynomials. Of course, the presentation benefits hugely from the many discoveries, generalizations, refinements, and simplifications that have occurred in the two centuries since Abel and Galois did their original work; and everything has been filtered through the author's own preferences and points of view. We will learn a lot along the way! The mathematically experienced reader may recognize my bias toward commutative algebra and fondness for number theory. I have tried to begin at the beginning, and to motivate each new topic with a specific question, until the subject has built up a momentum of its own. This approach has led to a fairly tight presentation, so I recommend proceeding in sequence through the chapters. Likewise, the exercises at the end of each chapter have been designed to support and extend the material in the preceding chapter, as well as prepare for the succeeding chapters. These exercises, with very few exceptions (and these have been noted), should be solvable by anyone who has completed the preceding material in the text.

The entire text can be covered in two or three 15-week semesters, depending on the number of instructional contact hours and the thoroughness of the in-class presentation. I found that I needed three semesters with four hours of meeting time each week to cover Chapters 2–22 completely, broken down as follows:

Semester 1	Chapters 2–11
Semester 2	Chapters 12–16
Semester 3	Chapters 17–22

At the least, Chapters 2–11 should provide sufficient content for a one-semester abstract algebra requirement, with coverage of the standard introductory topics in group theory and a glimpse of ring theory at the beginning (in the form of a number game) and end (when general rings are introduced). The two-semester sequence ends with the Fundamental Theorem of Galois Theory, which is proved for all finite Galois extensions in any characteristic. The student looking to undertake more advanced study should of course complete the entire book, including all of the exercises (and the projects in the final chapter). Some professional mathematicians may raise their eyebrows at my decision to include free groups, commutative diagrams, and universal properties so early in an introductory text (in Chapters 6, 7, and 17, respectively), as traditionally these topics have had to wait until graduate school. In fact, over the years, my undergraduate students have confirmed my feeling that this approach, far from presenting a technical barrier, is actually the most natural and easy to grasp.

The final chapter consists of a collection of projects, in which the reader is asked to supply nearly all of the proofs (in a sequence of "tasks"). These projects include stand-alone topics of specialized interest, such as kaleidoscopes and perfect numbers, as well as explorations of more standard topics such as Euclidean domains and power series, and a series of connected projects in linear algebra. All of the projects are designed to be accessible to the student who has completed the appropriate parts of the preceding chapters.

Abstract algebra is indeed a deep subject. It may require a year or more after a first course in the subject to fully absorb the material and perform an internal synthesis (I write this from personal experience). With this in mind, abstract algebra can transform not only the way one thinks about mathematics, but the way that one thinks—period.

Author

Steven J. Rosenberg is a professor in the Mathematics and Computer Science Department at the University of Wisconsin-Superior. He received his PhD from Ohio State University. As an educator, Dr. Rosenberg has both developed and taught a wide array of courses in mathematics and computer science. As a researcher, he has published results in the areas of algebraic number theory, cryptographic protocols, and combinatorial designs, among others. As a software developer, his clients included Coca-Cola Enterprises and the pension agency of Cook County Illinois. He has extensive experience in computer science and software engineering.

Acknowledgments

The author wishes to thank his former students: Aaron Anders, Josh Bentley, Ryan Bruner, Pavle Bulatovic, Armel Fetue, Joe Florestano, Dakota Jacobs, Nikola Kuzmanovski, Michael LaValley, Jounglag Lim, Zach Reiswig, Diwash Shrestha, Jeremy Syrjanen, and Hao Xu for providing corrections and suggestions in early drafts of this book; Chelsea Kowitz for giving me her notes from my lectures, on which most of the first ten chapters are based; Tom Lenosky, for suggesting that I might enjoy taking an abstract algebra class before I had declared a math major; and Herman and Minerva Katz for their unconditional support.

Symbols

Symbol	Description	Page
$\in$	is an element of	1
$\subseteq$	is a subset of	1
$\subset$	is a proper subset of	1
$|S|$	cardinality (size) of a set S	1
$|A|$	determinant of a square matrix A	45, 100
$|x|$	order of a group element x	77
$\mathbf{Z}$	the set of all integers	1
$\mathbf{Z}^+$	the set of all positive integers	1
$\mathbf{N}$	the set of all non-negative integers	2
$\mathbf{Q}$	the set of all rational numbers	2
$\mathbf{R}$	the set of all real numbers	2
$\mathbf{C}$	the set of all complex numbers	2
$\gcd(a, b)$	greatest common divisor of a and b	2, 235
$\gcd(A)$	greatest common divisor of a set A	235
$\operatorname{lcm}(a, b)$	least common multiple of a and b	2
$\notin$	is not an element of	2
$\nsubseteq$	is not a subset of	2
$\emptyset$	the empty set	2
:	such that	3
$|$	such that; divides	3; 114
$\cup$	union	4
$\cap$	intersection	4
$-$	set difference	4
$\times$	Cartesian product; componentwise product	4; 188
S^n	n^{th} Cartesian power of a set S	4
$f : D \to C$	f is a function from D to C	5
$f(x)$	output value of a function f at input x	6
$\mapsto$	elementwise mapping	6
id_S	identity function on a set S	6
$\circ$	composition of functions	6
$f(S)$	image of a set S under a function f	6
$f^{-1}(S)$	pre-image of a set S under a function f	7
f^{-1}	inverse function	7
$< \infty$	is finite	7
$f|_S$	restriction of a function f to a set S	7

Symbol	Description	Page
$\forall$	for all	8
$\exists$	there exists	8
$\implies$	implies	8
$\iff$	if and only if	8
$(f_i)_{i \in I}$	indexed collection	9, 195
$:=$ or $=:$	equals by definition	11
e (or e_G)	identity element of a group (named G)	23, 38
0	identity element of an abelian group	25
$\mathrm{Sym}(S)$	symmetric group on a set S	26
S_n	symmetric group on $\{1, 2, \ldots, n\}$	27
a^{-1}	inverse of group element a	30
a^n	n^{th} power of a in a group	31
$-a$	inverse of a in an abelian group	33
$n \cdot a$	n^{th} power of a in an abelian group	33
$!$	factorial	35
$\leq$	is a subgroup, subring, or subspace of	38, 103, 128
	or partial order on a set	296
$<$	is a proper subgroup, subring, or subspace of	38, 103, 128
$\nleq$	is not a subgroup of	38
$\langle S \rangle$	subgroup generated by a set S	41
	subspace spanned by S	128
$\langle s_1, \ldots, s_n \rangle$	subgroup generated by elements $s_1, \ldots, s_n$	41
P_n	regular n-sided polygon	48
D_{2n}	dihedral group of rigid symmetries of P_n	49
$\equiv$	is congruent to	54, 253
$\mathrm{Fr}(S)$	free group on a set S	56
$\ker(\sigma)$	kernel of σ	60, 104, 137
$\ln$	natural logarithm function	61
$\trianglelefteq$	is a normal subgroup of	64
zS	left coset of S	65
Sz	right coset of S	65
A/B	quotient object A mod B	67, 106
$\mathbf{Z}_n$	quotient group $\mathbf{Z}$ mod $n\mathbf{Z}$ under addition	68
$N_G(H)$	normalizer of a subgroup H in G	71
$\uplus$	disjoint union	73
$[A : B]$	index of B in A	76, 163
$\cong$	is isomorphic to	82, 137
$\hookrightarrow$	embeds in	87
$\mathrm{Aut}(A)$	group of automorphisms of A	88, 103
$\mathrm{Inn}(G)$	group of inner automorphisms of a group G	88
$Z(G)$	center of a group G	88
$\langle\langle S \rangle\rangle$	normal subgroup generated by a set S	92
G^{ab}	abelianization of a group G	94
1 or 1_R	identity element (of R) under multiplication	99

Symbol	Description	Page
$M_2(\mathbf{R})$	ring of all 2×2 matrices over $\mathbf{R}$	100
$\mathrm{GL}_n(F)$	general linear group of $n \times n$ matrices over F	100, 139
$C(R)$	center of a ring R	111
(S)	ideal generated by a set S	116
$(s_1, \ldots, s_n)$	ideal generated by elements $s_1, \ldots, s_n$	116
	or n-cycle in a symmetric group	213
$\dim_F(V)$	dimension of a vector space V over the field F	135
$M_{m,n}(F)$	ring of all $m \times n$ matrices over the field F	138
$M_n(F)$	ring of all $n \times n$ matrices over the field F	139
$R[x]$	polynomial ring in x with coefficients from R	141
$\deg(f)$	degree of a polynomial f	144
$\deg_v(f)$	degree of a polynomial f in the variable v	238
$\deg_{\mathrm{tot}}(f)$	total degree of a polynomial f	284
$f(\alpha)$	result of evaluating a polynomial f at α	145
ε_α	polynomial evaluation-at-α map	145
$R[\alpha]$	image of evaluation-at-α map, R adjoin α	146
$R[x_1, \ldots, x_n]$	multivariate polynomial ring over R	152, 238
$\mathrm{Irr}(\alpha, F, x)$	irreducible polynomial of α over F in x	155
$F[\rho_1, \ldots, \rho_n]$	F adjoin the elements $\rho_1, \ldots, \rho_n$	160
$\mathrm{Aut}(K/F)$	group of automorphisms of K over F	170
f'	derivative of a polynomial f	173
$\mathrm{char}(R)$	characteristic of a ring R	175
$\mathrm{Gal}(K/F)$	Galois group of K over F	178
K^H	fixed field of a group H in the field K	179
$\exists!$	there exists a unique	185
$\prod_{i=1}^n S_i$	repeated product	189
$\oplus$	direct sum	192
ϕ	Euler's totient function	194
$A \cdot B$	setwise product or ideal product	197, 233
$A + B$	setwise sum (in an abelian group)	197
$\sum_{i \in S} G_i$	repeated setwise sum	198
G_p	maximal p-subgroup of an abelian group G	200
kG	set of all k^{th} powers in an abelian group G	202
$\mathbf{Z}_p$	additive group of integers modulo p	203
$\mathbf{F}_p$	field of integers modulo p	203
$\|$	exactly divides	204
$\mathrm{exponent}(G)$	exponent of a group G	209
$\mathrm{Orb}_G(x)$	orbit of x under the action of a group G	212
$\mathrm{Stab}_G(x)$	stabilizer of x under the action of a group G	212
$R[U^{-1}]$	localization of a domain R at U	225
$k(R)$	field of fractions of a domain R	228
$F(x)$	field of rational functions over F in x	228
$>> 0$	is sufficiently large	230
$\exp_{\mathfrak{p}}(a)$	exponent of a prime ideal $\mathfrak{p}$ in a	234

Symbol	Description	Page
$\mathrm{cont}(f)$	content of a polynomial f	236
ζ_n	the primitive n^{th} root of unity $e^{2\pi i/n}$ in $\mathbf{C}$	250
$\Phi_n(x)$	n^{th} cyclotomic polynomial	251
$\mathcal{N}_{K/F}$	norm map from K down to F	260
$\det(M)$	determinant of the square matrix M	308
$O_n(F)$	orthogonal group of $n \times n$ matrices over F	324
$SO_n(F)$	special orthogonal group of $n \times n$ matrices over F	324
$\mathrm{Ann}_R(M)$	annihilator of the R-module M	359
$\mathrm{Tor}_R(M)$	torsion subset of the R-module M	361

1

Review of Sets, Functions, and Proofs

1.1 Sets

Of all the things studied in mathematics, the most fundamental is the *set*. Although the notion of sets as the basis for thinking about math was only introduced relatively recently (for mathematics!) by Georg Cantor in the 1870s, sets now occupy a central position at the heart of mathematics.

A set is to be thought of as a collection of things, called the *elements* of the set. Elements of a set can be any type of thing: numbers, functions (to be defined later), or even other sets. Two sets are considered equal when they have precisely the same elements. We use the ordinary equality symbol to indicate that two sets are equal, as in "$S = T$."

Notation 1.1. Let S be a set. The statement

$$x \in S,$$

read "x is in S" or "x in S," means that x is an element of the set S. The statement

$$T \subseteq S,$$

read "T is a subset of S," means that each element of T is also an element of S. The statement

$$T \subset S,$$

read "T is a proper subset of S," means that T is a subset of S and $T \neq S$.

Definition 1.2. The *cardinality* of a set is the number of distinct elements in the set. We denote the cardinality of the set S by $|S|$. We also speak of cardinality as *size*.

Definition 1.3. A *singleton* set is a set of cardinality 1.

1.1.1 Some Special Sets of Numbers

We have special notation for certain of the most important sets of numbers:

$\mathbf{Z}$ is the set of all *integers*, or whole numbers; we use the notation $\mathbf{Z}^+$ to denote the set of all *positive integers*;

DOI: 10.1201/9781003252139-1

N is the set of all *natural numbers*, or non-negative integers (note that we include 0 as an element of **N**, but some other authors do not);

Q is the set of all *rational numbers*, i.e., numbers which are ratios of two integers, where the denominator is not zero;

R is the set of all *real numbers*; and finally,

C is the set of all *complex numbers*.

As the first and most basic set of numbers in our list above, we remind the reader of some of the properties of **Z**. Given two integers a and b, we say that b *divides* a, written $b \mid a$, if there is some integer c such that $a = b \cdot c$, where $\cdot$ is ordinary multiplication.

The *greatest common divisor* of a and b, written $\gcd(a, b)$, is the largest integer d such that both $d \mid a$ and $d \mid b$; the gcd exists if $a \neq 0$ or $b \neq 0$.

The *least common multiple* of a and b, written $\text{lcm}(a, b)$, is the smallest positive integer m such that both $a \mid m$ and $b \mid m$; the lcm exists if $a \neq 0$ and $b \neq 0$.

An integer p is called *prime* if $p \geq 2$, and there is no integer a such that $1 < a < p$ and $a \mid p$. Two positive integers a and b are called *relatively prime* if we have $\gcd(a, b) = 1$.

1.1.2   Describing a Set

There are four common styles used to describe a set:

(1) We can list all the elements of the set inside of "curly braces," { and }. This style is used for small sets for which it is convenient (or at least possible!) to actually write down (symbols for) all the elements. We refer to this style as *explicit* description.

Example 1.4. Let $S = \{0, 7, 14, 21, 28, 35, 42, 49\}$. Then S is a set consisting of 8 distinct elements, so we have $|S| = 8$. We can describe S in words by saying that S is the set of all integers between 0 and 50, inclusive, which are integer multiples of 7. It is correct to write $21 \in S$ and $42 \in S$, for example. Since 3 is *not* an element of S, we indicate this fact by writing $3 \notin S$. It is correct to write $\{0, 21, 42\} \subseteq S$ and also to write $\{0, 21, 42\} \subset S$. However, since $\{0\} \notin S$ (all of the elements of S are *numbers*, not sets), the set $\{\{0\}\}$ is *not* a subset of S, which we can indicate by writing $\{\{0\}\} \nsubseteq S$.

Example 1.5. It is possible to have a set with no elements at all. There is only one such set (check the definition of when two sets are considered to be equal to each other, given above). This set is called the *empty set*, and denoted $\emptyset$. It is correct to write $\emptyset = \{\}$, which says that the empty set has no elements. However, note that $\emptyset \neq \{\emptyset\}$. We consider the empty set to be a thing itself (as opposed to nothing!), and so the set $\{\emptyset\}$ does have an element, namely $\emptyset$. Thus, $\emptyset \in \{\emptyset\}$ but $\emptyset \notin \emptyset$, so the two sets $\emptyset$ and $\{\emptyset\}$ are not equal. Note, however, that $\emptyset \subseteq \emptyset$. In fact, the empty set is a subset of *every* set.

(2) We may list several elements of the set, inside of curly braces, until a pattern is obvious, and then write the *ellipses* symbol "$\ldots$" (three dots),

followed by the final elements of the set (if the set is finite). In case the pattern also continues in the opposite direction, we write ellipses *before* the first explicitly listed elements as well. The reader should beware of possible ambiguity in using style (2): use this style sparingly if it all, and only use it when the pattern of set elements is quite clear.

Example 1.6. Let $U = \{0, 7, 14, \ldots, 49\}$. Then we have $U = S$, where S is the set from Example 1.4. At least, this is the most obvious interpretation of what happens in the omitted section of our description of U; there could possibly be other interpretations!

Example 1.7. We have
$$\mathbf{N} = \{0, 1, 2, 3, \ldots\}$$
and
$$\mathbf{Z} = \{\ldots, -2, -1, 0, 1, 2, \ldots\}.$$
Notice that the number of elements we choose to write explicitly is up to us; we just want to be sure that the pattern of elements is obvious.

(3) We can write a set in the *implicit* style
$$\{x \in S \; : \; C\}$$
or
$$\{x \in S \mid C\},$$
where:

x is a simple variable name (we may use x itself, or any other letter or symbol not already used for another purpose);

S is some already-known set; and

C is some *condition* involving the variable x.

In this context, ":" and "|" are each read "such that." The entire expression is read "the set of all x in S such that C" or "the set of all x in S satisfying C."

We note that the variable used inside the set description (called x here) is a so-called *dummy variable*: this means that it has no meaning outside of the curly braces describing this set; it is only a place-holder which is to be thought of as running through all elements of the given set S.

Example 1.8. We can describe the set of all natural numbers $\mathbf{N}$ as follows:
$$\mathbf{N} = \{n \in \mathbf{Z} \; : \; n \geq 0\}.$$

Example 1.9. We can describe the set S in Example 1.4 as
$$S = \{n \in \mathbf{N} \; : \; 7 \mid n \text{ and } n \leq 50\}.$$

Here, we have the compound condition "$7 \mid n$ and $n \leq 50$." We remind the reader that "$7 \mid n$" is read "7 divides n," which means that there is some integer m such that $n = 7 \cdot m$; in other words, 7 is a factor of n, or n is a multiple of 7 (by some *integer*). Notice that in this case, the symbol "|" means "divides," and not "such that." In general, we prefer to use the colon ":" for "such that," since the vertical bar, "|," has so many other uses (however, the colon is also used to mean something else, in function notation; see below).

(4) We can describe a set as the collection of all things of a certain form: all "formulas" whose variables come from certain specified sets, and which satisfy certain given conditions:

$$S = \{(\text{formula}) \ : \ (\text{conditions})\}.$$

Example 1.10. We can write the set $\mathbf{Q}$ of all rational numbers as

$$\mathbf{Q} = \{a/b \ : \ a, b \in \mathbf{Z} \text{ and } b \neq 0\}.$$

Example 1.11. Notice that style (3) can be replaced by style (4). For example, we can write the set S from Example 1.4 using style (4) as

$$S = \{n \ : \ n \in \mathbf{N}, 7 \mid n, \text{ and } n \leq 50\}.$$

1.1.3   Operations on Sets

There are several commonly used methods to produce a set from two or more given sets.

The *union* of two sets A and B is

$$A \cup B = \{x \ : \ x \in A \text{ or } x \in B\}.$$

The *intersection* is defined by

$$A \cap B = \{x \ : \ x \in A \text{ and } x \in B\}.$$

The *difference* (or *set difference*) is defined by

$$A - B = \{x \ : \ x \in A \text{ and } x \notin B\}.$$

The *Cartesian product* of two sets A and B is

$$A \times B = \{(a, b) \ : \ a \in A \text{ and } b \in B\}.$$

Here, we call (a, b) an *ordered pair* whose first component is a and whose second component is b. When we take the Cartesian product of a set with itself, we may use exponential notation: thus, A^2 is simply shorthand for $A \times A$.

Example 1.12. Two sets S and T are called *disjoint* if they have no elements in common. With set operations, we can write this condition as $S \cap T = \emptyset$.

Example 1.13. A point P in the ordinary Cartesian plane can be written using two coordinates x and y as $P = (x, y)$, where x and y are real numbers. The set S of all points in the Cartesian plane is thus

$$S = \{(x, y) \ : \ x, y \in \mathbf{R}\}.$$

But now we can realize that this set is merely the Cartesian product of $\mathbf{R}$ with itself:

$$S = \mathbf{R}^2.$$

So we can say that the Cartesian plane *is* $\mathbf{R}^2$. Extending this idea, we can write the set of all points in ordinary three-dimensional space as $\mathbf{R}^3$.

Remark 1.14. In the notation $\mathbf{R}^3$, we can ask whether we mean $A = (\mathbf{R} \times \mathbf{R}) \times \mathbf{R}$ or instead $B = \mathbf{R} \times (\mathbf{R} \times \mathbf{R})$. Formally, these are two different sets; an element of A looks like $((x, y), z)$, while an element of B looks like $(x, (y, z))$. However, in such situations we will agree to omit the inner parentheses and treat A and B as equal, so $A = B = \mathbf{R}^3 = \{(x, y, z) : x, y, z \in \mathbf{R}\}$.

The operations of union and intersection may be applied to more than two sets at a time; we define the union and intersection of any collection of sets, as follows:

Definition 1.15. Let $\mathfrak{C}$ be a collection of sets (that is, a set whose elements are sets). Then

$$\bigcup_{S \in \mathfrak{C}} S = \{x : \text{there exists } S \in \mathfrak{C} \text{ such that } x \in S\}$$

and

$$\bigcap_{S \in \mathfrak{C}} S = \{x : \text{for all } S \in \mathfrak{C}, x \in S\}.$$

1.2 Functions

Informally, a function is a rule which tells us how to process an input of a specified kind to produce an output. More formally, a function f from a set D to a set C is a way to unambiguously produce an element of C from any given element of D. We write

$$f : D \to C$$

to indicate that f is a function which accepts inputs which are elements of the set D and produces outputs which are elements of the set C. Here, D is called the *domain* of f and C is called the *codomain* of f. We read "$f : D \to C$" as "f is a function from D to C" or "f from D to C."

Even more formally, we can group each input and the corresponding output of a function f together in a single ordered pair, and consider f itself to be the set of all such ordered pairs:

Definition 1.16. A *function* f with domain D and codomain C is a set

$$f \subseteq D \times C$$

such that for every $x \in D$, there is a unique $y \in C$ for which $(x, y) \in f$.

Notation 1.17. Let f be a function. To write $f(x) = y$ means that $(x, y) \in f$. We may also write $f : x \mapsto y$ for $f(x) = y$.

Remark 1.18. The condition that for every $x \in D$ there is a unique $y \in C$ for which $(x, y) \in f$ simply says that for every possible input x, there is a unique output y determined by x using the function f; i.e, a unique y such that $f(x) = y$.

Remark 1.19. In the definition of a function given above, the codomain C is the least essential part. In fact, while the domain D of f is uniquely determined by f, the codomain is not. We are always free to replace C by any set which contains (at minimum) all the output values of our function. We note, however, that in some parts of mathematics it is convenient to include the codomain as part of the information about a function, i.e., to define our function to be the ordered pair (f, C) instead of just f.

Remark 1.20. Notice that, as promised, sets are the most basic objects around. Even a function was defined to be a special type of set. Actually, the formal definition of a function (as given above) says that a function f is exactly what we are used to calling the *graph* of f: namely, the set of all points (x, y) where $f(x) = y$.

If $f : D \to C$ is a function, then we say that f *maps* D to C, or that f is a *mapping* from D to C. Similarly, we read $f : x \mapsto y$ as "f maps x to y."

For each set, there is a distinguished function on that set which "does nothing"; or more accurately, which maps every element of the domain to itself, thus revealing the identity of each element:

Definition 1.21. Let S be a set. The *identity function* on S is the function

$$id_S : S \to S$$

defined by the formula

$$id_S(x) = x$$

for every x in S.

In case the codomain of one function is the domain of another function, it makes sense to perform *two* mappings of an input from the first function's domain:

Definition 1.22. Suppose that $f : A \to B$ and $g : B \to C$ are functions. Then the *composition of g with f* is the function $g \circ f : A \to C$ defined by the formula $(g \circ f)(x) = g(f(x))$ for every $x \in A$.

If $f(x) = y$, then we say that y is the *image* of x (under f). We also define images and pre-images of sets:

Definition 1.23. Let $f : D \to C$ be a function. For any set $A \subseteq D$, we define the image of A under f to be

$$f(A) = \{f(x) : x \in A\} \subseteq C.$$

For any set $B \subseteq C$, we define the pre-image of B under f to be

$$f^{-1}(B) = \{x \in D : f(x) \in B\} \subseteq D.$$

We distinguish three properties which a function may possess.

Definition 1.24. Let $f : D \to C$ be a function.

We say that f is *one-to-one*, or *injective*, if no two elements of D map to the same point of C. More formally, f is injective means that for all $x_1, x_2 \in D$, if $x_1 \neq x_2$ then $f(x_1) \neq f(x_2)$.

We say that f is *onto*, or *surjective*, if every element of C is the image of some element of D. More formally, f is surjective if for every $y \in C$, there is some $x \in D$ such that $f(x) = y$.

We say that f is *bijective*, or that f is a *bijection*, if f is both injective and surjective.

Remark 1.25. Two sets have the same cardinality precisely if there is a bijection from one to the other. In fact, this may be taken as a definition of what it means for two sets to have the same cardinality.

Remark 1.26. A set S is called *finite* if there is a natural number n such that there exists a bijection from S to $\{1, 2, \ldots, n\}$. (Notice that this latter set is empty if $n = 0$.) To indicate that S is finite, we sometimes write $|S| < \infty$ or $|S| = n < \infty$.

Remark 1.27. If $f : D \to C$ is a bijection, then there is a unique function $g : C \to D$ such that $f \circ g = id_D$ and $g \circ f = id_C$. We call g the *inverse* of f, and we write $g = f^{-1}$. This may cause some confusion with the notation for pre-images (introduced above). Even if f is a bijection, however, the notation is still unambiguous; just remember that we only take pre-images of *subsets* of the codomain, whereas the inverse function takes *elements* of the codomain as inputs. Also, in case $B \subseteq C$ and f is bijective, then the set $f^{-1}(B)$ is the same whether we interpret it as the pre-image of B under f or as the image of B under f^{-1}.

If $f : D \to C$ is a function and $S \subseteq D$, then f lets us compute a value in C for any given input from S. By limiting ourselves to input values from S, we get another function, which is technically different from f, because its domain is S instead of D:

Definition 1.28. Let $f : D \to C$ be a function, and let $S \subseteq D$. The *restriction of f to S* is the function

$$h : S \to C$$

defined by the formula

$$h(x) = f(x) \text{ for all } x \text{ in } S.$$

In this situation we may also say that f *extends* h, especially if we began with h and found f afterwards.

Notation 1.29. The restriction of f to S is denoted $f|_S$, read "f restricted to S."

1.3 Proofs

A *proof* is a chain of reasoning which uses definitions, logic, and previously established results in order to establish a new result. Ideally, every step in a proof should clearly follow from the previous steps, and should be justifiable as just indicated above.

One of the central skills necessary to succeed in higher mathematics is the ability to translate back and forth between informal and formal language. The informal language (usually a "natural" language, such as English) is valuable because it expresses our ideas in an easy-to-read and easy-to-absorb form, and also lets us express ideas when we are still not ready to express ourselves precisely. A more formal language (such as the symbolic language of logic) is very important because it lets us express our ideas very precisely, without the possibility of misinterpretation or ambiguity. The reader should keep in mind that when doing math, both points of view are useful; try to understand each new concept and each statement at both the informal level—the level of intuitive meaning—and the formal level, where it can be expressed most precisely. Much of doing math consists of "unpacking" formal statements to discover their meaning, then forming a strategy, and finally implementing the strategy back on the formal side of mathematics.

1.3.1 Logic

Here we only recall the four most important symbols from logic which have not yet been introduced in this text:

"$\forall$" is called the *universal quantifier*, and is read "for all" or "for every."

"$\exists$" is called the *existential quantifier*, and is read "there exists."

" $\implies$ " is called the *implication* or *conditional* symbol, and is read "implies."

" $\iff$ " is called the *biconditional* symbol, and is read "if and only if," sometimes abbreviated "iff."

In an informal context, or outside of formal set descriptions, we also use "s.t." to stand for "such that."

Example 1.30. We can write the definitions of injectivity and surjectivity formally, using the language of symbolic logic. Let $f : D \to C$ be a function. Then f is injective iff

$$\forall x_1, x_2 \in D, \ x_1 \neq x_2 \implies f(x_1) \neq f(x_2).$$

The definition of injectivity is often more convenient to work with in the logically equivalent form

$$\forall x_1, x_2 \in D, \ f(x_1) = f(x_2) \implies x_1 = x_2.$$

We can say that f is surjective iff

$$\forall y \in C, \ \exists x \in D \text{ s.t. } f(x) = y.$$

Example 1.31. We can formally state the definition of when two sets S and T are equal, as follows:

$$S = T \iff \forall x, (x \in S \iff x \in T).$$

Usually we break the right-hand statement down into two parts: $\forall x \in S, x \in T$ and $\forall x \in T, x \in S$. Note in particular that a set only "knows" whether or not a given object x is an element of it; being an element of a set is a yes-or-no property. There is no way to say that x occurs 42 times in the set S, for example. Likewise, elements of a set have no ordering by themselves; a set is an unordered collection of elements. If we want order, or repetition, then we need to use other constructions. One way to get these additional features is to use a Cartesian product; for example, in an ordered pair $(a, b) \in S^2$, we can say that a is the first component and b is the second component, and we may have $a = b$. Another option is given in the following example.

Example 1.32. Often, it is convenient use elements of some set I as labels or subscripts. Formally, let $f : I \to S$ be a function. Then we sometimes use the notation f_i instead of $f(i)$, and $(f_i)_{i \in I}$ instead of f. In this situation, we call I the *index set*, and an element of I is called an *index*. The function f itself is called an *indexed collection*, and we may also refer to f as the *indexing function*.

A familiar special case occurs when $I = \mathbf{Z}^+$ and $S = \mathbf{R}$. Let $a : \mathbf{Z} \to \mathbf{R}$ be a function. For every positive integer i, we have $a_i \in \mathbf{R}$. We usually refer to $(a_i)_{i \in \mathbf{Z}^+}$ as a *sequence* of real numbers. Notice that there may be repetitions among the a_i, and there is no way to recover the number of such repetitions or the order of the sequence from the set $\{a_i \ : \ i \in \mathbf{Z}^+\}$ of the a_i values alone.

As another example, consider the set $\mathfrak{C}$ of all closed intervals in $\mathbf{R}$ of length 2. For example, we have $[4, 6] \in \mathfrak{C}$, and $[\pi, \pi + 2] \in \mathfrak{C}$. If we know the left-hand endpoint of such an interval, then that interval is uniquely determined; this left endpoint is a real number. Therefore, it makes sense to use $\mathbf{R}$ as an index set here. So for each $x \in \mathbf{R}$ we let

$$C_x = [x, x + 2] = \{y \in \mathbf{R} \ : \ x \le y \le x + 2\},$$

and then observe that

$$\mathfrak{C} = \{C_x \ : \ x \in \mathbf{R}\}.$$

In this example, the indexing function C is one-to-one, and we may be content to use the subscript notation C_x without the rest of the notation for an indexed collection, since we only care about the resulting set $\mathfrak{C}$.

In the present text, we shall not delve deeply into the twin foundational subjects of set theory and mathematical logic; there are many books devoted to each of these subjects, which the interested reader may examine. But the reader should be aware that these subjects exist, and that they form the basis for what we may accept as "obvious" facts. For example, the substitution principle, which states that we may replace any expression with another so long as the two expressions are equal, belongs to mathematical logic. We make free use of the substitution principle. On the other hand, the historically controversial Axiom of Choice, which is equivalent to the statement that any non-empty collection of non-empty sets has a non-empty (Cartesian) product, turns out to have some very non-obvious consequences; we use this axiom less frequently, and make some effort to call attention to it when it is needed.

1.3.2   Proof Conventions

1. To prove a *universal statement*

$$\forall x \in A, P(x)$$

   start by choosing an *arbitrary* element of A:

   Let $x \in A$.

   Your goal is now to prove $P(x)$.

   Common mistakes: choosing x to have special properties instead of being arbitrary. Also, choosing x to be a compound expression (as opposed to a single variable symbol), as in "Let $A \cap E \in \mathcal{F}$."

2. To prove an *existential statement*

$$\exists x \in A \text{ s.t. } P(x)$$

   find a *specific* element x of A with the property $P(x)$.

   Common mistakes: leaving some details of x unspecified (as variables).

3. To prove an *implication*

$$P \implies Q$$

   start by assuming P:

   Suppose P.

   Your goal is now to prove Q.

   Sometimes it is convenient to prove the implication $P \implies Q$ by proving the logically equivalent statement (not Q) $\implies$ (not P), which is called the *contrapositive* of the original implication.

4. To prove a *biconditional statement*

$$P \iff Q$$

   break the proof into two sub-proofs: (1) Prove $P \implies Q$, (2) Prove $Q \implies P$.

5. To prove a *subset relationship*

$$A \subseteq B$$

convert this statement to the logically equivalent form

$$\forall x \in A, x \in B$$

and prove this statement (using Proof Convention (1)).

6. To prove an *equality of sets*

$$A = B$$

break the proof into two sub-proofs: (1) Prove $A \subseteq B$, (2) Prove $B \subseteq A$.

7. Always introduce a new variable (say x) by writing either

$$\text{Let } x = \ldots$$

or

$$\exists x \in A \text{ s.t. } \ldots$$

If you use the second method, then you need to justify the existence of x somehow. Sometimes when we introduce a new variable, we write "$x := \ldots$" or "$\ldots =: x$"; here, $:=$ and $=:$ both mean "is equal to by definition." Notice that the colon : always goes next to the new variable.

Common mistakes: using a variable before it is properly introduced; also, using a dummy variable outside its proper context, as in, "$\forall x \in A, P(x)$. So x satisfies ..." Here, x is a dummy variable that only makes sense inside the universal statement, not outside it in the second sentence. Notice that, according to our rule, x has *not* been introduced as a variable!

We briefly recall the main proof techniques here:

1. Direct Proof: In this technique, we start with what we are given (our hypotheses) and proceed directly to our goal.

2. Proof by Contradiction: To prove a statement P by contradiction, we start by writing

$$\text{Assume for a contradiction that } P \text{ is false.}$$

We proceed until we derive a statement which is clearly false, and conclude by writing

$$\text{This contradiction shows that } P \text{ is true.}$$

3. (a) Proof by Induction: This proof technique applies to statements
    of the form
    $$\forall n \in \mathbf{N}, P(n),$$
    where $P(n)$ is a statement which depends on the natural number n
    (note: any well-ordered set may be substituted for $\mathbf{N}$ here). To use
    induction, we must prove

    (i) $P(0)$ is true, *and*
    (ii) $\forall n \in \mathbf{N}, (P(n) \implies P(n+1))$.

    We first write

    Base Case: $n = 0$.

    and proceed to prove that $P(0)$ is true. Next we write

    Inductive Step: Suppose that $P(n)$ is true. (We call $P(n)$ the *in-
    ductive hypothesis*.)

    Our goal is now to prove that $P(n+1)$ is also true.

    (b) Proof by Strong Induction: As in ordinary induction, strong
    induction may be used to prove statements of the form

    $$\forall n \in \mathbf{N}, P(n).$$

    But in strong induction, our inductive hypothesis is (apparently)
    stronger: namely, $P(0)$ and $P(1)$ and $\cdots$ and $P(n)$. Thus the in-
    ductive step is (sometimes) easier when strong induction is used.

It is often useful to perform calculations and do specific examples in the
course of attempting to prove a general result. Similarly, it is a good idea to
keep track of one's current goal in a proof, and of the strategy to be used.
This "scratchwork" seldom appears in the final, clean version of a proof which
is printed in books. However, in order to give the reader an idea of how
and why a proof flows the way it does, we have taken the license to include
scratchwork inside of some of the proofs in this text. Such work is enclosed in
square brackets, [and], so it will not be confused with the proof itself.

1.4   How to Read This Book

Like any mathematics text, this book should be read slowly and carefully, with
plenty of paper at hand to take notes, and especially to work out and verify
the claims made in the text. One very good habit to form is to *pronounce*
everything as you read it (even if this takes place in your head): as you learn
each new symbol and new notation, make sure to also learn how to pronounce
it, and practice each time you see it. The alternative leads to skipping ahead

every time you encounter math notation, which is clearly not a good idea! And this is only the first step, of course. Make sure to work out all the details of each argument of a proof, of each new definition, and of each discussion, until *you* are satisfied with it. Take ownership of the math as you go.

Reading a mathematics text for understanding is a slow and sometimes difficult process. Math is denser than many subjects—just as gold is denser than air.

2

Introduction: A Number Game

2.1 A Game with Integers

Imagine a game that is played as follows. To set up the game, you gather a collection of scraps of paper—an *infinite* collection—and label each scrap with an integer. We require that each integer appears on one and only one scrap of paper. Then, while you aren't looking, your friend puts each scrap into its own separate bag.

There is no way for you to see inside the bags, and the bags look identical to each other from the outside; they are indistinguishable. But you are allowed to perform addition and multiplication on the bags: you may hand over (or otherwise point out) any two bags to your friend, and ask for the sum or product. Your friend will look inside the bags for you, add or multiply the enclosed numbers as requested, and retrieve for you the bag which contains the result. You may even ask for the sum or product of a bag with itself.

For example, we may hand over the bag containing 7 and the bag containing 4, and ask for their sum. Our friend will peek inside these bags, search for the bag containing 11—this may take a while!—and hand it over to us. But our friend never tells us the values of any of the numbers inside.

The goal of the game is to identify the contents of the bags. (More precisely, we would like to find a procedure which eventually identifies the contents of any given bag.)

Is there any strategy you can use to win this game? The answer is yes! To see this, think about numbers which have a special relationship to the operations of addition and multiplication. The leading candidates are 0 and 1.

Zero is the only integer which has no effect when added to another integer. If you can just find some pair of bags Z and X such that

$$Z + X = X,$$

then you will know that bag Z contains the number 0. By systematically searching for such a pair, you will eventually be able to find and label the bag Z containing 0.

After discovering 0, you can take advantage of the properties of the number 1 under multiplication. If you can find two bags U and Y such that

$$U \cdot Y = Y,$$

DOI: 10.1201/9781003252139-2

making sure that bag Y is *not* Z, then bag U must contain 1. Again, after enough attempts, you will be able to identify the bag U.

One final search is needed to find the bag containing -1. One way to do this is to look for a bag N such that

$$N + U = Z.$$

At this point, no more trial-and-error is required. You can add U to itself to obtain 2's bag, and keep adding U to identify the bags of all the positive integers. Similarly, starting with N, you will be able to take care of all negative integers. Thus we can discover the contents of all the bags!

For reference, let us speak of the number game we have just finished discussing as the number game on $\mathbf{Z}$ (recall that $\mathbf{Z}$ denotes the set of all integers).

2.2   A Bigger Game

Now that we know how to win the number game on $\mathbf{Z}$, let's make things more interesting by throwing in the irrational number $\sqrt{2}$. The number $\sqrt{2}$ gets its own scrap of paper (and its own bag), and you, the player, are allowed to add or multiply bags as before.

But there is a complication here: we cannot simply enlarge our collection of integer bags with one additional bag containing $\sqrt{2}$. For your friend would not be able to respond if you asked for the sum, say, of the bags containing $\sqrt{2}$ and 1. What is the smallest collection of numbers that we need to include before we can play our new and improved game?

First, observe that any number of the form $b\sqrt{2}$, where $b \in \mathbf{Z}$, needs to be included. Adding such numbers to ordinary integers, we see that any number of the form

$$a + b\sqrt{2}$$

(where a is an integer) also needs to be included. The unhappy prospect looms of having to repeat this procedure of multiplying, adding, multiplying, and adding without end. But remarkably, we can stop at this stage. For it is not hard to check (Exercise 2.2) that the sum or product of any two numbers of this form is again of the same form. Thus, we replace the set $\mathbf{Z}$ of our first game with the set

$$R := \{a + b\sqrt{2} \mid a, b \in \mathbf{Z}\} \tag{2.1}$$

to play our new game. We put each number from R in its own bag, and again the goal is to determine the contents of the bags.

The same reasoning as before will show that you can identify each ordinary integer within its own bag. For, even in this bigger set R, the numbers 0 and 1 still enjoy their unique properties with respect to addition and multiplication.

Now if you could only identify the bag containing $\sqrt{2}$, you would be able to finish and win the game (see Exercise 2.3).

One way you might think of to identify $\sqrt{2}$ is to use the fact that if bag T contains 2 and bag X contains $\sqrt{2}$, then

$$X \cdot X = T. \tag{2.2}$$

Since T can be found (2 is an integer!), we just need to look for such an X.

But there is a problem: the number $-\sqrt{2}$ has the same property! The two numbers $\sqrt{2}$ and $-\sqrt{2}$ are the only elements of R which satisfy equation 2.2, so your task boils down to distinguishing between these two numbers. How can you tell apart $\sqrt{2}$ from $-\sqrt{2}$? At this point, the curious reader is invited to ponder this question before reading on.

It turns out that there is a fundamental problem in distinguishing $\sqrt{2}$ from $-\sqrt{2}$: it is actually *impossible* to identify them, no matter how clever you are or how much time you have!

To convince ourselves of this, consider the following thought experiment. Suppose that you start playing the game one day, and at the end of the day, after having identified and labeled some of the bags, you go home. During the night, unbeknownst to you, your friend does some mischief: your friend switches the contents of the bags, replacing the scrap of paper labeled $\sqrt{2}$ with that labeled $-\sqrt{2}$, $1 + \sqrt{2}$ with $1 - \sqrt{2}$, and in general, $m + n\sqrt{2}$ with $m - n\sqrt{2}$ for arbitrary integers m and n.

The next day, you resume play, and start by checking your work of the previous day. One calculation at a time, you repeat the computations of yesterday's work. Suppose that, on the first day, bag A contained $a + \alpha\sqrt{2}$ and bag B contained $b + \beta\sqrt{2}$. Asking for their sum, you would be handed the bag C containing $(a + b) + (\alpha + \beta)\sqrt{2}$. On the second day, what will happen when you recalculate this sum? Since bag A will then contain $a - \alpha\sqrt{2}$ and B will contain $b - \beta\sqrt{2}$, when you compute $A + B$, you will be handed the bag containing $(a + b) - (\alpha + \beta)\sqrt{2}$. But this number is now inside bag C, so you will get the same result as on the first day.

Even more remarkably, computing the product of A and B gives the same result on the second day as on the first day: on the first day, the result is the bag D containing

$$(a + \alpha\sqrt{2}) \cdot (b + \beta\sqrt{2}) = (ab + 2\alpha\beta) + (a\beta + \alpha b)\sqrt{2},$$

while on the second day the result is the bag E containing

$$(a - \alpha\sqrt{2}) \cdot (b - \beta\sqrt{2}) = (ab + 2\alpha\beta) - (a\beta + \alpha b)\sqrt{2}.$$

But the bags D and E are the same!

Thus, the results of *all* computations we could possibly do with our bags will be the same on both days. If there were a way you could be sure that a certain bag X contained $\sqrt{2}$ on the first day, then the exact same calculations that led you to that conclusion, but performed on the second day, would lead you to conclude that this same bag X contains $\sqrt{2}$ again on day two. But on day two, X would contain $-\sqrt{2}$, and not $\sqrt{2}$. This contradiction shows that there can be no procedure which allows you do decide which bag contains $\sqrt{2}$.

2.3   Concluding Remarks

Our first number game, using the set of integers $\mathbf{Z}$, is winnable. We interpret this result as telling us that $\mathbf{Z}$ is completely determined by how its elements behave under the two operations of addition and multiplication.

The second game, using the set R, is not winnable; this tells us that R is *not* completely determined by its two operations of $+$ and $\cdot$. The key idea in establishing this fact was to consider switching $m + n\sqrt{2}$ and $m - n\sqrt{2}$. More formally, we can consider the function

$$f : R \to R$$

given by

$$f(m + n\sqrt{2}) = m - n\sqrt{2} \tag{2.3}$$

for $m, n \in \mathbf{Z}$. The reader may recognize the formula above from high school algebra: $m - n\sqrt{2}$ is sometimes called the *conjugate* of $m + n\sqrt{2}$, and it is useful, for example, to rationalize denominators in fractions of the form

$$\frac{a}{m + n\sqrt{2}}.$$

What makes the function f so useful in analyzing our game is that you cannot tell the difference between the bags before and after applying f; there is no way to tell that a switch was made. We can express this in a different way by saying that f is a *symmetry* of R. Algebraically speaking, f has a certain relationship (to be explored in the exercises) with the operations of addition and multiplication, and this relationship gives f its amazing usefulness.

Stepping back a bit further, these number games may start to give us a new perspective on number systems such as $\mathbf{Z}$ and R. At the start of a game, all the numbers are hidden inside of bags: we have removed any identifying information about the numbers. At this point in the game, there would be no way to tell apart two different number systems unless their *sizes* were different. Expressed another way, at the start of a game, we have stripped away all features of our number system except for the one feature it possesses as a mere *set*: namely, its size.

When play commences, we can take advantage of the additional features of the number system: namely, its *operations*. Using known properties of these operations, such as the existence of the additive identity element 0 and the multiplicative identity 1, and the rules for addition and multiplication, we can determine much about the hidden contents of the bags. In the case of $\mathbf{Z}$, we can determine the contents completely.

In traditional "high school" algebra, we start with a given number system such as $\mathbf{Z}$ or $\mathbf{R}$ and study its properties with respect to the usual operations of addition, subtraction, multiplication, etc. We then use these properties to solve equations and inequalities containing one (or more) unknown number. In *abstract algebra*, we do not start with a particular system such as the integers or the real numbers. Instead, we start with an unknown set and an unknown operation, and we specify the properties of the operation. (Later, we may allow two operations, or even more general extra structures on our set.) We proceed to determine the consequences of these properties; a typical question is, to what degree of uniqueness do these properties determine our set and operation?

This more abstract approach to algebra allows us to generalize the notion of a number system and to isolate those properties that are most important, together with the consequences of those properties. We can thus expand our vocabulary beyond that of number systems. Abstract algebra also sheds light on number systems themselves. Indeed, the power of abstract algebra is so great that it provides elegant, unifying proofs of previously known results and allows us to solve problems which can be formulated in terms of classical algebra and geometry, but which had defied solution for centuries. These are the hallmarks of a successful mathematical theory.

2.4 Exercises

Exercise 2.1. Describe a way of identifying the bag containing -1 which is different from the method used in the text.

Exercise 2.2. Let R be as in Equation 2.1. Prove that for all x and y in R, we have both $x + y \in R$ and $x \cdot y \in R$. (We say that R is *closed* under addition and multiplication.)

Exercise 2.3. Let R be as in Equation 2.1. In the number game for R, explain how you could identify the contents of all the bags if you could just determine which bag contained $\sqrt{2}$.

Exercise 2.4. A student wishes to consider playing the number game on the set of integers with the rational number $\frac{1}{2}$ thrown in. Imitating the example in the text, this student defines the set of numbers

$$S := \left\{ a + \frac{b}{2^n} \;\middle|\; a, b, n \in \mathbf{Z} \right\}.$$

(a) Suppose that the student attempts to define a function $f : S \to S$ by the formula

$$f\left(a + \frac{b}{2^n}\right) = a - \frac{b}{2^n} \text{ for } a, b, n \text{ in } \mathbf{Z}.$$

What is wrong with this definition of f?

(b) Find a description of S which is simpler than the one given at the beginning of this exercise. Prove that your description is correct.

(c) Prove that S is closed under addition and multiplication: that is, do Exercise 2.2 with S in place of R. (It will likely be convenient to use the description of S from part (b).)

(d) Is the number game on S winnable? Justify your answer.

Exercise 2.5. Let R be as in Equation 2.1.

(a) Prove that for all x in R, there are *unique* integers a and b such that $x = a + b\sqrt{2}$.

(b) Use part (a) of this exercise to explain why the definition of the function f given in Equation 2.3 makes sense. (We say that f is *well-defined* here.) Compare to Exercise 2.4 part (a).

Exercise 2.6. Consider playing the number game on some number system V. Here, V could be any set of numbers which is closed under addition and multiplication. Suppose that $g : V \to V$ is a bijective function.

(a) Suppose that whenever someone replaces x by $g(x)$ inside the bags, for all x in V, then there is no way for the player to tell the difference. Show that we must then have

$$g(x + y) = g(x) + g(y) \tag{2.4}$$

and

$$g(x \cdot y) = g(x) \cdot g(y) \tag{2.5}$$

for all x and y in V.

(b) Verify that the function f on R defined by Equation 2.3 satisfies both Equations 2.4 and 2.5 from part (a) of this exercise.

(c) Prove that there are exactly two functions on R which satisfy both equations, namely, the function f defined by equation 2.3 and the identity function on R.

3

Groups

3.1 Introduction

No one would dispute the usefulness of number systems. Our first task in abstract algebra is to *generalize* the concept of a number system by isolating their most important properties. We then declare that any system which has a certain subset of these specific properties is to be called a *group* or *ring* or *field*, etc., according to which subset of the properties it satisfies.

3.2 Binary Operations

From an algebraic point of view, what constitutes a "number system" such as **Z** is not just the *set* **Z**, but, even more importantly, the *operations* addition (+) and multiplication (·) Let us focus on addition. What kind of mathematical object is +?

To start with, + (on **Z**) is something that takes two integers as input and produces a single integer as output. This suggests that + may be a *function*. If so, what can its domain be? A single element of the domain of + must somehow provide us with *two* arbitrary integers.

We recall that elementary set theory has a device to capture two elements of a set in a single entity: namely, the Cartesian product. An element of $\mathbf{Z} \times \mathbf{Z}$ is an ordered pair (a, b) of integers. Thus we want to say that + is actually a function

$$+ \; : \; \mathbf{Z} \times \mathbf{Z} \to \mathbf{Z}.$$

In fact, this description of + is entirely accurate if we are only willing to allow the usual function notation

$$+(\,(a,b)\,)$$

to be written in the alternative format

$$a + b.$$

DOI: 10.1201/9781003252139-3

Thus, for example, the equation

$$3 + 4 = 7$$

can be rewritten as

$$+(\ (3, 4)\ ) = 7.$$

We say that $+$ is a *binary* operation on $\mathbf{Z}$ since it requires *two* inputs from $\mathbf{Z}$. Next we generalize this discussion by defining what we mean by a "binary operation" on an arbitrary set.

Definition 3.1. Let S be a set. A *binary operation on S* is a function

$$\triangle : S \times S \longrightarrow S.$$

Remark 3.2. In the definition above, S can be *any* set. S is not necessarily a set of numbers. We can and will allow S to be a set whose elements are numbers, or "grids" of numbers (known as *matrices*), or functions, for example.

Remark 3.3. Our notation for the output values of a binary operation will follow the example of the familiar operations of addition and multiplication of numbers. Instead of the usual function notation $\triangle(\ (a, b)\ )$ or its shorthand $\triangle(a, b)$, we use the "operator notation" $a \triangle b$.

Remark 3.4. We often use the symbol $\cdot$ ("dot") to denote a binary operation, as our symbol of choice. Sometimes we refer to $a \cdot b$ as a "product," and we may even refer to the operation $\cdot$ as "multiplication," even though $\cdot$ may have nothing to do with ordinary multiplication of numbers. Later, we will usually omit a symbol altogether, and simply write ab for $a \cdot b$. We often use $+$ ("plus" or "additive" notation) if our operation is *commutative* (see Definition 3.11 below).

Now that we have a general notion for the *kind* of thing that addition and multiplication of numbers are (namely, binary operations), we recall the fundamental properties of these familiar binary operations, so that we may define these properties, too, in generality. The properties that we wish to consider are associativity, commutativity, the existence of an identity element, and the existence of inverses.

Example 3.5. The *associative* property of $\mathbf{Z}$ under addition can be written as

$$\forall x, y, z \in \mathbf{Z},\ (x + y) + z = x + (y + z).$$

To define what it means for a general binary operation to be associative, we only need to replace $\mathbf{Z}$ and $+$ by a general set and a general binary operation on that set:

Definition 3.6. Let $\cdot$ be a binary operation on a set S. Then $\cdot$ is *associative* if

$$\forall x, y, z \in S,\ (x \cdot y) \cdot z = x \cdot (y \cdot z).$$

Notice in particular that the expressions $(x \cdot y) \cdot z$ and $x \cdot (y \cdot z)$ make sense: for $x \cdot y$ is an element of S, and so we can dot it with z.

We can also define the other properties of interest to us in generality:

Definition 3.7. An *identity element* for a binary operation $\cdot$ on a set S (also called an *identity element of S under $\cdot$*) is an element e of S such that

$$\forall x \in S, e \cdot x = x \cdot e = x.$$

Remark 3.8. We sometimes describe the defining property of an identity element informally by saying that an identity element is "absorbed" by other elements.

Definition 3.9. Let $\cdot$ be a binary operation on a set S which possesses an identity element e. An *inverse* of an element a of S (with respect to $\cdot$) is an element b of S such that

$$a \cdot b = b \cdot a = e.$$

Remark 3.10. The definition above is symmetric in a and b. Thus we can say that b is an inverse of a iff a is an inverse of b. Also, it appears that our definition of inverse depends on which identity element we choose, but we shall see later (Lemma 3.33) that there can be at most one identity element for a given binary operation.

Definition 3.11. Let $\cdot$ be a binary operation on a set S. Then $\cdot$ is *commutative* if

$$\forall x, y \in S, \ x \cdot y = y \cdot x.$$

Example 3.12. Let $S = \{a, b\}$ be a set with two elements, $a \neq b$. To form a binary operation $\cdot$ on S, we are free to decide what each of the "products" $a \cdot a$, $a \cdot b$, $b \cdot a$, and $b \cdot b$ should be—as long as each product is again in S. Some binary operations on S will be associative, while others will not be; and likewise for the other properties we defined above. Here is one way to define a binary operation $\cdot$ on S: $a \cdot a = b$, $a \cdot b = b$, $b \cdot a = b$, and $b \cdot b = a$. We observe that this operation is commutative, since $a \cdot b = b \cdot a = b$. However, our operation $\cdot$ is not associative, because $a \cdot (b \cdot b) = a \cdot a = b$, but $(a \cdot b) \cdot b = b \cdot b = a$. (See Example 3.25 for a "multiplication table" format for writing down a binary operation.)

3.3 Groups: Definition and Some Examples

Now it is time to define our first new type of object to study in depth.

Definition 3.13. A *group* is a set G together with a function $\cdot$, satisfying the following 4 axioms:

G0 · is a binary operation on G.

G1 The operation · is associative.

G2 G contains an identity element for the operation ·.

G3 Every element of G has an inverse in G.

Remark 3.14. By default, the symbol e will denote the identity element of a group (and not, for example, Euler's constant, which is the natural logarithm base). Sometimes we will use the notation e_G for e when we wish to indicate the group in question. Also, there are instances where other notation will be used for the identity element of a group, as dictated by circumstances and convention; these exceptions will be noted as they come up.

Remark 3.15. Formally, our group is the ordered pair $(G, \cdot)$. We call G the *underlying set* of the group, and we refer to · as the *group operation* or *group law*. If the group operation is understood, we often speak of G alone as a "group," even though the operation is more important than the underlying set G. The same set G may be a group in many different ways: G may admit many different group laws!

Remark 3.16. Notice that we did *not* require the commutative property in our list of axioms for a group. A group operation need not be commutative. This is important, because many algebra rules which are true for numbers fail in non-commutative groups. In general, we shall have to re-learn the rules of algebra from scratch for each new type of object we study, starting now with groups.

Definition 3.17. A group is called *commutative* or *abelian* (rhymes with "chameleon") if its operation is commutative.

Remark 3.18. When we say that a group is commutative, we mean that *all* pairs of elements in the group commute. We may also say that two particular elements commute (i.e., with each other), without requiring our group to be commutative. However, we never say that a pair of elements is "abelian": the term *abelian* is only used in a global sense, to mean that the group is commutative.

Generally speaking, the more properties (axioms) we require to be satisfied, the easier the resulting objects are to understand. In this vein, abelian groups are easy to work with compared to groups in general. We shall see instances of this fact below (see, for example, Warning 3.43 and Lemma 3.44).

Example 3.19. We claim that $(\mathbf{Z}, +)$ is a group. Let us verify the group axioms:

G0: Adding two integers gives an integer, so $+$ is a binary operation on $\mathbf{Z}$.

G1: Addition on $\mathbf{Z}$ is associative, as we have recalled earlier.

G2: We must find an integer e such that $e + x = x = x + e$ for every integer x. The choice $e = 0$ works, so 0 is an identity element for $(\mathbf{Z}, +)$. The reader may recall that 0 is sometimes called the *additive identity*.

G3: Let $x \in \mathbf{Z}$. We must show the existence of an integer y such that $x + y = 0 = y + x$. The choice $y = -x$ has the desired property. Again, the reader may have seen $-x$ referred to as the *additive inverse* of x.

Remark 3.20. For an abelian group, the group operation will frequently be denoted by $+$ ("addition"), even if our group is not an ordinary number system such as $\mathbf{Z}$. We also frequently use the symbol 0 instead of e to denote the identity element in an abelian group.

Example 3.21. The reader may verify that $(\mathbf{R}, +)$ is a group. The details are similar to those in the preceding example.

Example 3.22. In this example, let $\cdot$ be ordinary multiplication on $\mathbf{R}$. This operation is associative, and possesses an identity element, namely 1. Attempting to check Axiom **G3**, we need to solve the equation $xy = 1$ for an arbitrary real number x. The solution is $y = 1/x$. Unfortunately, this solution does not make sense when $x = 0$. Moreover, when $x = 0$, the equation $xy = 1$ has no solution. Therefore, 0 has no inverse under multiplication, and so $(\mathbf{R}, \cdot)$ is not a group.

But let's try to salvage the situation. Since the number 0 was the only problem, let us remove it! First, notice that when $\cdot$ is restricted to non-zero real numbers, we still get a binary operation (on $\mathbf{R} - \{0\}$), because the product of two non-zero real numbers is non-zero. Multiplication is still associative on this smaller set, and 1 still acts as an identity element. Also, if $x \in \mathbf{R} - \{0\}$ then $1/x \in \mathbf{R} - \{0\}$. Therefore $(\mathbf{R} - \{0\}, \cdot)$ is a group!

Example 3.23. Consider the set $S := \{\frac{1}{2}, 1, 2\}$ under multiplication. We know that multiplication of real numbers is associative. Further, 1 is an identity element for multiplication; and for each number in S, its multiplicative inverse is also in S. Yet S is *not* a group under multiplication. The reason is that multiplication is not an operation on S! For example, $2 \in S$, but $2 \cdot 2 \notin S$.

To get a group from S, we need to include all the integer powers of 2. The reader can verify that the set $\{2^n \mid n \in \mathbf{Z}\}$ is a group under multiplication; see also Example 4.11.

Example 3.24. Let $S = \{\alpha\}$ be a set with one element. Then there is only one possible binary operation $\cdot$ on S, namely, $\alpha \cdot \alpha = \alpha$. This operation makes S into a group! Of course, we have $e = \alpha$ here. A group with only one element is called a *trivial* group.

Example 3.25. Let $G = \{a, b, c\}$, and define a binary operation $\cdot$ on G by the following table:

$\cdot$	a	b	c
a	a	b	c
b	b	c	a
c	c	a	b

We interpret this table as follows. To compute the value of $x \cdot y$, we read the entry in the row labeled by x and the column labeled by y. For instance, we have $c \cdot b = a$, because the row labeled by c is the third row, the column labeled by b is the second column, and the entry in the third row, second column is a. Let us see what is involved in checking that $(G, \cdot)$ is a group.

G1: Checking associativity requires 27 verifications. This is because we must choose 3 arbitrary elements of G, with repetition allowed, where order matters; for example, choose b, b, and c. Each of these 27 verifications requires 4 computations using the table; in our example, we must check whether $b \cdot (b \cdot c) = (b \cdot b) \cdot c$, and this requires 4 "dot" computations. Therefore, 108 computations are required in all.

G2: We claim that a is an identity element for $(G, \cdot)$. Checking this only requires 5 verifications.

G3: The reader can verify that an inverse for a is a, an inverse for b is c, and an inverse for c is b. This requires at most 6 "dot" computations, with only 3 required if you take advantage of the symmetry in the definition of *inverse*. The 3 relevant computations are: $a \cdot a = a$, $b \cdot c = a$, and $c \cdot b = a$.

It turns out that $\cdot$ is associative, and so $(G, \cdot)$ is a group. But clearly we should not rely too heavily on brute-force computation or tables to understand group operations in general: even in this tiny example, well over 100 computations would be needed just to verify that G is a group!

Definition 3.26. A table whose rows and columns are labeled by the elements of a group G, such that the entry in row x and column y is the product $x \cdot y$, is called a *group table* or *Cayley table* for G.

Next we introduce an important family of groups called *symmetric groups*. First we need a preliminary definition.

Definition 3.27. Let S be a non-empty set. A *permutation of S* is a bijective function $f : S \to S$.

Definition 3.28. Let S be a non-empty set. Then

$$\text{Sym}(S) := \{f \mid f \text{ is a permutation of } S\}.$$

(Here, "Sym" stands for "symmetric" and is pronounced "sim.")

We would like to make $\text{Sym}(S)$ into a group by finding a suitable binary operation. Since the elements of $\text{Sym}(S)$ are *functions*, we look for an operation that produces a function from two given functions. A natural candidate is the operation of composition of functions. Recall that if we have two functions $\sigma : A \to B$ and $\tau : B \to C$ (so that the codomain of σ equals the domain of τ), then we can form their composition $\tau \circ \sigma : A \to C$ by using the formula $(\tau \circ \sigma)(x) = \tau(\sigma(x))$ for all x in A. Because of the following lemma, we call $\text{Sym}(S)$ the *symmetric group on S*.

Lemma 3.29. *If S is a non-empty set, then $(\text{Sym}(S), \circ)$ is a group, where $\circ$ is composition of functions.*

Proof. We verify each of the group axioms.

G0: [Check: $\forall \sigma, \tau \in \text{Sym}(S), \sigma \circ \tau \in \text{Sym}(S)$]

Suppose that $\sigma, \tau \in \text{Sym}(S)$. Then $\sigma, \tau : S \to S$, so $\sigma \circ \tau$ is defined, and is a function from S to S. It is not hard to show that the composition of two bijections is a bijection (Exercise 3.6), so we have $\sigma \circ \tau \in \text{Sym}(S)$, as desired.

G1: [We must show: $\forall f, g, h \in \text{Sym}(S), (f \circ g) \circ h = f \circ (g \circ h)$]

Let $f, g, h \in \text{Sym}(S)$. [Show: $(f \circ g) \circ h = f \circ (g \circ h)$]

[Two functions are equal if they have the same values at all points of their common domain. So we must show:

$$\forall x \in S, ((f \circ g) \circ h)(x) = (f \circ (g \circ h))(x). \ ]$$

Let $x \in S$. Then $((f \circ g) \circ h)(x) = (f \circ g)(h(x)) = f(g(h(x)))$ and $(f \circ (g \circ h))(x) = f((g \circ h)(x)) = f(g(h(x)))$, so $((f \circ g) \circ h)(x) = (f \circ (g \circ h))(x)$. Therefore, $\circ$ is associative.

G2: We need to find a permutation e of S such that

$$e \circ f = f \circ e = f$$

for every permutation f of S. In particular, for each permutation f and each x in S we need to have $e(f(x)) = f(x)$. Since permutations are surjective, if we take any element y of S, there is some x in S such that $f(x) = y$. This forces $e(y) = y$ for all y in S. So set $e = \text{id}_S$, the identity function on S. One can verify (Exercise 3.7) that this choice of e works.

G3: [We must show:

$$\forall f \in \text{Sym}(S) \ \exists g \in \text{Sym}(S) \text{ s.t. } f \circ g = g \circ f = e = \text{id}_S.$$

This should remind us of the properties of an *inverse function*. Strategy: show that f^{-1} exists, is in $\text{Sym}(S)$, and satisfies the equations for g above.]

Let $f \in \text{Sym}(S)$. Since f is a bijection from S to S, then f has an inverse function $f^{-1} : S \to S$. Now f^{-1} is also bijective (Exercise 3.8), so $f^{-1} \in \text{Sym}(S)$. We have $f \circ f^{-1} = f^{-1} \circ f = \text{id}_S$, so f^{-1} is an inverse of f in the group sense as well as the function sense, and we are done. $\square$

Notation 3.30. In the special case when $S = \{1, 2, 3, \ldots, n\}$ for some positive integer n, we denote the symmetric group on S by S_n. That is,

$$S_n := \text{Sym}(\{1, 2, 3, \ldots, n\}). \tag{3.1}$$

Example 3.31. For an element $f \in S_n$, let us use the *array notation*

$$f = \begin{bmatrix} 1 & 2 & 3 & \cdots & n \\ f(1) & f(2) & f(3) & \cdots & f(n) \end{bmatrix}$$

to represent f. For instance, let $\sigma = \begin{bmatrix} 1 & 2 & 3 & 4 \\ 3 & 1 & 4 & 2 \end{bmatrix}$ and let $\tau =$

$\begin{bmatrix} 1 & 2 & 3 & 4 \\ 4 & 3 & 1 & 2 \end{bmatrix}$. Then σ is the function $\sigma : \{1,2,3,4\} \to \{1,2,3,4\}$ such that $\sigma(1) = 3$, $\sigma(2) = 1$, $\sigma(3) = 4$, and $\sigma(4) = 2$. Let $\alpha = \sigma \circ \tau$, the product of σ with τ in S_4. Then we have $\alpha(1) = (\sigma \circ \tau)(1) = \sigma(\tau(1)) = \sigma(4) = 2$. The reader should check that, in array notation, we have

$$\alpha = \begin{bmatrix} 1 & 2 & 3 & 4 \\ 2 & 4 & 3 & 1 \end{bmatrix}.$$

That is, we have

$$\begin{bmatrix} 1 & 2 & 3 & 4 \\ 3 & 1 & 4 & 2 \end{bmatrix} \circ \begin{bmatrix} 1 & 2 & 3 & 4 \\ 4 & 3 & 1 & 2 \end{bmatrix} = \begin{bmatrix} 1 & 2 & 3 & 4 \\ 2 & 4 & 3 & 1 \end{bmatrix}.$$

Notice that we can compute a product of two elements like this in a symmetric group without ever leaving array notation. For instance, here we compute $1 \mapsto 4$ and $4 \mapsto 2$ to get $1 \mapsto 2$ in the result. Also notice that we end up computing this product "from right to left," which may seem counterintuitive until we remember that functions are composed by applying the right-hand function first.

3.4   First Results about Groups

In this section, we start to "tame" groups by proving that some of the familiar rules of algebra for numbers still hold true for groups. Just as importantly, we shall also see that some results which are true for number systems are *not* true for groups in general; in some cases, these failed results can still be shown to hold provided we are dealing with *abelian* groups.

As usual in mathematics, we should be careful not to make unspoken assumptions or make up our own rules when dealing with the algebra of groups. Our default position should be that nothing is true unless we can prove it!

Our first job will be to make sense of an expression like

$$a_1 \cdot a_2 \cdot a_3 \cdots \cdot a_n, \tag{3.2}$$

where the a_i's are elements of some group. We frequently write down such expressions with numbers: e.g.,

$$1 + 2 + 3 + 4. \tag{3.3}$$

In fact, our familiarity with ordinary arithmetic is perhaps so great that we may not see any problem here at all. What's the big deal?

Definition 3.38. Let G be a group, let $a \in G$, and let n be a positive integer. Then

$$a^n := \underbrace{a \cdot a \cdot \ldots \cdot a}_{n \text{ factors}},$$

$$a^{-n} := \underbrace{a^{-1} \cdot a^{-1} \cdot \ldots \cdot a^{-1}}_{n \text{ factors}},$$

$$a^0 := e.$$

Remark 3.39. Now that we have both multiplicative and exponential notation, it is natural to ask whether an expression such as $a \cdot b^2$ means $a \cdot (b^2)$ or instead $(a \cdot b)^2$. We follow the same order of operations as in traditional algebra, so powers have higher precedence than multiplication, and parentheses have the highest precedence of all. Thus $a \cdot b^2$ means $a \cdot (b^2)$. Later, when we study objects with two operations, both $+$ and $\cdot$, we will continue to abide by the standard order of operations.

Some of the rules for ordinary exponentiation of numbers are still true in arbitrary groups. Before reaching our next theorem, which states two such rules, we prove a lemma.

Lemma 3.40. *Let a be an element of a group, and let $n \in \mathbf{Z}$. Then $(a^n)^{-1} = a^{-n}$.*

Proof. Case 1: $n > 0$.

By Remark 3.36, it is enough to verify that $a^n \cdot a^{-n} = e$. When $n = 1$, this is immediate from the definition of a^{-1}. For $n > 1$, we have

$$a^n \cdot a^{-n} = \underbrace{a \cdot a \cdots a}_{n \text{ factors}} \cdot \underbrace{a^{-1} \cdot a^{-1} \cdots a^{-1}}_{n \text{ factors}}$$

$$= \underbrace{a \cdot a \cdots a}_{n-1 \text{ factors}} \cdot a \cdot a^{-1} \cdot \underbrace{a^{-1} \cdot a^{-1} \cdots a^{-1}}_{n-1 \text{ factors}}$$

$$= \underbrace{a \cdot a \cdots a}_{n-1 \text{ factors}} \cdot e \cdot \underbrace{a^{-1} \cdot a^{-1} \cdots a^{-1}}_{n-1 \text{ factors}}$$

$$= \underbrace{a \cdot a \cdots a}_{n-1 \text{ factors}} \cdot \underbrace{a^{-1} \cdot a^{-1} \cdots a^{-1}}_{n-1 \text{ factors}}.$$

We can continue to "cancel" the middle two terms until we reach the identity element, e, and we are done. More properly, our argument gives a proof by induction on n.

Case 2: $n < 0$.

Since $-n > 0$, Case 1 above gives $(a^{-n})^{-1} = a^{-(-n)} = a^n$. By Remark 3.36, we conclude that $(a^n)^{-1} = a^{-n}$, as desired.

Case 3: $n = 0$.

Then $a^n = a^0 = e$ and $a^{-n} = a^0 = e$, by Definition 3.38. But $e \cdot e = e$ by definition of identity, so we have $e^{-1} = e$, and we are done. $\square$

Theorem 3.41 (Laws of Exponents). *Let G be a group, $a \in G$, and $m, n \in \mathbf{Z}$. Then*

(i) $a^m \cdot a^n = a^{m+n}$ *and*

(ii) $(a^m)^n = a^{mn}$

Proof. Case 1: $m, n > 0$.

Then for (i) we have

$$a^m \cdot a^n = \underbrace{(a \cdot a \cdot ... \cdot a)}_{m \text{ factors}} \cdot \underbrace{(a \cdot a \cdot ... \cdot a)}_{n \text{ factors}} = \underbrace{a \cdot a \cdot ... \cdot a}_{m+n \text{ factors}} = a^{m+n}.$$

For (ii), we have

$$(a^m)^n = \underbrace{a^m \cdot a^m \cdot ... \cdot a^m}_{n \text{ factors of } a^m} = \underbrace{a \cdot a \cdot ... \cdot a}_{mn \text{ factors}} = a^{mn}.$$

Case 2: $m, n < 0$. Left to the reader in exercise 3.12.

Case 3: $m > 0$, $n < 0$.

[Idea: $a^5 \cdot a^{-2} = a^{3+2} \cdot a^{-2} = a^3 \cdot a^2 \cdot a^{-2} = a^3 \cdot e = a^3$. But this is justifiable based on our current knowledge only when $m + n > 0$. Otherwise, we model our proof on the following calculation: $a^2 \cdot a^{-5} = a^2 \cdot a^{-2-3} = a^2 \cdot a^{-2} \cdot a^{-3} = e \cdot a^{-3} = a^{-3}$.]

For (i), we first assume that $m + n > 0$ to get

$$
\begin{aligned}
a^m \cdot a^n &= a^{m+n-n} \cdot a^n \\
&= a^{m+n} \cdot a^{-n} \cdot a^n \quad &&\text{(by Case 1, since } m + n > 0 \text{ and } -n > 0) \\
&= a^{m+n} \cdot e \quad &&\text{(by Lemma 3.40)} \\
&= a^{m+n} \quad &&\text{(by definition of identity).}
\end{aligned}
$$

Next we prove (i) assuming that $m + n < 0$:

$$
\begin{aligned}
a^m \cdot a^n &= a^m \cdot a^{-m+m+n} \\
&= a^m \cdot a^{-m} \cdot a^{m+n} \quad &&\text{(by Case 2, since } -m < 0 \text{ and } m + n < 0)^{\bullet} \\
&= e \cdot a^{m+n} \quad &&\text{(by Lemma 3.40)} \\
&= a^{m+n} \quad &&\text{(by definition of identity).}
\end{aligned}
$$

If $m + n = 0$, then $m = -n$, so $a^m \cdot a^n = a^{-n} \cdot a^n = e = a^0 = a^{m+n}$.

This completes the proof of (i) in Case 3.

Now, for (ii): Let $b = a^m$. By Lemma 3.40, we have $b^n = (b^{-n})^{-1}$. On the other hand, we have $b^{-n} = (a^m)^{-n} = a^{-mn}$ by Case 1 above, since $m > 0$ and $-n > 0$. So $(a^m)^n = b^n = (b^{-n})^{-1} = (a^{-mn})^{-1} = a^{mn}$, as desired.

Case 4: $m < 0, n > 0$; Case 5: $m = 0$; Case 6: $n = 0$: Left to the reader in exercise 3.12. $\square$

Warning 3.42. We do not attempt to define a^r when a is a group element and $r \notin \mathbf{Z}$.

Warning 3.43. In general for group elements a and b, we may have $(a \cdot b)^n \neq a^n \cdot b^n$. Equality does, however, hold in an *abelian* group:

Lemma 3.44. *Let G be an abelian group and let $a, b \in G$. Then for all $n \in \mathbf{Z}$, we have*

$$(a \cdot b)^n = a^n \cdot b^n. \tag{3.4}$$

Proof. We prove the statement for $n \geq 0$, by induction on n. The case $n < 0$ is left to the reader as Exercise 3.13.

Base Case: $n = 0$. In this case, Equation 3.4 is true since $e \cdot e = e$.

Inductive Step: Suppose that $n \geq 0$ and that Equation 3.4 holds. [We must show: $(a \cdot b)^{n+1} = a^{n+1} \cdot b^{n+1}$]

Then

$$
\begin{aligned}
(a \cdot b)^{n+1} &= (a \cdot b)^n \cdot (a \cdot b) && \text{(by Theorem 3.41)} \\
&= a^n \cdot b^n \cdot a \cdot b && \text{(by inductive hypothesis)} \\
&= a^n \cdot a \cdot b^n \cdot b && \text{(since G is abelian and $b^n, a \in G$)} \\
&= a^{n+1} \cdot b^{n+1} && \text{(by Theorem 3.41).}
\end{aligned}
$$

$\square$

Notation 3.45 (Additive Notation). In an abelian group whose operation is written as $+$, we use a different notation for denoting inverses and exponents. Namely, we denote the inverse of a by $-a$ instead of a^{-1}; and we denote repeated addition using multiplicative notation, writing $n \cdot a$ instead of a^n. This "additive notation" for exponentiation thus conforms to our usage in ordinary number systems, and is especially important when we have a set with two different operations (for instance, both $+$ and $\cdot$).

We present one more result below which can help us solve equations in groups.

Lemma 3.46 (Cancellation Laws). *Let G be a group and let $a, b, c \in G$. Then:*
(i) If $a \cdot c = b \cdot c$, then $a = b$; and
(ii) If $c \cdot a = c \cdot b$, then $a = b$.

Proof. (i): Suppose that $a \cdot c = b \cdot c$. Then

$$
\begin{aligned}
(a \cdot c) \cdot c^{-1} &= (b \cdot c) \cdot c^{-1} \\
a \cdot (c \cdot c^{-1}) &= b \cdot (c \cdot c^{-1}) \\
a \cdot e &= b \cdot e \\
a &= b
\end{aligned}
$$

The proof of (ii) is similar. $\square$

Remark 3.47. The proof of the preceding lemma is probably more important than the result itself. Notice that it matters very much that c is on the right of the dot on both sides of equation (i), and likewise c is on the left of the dot on both sides of (ii). We recommend that the reader avoid thinking of "cancelling," but instead remember to apply the inverse to both sides of an equation, and to do so on the *same* side of the operation.

We conclude this chapter with some examples showing how to apply our knowledge of the algebra of groups. From now on, we shall make free use of the properties of inverses of group elements, of the laws of exponents, and of the absorption property of the identity element. We shall also find it convenient to omit the operation symbol (e.g. $\cdot$) sometimes; it should be understood to be present.

Example 3.48. Suppose that x, y, and z are elements of a group. Solve the following equation for x:

$$z \cdot x \cdot y^2 = z^2 \cdot y^5.$$

Solution:

$$
\begin{aligned}
zxy^2 &= z^2 y^5 \\
z^{-1} \cdot zxy^2 &= z^{-1} \cdot z^2 y^5 \\
xy^2 &= zy^5 \\
xy^2 \cdot y^{-2} &= zy^5 \cdot y^{-2} \\
x &= zy^3.
\end{aligned}
$$

Example 3.49. Suppose that x, y, and z are elements of a group. Solve the following equation for x:

$$zxy^2 = y^5 z^2.$$

Bad Solution:

$$
\begin{aligned}
\not{z}xy^2 &= y^5 z^{\not{2}1} \\
xy^2 &= y^5 z \\
x\not{y}^2 &= y^{\not{5}3} z \\
x &= y^3 z.
\end{aligned}
$$

Good Solution:

$$
\begin{aligned}
zxy^2 &= y^5 z^2 \\
z^{-1} \cdot zxy^2 &= z^{-1} \cdot y^5 z^2 \\
xy^2 &= z^{-1} y^5 z^2 \\
xy^2 \cdot y^{-2} &= z^{-1} y^5 z^2 \cdot y^{-2} \\
x &= z^{-1} y^5 z^2 y^{-2}.
\end{aligned}
$$

We cannot simplify this last expression (unless for example our group is abelian). Notice how careless use of "cancelling" led to an incorrect solution in our first attempt. We cannot cancel from the left and from the right at the same time!

3.5   Exercises

Exercise 3.1. Let $S = \{a, b\}$ with $a \neq b$.

(a) Write out the "multiplication tables" for *all possible* binary operations on S. How many are there? Can you give a general formula for the number of binary operations on a finite set as a function of the size of the set?

(b) Find at least one binary operation on S which possesses an identity element, and one which does not.

(c) Find at least one binary operation on S which is associative, and one which is not.

(d) Find at least one binary operation on S which is commutative, and one which is not.

(e) Find at least one binary operation on S which is a group law, and one which is not.

Exercise 3.2. Let G be a group with exactly 2 elements (we say that G has *order* 2). Thus we may write $G = \{e, x\}$ with $x \neq e$, where, as usual, e is the identity element of G.

(a) Can $x^{-1} = e$? Why or why not?

(b) What must x^{-1} be?

(c) Write out the group table for G.

(d) Prove that every group of order 2 is abelian.

Exercise 3.3. Prove that if e is the identity element of a group, then $e^n = e$ for all n in $\mathbf{Z}$. Suggestion: first use induction to prove the result for $n \geq 0$.

Exercise 3.4. Let $S = \left\{ \begin{bmatrix} a & b \\ c & d \end{bmatrix} : a, b, c, d \in \mathbf{R} \right\}$. Define a binary operation $\cdot$ on S by $\begin{bmatrix} a & b \\ c & d \end{bmatrix} \cdot \begin{bmatrix} u & v \\ w & x \end{bmatrix} = \begin{bmatrix} au + bw & av + bx \\ cu + dw & cv + dx \end{bmatrix}$.

(a) Prove that $\cdot$ is associative.

(b) Find an element $e \in S$ such that $m \cdot e = e \cdot m = m$ for all $m \in S$.

(c) Is $\cdot$ commutative?

Exercise 3.5 Let G be a group and let $a, b \in G$. Suppose that $b \cdot a = a^{-1} \cdot b$. Prove by induction on k that for all $k \in \mathbf{N}$, we have $b \cdot a^k = a^{-k} \cdot b$. Then extend this formula to all $k \in \mathbf{Z}$.

Exercise 3.6. Prove that if $\sigma : A \to B$ and $\tau : B \to C$ are two bijective functions, then the composition $\tau \circ \sigma : A \to C$ is also bijective.

Exercise 3.7. Prove that for any non-empty set S, the identity function on S is an identity element of $\mathrm{Sym}(S)$ under composition of functions.

Exercise 3.8. Prove that the inverse function of a bijection is also a bijection.

Exercise 3.9. Let $G = S_3$ be the symmetric group on $\{1, 2, 3\}$.

(a) Write out all of the elements of G using array notation (see Example 3.31). You should have 6 elements in all.

(b) Show that, in general, we have $|S_n| = n!$ for any positive integer n. Recall that $n! := 1 \cdot 2 \cdot 3 \cdots (n-1) \cdot n$ and is called the *factorial* of n.

(c) Write the group table for S_3. You may find it convenient to label each element with a single letter.

Exercise 3.10. Suppose that $a, b, c, d \in S$ and $\cdot$ is a binary operation on S. Consider the expression $a \cdot b \cdot c \cdot d$.

(a) Find all possible parenthesizations of this expression. How many are there?

(b) For each parenthesization you found in part (i) above, write down the expressions x and y as described in the proof of Lemma 3.32.

(c) Assuming that $\cdot$ is associative, show directly that all parenthesizations of this expression have the same value.

Exercise 3.11. Decide, with proof, whether each of the following is a group:

1. $\mathbf{Z}$ under ordinary multiplication;

2. $\mathbf{R}$ with the operation $\cdot$ given by $x \cdot y = \frac{x+y}{1+x^2 y^2}$;

3. $\mathbf{R}^2$ with the operation $\oplus$ given by $(a, b) \oplus (c, d) = (a+c, b+d)$;

4. The set of all irrational real numbers under ordinary multiplication;

5. The set $V = \{x \in \mathbf{R} : -c < x < c\}$ with the operation $\odot$ given by $x \odot y = \frac{x+y}{1+xy/c^2}$, where c is a fixed positive real number (this is the velocity addition law in special relativity theory, when c is the speed of light and the velocities are parallel).

Exercise 3.12. Finish the proof of Theorem 3.41 in Cases 2, 4, 5, and 6.

Exercise 3.13. Prove that if a and b are elements of an abelian group, then for all positive integers n, we have $(a \cdot b)^{-n} = a^{-n} \cdot b^{-n}$. (Use induction on n.)

Exercise 3.14. Generalize the result of Lemma 3.44 as follows. Prove that if $a_1, \ldots, a_r$ are elements of an abelian group, and $n \in \mathbf{Z}$, then we have $(a_1 \cdots a_r)^n = a_1^n \cdots a_r^n$.

Exercise 3.15. Let G be a group and let $x_1, x_2 \in G$. Find a formula for $(x_1 x_2)^{-1}$ by solving for z in the equation $x_1 x_2 \cdot z = e$. Then generalize your result by finding a formula for $(x_1 x_2 \cdots x_k)^{-1}$ for arbitrary elements $x_1, \ldots, x_k \in G$.

Exercise 3.16. We have seen in Example 3.22 how to shrink the set $\mathbf{R}$ to get a group under multiplication. In Example 3.23, on the other hand, we *enlarged* the underlying set that we started with in order to produce a group. Is it possible to enlarge $\mathbf{R}$ to get a group under multiplication, by adding a single new element and suitably extending the definition of multiplication? Prove your claim.

Exercise 3.17. In Example 3.25 we did not stipulate that a, b, and c are *distinct*. If, perversely, we assume that $a = b$ and that the table for the operation $\cdot$ is still valid, then show that we must in fact have $a = b = c$, so $|G| = 1$. Thus $(G, \cdot)$ is still a group!

Exercise 3.18. Suppose that $(S, \cdot)$ is a group, where $S \subseteq \mathbf{R}$ and $\cdot$ is ordinary multiplication. Can $0 \in S$? Prove your claim.

4

Subgroups

4.1 Groups Inside Groups

In this chapter, we begin to explore the relationships that groups can have with each other. In particular, we examine the situation in which one group can "live inside" another group. In addition to shedding light on relationships among different groups, we will find a process to actively seek groups inside of a given group.

We have seen that both $(\mathbf{Z}, +)$ and $(\mathbf{R}, +)$ are groups. They seem related. How exactly are they related to each other? Well, $\mathbf{Z} \subseteq \mathbf{R}$, and $+$ on $\mathbf{Z}$ is "inherited" from $+$ on $\mathbf{R}$. Let's formalize and generalize this idea. What needs to be clarified is the "inheritance" concept.

Recall that if $f : A \to B$ is a function, then we can *restrict* f to a subset of its domain. That is, given any subset C of A, we can consider the function

$$g : C \to B$$

given by the rule

$$g(x) = f(x) \text{ for all } x \text{ in } C.$$

We use the notation $g = f|_C$, read "g is the *restriction of f to C*." We also say in this situation that f is an *extension of g to A*. The point is that f and g produce the same output on any input that is in the intersection of their domains.

To make the relationship between $(\mathbf{Z}, +)$ and $(\mathbf{R}, +)$ clear, we should first find a way to distinguish between the two "$+$" operations, for they technically have different domains. Let us use the notation $+_{\mathbf{Z}}$ and $+_{\mathbf{R}}$ to distinguish these operations. Thus, we have

$$+_{\mathbf{Z}} : \mathbf{Z} \times \mathbf{Z} \longrightarrow \mathbf{Z}$$

is integer addition, and

$$+_{\mathbf{R}} : \mathbf{R} \times \mathbf{R} \longrightarrow \mathbf{R}$$

is real-number addition. These two operations are related by restriction: namely, $+_{\mathbf{Z}} = +_{\mathbf{R}}\big|_{\mathbf{Z} \times \mathbf{Z}}$. This discussion naturally leads to the following definition.

DOI: 10.1201/9781003252139-4

Definition 4.1. Let $(G, \cdot)$ be a group. A *subgroup* of $(G, \cdot)$ is a group $(H, \triangle)$ such that $H \subseteq G$ and $\triangle = \cdot|_{H \times H}$.

Notation 4.2. We write $(H, \triangle) \leq (G, \cdot)$, or simply $H \leq G$, to mean H is a subgroup of G. The notation $H < G$ means that H is a proper subgroup of G: that is, $H \leq G$ and $H \neq G$. We write $H \not\leq G$ to mean H is *not* a subgroup of G.

Example 4.3. $(\mathbf{Z}, +_{\mathbf{Z}}) \leq (\mathbf{R}, +_{\mathbf{R}})$. Indeed, this was our motivating example. On the other hand, $(\mathbf{R} - \{0\}, \cdot) \not\leq (\mathbf{R}, +)$, because multiplication and addition do not always give the same result when applied to non-zero real numbers; for example, $1 + 1 \neq 1 \cdot 1$.

To understand a given group G, one of the first questions we shall ask is, what are its subgroups? Given a group $(G, \cdot)$ and a subset H of G, there is at most one binary operation $\triangle$ on H that makes $(H, \triangle)$ a subgroup of $(G, \cdot)$: namely, $\triangle = \cdot|_{H \times H}$. Because of this, it makes sense to simply write $H \leq G$ to mean $(H, \cdot|_{H \times H}) \leq (G, \cdot)$ when the operation on G is understood, and there is seldom a need to use a separate symbol for the group law on H. A natural question is, for which subsets H of G is $H \leq G$? One way to answer this question will be presented shortly, in Theorem 4.8.

Let us consider Axiom (**G2**). If $H \leq G$, then each group, H and G, must contain an identity element, which we can denote e_H and e_G, respectively. The next result says that the identity elements of a group and a subgroup must coincide.

Lemma 4.4. *Let G be a group and suppose $H \leq G$. Then $e_G = e_H$.*

Proof. Since e_H is an identity element of H, we have $e_H \cdot e_H = e_H$. Also, since e_G is an identity element of G and $e_H \in G$, we have $e_H \cdot e_G = e_H$. So $e_H \cdot e_H = e_H \cdot e_G$. By the Cancellation Laws for G (Lemma 3.46), we have $e_H = e_G$. □

As a consequence of the previous lemma, whenever we consider a group together with some of its subgroups, we can simply use e to denote their shared identity element. Another convenience is that the notion of "inverse" coincides for a group and a subgroup:

Corollary 4.5. *Suppose that G is a group, $H \leq G$, and $x \in H$. Then the inverse of x considered as an element of the group H is equal to the inverse of x considered as an element of the group G.*

Proof. Suppose that $H \leq G$ and $x \in H$. Let a and b denote the inverse of x as an element of H and of G, respectively. Then $a \cdot x = e = b \cdot x$, so $a = b$ by Cancellation in G. □

Because of Corollary 4.5, it is safe to use the notation x^{-1} for the inverse of x even when there are multiple groups under consideration, so long as they are subgroups of a common "parent" group.

Certainly, the empty set cannot be a group, since it has no identity element; indeed, it has no elements at all. Two not so trivial things could prevent $H \leq G$. First, the operation of G may not restrict to a binary operation on H. For example, let $H = \{\frac{1}{2}, 1, 2\} \subseteq \mathbf{R} - \{0\}$. Then ordinary multiplication doesn't restrict well to H; see Example 3.23. The condition that is wanted here comes up so often that it has a name:

Definition 4.6. Let $(G, \cdot)$ be a group, and let $H \subseteq G$. Then H is *closed under* $\cdot$ if

$$\forall x, y \in H, \; x \cdot y \in H.$$

Second, H may not contain all the inverses of its own elements: for instance, $\mathbf{N} \subseteq \mathbf{Z}$ and $1 \in \mathbf{N}$, but $-1 \notin \mathbf{N}$, so $\mathbf{N}$ is not a subgroup of $\mathbf{Z}$ under addition. We next record the property that is missing here:

Definition 4.7. Let G be a group and let $H \subseteq G$. Then H is *closed under inverses* if

$$\forall x \in H, \; x^{-1} \in H.$$

The following theorem assures us that these are the only things that can go wrong:

Theorem 4.8 (Subgroup Test). *Let G be a group and let $H \subseteq G$. Then $H \leq G$ iff all three of the following conditions are satisfied:*

ST0 *H is non-empty;*

ST1 *H is closed under inverses;*

ST2 *H is closed under the group operation of G.*

Proof. ($\Rightarrow$) Suppose $H \leq G$. [We must show **ST0**, **ST1**, and **ST2**]

ST0: Since H is a subgroup of G, H contains the identity element e of G by Lemma 4.4, so H must be non-empty.

ST1: Let $x \in H$. By Axiom (**G3**) for H, there is an inverse y of x in H. By Corollary 4.5, we have $y = x^{-1}$, the inverse of x in G.

ST2: Let $x, y \in H$. Since $(H, \cdot|_{H \times H})$ is a group, then the restriction of $\cdot$ to $H \times H$ is a binary operation on H:

$$\cdot|_{H \times H} : H \times H \to H.$$

Thus we have $x \cdot y \in H$.

($\Leftarrow$) Suppose that **ST0**, **ST1**, and **ST2** are satisfied.

[We must show $H \leq G$, which means $(H, \cdot|_{H \times H})$ is a group]

Let $\Delta = \cdot|_{H \times H}$. We proceed to verify the group axioms for (H, Δ).

G0: We must show that the range of Δ is a subset of H. This is exactly the condition **ST2**.

G1: Let $x, y, z \in H$. Then

$$
\begin{aligned}
(x\Delta y)\Delta z &= (x \cdot y) \cdot z & \text{(since } \Delta \text{ is the restriction of } \cdot) \\
&= x \cdot (y \cdot z) & \text{(by Axiom \textbf{G1} for } G) \\
&= x\Delta(y\Delta z) & \text{(since } \Delta \text{ is the restriction of } \cdot).
\end{aligned}
$$

G2: By **ST0**, there is some element x in H. We have $x^{-1} \in H$ (by **ST1**), and so $x \cdot x^{-1} \in H$ (by **ST2**). Thus $e_G \in H$. We verify that e_G is an identity element for Δ. Let $a \in H$. Then $a\Delta e_G = a \cdot e_G = a$ and $e_G \Delta a = e_G \cdot a = a$. This establishes Axiom **G2**.

G3: Let $x \in H$. Then by **ST1**, we have $x^{-1} \in H$. Now $x\Delta x^{-1} = x \cdot x^{-1} = e_G$ and $x^{-1}\Delta x = x^{-1} \cdot x = e_G$. From the paragraph above, e_G is an identity element for Δ, so we are done. $\qquad\square$

Remark 4.9. When we use the Subgroup Test in the "backwards" direction, to prove $H \leq G$, we usually establish **ST0** by showing $e \in H$. Lemma 4.4 assures us that this strategy is a good one.

Remark 4.10. Again, when used to prove $H \leq G$, the Subgroup Test never requires us to verify Axiom **G1**. This is fortunate, as associativity is in some sense the toughest of the group axioms to check: see Example 3.25. Notice how, in the proof of the Subgroup Test, associativity for Δ followed automatically from associativity for $\cdot$.

Example 4.11. We will use the Subgroup Test to verify that the set

$$
H := \{2^n \mid n \in \mathbf{Z}\}
$$

is a subgroup of $G := \mathbf{R} - \{0\}$ under ordinary multiplication.

ST0: We have $2^0 \in H$, so H is non-empty.

ST1: Let $x \in H$. Then $x = 2^n$ for some n in $\mathbf{Z}$. In the group G, the inverse of x is 2^{-n}. But $-n \in \mathbf{Z}$, so $2^{-n} \in H$.

ST2: Let $x, y \in H$. Then $x = 2^m$ and $y = 2^n$ for some $m, n \in \mathbf{Z}$. So $x \cdot y = 2^{m+n}$, and $2^{m+n} \in H$ since $m + n \in \mathbf{Z}$.

4.2 The Subgroup Generated by a Set

We have seen (e.g. in Example 3.23) that not every subset of a group is a subgroup; and now we have a test, the Subgroup Test, that lets us check whether a given subset is a subgroup. Next we ask, Given a group G and a subset $S \subseteq G$, what is the *smallest* subgroup H of G which contains S?

Questions such as this are often fruitful in mathematics, but they do not always have answers! As a simple example, we could ask, What is the smallest positive real number? Of course, there isn't one. In the case of our question about subgroups, however, we will find that there is always an answer.

Suppose for the moment that we know some elements $s_1, s_2, \ldots, s_k$ of S (not assumed to be distinct). Since H is required to contain S, we certainly must have $s_i \in H$ for each i. Since $H \leq G$, the Subgroup Test forces $s_i^{-1} \in H$ for each i, since H must be closed under inverses by **ST1**. We can summarize what we now know by stating that $s_i^{\varepsilon} \in H$ for each i in $\{1, \ldots, k\}$ and each ε in $\{1, -1\}$.

Next, using the closure of H under the group operation (**ST2**), we can say that the product of any two such terms must lie in H. Using closure under $\cdot$ repeatedly, we conclude that $s_1^{\varepsilon_1} s_2^{\varepsilon_2} \cdots s_k^{\varepsilon_k} \in H$ for any choice of $\varepsilon_1, \varepsilon_2, \ldots, \varepsilon_k$ in $\{1, -1\}$.

The next result says that we need seek no further:

Theorem 4.12. *Let G be a group and let $S \subseteq G$ with S non-empty. Let*

$$H = \{s_1^{\varepsilon_1} s_2^{\varepsilon_2} \cdots s_k^{\varepsilon_k} \mid s_1, \ldots, s_k \in S, \varepsilon_1, \ldots, \varepsilon_k \in \{1, -1\}\}.$$

Then $H \leq G$. Further, H is the smallest subgroup of G which contains S, in the following sense: if K is any subgroup of G which contains S, then $H \subseteq K$.

Remark 4.13. The elements $s_1, \ldots, s_k$ do not need to be distinct! There may be repetitions in this list.

Proof. We use the Subgroup Test.

ST0: Since S is non-empty, $\exists s \in S$. So $s \in H$ (taking $k = 1$, $s_1 = s$, and $\varepsilon_1 = 1$), and H is non-empty. (This argument also shows that $S \subseteq H$.)

ST1: Let $x \in H$. [Show $x^{-1} \in H$] Then we can write $x = s_1^{\varepsilon_1} s_2^{\varepsilon_2} \cdots s_k^{\varepsilon_k}$ with $s_i \subset S$ and $\varepsilon_i \in \{1, -1\}$ (by definition of H). Now the inverse of x is given by the formula

$$
\begin{aligned}
x^{-1} &= (s_k^{\varepsilon_k})^{-1} \cdots (s_2^{\varepsilon_2})^{-1} (s_1^{\varepsilon_1})^{-1} \quad \text{(by Exercise 3.15)} \\
&= s_k^{-\varepsilon_k} \cdots s_2^{-\varepsilon_2} s_1^{-\varepsilon_1} \quad\quad\quad\;\; \text{(by Theorem 3.41)}
\end{aligned}
$$

Since $\varepsilon_i \in \{1, -1\}$, we also have $-\varepsilon_i \in \{1, -1\}$, and so x^{-1} is of the right form to belong to H: namely, a product of elements of S raised to powers of ± 1.

ST2: Let $x, y \in H$. [Show $xy \in H$] Then we can write $x = s_1^{\varepsilon_1} s_2^{\varepsilon_2} \cdots s_k^{\varepsilon_k}$ and $y = t_1^{\alpha_1} t_2^{\alpha_2} \cdots t_\ell^{\alpha_\ell}$ with $s_i, t_j \in S$ and $\varepsilon_i, \alpha_j \in \{1, -1\}$. So we have

$$xy = s_1^{\varepsilon_1} s_2^{\varepsilon_2} \cdots s_k^{\varepsilon_k} t_1^{\alpha_1} t_2^{\alpha_2} \cdots t_\ell^{\alpha_\ell},$$

which is again of the right form to be in H.

The last statement in the Theorem follows from the fact that a subgroup is closed under inverses and the group operation, as outlined in the argument immediately preceding this proof. $\qquad\square$

Notation 4.14. The group H defined in the statement of Theorem 4.12 is denoted $\langle S \rangle$, and called the *group generated by S*. If $S = \{s_1, \ldots, s_n\}$ is a finite set, then we also use the notation $\langle s_1, \ldots, s_n \rangle$ instead of $\langle \{s_1, \ldots, s_n\} \rangle$ for the group generated by S. In case $S = \emptyset$, we interpret $\langle S \rangle$ to mean $\langle e \rangle$, which is the smallest subgroup of all (see Exercise 4.1).

Example 4.15. The simplest application of Theorem 4.12 occurs when S contains just one element. Let G be a group and let $a \in G$. Set $S = \{a\}$. Then we have

$$
\begin{aligned}
\langle S \rangle &= \{s_1^{\varepsilon_1} s_2^{\varepsilon_2} \cdots s_k^{\varepsilon_k} \mid s_i \in S, \varepsilon_i \in \{1, -1\}\} &\text{(by def. of } \langle S \langle \text{)} \\
&= \{a^{\varepsilon_1} a^{\varepsilon_2} \cdots a^{\varepsilon_k} \mid \varepsilon_i \in \{1, -1\}\} &\text{(since } S = \{a\}) \\
&= \{a^{\varepsilon_1 + \varepsilon_2 + \cdots + \varepsilon_k} \mid \varepsilon_i \in \{1, -1\}\} &\text{(by Laws of Exponents)} \\
&= \{a^n \mid n \in \mathbf{Z}\},
\end{aligned}
$$

this last description coming from the fact that an arbitrary finite sum $\varepsilon_1 + \varepsilon_2 + \cdots + \varepsilon_k$ of 1's and -1's can give any integer n as a result. According to Theorem 4.12, this set is a subgroup of G; according to our notational rules, we may also denote it by $\langle a \rangle$, which we generally prefer.

The case of a group generated by a single element is so important that it has its own terminology:

Definition 4.16. Let H be a group. If $H = \langle a \rangle$ for some $a \in H$, then H is called *cyclic*, and we say that a is a *generator* for H.

Warning 4.17. In Definition 4.16, we need $a \in H$, as opposed to $a \subseteq H$. In the notation $\langle x \rangle$, it is important to know whether x represents a *set* or an *element*. In both cases, $\langle x \rangle$ will be a group. When x is a single element of a group, then $\langle x \rangle$ is a cyclic group; but if x is a *set* of size greater than 1, then $\langle x \rangle$ will in general not be cyclic. To help avoid confusion, we try to use lower-case letters to stand for elements and upper-case letters to stand for sets.

Notice that in Definition 4.16 we did not say that H was a subgroup of some other group. Even though we began this section with the notion of constructing subgroups within a given group, we have arrived at an "intrinsic" or "absolute" (not relative) notion of what it means to be cyclic.

Nevertheless, we can also apply our relative point of view too, and assert that every group G must contain (usually a lot of) cyclic subgroups: just pick any $g \in G$ and then we can form the cyclic subgroup $\langle g \rangle$ of G.

Example 4.18. Consider the group $(\mathbf{Z}, +)$. Notice that "3^2" actually means $3+3$ here, since the group operation is addition. Thus we use additive notation here, and write $2 \cdot 3$ instead of "3^2" in this group. Conveniently, this notation agrees with the interpretation of $\cdot$ as ordinary multiplication.

More generally, we write $n \cdot x$ instead of "x^n" in the group $(\mathbf{Z}, +)$, for x and n in $\mathbf{Z}$. Thus for any integer x, the group generated by x is

$$
\langle x \rangle = \{n \cdot x \ : \ n \in \mathbf{Z}\},
$$

the set of all integer multiples of x. Since every integer is an integer multiple of 1, we have $\langle 1 \rangle = \mathbf{Z}$, so $\mathbf{Z}$ is a cyclic group under addition. We also have $\langle -1 \rangle = \mathbf{Z}$. In general, a cyclic group will have more than one generator, unless the group has order 1 or 2.

Example 4.19. Consider the group $(\mathbf{Q}^\times, \cdot)$, where $\mathbf{Q}^\times = \mathbf{Q} - \{0\}$ and $\cdot$ is ordinary multiplication. If $\mathbf{Q}^\times$ were cyclic, then we would have $\mathbf{Q}^\times = \langle \frac{a}{b} \rangle = \{(\frac{a}{b})^n \mid n \in \mathbf{Z}\}$ for some non-zero integers a and b. But if p is any prime number which does not divide a or b, then the equality $(\frac{a}{b})^n = p$ is impossible for any integer n. Therefore, $(\mathbf{Q}^\times, \cdot)$ is not cyclic.

Next we come to an important general property of subgroups: the intersection of any collection of subgroups of a given group is a subgroup of that group. The proof is entirely formal.

Lemma 4.20. *Let G be a group, and let $\mathfrak{C}$ be a non-empty collection of subgroups of G. Then $\cap_{H \in \mathfrak{C}} H \leq G$.*

Proof. Set $I = \cap_{H \in \mathfrak{C}} H$. We shall use the Subgroup Test to verify that $I \leq G$.

ST0: [Show $e \in I$; this means $\forall H \in \mathfrak{C}, e \in H$] Let $H \in \mathfrak{C}$. By Lemma 4.4, $e \in H$. Since H was an arbitrary member of $\mathfrak{C}$, we have $e \in I$. So I is non-empty.

ST1: Let $x \in I$. Then $\forall H \in \mathfrak{C}, x \in H$. [Show $x^{-1} \in I$; this means $\forall H \in \mathfrak{C}, x^{-1} \in H$] Let $H \in \mathfrak{C}$. Then $x \in H$ and $H \leq G$. Now H is closed under inverses (by **ST1** for H), so we have $x^{-1} \in H$. Since H was an arbitrary member of $\mathfrak{C}$, we have $x^{-1} \in I$.

ST2: Let $x, y \in I$. [Show $x \cdot y \in I$] Let $H \in \mathfrak{C}$. Then $x, y \in H$ and $H \leq G$. Now H is closed under the group operation (by **ST2** for H), so $x \cdot y \in H$. Since H was an arbitrary member of $\mathfrak{C}$, we have $x \cdot y \in I$. □

Armed with Lemma 4.20, we return to the question of finding the smallest subgroup of a group G which contains a given subset S of G. Consider the collection

$$\mathfrak{C} := \{H \mid H \leq G \text{ and } S \subseteq H\}$$

of all subgroups of G which contain S. We are guaranteed by Lemma 4.20 that the intersection of all subgroups of G containing S,

$$I := \bigcap_{H \in \mathfrak{C}} H,$$

is a subgroup of G. A bit of thought shows that I is the smallest possible subgroup of G which contains S. Comparing this result with Theorem 4.12, we conclude that $I = \langle S \rangle$. We thus have two ways of viewing the smallest subgroup of G containing S: the "bottom-up" view of Theorem 4.12, where we build $\langle S \rangle$ up from elements of S and their inverses, and the "top-down" view we just saw, where we arrive at $\langle S \rangle$ by starting with G and paring away the excess, intersecting all the subgroups containing S.

4.3 Exercises

Exercise 4.1. Let G be a group.

(a) Prove that $G \leq G$.

(b) Prove that $\{e\} \leq G$.

Exercise 4.2 (The Two-Thirds Rule). Let $H \leq G$, and suppose that $x, y, z \in G$ and $x \cdot y = z$. Prove that if any two of these elements are in H, then so is the third.

Exercise 4.3. Let H, K, and G be groups.

(a) Prove that if $H \leq K$ and $K \leq G$, then $H \leq G$.

(b) Prove that if $H \leq G$, $K \leq G$, and $H \subseteq K$, then $H \leq K$.

Exercise 4.4. Decide, with proof, whether each of the following sets is a subgroup of **R** under ordinary addition:

1. **Q**

2. $\{a + b\sqrt{2} \ : \ a, b \in \mathbf{Z}\}$

3. The set of all irrational numbers together with 0

4. $\{x^3 \ : \ x \in \mathbf{R}\}$ (Note: "x^3" means $x \cdot x \cdot x$, and not $x + x + x$ here.)

Exercise 4.5. Prove that any group of order 2 is cyclic. (See Exercise 3.2.)

Exercise 4.6. Let $G = S_3$.

(a) Find all of the cyclic subgroups of G. How many are there? What are their orders?

(b) Find two elements $x, y \in G$ which do not commute.

(c) Show that the x and y that you found in part (b) generate G; that is, show that $G = \langle x, y \rangle$. (How did I know this would happen?)

Exercise 4.7. Consider the group $(\mathbf{Z}, +)$.

(a) Describe the subgroup $\langle n \rangle$ as simply as possible, where n is an arbitrary integer.

(b) What is your first reaction to the question: Is $\langle 4, 6 \rangle$ cyclic?

(c) Prove that $\langle 4, 6 \rangle = \langle 2 \rangle$. Is $\langle 4, 6 \rangle$ cyclic?

(d) Make a conjecture about the nature of the subgroup $\langle a, b \rangle$ for general integers a and b. What is the simplest way to describe this subgroup? Can you prove your conjecture?

Exercise 4.8. Let S be a set and let $\cdot$ be an associative binary operation on S with an identity element e. Thus we are saying that $(S, \cdot)$ satisfies Group Axioms **G0**, **G1**, and **G2** but *not* necessarily **G3**. Let $T = \{x \in S \mid \exists y \in S \text{ s.t. } x \cdot y = y \cdot x = e\}$. (That is, T consists of all elements in S with inverses in S.)

(a) Let $\Delta = \cdot|_{T \times T}$. Prove that (T, Δ) is a group.

(b) Why is it technically incorrect to say that $(T, \cdot)$ is a group? (We will abuse notation this way sometimes in the future, however.)

Exercise 4.9. Must the union of two subgroups of a group G be a subgroup of G?

Exercise 4.10. This exercise refers to Exercise 3.4. For an element $A = \begin{bmatrix} a & b \\ c & d \end{bmatrix}$ in S, define $|A| = ad - bc$.

(a) Prove that for all $A, B \in S$, we have $|A \cdot B| = |A| \cdot |B|$. (Note that the $\cdot$ on the left side is the operation on S defined in Exercise 3.4, while the $\cdot$ on the right is ordinary multiplication in $\mathbf{R}$.)

(b) For an arbitrary element A of S such that $|A| \neq 0$, find an element C of S such that $A \cdot C = e$, where e is the identity element found in Exercise 3.4.

(c) Compute $|e|$.

(d) Prove that if $A \in S$ and $|A| = 0$, then there is no $C \in S$ such that $A \cdot C = e$.

(e) Set $G = \{A \in S \ : \ |A| \neq 0\}$. Use the previous parts of this exercise together with Exercise 4.8 to prove that $(G, \cdot)$ is a group.

Exercise 4.11. Let G and S be as in Exercise 4.10.

(a) Let $H = \{A \in S \ : \ |A| = 1\}$. Prove that $H \leq G$.

(b) Find a subgroup of G of order 2. Hint: see Exercise 4.5.

Exercise 4.12. Let G be a group and let $g \in G$. Set $H = \{x \in G \ : \ g \cdot x = x \cdot g\}$. Note: we often denote H by $C_G(g)$, and call it the *centralizer* of g in G.

(a) Prove that $H \leq G$.

(b) Prove that $\langle g \rangle \leq H$.

Exercise 4.13. Suppose that G is a group containing two elements x and y such that $x \cdot y = y \cdot x$.

(a) Convince yourself that $\langle x, y \rangle$ must be abelian, and explain why.

(b) Prove the result from part (a) formally.

(c) Generalize the result from part (b) as follows: let $x_1, \ldots, x_n \in G$ and suppose that these elements commute pairwise (that is, $x_i x_j = x_j x_i$ for all i, j). Prove that $\langle x_1, \ldots, x_n \rangle$ is abelian.

5

Symmetry

5.1 What is Symmetry?

One application of group theory is to the study of symmetries of objects. But what do we mean by a "symmetry"? Consider a blank rectangular sheet of paper lying on my desk in front of you. If I wait until your back is turned and then sneakily turn the paper by exactly 180 degrees about its center, when you look again you will have no way to tell that I changed anything! The success of this parlor trick has something to do with the symmetry of the sheet.

We proceed to give a mathematical treatment to the example above. Let us agree to put the origin $(0,0)$ of a Cartesian coordinate system at the paper's center. The action of rotating the sheet of paper by 180 degrees about its center can be captured by a *function*. Namely, consider the function $f : \mathbf{R}^2 \rightarrow \mathbf{R}^2$ which takes a point P as input and rotates P by 180 degrees about the origin. It is not hard to discover a formula describing this function: namely, $f : (x, y) \mapsto (-x, -y)$.

We would like to say that this function f has no discernible effect on the sheet of paper. Clearly, f moves around individual points on the sheet, but, as a whole, f does not change the *set* of points of the sheet.

The reader may have heard symmetry described as a property of an object, as in, "a human has bilateral symmetry" or "the function $y = x^2$ is symmetric about the y-axis." Our goal is to specify a "symmetry" of an object as a freestanding thing unto itself, and moreover a standard type of mathematical entity: we will say that a symmetry of an object *is* a function on the underlying space which leaves our object, as a whole, unchanged.

Definition 5.1. Let $X \subseteq S$ be two sets. Then a *symmetry* of X with respect to S is a bijective function $f : S \rightarrow S$ such that $f(X) = X$.

Remark 5.2. Recall how we define the image of a set under a function: namely, $f(X) := \{f(a) \mid a \in X\}$, so the image of X under f is the set of images of all points of X. Thus, the definition of symmetry given above does *not* assert that f takes every point of X to itself, but rather that f takes X as a whole to itself.

Example 5.3. Let $S = \mathbf{R}^2$, let $a, b > 0$, and let $X = [-a, a] \times [-b, b]$. Define a function $f : S \rightarrow S$ by $f : (x, y) \mapsto (-x, -y)$. Then f is bijective, and we have

DOI: 10.1201/9781003252139-5

$P \in X$ iff $f(P) \in X$. So $f(X) = X$, and f is a symmetry of X with respect to $\mathbf{R}^2$. Notice that X is just the rectangle of width $2a$ and height $2b$ centered at the origin, and f is the symmetry of X described above in our discussion about the sheet of paper.

Example 5.4. Let S be any non-empty set, and let f be any bijective function from S to S. Then we have $f(S) = S$, since f is surjective. Therefore f is a symmetry of S with respect to itself. We usually say simply that f is a *symmetry of S* in this case. But recall that we defined a *permutation* of S to be a bijective function from S to S. We conclude that Sym(S), the symmetric group on S, is exactly the set of all symmetries of S!

In general, the set of all symmetries of X with respect to S is a subgroup of Sym(S) (Exercise 5.1). Often, we are interested only in those symmetries of an object which satisfy some additional conditions. For instance, we may desire that our sheet of paper should not have to be stretched, shrunk, or torn into pieces and re-assembled in order to apply a symmetry; something like rotation, on the other hand, would be reasonable. One way to capture this kind of requirement is to stipulate that a symmetry should preserve distances and angles; a symmetry with this property is sometimes called a *rigid* symmetry. A looser requirement of this nature would be simply to require a symmetry to be continuous.

5.2   Dihedral Groups

Fix an integer $n \geq 3$, and let P_n be the regular n-sided polygon (or "n-gon" for short) in $\mathbf{R}^2$ centered at the origin and with one vertex at $(1, 0)$. What are the rigid symmetries of P_n with respect to $\mathbf{R}^2$?

One rigid symmetry of P_n is the rotation R about the origin by an angle of $2\pi/n$ radians in the counterclockwise direction. The function R moves each vertex of P_n to the vertex next to it in counterclockwise order. Let us derive a formula for R with respect to Cartesian coordinates.

Let $P = (x, y)$ be a point in $\mathbf{R}^2$. To describe the action of R on P, it is convenient to make use of polar coordinates. Supposing that the polar coordinates of P are (r, θ), we have $x = r\cos(\theta)$, $y = r\sin(\theta)$. For convenience, set $\alpha = 2\pi/n$.

Then the polar coordinates of $R(P)$ are $(r, \ \theta + \alpha)$. We are interested in the Cartesian coordinates (x', y') of $R(P)$. We have $x' = r\cos(\theta + \alpha) = r\cos(\theta)\cos(\alpha) - r\sin(\theta)\sin(\alpha) = x\cos(\alpha) - y\sin(\alpha)$, and $y' = r\sin(\theta + \alpha) = r\sin(\theta)\cos(\alpha) + r\cos(\theta)\sin(\alpha) = y\cos(\alpha) + x\sin(\alpha)$. Therefore we have

$$R(x, y) = (x\cos(\alpha) - y\sin(\alpha), x\sin(\alpha) + y\cos(\alpha)). \qquad (5.1)$$

In addition to its rotational symmetry, the polygon P_n is also symmetric about the x-axis. The corresponding symmetry of P_n with respect to $\mathbf{R}^2$ is the "flip" F about the x-axis, given by the formula

$$F(x,y) = (x, -y). \tag{5.2}$$

At this point, the reader should think about what other rigid symmetries of P_n may exist.

In fact, we assert (without proof; the reader is invited to confirm it) that all of the rigid symmetries of P_n can be obtained by composing some number of rotations and flips in some order. What is the set of all symmetries that can be obtained in this manner? Recall that in a symmetric group, the group operation is composition of functions. Thus, our answer is exactly the subgroup of $\text{Sym}(\mathbf{R}^2)$ generated by F and R.

Definition 5.5. Let $n \geq 3$ be an integer. The *dihedral group* D_{2n} is the subgroup of $\text{Sym}(\mathbf{R}^2)$ generated by F and R, where F is the flip about the x-axis and R is counterclockwise rotation about the origin by $2\pi/n$ radians. That is,

$$D_{2n} := \langle F, R \rangle.$$

We shall see shortly the reason for using $2n$ instead of n in our notation.

Our immediate goal is to understand the group D_{2n}. A typical element of D_{2n} is a product of terms of the form F, R, F^{-1}, and R^{-1}. Thus, some examples of elements of D_{2n} are F, R, $F \circ R$, $F \circ R^{-1}$, $R \circ R \circ F \circ R \circ F \circ R^{-1}$, etc.

As a start, let us analyze F^{-1} and R^{-1}. These are just the inverse functions of F and R, respectively. So R^{-1} is the *clockwise* rotation about the origin by $2\pi/n$ radians; this is the same as the counterclockwise rotation by $-2\pi/n$ radians. An easy way to obtain a formula for R^{-1} is to substitute $-\alpha$ for α in the formula for R. We find $R^{-1}(x,y) = (x\cos(-\alpha) - y\sin(-\alpha), x\sin(-\alpha) + y\cos(-\alpha))$. Taking advantage of the fact that sine is odd and cosine is even, we get

$$R^{-1}(x,y) = (x\cos(\alpha) + y\sin(\alpha), y\cos(\alpha) - x\sin(\alpha)). \tag{5.3}$$

Notice that applying F twice brings every point of $\mathbf{R}^2$ back to its original position. In other words, $F \circ F$ is the identity function on $\mathbf{R}^2$; but this is also the identity element of $\text{Sym}(S)$. Therefore we have $F^{-1} = F$, and

$$F^2 = e. \tag{5.4}$$

Next, we consider the cyclic subgroup generated by each element F and R separately. Since $F^2 = e$, then $F^3 = F^2 \circ F = e \circ F = F$, $F^4 = F^3 \circ F = F \circ F = F^2 = e$, etc. We also have $F^0 = e$ (by Definition 3.38). Further, for every positive integer k, we have $F^{-k} = (F^{-1})^k$ (by the Laws of Exponents) $= F^k$ (Since $F^{-1} = F$). We conclude that for any integer k, we have

$$F^k = \begin{cases} e, & \text{if } k \text{ is even;} \\ F, & \text{if } k \text{ is odd.} \end{cases}$$

In particular, $\langle F \rangle = \{e, F\}$.

We now consider $\langle R \rangle = \{R^k \mid k \in \mathbf{Z}\}$. First let k be a positive integer. Then R^k, the composition of k factors of R, is just a counterclockwise rotation by $2k\pi/n$ radians. Also, R^{-k} is the inverse of R^k, which is a counterclockwise rotation by $-2k\pi/n$ radians.

We conclude that for all integers k, R^k is a counterclockwise rotation by $2k\pi/n$ radians. These powers of R start repeating when $k = n$. Indeed, R^n is a rotation by exactly 2π radians or 360 degrees, so we have

$$R^n = e. \tag{5.5}$$

(Exercise 5.4 asks you to formally verify this assertion.)

It follows that $R^{n+1} = R^n \circ R = e \circ R = R$, $R^{n+2} = R^n \circ R^2 = e \circ R^2 = R^2$, etc.; so we only need the first n non-negative powers of R to capture all positive powers of R. For negative powers, we observe that $R \circ R^{n-1} = R^n = e$, and so we must have $R^{-1} = R^{n-1}$. From this, we see that all the negative powers of R can be written as positive powers of R. We conclude that $\langle R \rangle = \{e, R, R^2, R^3, \ldots, R^{n-1}\}$.

We interrupt this discussion of D_{2n} to record and prove a general version of what we have just learned about $\langle R \rangle$.

Lemma 5.6. *Let G be a group, let $x \in G$, and suppose that $x^k = e$ for some positive integer k. Then we have $\langle x \rangle = \{e, x, x^2, x^3, \ldots, x^{k-1}\}$.*

Proof. Set $H = \{e, x, x^2, x^3, \ldots, x^{k-1}\}$. Since $\langle x \rangle = \{x^j \mid j \in \mathbf{Z}\}$, we certainly have $H \subseteq \langle x \rangle$. On the other hand, if $y \in \langle x \rangle$, then we can write $y = x^j$ for some integer j; and in turn, by the Remainder Theorem for integer division, we can write $j = qk + r$ for integers q and r such that $0 \leq r \leq k - 1$, to get $y = x^j = x^{qk+r} = x^{qk}x^r = (x^k)^q x^r = e^q x^r = ex^r = x^r$, so $y \in H$. (This last series of equalities used the Laws of Exponents and also Exercise 3.3).   □

Warning 5.7. Lemma 5.6 shows that the order of $\langle x \rangle$ is *at most* k. But the elements e, x, x^2, $x^3, \ldots, x^{k-1}$ need not be distinct. Thus we may have $|\langle x \rangle| < k$. However, if we take k to be the *smallest* positive integer with the property of the Lemma, then we do have $|\langle x \rangle| = k$; see Exercise 5.7.

The next step in our quest to understand D_{2n} is to find a relationship between F and R. Suppose that we take a regular pentagon ($n = 5$) and label its vertices in counterclockwise order 0, 1, 2, 3, 4, starting with the vertex at $(1, 0)$. After applying the rotation R, vertex 4 will be at $(1, 0)$, and the new vertex order is 4, 0, 1, 2, 3. If we next apply the flip F, the final vertex order will be 4, 3, 2, 1, 0.

On the other hand, suppose we start with the original order 0, 1, 2, 3, 4, but first apply F. We get the new order 0, 4, 3, 2, 1. We can achieve the order 4, 3, 2, 1, 0 from here by rotating once in the *clockwise* direction: that is, by applying R^{-1}.

This discussion seems to indicate that

$$FR = R^{-1}F. \tag{5.6}$$

We proceed to prove this equation rigorously in the general case.

Claim 5.8. *For any $n \geq 3$, we have $FR = R^{-1}F$ in the dihedral group D_{2n}.*

Proof. [FR and $R^{-1}F$ are *functions*] [Two functions are equal iff they have the same domain and they give the same output for each input: so we must prove $\forall P \in \mathbf{R}^2, (FR)(P) = (R^{-1}F)(P)$] Let $P \in \mathbf{R}^2$, and write $P = (x, y)$. Then

$$
\begin{aligned}
(FR)(P) &= F(R(P)) \\
&= F(x\cos(\alpha) - y\sin(\alpha), x\sin(\alpha) + y\cos(\alpha)) \\
&= (x\cos(\alpha) - y\sin(\alpha), -x\sin(\alpha) - y\cos(\alpha)),
\end{aligned}
$$

and

$$
\begin{aligned}
(R^{-1}F)(P) &= R^{-1}(F(P)) \\
&= R^{-1}(x, -y) \\
&= (x\cos(\alpha) - y\sin(\alpha), -y\cos(\alpha) - x\sin(\alpha)).
\end{aligned}
$$

$\square$

We will say that Equations 5.4, 5.5, and 5.6 are "relations" involving F and R. More generally, we make the following definition, which simply says that we will use the term "relation" to mean any equation involving group elements. This use of the word "relation" should not be confused with that in Definition 8.3.

Definition 5.9. Let G be a group. A *relation* in G is an equation of the form

$$
x_1 x_2 \cdots x_m = y_1 y_2 \cdots y_n
$$

where $x_1, \ldots, x_m, y_1, \ldots, y_n \in G$.

The next result and its corollary tell us a lot about the structure of D_{2n}. We state these results at a level of generality that goes somewhat beyond the dihedral groups, in consideration of a future application. (We keep the notation F and R, so that these symbols could, but do not necessarily, represent the flip and rotation of D_{2n} in the result below.)

Proposition 5.10. *Suppose that G is a group generated by two elements F and R which satisfy the relations $F^2 = e$, $R^n = e$, and $FR = R^{-1}F$ (where n is a positive integer). Then*

$$
\begin{aligned}
G = \{ \quad e, \quad & R, \quad R^2, \quad R^3, \quad \ldots, \quad R^{n-1}, \\
F, \quad & FR, \quad FR^2, \quad FR^3, \quad \ldots, \quad FR^{n-1}\}.
\end{aligned}
$$

Proof. Let $S = \{e, R, R^2, \ldots, R^{n-1}, F, FR, FR^2, \ldots, FR^{n-1}\}$. Then certainly $S \subseteq G$. Our strategy will be to show that $S \leq G$. Once we know this, we can argue as follows: S is a subgroup of G which contains F and R; but $G = \langle F, R \rangle$, so G is the smallest subgroup of G which contains F and R; therefore, $G \subseteq S$, and so $S = G$.

It remains to show that $S \leq G$. We use the Subgroup Test.

ST0: $e \in S$, so S is non-empty.

ST1: Let $x \in S$. [Show $x^{-1} \in S$] Then we can write $x = F^a R^b$ for some $a \in \{0,1\}$ and $b \in \{0,1,\ldots,n-1\}$. So $x^{-1} = R^{-b}F^{-a}$ (by Exercise 3.15). We consider two cases:

Case 1: $a = 0$.

Then $x^{-1} = R^{-b} \in \langle R \rangle$. But $\langle R \rangle = \{e, R, R^2, \ldots, R^{n-1}\}$ (by Lemma 5.6). So $\langle R \rangle \subseteq S$, and thus $x^{-1} \in S$.

Case 2: $a = 1$.

Then $x^{-1} = R^{-b}F^{-1} = R^{-b}F$ (since $F = F^{-1}$) $= FR^b$ (by Exercise 3.5). So $x^{-1} = x \in S$.

ST2: Let $x, y \in S$. [Show $xy \in S$] Then we can write $x = F^a R^b$ and $y = F^c R^d$ for some integers a, b, c, d with $a, c \in \{0,1\}$ and $b, d \in \{0,1,2,\ldots,n-1\}$. So $xy = F^a R^b F^c R^d$. We break into two cases according to the value of c:

Case 1: $c = 0$.

Then $xy = F^a R^b R^d = F^a R^{b+d}$. Now $R^{b+d} \in \langle R \rangle$, so, by Lemma 5.6, we can write $R^{b+d} = R^j$ for some integer j with $0 \leq j \leq n - 1$. Thus $xy = F^a R^j \in S$.

Case 2: $c = 1$.

Then $xy = F^a R^b F R^d = F^a F R^{-b} R^d = F^{a+1} R^{d-b}$. Using Lemma 5.6 twice, we can write $F^{a+1} = F^k$ for some $k \in \{0,1\}$ and $R^{d-b} = R^j$ for some $j \in \{0,1,2,\ldots,n-1\}$. Therefore $xy = F^k R^j \in S$.   □

Corollary 5.11. *Suppose G, F, and R are as in Proposition 5.10. Then the group operation of G is given by*

$$(F^a R^b)(F^c R^d) = F^{a+c} R^{d+(-1)^c b} \tag{5.7}$$

for $a, c \in \{0,1\}$ and $b, d \in \{0,1,2,\ldots,n-1\}$.

Proof. This follows by inspection from the two cases $c = 0$ and $c = 1$ of the proof of **ST2** above.   □

The next result completes our description of D_{2n}, for the moment.

Theorem 5.12. *For any integer $n \geq 3$, the order of D_{2n} is $2n$.*

Proof. By Proposition 5.10, the order of D_{2n} is at most $2n$. [Strategy: Prove that the elements e, R, R^2, R^3, $\ldots$, R^{n-1}, F, FR, FR^2, FR^3, $\ldots FR^{n-1}$ are distinct in D_{2n}. To prove that a list has distinct entries, we prove that if two list entries are equal then their indexes are equal] Suppose that $F^a R^b = F^c R^d$ with $a, c \in \{0,1\}$ and $b, d \in \{0,1,2,\ldots,n-1\}$, and assume without loss of generality that $b \leq d$. [Show: $a = c$ and $b = d$] Then we have $F^{a-c} = R^{d-b}$. We break into cases based on the value of F^{a-c}. The two cases $F^{a-c} = e$ and $F^{a-c} = F$ are exhaustive, by Lemma 5.6.

Case 1: $F^{a-c} = e$.

Then $R^{d-b} = e$. Since $0 \leq b \leq d \leq n-1$, we have $0 \leq d-b < n$. It follows that $b = d$. (For a formal algebraic proof of this last implication, see Exercise 5.4.)

Also, we have $F^a = F^c$, and $a, c \in \{0, 1\}$. We suppose for a contradiction that $a \neq c$. Then we have $F^0 = F^1$, so $F = e$. But F is not the identity function on $\mathbf{R}^2$. Therefore, $a = c$.

Case 2: $F^{a-c} = F$.

Then $R^{d-b} = F$. Let $k = d - b$. Then $0 \leq k < n - 1$. Evaluating both functions in the equation $R^k = F$ at the point $(1, 0)$ gives $(\cos(2k\pi/n), \sin(2k\pi/n)) = (1, 0)$ (see Exercise 5.4). But $0 \leq 2k\pi/n < 2\pi$, so we must have $k = 0$, and $b = d$. But now we have $F = R^{d-b} = R^0 = e$, a contradiction. Therefore, Case 2 is impossible. $\qquad \square$

Remark 5.13. At last we see the reason for the $2n$ in the notation: D_{2n} is the *dihedral group of order* $2n$.

We conclude this section with a meditation on the dihedral groups. Originally, we conceived of D_{2n} as a (mysterious) collection of symmetries of a regular n-gon, whose elements are functions from $\mathbf{R}^2$ to $\mathbf{R}^2$. Now we have come to realize that D_{2n} can be described as a collection of certain combinations of the functions R and F, along with a certain group operation. We shall see that there is a sense in which D_{2n} can be described as the group generated by the *symbols* R and F subject to the three relations $F^2 = e$, $R^n = e$, and $FR = R^{-1}F$. The next sections will allow us to make this idea precise.

5.3 Exercises

Exercise 5.1. Let S be a non-empty set and let $X \subseteq S$. Let $\text{Sym}_X(S)$ denote the set of all symmetries of X with respect to S. Prove that $\text{Sym}_X(S)$ is a subgroup of $\text{Sym}(S)$.

Exercise 5.2. Suppose that T is a subset of the x, y-plane which is symmetric about the line $y = x$ (in the language of high-school algebra).

(a) Find a function $f : \mathbf{R}^2 \to \mathbf{R}^2$ such that f is a symmetry of T with respect to $\mathbf{R}^2$ and f is not the identity function on $\mathbf{R}^2$.

(b) What is $\langle f \rangle$?

Exercise 5.3. Using Equations 5.1 and 5.3, verify algebraically that $R \circ R^{-1}$ and $R^{-1} \circ R$ are both equal to the identity function on $\mathbf{R}^2$.

Exercise 5.4. (a) Starting with equation 5.1, use induction to prove the formula $R^k(x, y) = (x\cos(k\alpha) - y\sin(k\alpha), x\sin(k\alpha) + y\cos(k\alpha))$ for all positive integers k.

(b) Use the result of part (a) to show that $R^n = e$, but $R^k \neq e$ if $1 \leq k < n$.

Exercise 5.5. Let $x \in D_{2n}$ for some integer $n \geq 3$.

(a) Prove that if $x = FR^k$ for some $k \in \{0, 1, 2, \ldots, n-1\}$, then $\langle x \rangle$ has order 2.

(b) Prove that if $x = R^k$ for some $k \in \{0, 1, 2, \ldots, n-1\}$, then $\langle x \rangle$ has order at most n.

Exercise 5.6. Use the following steps to prove that every subgroup of a cyclic group is also cyclic. Suppose that $G = \langle a \rangle$ is a cyclic group generated by an element $a \in G$. Let $H \leq G$ with $H \neq \{e\}$. Set $E = \{j \in \mathbf{Z} : j \geq 1 \text{ and } a^j \in H\}$.

(a) Prove that E is non-empty.

(b) Why must E have a smallest element?

(c) Let m be the smallest element of E. Prove that $H = \langle a^m \rangle$. Hint: Use the Remainder Theorem for integer division.

Exercise 5.7. Let G be a group, let $a \in G$, and suppose that $a^k = e$ for some positive integer k. Let $m = \min\{n \in \mathbf{Z} : n \geq 1 \text{ and } a^n = e\}$ be the smallest positive integer with this property.

(a) Prove that $|\langle a \rangle| = m$.

(b) Prove that for any integer j, we have $a^j = e$ iff $m \mid j$.

(c) Deduce that for two integers i and j, we have $a^i = a^j$ iff $i \equiv j \pmod{m}$. Recall that we write $i \equiv j \pmod{m}$ to mean m divides $i - j$, and we read the former expression as "i is congruent to j modulo m."

Exercise 5.8. Let G be a group and let $a \in G$. Prove that if G is *finite*, then there exists a positive integer k such that $a^k = e$.

Exercise 5.9. Let S be a non-empty set. Let $X \subseteq S$ and let $H \leq \mathrm{Sym}(S)$. Let $Y = \{f(a) \mid f \in H \text{ and } a \in X\}$. Let G be the group of symmetries of Y with respect to S.

(a) Prove that G contains H.

(b) Show by an example that G may be strictly bigger than H.

(c) Consider the special case $S = \mathbf{R}^2$, and let X be the line segment from $(0, 1)$ to $(1, 0)$. Make a sketch of the corresponding set Y when $H = D_6$. Note: We view X as an arbitrary "seed" (or starting point) object, and H as a group of symmetries that we want to force upon X, resulting in the symmetric object Y. This process reminds us of the operation of a kaleidoscope, where X is a real object with no particular symmetries (say, a collection of bits of colored plastic), and Y is the image we actually see in the kaleidoscope. The principle that by applying every possible symmetry in some group H to an arbitrary object we get a highly symmetric object (invariant under H) we refer to as the *Kaleidoscope Principle*.

6

Free Groups

6.1 The Free Group Generated by a Set

Suppose we have a non-empty set S. For the sake of definiteness, let us take a 2-element set, $S = \{x, y\}$. What does $\langle S \rangle$ look like? Well, $\langle S \rangle$ contains x, y, x^{-1}, y^{-1}, as well as xy^{-1}, $x^{-1}y^{-1}$, and so on: that is, $\langle S \rangle$ is the set of all "words" formed out of the symbols x, y, x^{-1}, and y^{-1}.

But wait a moment—we never said that S was a subset of any group!

Well, we can run with this description of $\langle S \rangle$, and actually *create* a group out of these "words." Of course, in the resulting group, we expect inverses to cancel each other; but otherwise, we will not impose any relations among the elements.

Definition 6.1. Let S be a non-empty set. A *word* on S is an ordered sequence w of the form $w = s_1^{\varepsilon_1} s_2^{\varepsilon_2} \cdots s_k^{\varepsilon_k}$, where k is a non-negative integer (called the *length* of w), $s_i \in S$, and $\varepsilon_i \in \{1, -1\}$ for each i. We also write s_i for s_i^1. The word w is called *reduced* if it does not contain two adjacent terms of the form tt^{-1} or $t^{-1}t$.

Remark 6.2. A word is to be thought of as merely a succession of symbols. Two reduced words are equal if and only if they *appear* equal, i.e., if they consist of exactly the same elements of S to the same powers, written in the same order.

Remark 6.3. Notice that we allowed a word to consist of *zero* terms ($k = 0$). We call such a word the *empty* word, and denote it by e.

Example 6.4. The word $xyx^{-1}y^{-1}yyy$ is not reduced, but we can "reduce" it to get the word $xyx^{-1}yy$.

In general, reducing a word consists of eliminating all adjacent pairs which prevent the word from being reduced: that is, eliminating everything of the form tt^{-1} and $t^{-1}t$, repeatedly, until no more such pairs remain.

To "multiply" two words, we simply write them one after the other, and then reduce. For example, we have $(xyx^{-1}y^{-1}) \cdot (yyy) = xyx^{-1}y^{-1}yyy = xyx^{-1}yy$. The formal name for placing two words side by side to make a new word is *concatenation* (from the Latin word for "chain").

DOI: 10.1201/9781003252139-6

Definition 6.5. Let S be a non-empty set. The *free group on* S is the pair $(\mathrm{Fr}(S), \cdot)$, where $\mathrm{Fr}(S)$ is the set of all reduced words on S, and $\cdot$ is the operation of concatenation followed by reduction.

Remark 6.6. As we would expect, we have $xyx^{-1}y^{-1} \cdot yyy = xyx^{-1}yy$ in the free group on $\{x, y\}$. However, if we are careful to always write the $\cdot$ when multiplying, then we technically *cannot* write $xyx^{-1}y^{-1}yyy = xyx^{-1}yy$ in this group because $xyx^{-1}y^{-1}yyy$ is not even an element of the group: it is not reduced!

Note that $(\mathrm{Fr}(S), \cdot)$ really is a group; we give an informal demonstration that the group axioms are satisfied:

G1: Concatenation is associative, and reductions can be performed in any order without affecting the final result. To compute either $(w_1 \cdot w_2) \cdot w_3$ or $w_1 \cdot (w_2 \cdot w_3)$, the concatenation $w_1w_2w_3$ may as well be done first, followed by all the reductions. Thus the two expressions are equal.

G2: The empty word, e, is an identity element of $(\mathrm{Fr}(S), \cdot)$. For concatenating a word with the empty word does not change the original word.

G3: The inverse of a reduced word is another reduced word: we have

$$(s_1^{\varepsilon_1} s_2^{\varepsilon_2} \cdots s_k^{\varepsilon_k})^{-1} = s_k^{-\varepsilon_k} \cdots s_2^{-\varepsilon_2} s_1^{-\varepsilon_1};$$

so if $s_i^{\varepsilon_i}$ and $s_{i+1}^{\varepsilon_{i+1}}$ are never inverses of each other, then $s_{i+1}^{-\varepsilon_{i+1}}$ and $s_i^{-\varepsilon_i}$ are never inverses of each other.

Note that the free group on S is generated by the elements of S. So we really did find $\langle S \rangle$ in some sense, just as we originally set out to do, even though S was not originally assumed to be inside of any group! Namely, $\langle S \rangle = \mathrm{Fr}(S)$.

Remark 6.7. A free group such as $\mathrm{Fr}(\{x, y\})$ is "free" in the sense that its elements have no non-trivial relations among themselves: that is, no relations that are not forced by cancellation of symbols with their inverses. A free group is "relation-free."

By contrast, the dihedral group $D_{2n} = \langle F, R \rangle$ does have non-trivial relations, such as $FR = R^{-1}F$ and $FF = e$.

6.2   Exercises

Exercise 6.1. Let w be a word on the set $\{x, y\}$. Prove that the reduced form of w is independent of the order in which reductions are performed.

Exercise 6.2. Give a detailed proof that $\mathrm{Fr}(\{x, y\})$ is a group with the operation of concatenation followed by reduction.

Exercise 6.3. Let S be a finite set of size n. Let $k \in \mathbf{N}$.
   (a) How many words on S have length k?
   (b) How many reduced words on S have length k?

7

Group Homomorphisms

7.1 Relationships between Groups

Even though the two groups $\text{Fr}(\{x,y\}) = \langle x, y \rangle$ and $D_{2n} = \langle F, R \rangle$ are very different from each other, we sense that a relationship exists between them, since they can both be generated by two elements. To be specific, consider the correspondence $e \leftrightarrow e$, $x \leftrightarrow F$, $y \leftrightarrow R$, $xy \leftrightarrow FR$, $xyx^{-1} \leftrightarrow FRF^{-1}$, $xy^3 \leftrightarrow FR^3$, etc., where we associate a word on $\{x, y\}$ with the corresponding "word" on $\{F, R\}$.

This correspondence is not one-to-one, since different elements of $\text{Fr}(\{x, y\})$ can correspond to the same element of D_{2n}: for instance, we have $e \leftrightarrow e$ and $x^2 \leftrightarrow F^2 = e$, but $e \neq x^2$ in $\text{Fr}(\{x, y\})$. The correspondence does, however, give us a *function*

$$\omega \; : \; \text{Fr}(\{x, y\}) \to D_{2n} \tag{7.1}$$

sending a word on $\{x, y\}$ to the element of D_{2n} obtained by replacing x with F and y with R. This function expresses a natural relationship between these two groups. For ease of notation, set $H = \text{Fr}(\{x, y\})$.

Let us explore what makes this function so natural. We could say that ω is natural because it is simply a substitution rule that replaces x by F and y by R. While this is true, we seek a more "functional" explanation, which involves the relationship of ω to the group laws of the two groups in question.

To apply a group law, we need two group elements. So let us choose two elements a and b in H. For example, set $a = xyx^{-1}$ and $b = yyx$. Then we have $\omega(a) = FRF^{-1}$ and $\omega(b) = RRF$. Playing around with products, we see that $\omega(a \cdot b) = \omega(xyx^{-1}yyx) = FRF^{-1}RRF$, which is the same as $\omega(a) \cdot \omega(b)$. This phenomenon occurs for arbitrary elements a and b in H, as you should convince yourself; it forms the basis of the following definition.

Definition 7.1. Let $(G_1, \cdot_1)$ and $(G_2, \cdot_2)$ be groups. A *group homomorphism* from G_1 to G_2 is a function $\sigma \; : \; G_1 \to G_2$ such that

$$\forall a, b \in G_1, \; \sigma(a \cdot_1 b) = \sigma(a) \cdot_2 \sigma(b).$$

Remark 7.2. We arrived at the notion of group homomorphism by noting a relationship which appeared to exist between two particular groups. In general, the fundamental way (some might say the *only* way) that two groups can be

DOI: 10.1201/9781003252139-7

related is via a group homomorphism from one to the other. In the same vein, we seldom if ever care about any function from a group to another group unless that function is a group homomorphism.

Example 7.3. The function ω in Equation 7.1 is a group homomorphism from $\mathrm{Fr}(\{x,y\})$ to D_{2n}.

We shall see several more examples in the following section.

Next we establish the result which says that a group homomorphism respects identity elements, inverses, and products of finitely many factors to arbitrary integer powers.

Proposition 7.4. *Let* $\sigma : G_1 \to G_2$ *be a group homomorphism, and let* e_i *be the identity element of* G_i. *Then we have:*

(i) $\sigma(e_1) = e_2$.

(ii) For all $a \in G_1$, $\sigma(a^{-1}) = (\sigma(a))^{-1}$.

(iii) $\sigma(a_1^{n_1} a_2^{n_2} \cdots a_k^{n_k}) = (\sigma(a_1))^{n_1}(\sigma(a_2))^{n_2} \cdots (\sigma(a_k))^{n_k}$ *for all* a_1, a_2, ..., a_k *in* G_1 *and all* n_1, n_2, ..., n_k *in* $\mathbf{Z}$.

Proof. (i) We have $\sigma(e_1 \cdot e_1) = \sigma(e_1) \cdot \sigma(e_1)$ since σ is a group homomorphism. Also, $e_1 \cdot e_1 = e_1$ since e_1 is an identity element. Therefore,

$$\sigma(e_1) = \sigma(e_1) \cdot \sigma(e_1). \tag{7.2}$$

Since $\sigma(e_1) \in G_2$ and e_2 is the identity element of G_2, we also have

$$\sigma(e_1) = e_2 \cdot \sigma(e_1). \tag{7.3}$$

Using the Right Cancellation Law, we find $\sigma(e_1) = e_2$, as desired.

(ii) Let $a \in G_1$. Then we have $\sigma(a \cdot a^{-1}) = \sigma(a) \cdot \sigma(a^{-1})$ by the fact that σ is a group homomorphism. Also, $a \cdot a^{-1} = e_1$, so $\sigma(a \cdot a^{-1}) = \sigma(e_1)$, $= e_2$ by part (i) above. Therefore, $\sigma(a) \cdot \sigma(a^{-1}) = e_2$, so $\sigma(a^{-1})$ is the inverse of $\sigma(a)$.

(iii) First, the case $k = 1$ follows when $n_1 = 0$ from part (i) above; for positive powers n_1, it follows from the defining property of a group homomorphism by induction on n_1; for negative powers n_1, the proof is similar, but also uses the result of part (ii) above. Then the general case $k > 1$ follows by induction on k, again using the definition of group homomorphism. The reader is asked to give a detailed proof in Exercise 7.3. $\qquad\square$

The defining property of a group homomorphism $\sigma : G_1 \to G_2$ can also be expressed as follows:

To compute σ of the product $a \cdot b$, we can either compute $\sigma(a)$ and $\sigma(b)$, then take the product of the results; or compute the product $a \cdot b$, and then take σ of this product. The result is the same in both cases.

It seems that we are saying that the order in which we apply the group law and apply σ doesn't matter. In other words, σ "commutes" with the group law. But in fact, there are actually two different group laws here, so we need to clarify what we mean. It will help to first rewrite the defining property of a group homomorphism in function notation instead of operator notation.

Recall that the group law $\cdot_1$ is technically a function

$$\cdot_1 \ : \ G_1 \times G_1 \to G_1.$$

Further, recall that in our operator notation for group laws, $a \cdot_1 b$ is just another way to write $\cdot_1(a, b)$; and similarly for the group law of G_2. Thus we can rewrite the property that σ is a group homomorphism as

$$\sigma(\cdot_1(a, b)) = \cdot_2(\sigma(a), \sigma(b)). \tag{7.4}$$

A diagram of functions will help us explain a precise sense in which σ "commutes" with the two group laws.

$$
\begin{array}{ccc}
G_1 \times G_1 & \xrightarrow{\ \sigma \times \sigma\ } & G_2 \times G_2 \\
\downarrow{\scriptstyle \cdot_1} & & \downarrow{\scriptstyle \cdot_2} \\
G_1 & \xrightarrow{\ \sigma\ } & G2
\end{array}
\tag{7.5}
$$

In Diagram 7.5, four sets and four functions are represented. The bottom row is simply the homomorphism σ. The two sides are exactly the group operations of G_1 and G_2. Finally, the function

$$\sigma \times \sigma \ : \ G_1 \times G_1 \to G_2 \times G_2$$

is defined by the natural-seeming formula

$$(\sigma \times \sigma)(a, b) = (\sigma(a), \sigma(b)).$$

Starting with an element (a, b) in the upper-left corner of the diagram, there are two possible ways to reach the lower-right corner: we can apply $\sigma \times \sigma$ followed by $\cdot_2$, or we can apply $\cdot_1$ followed by σ. Equation 7.4 says that we get the same result no matter which of these two paths we take. Another way to say this is:

$$(\cdot_2) \circ (\sigma \times \sigma) = \sigma \circ (\cdot_1). \tag{7.6}$$

We may describe the property of "commutativity" by saying that the result of an operation is independent of the order in which we choose the inputs. Equation 7.6 says that the result in G_2 of any input from $G_1 \times G_1$ is independent of the order in which we choose to move in Diagram 7.5: it doesn't matter whether we first move right and then down, or first down and then right. Thus, we would like to say that there is a sense in which Diagram 7.5 is "commutative." We make this notion precise in the context of a general function diagram by means of the following definition.

Definition 7.5. A function diagram is *commutative* if, for any two paths from a set A in the diagram to another set B in the diagram, the corresponding composite functions are equal. We also express this by saying that the diagram *commutes*.

Thus, our final restatement of the group homomorphism property is: A function $\sigma \ : \ G_1 \to G_2$ is a group homomorphism iff Diagram 7.5 commutes.

7.2   Kernels: How Much Did We Lose?

We now turn to the subject of relations within a group. First, notice that every relation can be rewritten in an equivalent form such that the right-hand side of the relation is e:

$$x_1 x_2 \cdots x_m = y_1 y_2 \cdots y_n \iff x_1 x_2 \cdots x_m y_n^{-1} \cdots y_2^{-1} y_1^{-1} = e. \qquad (7.7)$$

Definition 7.6. We shall refer to the right-hand side of the biconditional in Statement 7.7 as the *standard form* of the relation on the left-hand side.

Let us return to the group homomorphism ω of Equation 7.1. We will interest ourselves in the relations of D_{2n}. As we have seen, three of the relations in D_{2n} are $F^2 = e$, $R^n = e$, and $FR = R^{-1}F$; the last relation can be rewritten as $FRF^{-1}R = e$. Now, we have defined a "relation" to be an *equation*; and equations are useful tools in algebra, but we prefer not to make a study of equations as a class of objects (we leave this to the logicians). Luckily, every relation has an equivalent standard form in which the right-hand side is just e, and so it might appear that every relation can be given by a single group element, namely, the left-hand side of this standard form. For instance, the group elements corresponding to the three relations in D_{2n} stated above are F^2, R^n, and $FRF^{-1}R$.

But wait a moment: each of these three group elements is just e; that was the whole point! So it is not accurate to say that a relation is determined by the left-hand side of its equivalent standard form *thought of as a single group element*. Rather, we want to think of a relation as a *formal list of symbols*. This seems familiar: it reminds us of words in a free group.

Indeed, consider the elements x^2, y^n, and $xyx^{-1}y$ in the free group on $\{x, y\}$. These three elements are certainly *not* equal to e (nor to each other). The fact that their images under ω are all e is the telling point. We realize that the elements of $\mathrm{Fr}(\{x, y\})$ which map to e under ω exactly correspond to relations in D_{2n}. To further our study of relations in a group, we therefore make the following definition.

Definition 7.7. Let G_1 and G_2 be groups, and let e_i denote the identity element of G_i. The *kernel* of a group homomorphism $\sigma : G_1 \to G_2$ is the set $\ker(\sigma) := \sigma^{-1}(\{e_2\}) = \{a \in G_1 : \sigma(a) = e_2\}$.

Remark 7.8. In the definition of kernel, the group G_1 need not be free; the definition applies to any group homomorphism between any two groups.

We will now describe the connection between kernels and relations in the general case; the reader should keep in mind the example of the map from $\mathrm{Fr}(\{x, y\})$ to D_{2n} given above. We see from Proposition 7.4 that if some elements $a_1, \ldots, a_k$ of G_1 satisfy a relation in G_1, then their images $\sigma(a_1), \ldots, \sigma(a_k)$ satisfy the same relation, but in G_2. For example, if $a_1 a_2^{-1} a_3^4 = e_1$, then we must also have $\sigma(a_1)\sigma(a_2)^{-1}\sigma(a_3)^4 = e_2$. But the converse does not hold: $\sigma(a_1), \ldots, \sigma(a_k)$ may satisfy additional relations not satisfied by $a_1, \ldots, a_k$. The extent to which we add new relations in going from G_1 to G_2 is measured by the kernel $\ker(\sigma)$. A more informal way to express this is to say that the kernel measures how much G_1 "collapses" in being transported to G_2.

Example 7.9. Let $G_1 = \mathbf{R}^\times := \mathbf{R} - \{0\}$ under ordinary multiplication. Let $G_2 = \{1, -1\}$, also under multiplication. Consider the "sign" function σ : $G_1 \to G_2$ defined by

$$\sigma(x) = \begin{cases} 1, & \text{if } x > 0; \\ -1, & \text{if } x < 0. \end{cases}$$

To check that σ is a group homomorphism, we must check that $\sigma(x) \cdot \sigma(y) = \sigma(xy)$ for all non-zero real numbers x and y. But this statement merely expresses the familiar rules about signs of products: positive times positive is positive, positive times negative is negative, and so on. We have $\ker(\sigma) = \{x \in \mathbf{R}^\times : \sigma(x) = 1\} = \mathbf{R}^+$, the set of all positive real numbers. The map σ collapses all positive numbers onto 1 and all negative numbers onto -1, thus simplifying the structure of the group $\mathbf{R}^\times$ down to the sign rules referred to above. In fact, it may be instructive to replace the set $\{1, -1\}$ with the set $\{+, -\}$ for G_2 in this example.

Example 7.10. Consider the absolute value function σ : $\mathbf{R}^\times \to \mathbf{R}^\times, x \mapsto |x|$. The well-known identity

$$|x \cdot y| = |x| \cdot |y| \tag{7.8}$$

says precisely that σ is a group homomorphism. We have $\ker(\sigma) = \{x \in \mathbf{R}^\times : |x| = 1\} = \{1, -1\}$. This corresponds to the fact that in taking the absolute value of a number, we lose the information about its sign, but nothing else is lost.

Example 7.11. Let $\mathbf{R}^+$ denote the set of all positive real numbers. Consider the natural logarithm function $\ln$: $\mathbf{R}^+ \to \mathbf{R}$, where we take the group operation on $\mathbf{R}^+$ to be multiplication and that on $\mathbf{R}$ to be addition. The identity

$$\ln(x \cdot y) = \ln(x) + \ln(y) \tag{7.9}$$

is exactly the condition we need for ln to be a group homomorphism; notice that the operation on the left-hand side of Equation 7.9, inside the logarithm, is multiplication in $\mathbf{R}^+$, while that on the right is addition in $\mathbf{R}$. We have $\ker(\ln) = \{x \in \mathbf{R}^+ : \ln(x) = 0\} = \{1\}$. This is the smallest of any of the kernels we have seen so far. We know in general (by Proposition 7.4) that every group homomorphism takes the identity element of its domain to the identity element of its codomain, so the kernel in the present example is as small as it could be. This tells us that the natural logarithm function loses *no* information when transporting the positive real numbers under multiplication to the real numbers under addition. Indeed, the natural logarithm function is invertible. The fact that the natural logarithm is a group homomorphism with kernel $\{1\}$

says that ln faithfully translates multiplication into addition. Because addition is easier to perform than multiplication, logarithms were used in former times to ease the computational burden of human calculators (see Exercise 7.4). This is the reason for the explosive popularity of the logarithm upon its original discovery, although the idea of a "group" was then unknown.

Example 7.12. Let G be any group, and let $S = \{x_g \; : \; g \in G\}$; that is, S is a set containing one symbol for each element of G. Let $F = \text{Fr}(S)$ be the free group on S. (The reader should think about why we choose not to simply consider $\text{Fr}(G)$.) We can define a function $\tau \; : \; S \to G$ by the formula $\tau(x_g) = g$. There is a unique way to extend τ to a group homomorphism $\sigma \; : \; F \to G$, namely, by taking a word in the x_g's to the corresponding term in the g's. Clearly, σ is surjective; for the image of τ is already all of G. We say that every group is the homomorphic image of a free group.

One theme in group theory is that many "naturally defined" subsets of a group turn out to be subgroups. The next result confirms this principle in two instances of sets which are naturally associated with a group homomorphism: the image and the kernel.

Theorem 7.13 (Images and kernels are subgroups). *Let $\sigma \; : \; G_1 \to G_2$ be a group homomorphism. Then*
   (i) $\sigma(G_1) \leq G_2$, and
   (ii) $\ker(\sigma) \leq G_1$.

Proof. We use the Subgroup Test for both parts.
   (i) **ST0:** $\sigma(G_1)$ is non-empty since $\sigma(e_1) \in \sigma(G_1)$.
   ST1: Let $y \in \sigma(G_1)$. [Show $y^{-1} \in \sigma(G_1)$] Then $\exists x \in G_1$ such that $y = \sigma(x)$ (by definition of $\sigma(G_1)$). So $\sigma(x^{-1}) = \sigma(x)^{-1}$ (by Proposition 7.4 part (ii)) $= y^{-1}$. But $x^{-1} \in G_1$, so $y^{-1} \in \sigma(G_1)$.
   ST2: Let $y, z \in \sigma(G_1)$. [Show $yz \in \sigma(G_1)$] Then $\exists w, x \in G_1$ such that $y = \sigma(w)$ and $z = \sigma(x)$ (by definition of $\sigma(G_1)$). So $yz = \sigma(w)\sigma(x) = \sigma(wx)$ (by definition of group homomorphism). But $wx \in G_1$, so $yz \in \sigma(G_1)$.
   (ii) **ST0:** We have $\sigma(e_1) = e_2$ by Proposition 7.4 part (i). Therefore, $e_1 \in \ker(\sigma)$ (by definition of kernel). So $\ker(\sigma)$ is non-empty.
   ST1: Let $x \in \ker(\sigma)$. [Show $x^{-1} \in \ker(\sigma)$] Then $\sigma(x) = e_2$ (by definition of kernel). So $\sigma(x^{-1}) = \sigma(x)^{-1}$ (by Proposition 7.4 part (ii)) $= e_2^{-1} = e_2$. Thus we have $x^{-1} \in \ker(\sigma)$.
   ST2: Let $w, x \in \ker(\sigma)$. [Show $wx \in \ker(\sigma)$] Then $\sigma(w) = \sigma(x) = e_2$ (by definition of kernel). So $\sigma(wx) = \sigma(w)\sigma(x)$ (by definition of group homomorphism) $= e_2 \cdot e_2 = e_2$. Therefore, $wx \in \ker(\sigma)$.   □

As a first application of Theorem 7.13, we refine Example 7.12.

Example 7.14. Let G be any group, and suppose that T is a subset of G which generates G: that is, $\langle T \rangle = G$. Instead of taking one symbol for each element of G as we did in Example 7.12, it is enough to take one symbol for each element of T. So let $S = \{x_g \; : \; g \in T\}$. There is again a unique group homomorphism σ : $\text{Fr}(S) \to G$ such that $\sigma(x_g) = g$ for each g in T. Since the image of σ contains T, and is a subgroup of G by Theorem 7.13, then it contains $\langle T \rangle$, which is all of G; so σ is surjective.

Theorem 7.13 says that the image and the kernel of a group homomorphism are subgroups of the codomain and the domain, respectively. This result raises the question of whether a sort of converse might be true: given a group G and a subgroup $H \leq G$, we ask (1) Is H the image of some group homomorphism with codomain G?, and (2) Is H the kernel of some group homomorphism with domain G?

To answer question (1), consider the function σ : $H \to G$, $\sigma(x) = x$; that is, σ is just the identity function on H. It is easy to see that σ is a group homomorphism with image H. So the answer to our first question is Yes.

The second question asks, is every subgroup a kernel of a group homomorphism defined on its parent group? Let us consider a small example. Let $G_1 = S_3$, and let $\alpha = \begin{bmatrix} 1 & 2 & 3 \\ 2 & 1 & 3 \end{bmatrix} \in S_3$. Set $H = \langle \alpha \rangle$. Then H is a group of order 2. Suppose that there exists a group homomorphism σ : $G_1 \to G_2$, for some group G_2, such that $\ker(\sigma) = H$. Then we have $\sigma(\alpha) = e_2$, $\sigma(e_1) = e_2$, and no other elements of G_1 are taken to e_2 by σ. If β is any element of G_1, then we must have $\sigma(\alpha\beta) = \sigma(\alpha)\sigma(\beta) = e_2 \cdot \sigma(\beta) = \sigma(\beta)$; similarly, $\sigma(\beta\alpha) = \sigma(\beta)\sigma(\alpha) = \sigma(\beta) \cdot e_2 = \sigma(\beta)$. In fact, using Proposition 7.4, we can see that all of the α's "go away" when we apply σ: for example, we have

$$\sigma(\beta^3 \alpha \beta \alpha^2) = \sigma(\beta^3 \beta) = \sigma(\beta^4).$$

One of the simplest such formulas we can obtain in this manner is

$$\sigma(\beta\alpha\beta^{-1}) = \sigma(\beta\beta^{-1}) = \sigma(e_1) = e_2.$$

Therefore, we have

$$\beta\alpha\beta^{-1} \in \ker(\sigma). \tag{7.10}$$

Now let us take $\beta = \begin{bmatrix} 1 & 2 & 3 \\ 1 & 3 & 2 \end{bmatrix}$. Calculating in S_3, we find

$$\beta\alpha\beta^{-1} = \begin{bmatrix} 1 & 2 & 3 \\ 3 & 2 & 1 \end{bmatrix} \notin \{e_1, \alpha\} = \ker(\sigma). \tag{7.11}$$

This contradiction proves that H cannot be the kernel of *any* group homomorphism whose domain is S_3.

Thus, the answer to our second question is No. The discussion that led to our negative answer will also provide insight into the reason behind that answer. Equation 7.10 was the obstacle which prevented H from being a kernel. Motivated by this condition, we make two definitions:

Definition 7.15. Let G be a group, and let $\alpha, \gamma \in G$. We say that α and γ are *conjugates* of each other (in G) if there exists $\beta \in G$ such that $\gamma = \beta\alpha\beta^{-1}$.

In this case we also say that γ is the *conjugate* of α by β. The reader should check that the relationship of being conjugates is indeed symmetric: α and γ are conjugates iff γ and α are conjugates.

Definition 7.16. Let G be a group and let $H \leq G$. Then H is *normal* in G if

$$\forall \alpha \in H \; \forall \beta \in G, \; \beta \alpha \beta^{-1} \in H.$$

Notation 7.17. We write $H \trianglelefteq G$ to denote that H is normal in G.

Remark 7.18. The definition of normal says that H is normal in G if H is a subgroup of G which is "closed under conjugation by arbitrary elements of G."

Theorem 7.19. *Let $\sigma : G_1 \to G_2$ be a group homomorphism. Then $\ker(\sigma) \trianglelefteq G_1$.*

Proof. Let $K = \ker(\sigma)$. From Theorem 7.13, we know that $K \leq G_1$. It remains to show that K is closed under conjugation by arbitrary elements of G_1. So let $\alpha \in K$ and let $\beta \in G_1$. [Show: $\beta \alpha \beta^{-1} \in K$; that is, show $\sigma(\beta \alpha \beta^{-1}) = e_2$] We compute

$$
\begin{aligned}
\sigma(\beta \alpha \beta^{-1}) &= \sigma(\beta)\sigma(\alpha)(\sigma(\beta))^{-1} && \text{(by Proposition 7.4)} \\
&= \sigma(\beta) \cdot e_2 \cdot (\sigma(\beta))^{-1} && \text{(since } \alpha \in K = \ker(\sigma)) \\
&= \sigma(\beta) \cdot (\sigma(\beta))^{-1} \\
&= e_2.
\end{aligned}
$$

This shows that $\beta \alpha \beta^{-1} \in K$, as desired. $\square$

7.3   Cosets

Being a subgroup was not enough to be a kernel; is it too much to hope that every *normal* subgroup is a kernel? Is there a converse to Theorem 7.19? Given a normal subgroup $K \trianglelefteq G$, we would like to dream up a group H and a group homomorphism $\sigma : G \to H$ such that $\ker(\sigma) = K$.

Suppose that $x \in G$, $y \in H$, and $\sigma(x) = y$. Our starting point will be to treat x as an unknown and solve this equation: that is, to investigate the pre-image of y under σ, i.e. $\sigma^{-1}(\{y\})$. Because we will write this so often, we will take the liberty of dropping the set braces and writing simply $\sigma^{-1}(y)$; this does *not* imply that σ has an inverse function! Well, what is $\sigma^{-1}(y)$? Suppose to start with that we have some solution $z \in \sigma^{-1}(y)$. Let us choose an arbitrary $w \in \sigma^{-1}(y)$. Then our idea is that w and z are treated the same way by σ; so w and z are "equivalent" according to σ; thus $z^{-1}w$ should be equivalent to e_G according to σ, since a group homomorphism respects group relations. But a group homomorphism sends the identity element to the identity element, so we should have $\sigma(z^{-1}w) = e_H$.

More precisely, we calculate $\sigma(z^{-1}w) = \sigma(z)^{-1}\sigma(w) = y^{-1}y = e_H$. Thus, we can say that $z^{-1}w \in \ker(\sigma) = K$. By setting $k = z^{-1}w$, we can write $w = zk$.

Conversely, it is not hard to check that any element of G of the form $w = zk$, where $k \in K$, has the property that $w \in \sigma^{-1}(y)$. Thus we want to assert that

$$\sigma^{-1}(y) = \{zk \mid k \in K\}. \tag{7.12}$$

Because of the natural importance of this type of set, we make a definition:

Definition 7.20. Let T be a set with an associative binary operation (written as multiplication), and let $S \subseteq T$. Let $z \in T$. Then

$$zS := \{zs \mid s \in S\}$$

and

$$Sz := \{sz \mid s \in S\}.$$

We call zS a *left coset of S in T* and Sz a *right coset of S in T*.

Remark 7.21. In the definition of coset above, we only required S to be a *subset* of the *set T*; in our previous discussion which motivated this definition, the place of S was taken by a *normal subgroup* of a *group*. In practice, the concept of a coset of S in T is most useful when T is a group and S is a subgroup of T, though not necessarily a normal subgroup.

We summarize our discussion so far by stating the following lemma and corollary; the reader is asked to supply a complete proof of the lemma in Exercise 7.5.

Lemma 7.22. *Let $\sigma : G \to H$ be a group homomorphism with kernel K. Then the pre-image of an element in the range of σ is a coset of K in G. More precisely, let $x \in G$ and let $y = \sigma(x)$; then $\sigma^{-1}(y) = xK = Kx$.* $\square$

Corollary 7.23. *If $\sigma : G \to H$ is a group homomorphism with kernel K, then there is a bijective correspondence ψ from the set of all left cosets of K in G to the image of σ, given by $\psi(xK) = \sigma(x)$ for x in G.*

Proof. First, we check that ψ is well-defined. Suppose that a given coset C of K in G can be written in two ways, as $C = xK = yK$, with $x, y \in G$. [We must show that $\sigma(x) = \sigma(y)$.] Then we have $e_G \in K$, so $x = x \cdot e_G \in xK$. Since $xK = yK$, we have $x \in yK$, so $x = yk$ for some $k \in K$. Thus $\sigma(x) = \sigma(yk) = \sigma(y)\sigma(k) = \sigma(y)e_H = \sigma(y)$, as required.

Next, we show that ψ is injective. Suppose that $\psi(C) = \psi(D)$ where C and D are left cosets of K in G. Then we can write $C = xK$ and $D = wK$ for some $x, w \in G$, by definition of left coset; and using the definition of ψ, we have $\sigma(x) = \sigma(w) =: y$. By Lemma 7.22, we have $\sigma^{-1}(y) = xK$ and also $\sigma^{-1}(y) = wK$, so $C = D$.

Finally, we show that ψ is surjective. Let $y \in \sigma(G)$. Then $y = \sigma(x)$ for some $x \in G$. So $y = \psi(xK)$, and y is in the image of ψ. $\square$

Before resuming our quest for a converse to Theorem 7.19, we establish some natural properties of the multiplication of group elements with cosets.

Lemma 7.24. *Let G be a group. Let $S \subseteq G$ and $x, y \in G$. Then*
  (i) $x(yS) = (xy)S$, and
  (ii) $x(Sy) = (xS)y$.

Proof. We prove (i); the proof of (ii) is similar. Let $T = \{xyz \ : \ z \in S\}$. We will show that $T = x(yS)$.
  [Show $T \subseteq x(yS)$] Let $t \in T$. Then $t = xyz$ for some $z \in S$. Now $yz \in yS$, so $t = xyz = x(yz) \in x(yS)$.
  [Show $x(yS) \subseteq T$] Let $w \in x(yS)$. Then $w = xa$ for some $a \in yS$, and $a = yb$ for some $b \in S$. So $w = x(yb) = xyb \in T$.
  We have shown that $T = x(yS)$; similarly, it can be shown that $T = (xy)S$. Thus $x(yS) = (xy)S$. $\square$

Lemma 7.22 says, in part, that the left and right cosets aK and Ka are the same, when K is the kernel of a group homomorphism. Since we are wondering whether every normal subgroup is also a kernel, it makes sense to ask whether aN and Na are equal when N is a normal subgroup; the affirmative answer, given in the following lemma, brings us another step closer to our goal:

Lemma 7.25. *If $N \trianglelefteq G$ and $a \in G$, then $aN = Na$.*

Proof. Suppose that $N \trianglelefteq G$ and $a \in G$. [Show $aN \subseteq Na$] Let $x \in aN$. Then $x = ay$ for some $y \in N$. [Show $x = za$ for some $z \in N$] [Then $z = xa^{-1}$] Let $z = xa^{-1}$. Then $z = aya^{-1}$, the conjugate of the element $y \in N$ by the element $a \in G$. By definition of normal, we have $z \in N$. Now $x = ay = aya^{-1}a = za \in Na$. Therefore, $aN \subseteq Na$. Similarly, $Na \subseteq aN$, which concludes the proof. $\square$

7.4   Quotient Groups

Next, we shift our creativity into high gear to complete our program of constructing a group homomorphism with prescribed domain and kernel. So suppose again that G is a group and $N \trianglelefteq G$. Suppose that $\sigma : G \to H$ is a group homomorphism with kernel N, as desired. Observe that we are free to replace H by any subgroup L of H, as long as L contains the image $\sigma(G)$. Indeed, the codomain of a function is arbitrary except for the requirement that it must contain the image of the domain. We know from Theorem 7.13 that the image

$\sigma(G)$ itself is a subgroup of H. So we may take $L = \sigma(G)$ and consider σ as a group homomorphism from G to the group $\sigma(G)$,

$$\sigma \; : \; G \to \sigma(G).$$

Note that we have not changed the map σ at all; we have merely changed our point of view about the codomain of σ. From this point of view, σ is *surjective*. So our first insight is that we may as well look for a *surjective* group homomorphism σ with domain G and kernel N.

Thus, to further our discussion, we assume that $\sigma : G \to H$ is a surjective group homomorphism. Now, σ may not be one-to-one, but the Corollary to Lemma 7.22 says that we do have a one-to-one correspondence between the *cosets* of N in G and the elements of H. To be specific, let us use left cosets. Then σ carries each left coset xN (for $x \in G$) to a single element $\sigma(x)$ of H. The left cosets of N in G mirror the elements of H.

Now recall that we were given G and N, with $N \trianglelefteq G$, and we had to find H and σ (if possible) such that $\sigma : G \to H$ is a group homomorphism with kernel N. Our final idea in this program is quite radical. We have seen that *if* a suitable H and σ exist, then we may assume that σ is surjective, and then the elements of H must be in bijective correspondence, via σ^{-1}, with the left cosets of N in G. Our idea is this: We will make a new group Q *out of* the left cosets of N in G, such that $\sigma^{-1} : H \to Q$ is a bijective group homomorphism.

We proceed to deduce what the group law on these cosets must be. First, the reader should verify that the inverse of σ^{-1} isn't actually σ, but instead the function $\tau : Q \to H$ such that $\tau(xN) = \sigma(x)$ for x in G. By Exercise 7.10, τ must be a group homomorphism. For $xN, yN \in Q$, we have $\tau(xN \cdot yN) = \tau(xN)\tau(yN) = \sigma(x)\sigma(y) = \sigma(xy)$ (since σ is a homomorphism), $= \tau((xy)N)$. Now τ is bijective, so in particular τ is injective, and we must have $xN \cdot yN = (xy)N$.

We have shown that, given groups N and G with $N \trianglelefteq G$, if there is a group homomorphism with domain G and kernel N, then the set of all left cosets of N in G is itself a group, with multiplication defined in a very natural way, namely $xN \cdot yN = (xy)N$. This motivates the following result.

Theorem 7.26. *Let G be a group and let N be a normal subgroup of G. Let*

$$Q = \{xN \; : \; x \in G\},$$

the set of all left cosets of N in G. Define a binary operation $\cdot$ on Q by

$$xN \cdot yN = (xy)N.$$

Then Q is a group, called the quotient group of G by N, and written G/N (pronounced "G mod N").

Proof. First, we must show that $\cdot$ is well-defined on Q. Suppose that $x_1, y_1, x_2,$ $y_2 \in G$ and that $x_1 N = x_2 N$ and $y_1 N = y_2 N$. [Show $(x_1 y_1)N = (x_2 y_2)N$] Then $(x_1 y_1)N = x_1(y_1 N) = x_1(y_2 N) = x_1(N y_2)$ (by Lemma 7.25) $= (x_1 N)y_2$ $= (x_2 N)y_2 = (N x_2)y_2 = N(x_2 y_2) = (x_2 y_2)N$.

Next, we verify the group axioms.

G1: Let $x, y, z \in Q$. Then we can write $x = aN$, $y = bN$, $z = cN$ for some $a, b, c \in G$. We have $(xy)z = (aN \cdot bN) \cdot cN = (ab)N \cdot cN = ((ab)c)N$ $= (a(bc))N$ (by Axiom G1 for the group G) $= aN \cdot (bc)N = aN \cdot (bN \cdot cN)$ $= x(yz)$.

G2: We claim that N is an identity element for $\cdot$ on Q. Notice that $N = e_G N$, so N is a left coset of N in G, hence $N \in Q$. Further, if $a \in G$, then we have $aN \cdot e_G N = (a e_G)N = aN$, and similarly, $e_G N \cdot aN = aN$.

G3: We can see that $a^{-1}N$ is an inverse for aN, since $a^{-1}N \cdot aN = e_G N = aN \cdot a^{-1}N$. $\qquad\square$

Remark 7.27. Theorem 7.26 requires N to be a normal subgroup of G; the theorem is not true otherwise. The reader should examine the proof to see where the assumption of normality is used: it is only used to verify that the group law on Q is well-defined.

Remark 7.28. We reiterate that an element of G/N is a *set*: more precisely, a left coset of N in G. The set N itself is the identity element of G/N.

Example 7.29. Let $n \in \mathbf{Z}$. We claim that $\langle n \rangle \trianglelefteq (\mathbf{Z}, +)$. First, recall that $\langle n \rangle$ is just the set of all integer multiples of n, so that we can write $\langle n \rangle = n\mathbf{Z}$. To show that $n\mathbf{Z} \trianglelefteq \mathbf{Z}$, we must show that for all $x \in \mathbf{Z}$ and $y \in n\mathbf{Z}$, we have $xyx^{-1} \in n\mathbf{Z}$. Since we use additive notation in $(\mathbf{Z}, +)$, xyx^{-1} really means $x + y - x$, which is just y, and hence is in $n\mathbf{Z}$, as desired.

The elements of $\mathbf{Z}/n\mathbf{Z}$ are the cosets of $n\mathbf{Z}$ in $\mathbf{Z}$. They look like $a + n\mathbf{Z}$ where $a \in \mathbf{Z}$. This is exactly the *congruence class* of all integers congruent to a modulo n.

As a special case, when $n = 3$, the elements of $\mathbf{Z}/n\mathbf{Z}$ are:

$$0 + 3\mathbf{Z} = \{\ldots, -6, -3, 0, 3, 6, \ldots\}$$

$$1 + 3\mathbf{Z} = \{\ldots, -5, -2, 1, 4, 7, \ldots\}$$

$$2 + 3\mathbf{Z} = \{\ldots, -4, -1, 2, 5, 8, \ldots\}$$

These three cosets together contain all the integers. Any coset of $\langle 3 \rangle$ in $\mathbf{Z}$ can be realized as one of these three. For instance, $8 + 3\mathbf{Z}$ contains $8 + 3 \cdot 0 = 8$, $8 + 3 \cdot -1 = 5$, $8 + 3 \cdot 1 = 11$, etc.; that is, all integers congruent to 8 modulo 3, which is the same as all integers congruent to 2 modulo 3. So $8 + 3\mathbf{Z} = 2 + 3\mathbf{Z}$.

In general, when $n > 0$, the order of $\mathbf{Z}/n\mathbf{Z}$ is the number of congruence classes modulo n in $\mathbf{Z}$, which is n: these congruence classes are $0 + n\mathbf{Z}, 1 + n\mathbf{Z}$, $\ldots, (n-1) + n\mathbf{Z}$. We also use the notation $\mathbf{Z}_n$ for $\mathbf{Z}/n\mathbf{Z}$ when $n > 0$. The notation $\mathbf{Z}_n$ is used primarily in elementary texts on abstract algebra to denote the quotient group $\mathbf{Z}/n\mathbf{Z}$. In the branch of mathematics known as number theory, the notation $\mathbf{Z}_p$ (when p is a prime integer) denotes an entirely different object, the set of so-called "p-adic integers," which we shall not study here.

We are finally ready to prove a converse to Theorem 7.19. Given a group G and a subgroup $N \trianglelefteq G$, the quotient group G/N provides a target for a group homomorphism with domain G and kernel N:

Theorem 7.30. *Let G be a group and let $N \trianglelefteq G$. Define a function σ : $G \to G/N$ by $a \mapsto aN$. Then σ is a surjective group homomorphism, and $\ker(\sigma) = N$. We call σ the* natural map *from G to G/N.*

Proof. Suppose that $N \trianglelefteq G$, and define σ as above. [Show σ is a homomorphism] Let $a, b \in G$. Then $\sigma(a \cdot b) = (a \cdot b)N$ (by definition of σ) $= aN \cdot bN$ (by definition of the group law on G/N) $= \sigma(a) \cdot \sigma(b)$. So σ is a group homomorphism.

[Show σ is surjective] Let $x \in G/N$. Then $x = aN$ for some $a \in G$. So $x = \sigma(a)$. Thus, σ is surjective.

[Show $\ker(\sigma) = N$]

[Show $\ker(\sigma) \subseteq N$] Let $a \in \ker(\sigma)$. Then $a \in G$ and $\sigma(a) = e_{G/N}$, by definition of kernel. But $e_{G/N} = N$, and $\sigma(a) = aN$. Thus we have $aN = N$. Since $N \leq G$, we have $e_G \in N$, which gives $a \cdot e_G \in aN = N$. Thus, $a \in N$. We have shown that $\ker(\sigma) \subseteq N$.

[Show $N \subseteq \ker(\sigma)$] Let $a \in N$. [Show $a \in \ker(\sigma)$; this means $\sigma(a) = e_{G/N} = N$] Then $\sigma(a) = aN$ (by definition of σ). [Show $aN = N$] Let $x \in aN$. Then $x = ab$ for some $b \in N$. So $x \in N$ (by ST2). Conversely, if $y \in N$, then $y = aa^{-1}y$, and $a^{-1}y \in N$ (by ST1 and ST2), so $y \in aN$. Thus $aN = N$, so $\sigma(a) = N = e_{G/N}$. Hence $a \in \ker(\sigma)$, and we have shown that $N \subseteq \ker(\sigma)$.

Therefore, $\ker(\sigma) = N$, which completes the proof. $\square$

Corollary 7.31 (kernel $\longleftrightarrow$ normal). *Let G be a group and let $S \subseteq G$. Then the following are equivalent:*

(i) there exists a group homomorphism with domain G and kernel S;
(ii) $S \trianglelefteq G$.

Proof. Combine Theorem 7.19 and Theorem 7.30. $\square$

7.5 Exercises

Exercise 7.1. Suppose that we tried to define a function

$$\theta : D_{2n} \to \mathrm{Fr}(\{x, y\})$$

like the function ω of Equation 7.1, but in the opposite direction, sending a word on $\{F, R\}$ to the corresponding word on $\{x, y\}$. What is wrong with such a definition?

Exercise 7.2. Consider a group $(G, \cdot)$. Create a function diagram with the property that G is a commutative group iff the diagram commutes. Hint: Use G and $G \times G$ at least once each, and find appropriate functions.

Exercise 7.3. Provide a detailed proof of Proposition 7.4 part (iii) by following the steps outlined in the proof given in the text.

Exercise 7.4. While not strictly an exercise in abstract algebra, this problem explores the practical significance of the logarithm function stemming from the fact that it is a bijective group homomorphism. Let us consider three different methods for multiplying two real numbers where each number is known to d significant digits in base ten.

(a) In the "Grade-School Multiplication" method, we multiply the first number by each digit of the second number, shifting to adjust for the place of the digits; then we add the resulting rows to get the final answer. Show that this method takes at least d^2 operations to complete.

(b) In the "Exhaustive Pre-computation" method, we buy a book containing the answer to every multiplication problem involving two d-digit base-ten numbers. Show that this book must contain 10^{2d} entries.

(c) In the "Abstract Algebra" method, we buy a book containing the base-ten logarithm of every d-digit base-ten number. To compute a product $x \cdot y$, we look up the logarithms of x and y, add them together, and finally we look up the number z whose logarithm is this sum. Show that our book only needs to contain 10^d entries, and we only need to perform about d operations, although our answer will only be an approximation. (More accurately, each lookup takes linear time in d, and so does the addition.) Thus in the case $d = 5$, method (b) is already infeasible due to the size of the book, and method (c) is (naively speaking) already 5 times faster than method (a).

Exercise 7.5. Prove Lemma 7.22.

Exercise 7.6. Prove that if $H \leq G$ and for all $a \in G$ we have $aH = Ha$, then $H \trianglelefteq G$. (Note that this is a sort of converse to Lemma 7.25.)

Exercise 7.7. Suppose that $\sigma : G_1 \to G_2$ is a group homomorphism and $H \leq G_1$. Give a simple argument showing that $\sigma(H) \leq G_2$.

Exercise 7.8. (a) Let $\sigma : G_1 \to G_2$ be a surjective group homomorphism. Prove that if $N \trianglelefteq G_1$, then $\sigma(N) \trianglelefteq G_2$.

(b) Show that the result of part (a) above fails in general if σ is not surjective.

Exercise 7.9. Suppose that $\sigma : G \to H$ and $\tau : H \to L$ are group homomorphisms. Prove that $\tau \circ \sigma$ is a group homomorphism, and that $\ker(\tau \circ \sigma) \supseteq \ker(\sigma)$.

Exercise 7.10. Prove that if $\sigma : G \to H$ is a bijective group homomorphism, then the function $\sigma^{-1} : H \to G$ is also a bijective group homomorphism.

Exercise 7.11. Let G be the group of Exercise 4.10 part (e). Set

$$H = \left\{ \begin{bmatrix} a & b \\ -b & a \end{bmatrix} : a, b \in \mathbf{R}, \text{ not both } 0 \right\}.$$

Find an injective group homomorphism from $\mathbf{C}^\times$ to G whose image is H, where $\mathbf{C}^\times$ is the group of all non-zero complex numbers under ordinary multiplication. Why does this immediately imply that $H \leq G$?

Exercise 7.12. Let $\sigma : G_1 \to G_2$ be a group homomorphism. Let $H = \sigma(G_1)$ be the image of G_1 under σ, and let $K = \ker(\sigma)$. Prove that H is abelian iff for all a, b in G_1 we have $aba^{-1}b^{-1} \in K$.

Exercise 7.13. (a) For each cyclic subgroup of S_3, compute (i) all of its right cosets in S_3, and (ii) all of its left cosets in S_3.
   (b) Which cyclic subgroups of S_3 are normal in S_3?

Exercise 7.14. Construct group tables for the following groups: (a) $\mathbf{Z}_3$ (b) $\mathbf{Z}_4$ (c) $\mathbf{Z}_5$

Exercise 7.15. Let G be a group and let $N \trianglelefteq G$. Let $C, D \in G/N$. Let $C \cdot D$ denote the product of C with D computed using the group law on G/N given in the statement of Theorem 7.26. Let $C * D$ denote the "setwise product" of C with D, namely $C * D := \{x \cdot y : x \in C \text{ and } y \in D\}$. Prove that $C \cdot D = C * D$.

Exercise 7.16. Let G be a group and let $H \leq G$. Define the *normalizer of H in G* to be
$$N_G(H) = \{g \in G : gHg^{-1} = H\}.$$

   (a) Prove that $N_G(H) \leq G$.
   (b) Prove that $H \trianglelefteq N_G(H)$.
   (c) Prove that if $L \leq G$ and $H \trianglelefteq L$, then $L \subseteq N_G(H)$. Thus, $N_G(H)$ is the largest subgroup of G in which H is normal.

Exercise 7.17. Let G be a group, and let $N \trianglelefteq G$. Set $Q = G/N$. Let
$$S = \{H : H \leq G \text{ and } H \supseteq N\}$$

and let
$$T = \{L : L \leq Q\}.$$

   (a) For $H \in S$, define $\sigma(H) = \{xN : x \in H\} \subseteq Q$. Prove that for all $H \in S$, we have $\sigma(H) \in T$.
   (b) For $L \subset T$, define $\tau(L) = \{x \in G : xN \in L\} \subseteq G$, called the *lift* of L to G. Prove that τ is a function from T to S.
   (c) Prove that σ and τ are inverse to each other. Thus, there is a natural one-to-one correspondence between subgroups of G/N and subgroups of G which contain N.
   (d) Prove that $\tau(L) = \bigcup_{C \in L} C = \{xy : x \in G, y \in N, \text{ and } xN \in L\}$.

8

Lagrange's Theorem

8.1 Cosets and Partitions

The concepts and definitions which naturally arose when we studied group homomorphisms are so fruitful that we have yet to exhaust their potential. Among these is the concept of a *coset*, which we next examine in greater detail. Recall that we defined the left coset xS for a subset S of a group G and an element x of G to be

$$xS = \{xs \mid s \in S\}.$$

Remark 8.1. The discussion and results in this section deal mostly with left cosets, but our results apply equally well in the case of right cosets. In particular, Theorem 8.7 and Lemma 8.8 are both true with left cosets replaced by right cosets.

We were motivated to define the concept of a coset by the fact that, given a group homomorphism $\sigma : G \to H$, the pre-image $\sigma^{-1}(y)$ of an element $y \in \sigma(G)$ is a left coset of the kernel $K - \ker(\sigma)$. Specifically, we have $\sigma^{-1}(y) - xK$ for any element $x \in G$ such that $\sigma(x) = y$ (Lemma 7.22).

Now, if $f : D \to C$ is *any* function, from any set D to any set C, then the pre-images of the points of C always form a *partition* of D; that is, these pre-image sets cover all of D and are mutually disjoint. Because the concept of a partition is an important one which will occur again, we take a brief digression to develop these ideas; we start with a formal definition:

Definition 8.2. Let S be a set. A *partition* of S is a set C consisting of subsets of S such that

$$S = \underset{B \in C}{\dot\bigcup} B.$$

Here, the dot in the union symbol indicates a *disjoint* union; that is, a union of disjoint sets.

One way to think about a partition C of a set S is that the sets in C neatly carve S into separate regions or "classes." We may say that two elements x and y of S are "related" to each other if they are in the same class; that is, if there exists $B \in C$ such that $x, y \in B$. Going in the other direction, we will define the general notion of a "relation" on a set. The intention is to define what it means for two elements to be "related" to each other in some way. Given

DOI: 10.1201/9781003252139-8

two elements $x, y \in S$ (possibly identical), we just need to know whether x is related to y; that is, we need to know whether this pair of elements satisfies our relation. Thus, in the extreme economy typical of mathematics, we define a relation on S to *be* a set of ordered pairs from S:

Definition 8.3. A *relation* on a set S is a set $\sim \subseteq S^2$. We usually use operator notation for a relation; hence, writing "$a \sim b$" means $(a, b) \in \sim$. We read this as "a is related to b under $\sim$."

If we try to form a partition of S from a given relation $\sim$ on S, then we can run into problems; to start with, we need to be able to describe the classes carved out by such a partition. For $a \in S$, we would like to define the "class of a" to be the set

$$[a] := \{b \in S \ : \ b \sim a\}.$$

We would like to know when these sets $[a]$ (for $a \in S$) will form a partition of S whose classes correspond to the sets $[a]$. For this to happen, we certainly want a to be in its own class, i.e. $a \in [a]$; likewise, we would like two classes $[a]$ and $[b]$ never to "partially overlap": that is, we should not have $[a] \cap [b] \neq \emptyset$ unless $[a] = [b]$. To get a relation whose classes behave in such a nice way, we will impose extra properties on our relation. The reader should verify that the following three properties will be true for any relation defined from the classes of a partition:

Definition 8.4. A relation $\sim$ on a set S is called *reflexive* if $\forall a \in S, a \sim a$; *symmetric* if $\forall a, b \in S, a \sim b \implies b \sim a$; and *transitive* if $\forall a, b, c \in S, (a \sim b$ and $b \sim c) \implies a \sim c$.

A relation satisfying all three of these properties is singled out for special status:

Definition 8.5. We say that a relation $\sim$ is an *equivalence relation* if $\sim$ is reflexive, symmetric, and transitive. If $\sim$ is an equivalence relation on S and $a \in S$, then the *equivalence class of a* (under $\sim$) is the set $[a] := \{b \in S \ : \ b \sim a\}$.

It turns out that we now have enough to realize our hope and go backwards from a relation to a partition:

Lemma 8.6. *Suppose that $\sim$ is an equivalence relation on a set S. Then the equivalence classes of S under $\sim$ partition S.*

Proof. Exercise 8.3. $\qquad\qquad\qquad\qquad\qquad\qquad\qquad\qquad\qquad\qquad\qquad\qquad\square$

Returning to our main discussion, we know from Exercise 8.4 that the pre-images of points in the image of a function $f : D \to C$ will partition the function's domain: that is, we always have

$$D = \bigcup_{y \in C} f^{-1}(y).$$

It follows that if K is the kernel of some group homomorphism with domain G, then the left cosets of K in G should partition G. By Corollary 7.31, to be a kernel is the same thing as to be a normal subgroup. Thus, the left cosets of a normal subgroup must partition the group. The next result shows that we can remove the normality condition from the previous statement:

Theorem 8.7. *Let $H \leq G$. Then:*
  (i) For all $a, b \in G$, either $aH = bH$ or $aH \cap bH = \emptyset$.
  (ii) The left cosets of H in G partition G.

Proof. (i) Let $a, b \in G$. Suppose that $aH \cap bH \neq \emptyset$. [Show $aH = bH$] [Show $aH \subseteq bH$ and $bH \subseteq aH$] Then $\exists x \in aH \cap bH$, so we can write $x = ah_1 = bh_2$ for some $h_1, h_2 \in H$.

Let $y \in aH$. Then $y = ah$ for some $h \in H$. Now $a = bh_2h_1^{-1}$, so $y = bh_2h_1^{-1}h$. We have $h_2h_1^{-1}h \in H$ by ST1 and ST2, so $y \in bH$. Thus $aH \subseteq bH$.

Similarly, we can show that $bH \subseteq aH$. So $aH = bH$, as desired.

(ii) Formally, we must show that

$$G = \bigcup_{C \in \mathfrak{L}} C,$$

where $\mathfrak{L}$ is the set of all left cosets of H in G.

Let $x \in G$. Then $x = x \cdot e \in xH$ (since $e \in H$). This shows that $G \subseteq \bigcup_{C \in \mathfrak{L}} C$.

To show that $\bigcup_{C \in \mathfrak{L}} C \subseteq G$, we simply observe that each left coset C of H in G is a subset of G. (Indeed, everything in our discussion is inside of G!)

By part (i), the union is disjoint. This completes the proof of (ii). □

8.2 The Size of Cosets

The next result implies that cosets of the same set (within a group) always have the same size, even if that set is not a subgroup:

Lemma 8.8. *Let G be a group and let $S \subseteq G$. Let $a \in G$. Then the function $f : S \to aS$ given by $s \mapsto as$ is a bijection. Thus all left cosets of S in G have the same size as S, and hence the same size as each other.*

Proof. [Show f is injective] Let $x, y \in S$, and suppose that $f(x) = f(y)$. Then $ax = ay$, so $x = y$ by Lemma 3.46.

[Show f is surjective] Let $z \in aS$. Then $z = as$ for some $s \in S$, and $f(s) = z$. □

Definition 8.9. Let G be a group, let $H \leq G$, and let C be a coset of H in G. An element of C is called a *representative* of C, or a *coset representative* for C.

Remark 8.10. Suppose that $H \leq G$ and $a \in G$. Let $C = aH$. Since $e \in H$, we always have $a = ae \in aH$, so a is a coset representative for C. Also, if $b \in aH$, then likewise $b = be \in bH$, so we must have $bH = aH$ by Theorem 8.7. Thus, any coset representative b for C has the property that $C = bH$.

Warning 8.11. Cancellation does *not* work for coset representatives: $aH = bH$ does not imply $a = b$ in general. This is clear from Remark 8.10, since we have $aH = bH$ for *every* $b \in aH$!

Definition 8.12. Let $H \leq G$. The *index of H in G* is the number of left cosets of H in G.

Notation 8.13. The index of H in G is denoted $[G : H]$.

Remark 8.14. The index of a subgroup in a group is either a positive integer, or else it is infinite. A finite group can only have subgroups of finite index, but an infinite group can have subgroups of finite or infinite index. The bigger a subgroup is, the smaller the index of that subgroup will be. At one extreme, the index of H in G is one if and only if $H = G$; this is Exercise 8.1.

Definition 8.15. Let G be a group and let $H \leq G$. A *complete set of (left) coset representatives* for H in G is a set $R \subseteq G$ such that every left coset of H in G contains exactly one element of R.

Remark 8.16. We can form a complete set of coset representatives for H in G simply by choosing one element from every left coset of H in G; this must work by Theorem 8.7. (The innocuous-seeming statement that it is possible in general to simultaneously choose an element from each left coset of H in G is actually equivalent to the Axiom of Choice. We shall only do this, however, in cases where the index of H in G is finite, so we do not need to worry about this axiom here.) Thus, the size of any complete set of coset representatives for H in G is equal to the index of H in G.

We have remarked that results which are true for left cosets are generally also true for right cosets. Usually it makes no difference whether we work with left as opposed to right cosets. At this point, however, it is reasonable to ask whether we always get the same result by using right cosets instead of left cosets in the definition of the index of a subgroup in a group. The following result implies that we do. We note that this result follows by a simple counting argument in case the group in question is finite, but our result is not limited to finite groups.

Lemma 8.17. *If G is a group and $H \leq G$, then the number of left cosets of H in G equals the number of right cosets of H in G.*

Proof. Let $\mathfrak{L}$ and $\mathfrak{R}$ denote the sets of all left cosets and right cosets of H in G, respectively. Let $f : G \to G$ be the inverse function, defined by $f(x) = x^{-1}$. Then by Exercise 8.6 (a), f induces two functions $\sigma : \mathfrak{L} \to \mathfrak{R}$, $\sigma(L) = f(L)$ and $\tau : \mathfrak{R} \to \mathfrak{L}$, $\tau(R) = f(R)$. By Exercise 8.6 (b), f is its own inverse, so σ and τ are inverses of each other, hence bijective. This completes the proof. $\quad\square$

8.3 Reaping the Consequences

Our previous work has rich implications for the structure of finite groups. Many of these implications follow immediately from the next result.

Theorem 8.18 (Lagrange's Theorem). *Let G be a finite group. If $H \leq G$, then $|H| \cdot [G : H] = |G|$, and, in particular, $|H|$ divides $|G|$.*

Proof. Suppose that G is a finite group and $H \leq G$. By Theorem 8.7 part (ii), we have $G = \bigcup_{C \in \mathfrak{L}} C$, where $\mathfrak{L}$ is the set of all left cosets of H in G. Therefore, $|G| = \sum_{C \in \mathfrak{L}} |C|$. By Lemma 8.8, we have $|C| = |H|$ for all C in $\mathfrak{L}$. It follows that $|G| = |H| \cdot |\mathfrak{L}|$. The result now follows from the definition of $[G : H]$. $\square$

Corollary 8.19. *If G is a finite group and $N \trianglelefteq G$, then $|G/N| = |G| / |N|$.*

Proof. By the definitions of G/N (in Theorem 7.26) and $[G : N]$ (in Definition 8.12 and Notation 8.13), we have $|G/N| = [G : N]$. The result now follows from Lagrange's Theorem. $\square$

Lagrange's Theorem provides one of the most basic and important tools in the study of finite groups. As an example of its power, we next show that any finite group of prime order has a very special form.

Theorem 8.20. *Every group of prime order is cyclic.*

Proof. Suppose that G is a group of order p, where p is prime. Then $|G| > 1$, so $\exists x \in G - \{e\}$. Let $H = \langle x \rangle$ and let $n = |H|$. By Lagrange's Theorem, we have $n \mid p$. Also, we have $n > 1$, since $x \neq e$ and both e and x belong to H. Therefore, $n = p = |G|$. Since $H \subseteq G$, it follows that $H = G$. Thus $\langle x \rangle = G$, so G is cyclic. $\square$

Another corollary of Lagrange's theorem involves cyclic subgroups of a finite group. Before presenting this result, we make a standard definition:

Definition 8.21. Let G be a group and let $x \in G$. The *order* of x is the order of the cyclic subgroup generated by x.

Notation 8.22. We denote the order of x by $|x|$.

Remark 8.23. By definition of the order of x, we have $|x| = |\langle x \rangle|$. Just as with the index of a subgroup, the order of a group element may be a positive integer or it may be infinite.

Here finally is the promised result:

Lemma 8.24. *Let G be a finite group, and let $x \in G$. Then*

$$|x| = \min\{n \in \mathbf{Z} \mid n > 0 \text{ and } x^n = e\},$$

and $|x|$ divides $|G|$.

Proof. Since G is finite, then by Exercise 5.8, there is a positive integer k such that $x^k = e$. Now the formula for $|x|$ follows from Exercise 5.7 and the definition of $|x|$. That $|x|$ divides $|G|$ follows from the definition of $|x|$ and Lagrange's Theorem.   □

Corollary 8.25. *If G is a finite group and $x \in G$, then $x^{|G|} = e$.*

Proof. By Lemma 8.24, we can write $|G| = |x| \cdot k$ for some $k \in \mathbf{Z}$. Thus, $x^{|G|} = x^{|x| \cdot k} = (x^{|x|})^k = e^k = e$.   □

Example 8.26. Suppose that G is a group of order 7. Since 7 is prime, Theorem 8.20 implies that $G = \langle x \rangle$ for some $x \in G$. So we must have $|x| = |G| = 7$, by definition of $|x|$. Now by Lemma 8.24, we have $x^7 = e$. Thus, $G = \{e, x, x^2, x^3, x^4, x^5, x^6\}$ by Lemma 5.6. The reader can check that to multiply two elements of G written as powers of x, we add the powers modulo 7: $x^a \cdot x^b = x^c$, where $c \equiv a + b \pmod 7$. Thus, knowing only that a group has order 7, we have been able to essentially write down a formula for the group law! The main result behind this analysis is Lagrange's Theorem.

8.4   Exercises

Exercise 8.1. Let G be a group and let $H \leq G$. Prove that $[G : H] = 1$ iff $H = G$. (Prove the general case; do not assume that G is finite!)

Exercise 8.2. Let G be a group, and let $H \leq G$. Let $a, b \in G$. Prove that $aH = bH$ iff $a^{-1}b \in H$.

Exercise 8.3. Prove Lemma 8.6.

Exercise 8.4. Let $f : D \to C$ be a function. Define a relation $\sim$ on D by $a \sim b$ iff $f(a) = f(b)$.
   (a) Prove that $\sim$ is an equivalence relation on D.
   (b) Prove that the equivalence classes of D under $\sim$ are the pre-image sets $f^{-1}(y)$ for $y \in f(D)$.

Exercise 8.5. Let $\sim$ be an equivalence relation on a set S. Prove that there exists a function f such that $\sim$ is the relation obtained from f via the construction of Exercise 8.4.

Exercise 8.6. Let G be a group, and suppose $H \leq G$. Define a function $f : G \to G$ by $f(x) = x^{-1}$ for $x \in G$.
   (a) Prove that if C is a left coset of H in G, then $f(C)$ is a right coset of H in G, and vice versa.
   (b) Prove that $f \circ f = \mathrm{id}_G$, the identity function on G.

Exercise 8.7. Suppose that $H \leq G$ and $[G : H] = 2$. Prove that $H \trianglelefteq G$.

Exercise 8.8. Suppose that we tried to prove Lemma 8.17 by defining a function $\sigma : \mathfrak{L} \to \mathfrak{R}$ using the formula $\sigma(aH) = Ha$ for $a \in G$, with the same notation as in the proof of that lemma.

(a) Find an example of a group G and subgroup $H \leq G$ such that σ is not well-defined. Hint: take $G = S_3$.

(b) Prove that in general σ is well-defined iff $H \trianglelefteq G$.

Exercise 8.9. (a) Why must every subgroup of S_3 have order 1, 2, 3, or 6?

(b) Let $\alpha = \begin{bmatrix} 1 & 2 & 3 \\ 2 & 3 & 1 \end{bmatrix} \in S_3$, and let $H = \langle \alpha \rangle$. Prove that $H \trianglelefteq S_3$.

Exercise 8.10. Let $\beta = \begin{bmatrix} 1 & 2 & 3 \\ 2 & 1 & 3 \end{bmatrix} \in S_3$, and let $L = \langle \beta \rangle$.

(a) Prove that $L \ntrianglelefteq S_3$.

(b) Without doing any more "real" work, deduce what the normalizer of L in S_3 must be (see Exercise 7.16).

Exercise 8.11. Let G be a finite cyclic group of order n, and let $a \in G$ be a generator of G. Suppose that $H \leq G$, and set $E = \{j \in \mathbf{Z} : j \geq 1 \text{ and } a^j \in H\}$. Let $r = |H|$.

(a) In Exercise 5.6, assuming $r > 1$ it was shown that E is non-empty and that $H = \langle a^m \rangle$, where $m = \min(E)$. Prove this is also true under our present hypotheses even if $r = 1$.

(b) Why must r divide n?

(c) Prove that n divides mr. (Hint: see Exercise 5.7.)

(d) Let $q = n/r$, and set $\tilde{H} = \langle a^q \rangle$. Show that $q \mid m$, and deduce that $a^m \in \tilde{H}$, hence $H \leq \tilde{H}$.

(e) Prove that $|a^q| = r$, and deduce that $\tilde{H} = H$. Thus, $\langle a^{n/r} \rangle$ is the unique subgroup of G of order r.

Exercise 8.12. Once upon a time, a mathematician attempted to recall the definition of the normalizer of a subgroup H in a group G, but made a mistake, and instead defined

$$B_G(H) = \{g \in G : gHg^{-1} \subseteq H\}.$$

(Compare this to the correct definition of the normalizer $N_G(H)$ in Exercise 7.16.) In this exercise, you will show that in fact $B_G(H)$ is not always a subgroup of G, much less a normalizer subgroup. The idea is to find a group G containing two elements x and y such that

$$yxy^{-1} = x^2, \tag{8.1}$$

so that y will be in $B_G(\langle x \rangle)$. Then we will take $H = \langle x \rangle$. Equation 8.1 seems to say that conjugating x by y will double the power of x, and so conjugating x by y^{-1} should cut the power of x in half. But if the group H does not contain a "square root" of x, then y^{-1} will *not* lie in $B_G(H)$!

(a) Suppose that $x, y \in G$ and Equation 8.1 holds. Prove by induction that we have

$$y^k x = x^{2^k} y^k \tag{8.2}$$

for every non-negative integer k.

(b) Making use of Equation 8.2, prove that we have

$$y^b x^a = x^{a \cdot 2^b} y^b \tag{8.3}$$

for all integers a, b with $b \geq 0$.

(c) Use Equation 8.3 to show that we have

$$\left(x^a y^b\right)\left(x^\alpha y^\beta\right) = x^{a + \alpha \cdot 2^b} y^{b + \beta} \tag{8.4}$$

for all integers a, b, α, β with $b \geq 0$.

(d) Motivated by Equation 8.3, define

$$G = \mathbf{Q} \times \mathbf{Z}$$

with binary operation $\cdot$ given by

$$(a, b) \cdot (\alpha, \beta) = (a + \alpha \cdot 2^b, b + \beta).$$

Prove that $(G, \cdot)$ is a group.

(e) With G as in part (d) above, set $x = (1, 0)$ and $y = (0, 1)$. Let $H = \langle x \rangle$. Prove that $H = \{(n, 0) : n \in \mathbf{Z}\}$.

(f) Using the notation of part (e) above, prove that $y \in B_G(H)$ but $y^{-1} \notin B_G(H)$. Conclude that $B_G(H)$ is not a group.

(g) Prove that if G is any *finite* group, and $H \leq G$, then $B_G(H) = N_G(H)$, and thus, in particular, $B_G(H) \leq G$. Hint: Lemma 8.8 and Remark 8.1 can be used here.

9

Special Types of Homomorphisms

9.1 Isomorphisms

Consider the two groups

$$\mathbf{Z}_3 = \{0 + 3\mathbf{Z}, 1 + 3\mathbf{Z}, 2 + 3\mathbf{Z}\}$$

and

$$G = \langle \begin{bmatrix} 1 & 2 & 3 \\ 3 & 1 & 2 \end{bmatrix} \rangle \leq S_3.$$

For convenience, let us use the notation $\bar{0} = 0 + 3\mathbf{Z}$, $\bar{1} = 1 + 3\mathbf{Z}$, $\bar{2} = 2 + 3\mathbf{Z}$, and set $\alpha = \begin{bmatrix} 1 & 2 & 3 \\ 3 & 1 & 2 \end{bmatrix}$. Let $\beta = \alpha^2$. Then we have $\mathbf{Z}_3 = \{\bar{0}, \bar{1}, \bar{2}\}$ and $G = \{\alpha, \beta, e\}$. Constructing group tables for each group, we find:

$$\mathbf{Z}_3 : \quad \begin{array}{c|ccc} + & \bar{0} & \bar{1} & \bar{2} \\ \hline \bar{0} & \bar{0} & \bar{1} & \bar{2} \\ \bar{1} & \bar{1} & \bar{2} & \bar{0} \\ \bar{2} & \bar{2} & \bar{0} & \bar{1} \end{array}$$

$$G : \quad \begin{array}{c|ccc} \circ & e & \alpha & \beta \\ \hline e & e & \alpha & \beta \\ \alpha & \alpha & \beta & e \\ \beta & \beta & e & \alpha \end{array}$$

First of all, each group has order 3. More than this, the patterns in the two tables are identical: only the labels of the elements are different. If we were to re-label $\bar{0}$ as e, $\bar{1}$ as α, and $\bar{2}$ as β, then the group table for $\mathbf{Z}_3$ would be *the same* as the group table for G. *Except for how we happened to label the elements, the two groups are identical.*

Let us find a precise mathematical statement which captures this notion of two groups $(G_1, \cdot_1)$ and $(G_2, \cdot_2)$ being the same except for how their elements are labeled. First, we need to formalize the notion of "re-labeling" the elements of G_1 to match those of G_2. Actually, all we need to do is to associate each element of G_1 with a unique element of G_2 and vice versa, and this can be accomplished by giving a bijection

$$\sigma \; : \; G_1 \to G_2.$$

DOI: 10.1201/9781003252139-9

In our example above, we would take $\sigma(\bar{0}) = e$, $\sigma(\bar{1}) = \alpha$, and $\sigma(\bar{2}) = \beta$. With this identification of the elements of the two groups, what does it mean to say that the group tables are identical? Informally, in our example above, notice that we have "σ(Table for G_1) = Table for G_2." More formally, suppose that the table for G_1 contains a row labeled by the element x and a column labeled by y:

$$G_1 : \quad \begin{array}{c|c} \cdot_1 & y \\ \hline x & z \end{array}$$

There is some element z of G_1 in row x and column y. Meanwhile, the corresponding row and column in the table for G_2 are labeled by some elements a and b of G_2, and the corresponding entry is some element c of G_2:

$$G_2 : \quad \begin{array}{c|c} \cdot_2 & b \\ \hline a & c \end{array}$$

Now, to say that the row for a corresponds to the row for x just means that $\sigma(x) = a$. Similarly, $\sigma(y) = b$. To say that the group tables are identical after re-labeling the elements of G_1 via σ just means that, in addition, $\sigma(z) = c$. But we also have $x \cdot_1 y = z$ and $a \cdot_2 b = c$, because of how a group table is constructed. Therefore, $\sigma(x \cdot_1 y) = \sigma(z) = c = a \cdot_2 b = \sigma(x) \cdot_2 \sigma(y)$. This equation exactly says that σ is a group homomorphism! (We rely on the fact that x and y are arbitrary elements of G_1.) In fact, the only difference between our requirements for σ and our requirements for a general group homomorphism from G_1 to G_2 is that we also required σ to be a bijection.

Definition 9.1. Let G_1 and G_2 be groups. An *isomorphism* from G_1 to G_2 is a bijective group homomorphism from G_1 to G_2.

Notation 9.2. We write $G_1 \cong G_2$ if there exists an isomorphism from G_1 to G_2, and we say in this case that G_1 *is isomorphic to* G_2, or that G_1 and G_2 *are isomorphic* (to each other).

Remark 9.3. Again, we interpret the condition $G_1 \cong G_2$ as saying that G_1 and G_2 are identical as groups, except for how their elements happen to be labeled. An isomorphism $\sigma : G_1 \to G_2$ tells us how to identify elements of G_1 with elements of G_2 so that the group laws on G_1 and G_2 are the same.

Remark 9.4. Since an isomorphism is bijective, a necessary condition for two groups to be isomorphic is that they have the same order. However, this condition is not sufficient: see Exercise 9.2.

Isomorphism expresses the idea of "sameness" in a way that renders the labeling of group elements irrelevant. The following result says that this notion of sameness enjoys three properties that any good notion of sameness should have, namely, those properties which make isomorphism into an *equivalence relation* (see Definition 8.5).

Lemma 9.5. *Let G_1, G_2, G_3 be groups. Then:*

(1) id_1 : $G_1 \to G_1$ is an isomorphism, where id_1 is the identity map on G_1;

(2) If σ : $G_1 \to G_2$ is an isomorphism, then σ^{-1} : $G_2 \to G_1$ is an isomorphism;

(3) If σ : $G_1 \to G_2$ and τ : $G_2 \to G_3$ are isomorphisms, then $\tau \circ \sigma$: $G_1 \to G_3$ is an isomorphism.

In particular, we have:

$G_1 \cong G_1$;

$G_1 \cong G_2 \implies G_2 \cong G_1$;

$G_1 \cong G_2$ *and* $G_2 \cong G_3 \implies G_1 \cong G_3$.

Proof. (1) Left to the reader as Exercise 9.1.

(2) This is the content of Exercise 7.10.

(3) This follows from Exercise 7.9 and the fact that the composition of two bijections is a bijection. □

Remark 9.6. The equivalence class of a group according to the relation of isomorphism is known as the *isomorphism class* of the group. When we speak of identifying a group *up to isomorphism*, we mean identifying the isomorphism class of the group. Often, we accomplish this task by finding a particular "known" group which is isomorphic to the group in question.

In Theorem 7.30, we saw that there is a special "natural" group homomorphism from a group to a quotient of that group. Now that we understand the concept of an isomorphism of groups, we can revisit the relationship between group homomorphisms and quotient groups, and prove that the image of an arbitrary group homomorphism "is" (is isomorphic to) a quotient group.

Theorem 9.7 (The Fundamental Theorem of Group Homomorphisms). *If σ : $G_1 \to G_2$ is any group homomorphism, then $\sigma(G_1) \cong G_1/\ker(\sigma)$. More specifically, the function τ : $G_1/K \to \sigma(G_1)$ given by the formula $\tau(aK) = \sigma(a)$ for $a \in G_1$ is a well-defined group isomorphism, where $K = \ker(\sigma)$.*

Proof. Suppose that σ : $G_1 \to G_2$ is a group homomorphism, and set $K = \ker(\sigma)$, $H = \sigma(G_1)$. We attempt to define a function τ : $G_1/K \to H$ by the formula $aK \mapsto \sigma(a)$ for $a \in G_1$. Because a coset representative a is *not* uniquely determined by the coset it's in, we must check that τ is well-defined. So suppose that $aK = bK$ where $a, b \in G_1$. Then $a \in bK$ (why?), so $a = bx$ for some $x \in K$. Therefore, $\sigma(a) = \sigma(bx) = \sigma(b)\sigma(x)$ (since σ is a group homomorphism) $= \sigma(b)$ (since $x \in \ker(\sigma)$). This shows that our formula for τ gives the same result whether we use a or b as a coset representative. So τ is well-defined.

We next verify that τ is a group homomorphism. For two cosets cK and dK of K in G_1, we compute $\tau(cK \cdot dK) = \tau((cd)K) = \sigma(cd) = \sigma(c)\sigma(d)$ (since σ is a group homomorphism) $= \tau(cK)\tau(dK)$.

Finally, we check that τ is bijective.

[Show τ is surjective] Let $h \in H$. Then $h = \sigma(a)$ for some $a \in G_1$, by definition of H as $\sigma(G_1)$. So $\tau(aK) = \sigma(a) = h$. Thus, τ is surjective.

[Show τ is injective] Suppose that $C, D \in G_1/K$ and $\tau(C) = \tau(D)$. Then we can write $C = aK$ and $D = bK$ for some $a, b \in G_1$. So $\tau(aK) = \tau(bK)$. [We must show that $aK = bK$] Then $\sigma(a) = \tau(aK) = \tau(bK) = \sigma(b)$. Now $\sigma(a) = \sigma(b) =: c \in G_2$. By Lemma 7.22, we have $\sigma^{-1}(c) = aK = bK$. So τ is injective.

Now we know that τ is an isomorphism from G_1/K to H, so $G_1/\ker(\sigma) \cong \sigma(G_1)$. By Lemma 9.5, we can also write $\sigma(G_1) \cong G_1/\ker(\sigma)$. $\qquad\square$

Remark 9.8. We refer to τ in the proof above as the homomorphism *induced by* σ on G_1/K.

Remark 9.9. One way to view Theorem 9.7 is that an arbitrary group homomorphism

$$G_1 \overset{\sigma}{\to} G_2$$

"factors" as a composition of two group homomorphisms

$$G_1 \overset{\nu}{\to} G_1/\ker(\sigma) \overset{\tilde{\sigma}}{\to} G_2,$$

where ν is the natural map, which is surjective, and $\tilde{\sigma}$ is the map induced by σ on $G_1/\ker(\sigma)$, which is injective. By saying that σ "factors," we mean that $\sigma = \tilde{\sigma} \circ \nu$. Just as integers can factor under the binary operation of multiplication, functions—in this case, group homomorphisms—can factor under the binary operation of composition!

A typical use of the isomorphism concept is to try to understand some group under study by showing that it is isomorphic to a well-understood group. The following results, culminating in Corollary 9.12, illustrate one such situation.

Lemma 9.10. $\mathrm{Fr}(\{x\}) \cong (\mathbf{Z}, +)$.

Proof. Let $F = \mathrm{Fr}(\{x\})$. Then $F = \{x^j \; : \; j \in \mathbf{Z}\}$. Define $\sigma \; : \; F \to \mathbf{Z}$ by $x^j \mapsto j$ for each j in $\mathbf{Z}$. We verify that σ is a group homomorphism: for all $j, k \in \mathbf{Z}$, we have $\sigma(x^j \cdot x^k) = \sigma(x^{j+k}) = j + k = \sigma(x^j) + \sigma(x^k)$. It is straightforward to verify that σ is bijective. $\qquad\square$

Theorem 9.11. *Every cyclic group is isomorphic to a quotient of* $(\mathbf{Z}, +)$.

Proof. Let G be a cyclic group. Then $G = \langle a \rangle$ for some $a \in G$. Let $F = \mathrm{Fr}(\{x\})$. By Example 7.14, with $T = \{a\}$, we have a surjective group homomorphism $\tau \; : \; F \to G$ with $\tau(x) = a$. By Lemma 9.10, we have $F \cong \mathbf{Z}$;

so by Lemma 9.5, we also have $\mathbf{Z} \cong F$ via some isomorphism σ. Thus we have a sequence of two surjective group homomorphisms

$$\mathbf{Z} \xrightarrow{\sigma} F \xrightarrow{\tau} G.$$

Let $h = \tau \circ \sigma$. Then h is surjective since both τ and σ are surjective; and by Exercise 7.9, $h : \mathbf{Z} \to G$ is a group homomorphism. So by Theorem 9.7, we have $G \cong \mathbf{Z}/\ker(h)$. Thus, G is isomorphic to a quotient of $\mathbf{Z}$ under addition, as required. $\qquad\qquad\square$

Corollary 9.12. *Every cyclic group is isomorphic to either* $(\mathbf{Z}, +)$ *or to* $\mathbf{Z}_n$ *for some positive integer* n.

Proof. Let G be a cyclic group. Then by Theorem 9.11, $G \cong \mathbf{Z}/K$ for some normal subgroup $K \trianglelefteq (\mathbf{Z}, +)$. Now $(\mathbf{Z}, +)$ is cyclic (generated by 1, for example). So by Exercise 5.6, we know that K is cyclic; so $K = \langle n \rangle = n\mathbf{Z}$ for some $n \in \mathbf{Z}$. Since $n\mathbf{Z} = (-n)\mathbf{Z}$, we may assume that $n \geq 0$. Now if $n > 0$, then $G \cong \mathbf{Z}/n\mathbf{Z} = \mathbf{Z}_n$ by definition of $\mathbf{Z}_n$; while if $n = 0$, then $G \cong \mathbf{Z}/\{0\} \cong \mathbf{Z}$ by Exercise 9.1. $\qquad\qquad\square$

Corollary 9.12 may be viewed as a "classification theorem": it identifies each cyclic group as a well-understood group (either $\mathbf{Z}$ or $\mathbf{Z}_n$), up to isomorphism. For each possible order 1, 2, 3, ..., or ∞, there is exactly one corresponding isomorphism class of cyclic group (compare Remark 9.4).

Example 9.13. Let G be a group of order 7; as in Example 8.26, we know that G must be cyclic. Using Corollary 9.12, we can now say that $G \cong \mathbf{Z}_7$. Thus, there is only one group of order 7, up to isomorphism! By contrast, we could try to classify all the groups of order 7 by starting with a set with 7 elements and writing out every possible 7×7 binary operation table. The number of distinct binary operations on a set of order 7 is equal to

$$7^{49} = 256923577521058877808861147722423562131607.$$

After writing out these tables, we would proceed to eliminate those which fail to satisfy at least one of the group axioms. Finally, we would identify any two group tables which represent isomorphic groups. That would leave us with exactly one table, equivalent to the table for $\mathbf{Z}_7$.

9.2 Automorphisms

Consider again the group table for $\mathbf{Z}_3$, using the notation of Section 9.1. We had

$$\mathbf{Z}_3 : \quad
\begin{array}{c|ccc}
+ & \bar{0} & \bar{1} & \bar{2} \\
\hline
\bar{0} & \bar{0} & \bar{1} & \bar{2} \\
\bar{1} & \bar{1} & \bar{2} & \bar{0} \\
\bar{2} & \bar{2} & \bar{0} & \bar{1}
\end{array}$$

Now consider the result of swapping $\bar{1}$ and $\bar{2}$ everywhere in this table. We get a new table:

	$\bar{0}$	$\bar{2}$	$\bar{1}$
$\bar{0}$	$\bar{0}$	$\bar{2}$	$\bar{1}$
$\bar{2}$	$\bar{2}$	$\bar{1}$	$\bar{0}$
$\bar{1}$	$\bar{1}$	$\bar{0}$	$\bar{2}$

The pattern of labels in the second table is identical to the pattern of labels in the first table. This leads us to the conclusion that the function $\sigma \; : \; \mathbf{Z}_3 \to \mathbf{Z}_3$ given by $\bar{0} \mapsto \bar{0},\, \bar{1} \mapsto \bar{2},\, \bar{2} \mapsto \bar{1}$ is an isomorphism from $\mathbf{Z}_3$ to $\mathbf{Z}_3$.

From another point of view, we can construct the second table from the first by swapping the rows labeled by $\bar{1}$ and $\bar{2}$, and also swapping the columns labeled by $\bar{1}$ and $\bar{2}$. Even though the second table itself is different from the first table, there is some sense in which the swaps have no effect: namely, if we construct a binary operation from the second table, it will be *identical* to the binary operation given by the first table. This suggests that we are dealing with a kind of *symmetry* here.

Indeed, we can formally acknowledge our swap to be a symmetry according to Definition 5.1. Namely, let (for convenience) $(G, \cdot) = (\mathbf{Z}_3, +)$, let $S = G^3 = G \times G \times G$, and let $X = \{(a,\ b,\ a \cdot b) \mid a, b \in G\}$. Then the isomorphism σ induces a permutation $\tilde{\sigma}$ of S via the formula

$$\tilde{\sigma} \; : \; (a, b, c) \mapsto (\sigma(a), \sigma(b), \sigma(c)), \tag{9.1}$$

and $\tilde{\sigma}$ is a symmetry of X with respect to S (you are asked to prove this in Exercise 9.3). Now X just corresponds to the group table for G, so we may view $\tilde{\sigma}$ as a symmetry of the group table. This entire discussion generalizes to the situation when we have an isomorphism from any group G to itself, and it is important enough to be given a name:

Definition 9.14. Let G be a group. An *automorphism* of G is an isomorphism from G to G.

Remark 9.15. An automorphism of G corresponds to a symmetry of the group G: a permutation of the elements of G which leaves the group law for G unchanged. Just as a group element can represent a symmetry of an object, an automorphism represents a symmetry of a group! This suggests that the set of all automorphisms of a given group may itself be a group, and this is in fact the case (see Exercise 9.8).

9.3 Embeddings

We have seen that when we require a group homomorphism to be bijective, then it acquires a special status, that of an *isomorphism*, which guarantees us that the domain and codomain have the same structure as groups, although the elements are perhaps labeled differently. Yet the concept of "codomain," as we have observed, is somewhat artificial: it is not really intrinsic to a function. For this reason, of the two conditions, injectivity and surjectivity, which combine to make bijectivity, the condition of injectivity is the more essential.

Definition 9.16. An *embedding* is an injective homomorphism.

Notation 9.17. We write $\sigma : H \hookrightarrow G$ to indicate that σ is an embedding of H into G. In this situation, we say that H *embeds in* G or that H can be embedded in G. Sometimes we do not name our embedding, but write merely $H \hookrightarrow G$ to mean that H embeds in G.

The significance of this notion is that when we consider an embedding $\sigma : H \hookrightarrow G$, we can let $L = \sigma(H)$, so that σ gives a bijective homomorphism— an *isomorphism*—from H to L. Note that $L \leq G$ by Theorem 7.13. Thus, to say that H embeds in G is to say that H is isomorphic to a subgroup of G.

We have noted that the kernel of a group homomorphism measures how much information is lost in passing from the domain to the image. One manifestation of this principle is the following lemma, which represents the extreme case when a kernel is as small as possible: namely, the case of an embedding, in which *no* information is lost.

Lemma 9.18. *A group homomorphism is an embedding iff its kernel is the trivial subgroup,* $\{e\}$.

Proof. Exercise 9.14. □

Earlier, we remarked that "the fundamental way (some might say the only way) that two groups can be related is via a group homomorphism from one to the other." To those who objected that another way two groups can be related is that one group could be a subgroup of another, we can now respond as follows: The subgroup relation is essentially the relation of one group embedding in another group, and is thus a special kind of homomorphism relation. More precisely, one group is a subgroup of a second group if and only if their underlying sets have a subset relationship which gives rise to a group homomorphism (see Exercise 9.4).

9.4   Exercises

Exercise 9.1. Let G be a group.
  (a) Prove that the identity map from G to itself is an isomorphism.
  (b) Let $H = \{e\}$. Prove that $H \trianglelefteq G$ and that $G/H \cong G$.

Exercise 9.2. (a) Pat thinks that two groups G and H are not isomorphic iff $\exists \, \sigma \, : \, G \to H$ such that σ is not an isomorphism. Is Pat correct?

(b) Find, with proof, two groups of order 4 which are not isomorphic.

Exercise 9.3. Prove that the map $\tilde{\sigma}$ defined by Equation 9.1 is indeed a symmetry of X with respect to S (see the context of this equation for definitions of X and S). Make your proof general: only use the fact that σ is an automorphism of a group G; do not assume that $G = \mathbf{Z}_3$.

Exercise 9.4. Let $(G, \cdot)$ be a group, and let $S \subseteq G$. Suppose that $(S, \triangle)$ forms a group. Let $\sigma \, : \, G \to G$ be the identity function on G, and let $\tau \, : \, S \to G$ be the restriction of σ to S. Prove that $(S, \triangle) \leq (G, \cdot)$ iff τ is a group homomorphism from $(S, \triangle)$ to $(G, \cdot)$.

Exercise 9.5. Prove that if n and m are positive integers with $n \leq m$, then $S_n \hookrightarrow S_m$ (see Notation 3.30).

Exercise 9.6. Let S and T be two sets such that $|S| = |T|$; recall that this means there is a bijection from S to T. Prove that $\mathrm{Sym}(S) \cong \mathrm{Sym}(T)$. (You should *not* assume that S and T are finite.)

Exercise 9.7. Use group tables for the groups D_6 and S_3 to construct an isomorphism between these groups.

Exercise 9.8. Let G be a group. Define $\mathrm{Aut}(G)$ to be the set of all automorphisms of G, equipped with the binary operation given by composition of functions. Prove that $\mathrm{Aut}(G) \leq \mathrm{Sym}(G)$. In particular, $\mathrm{Aut}(G)$ is a group.

Exercise 9.9. Let G be a group and let $g \in G$. Define a function $\phi_g \, : \, G \to G$ by the formula $\phi_g(x) = gxg^{-1}$ for $x \in G$. Prove that ϕ_g is an automorphism of G. We call ϕ_g the *inner automorphism* determined by g, or the *conjugation by g* map.

Exercise 9.10. Let G be a group. Let

$$\mathrm{Inn}(G) = \{\phi_g \, : \, g \in G\},$$

the set of all inner automorphisms of G. Prove that $\mathrm{Inn}(G) \leq \mathrm{Aut}(G)$. (Refer to Exercises 9.8 and 9.9.)

Exercise 9.11. Let G be a group. Define

$$Z(G) = \{x \in G \, : \, \forall g \in G, \; gx = xg\}.$$

$Z(G)$ is called the *center* of G.

(a) Prove that $Z(G) \trianglelefteq G$.
(b) Generalize part (a) by proving that for all $H \leq Z(G)$, we have $H \trianglelefteq G$.
(c) Complete the following statement: G is abelian iff $Z(G) = \underline{\quad}$.

Exercise 9.12. Let G be a group and let $Z(G)$ be the center of G. Define a function $f \, : \, G \to \mathrm{Inn}(G)$ by the formula $g \mapsto \phi_g$. (Refer to Exercises 9.10 and 9.11.)

(a) Prove that f is a group homomorphism.
(b) Prove that $\mathrm{Inn}(G) \cong G/Z(G)$.

Exercise 9.13. This exercise refers to Exercise 4.10. Let $H = \{A \in S : |A| = 1\}$. Prove that $H \trianglelefteq G$ and $G/H \cong \mathbf{R}^\times$.

Exercise 9.14. Prove Lemma 9.18.

Exercise 9.15. Let $\sigma : G \to H$ be a group homomorphism, and set $K = \ker(\sigma)$. Suppose that $N \trianglelefteq G$ and $N \subseteq K$.

(a) Prove that for $a \in G$, the formula $\tilde{\sigma}(aN) = \sigma(a)$ gives a well-defined group homomorphism $\tilde{\sigma} : G/N \to H$.

(b) Prove that we can write $\sigma = \tilde{\sigma} \circ \nu$ where $\nu : G \to G/N$ is the natural map. We say that σ *factors through* G/N.

Exercise 9.16 (Automorphisms of $\mathbf{Z}_n$). Let n be a positive integer, and let $G = \mathbf{Z}_n$. Let $u = 1 + n\mathbf{Z} \in G$. Let $a \in \mathbf{Z}$, and set $\alpha = a + n\mathbf{Z} \in G$.

(a) Suppose that $\sigma \in \text{Aut}(G)$ and $\sigma(u) = \alpha$. Prove that we have $\sigma(ku) = \sigma(k + n\mathbf{Z}) = k\alpha = (ka) + n\mathbf{Z}$ for each $k \in \mathbf{Z}$. Conclude that there is at most one automorphism of G which sends u to α.

(b) Prove that the function $f : G \to G$ given by the formula $f(k + n\mathbf{Z}) = (ka) + n\mathbf{Z}$ is a well-defined group homomorphism.

(c) Prove that the map f from part (b) above is an automorphism of G if and only if $\gcd(a, n) = 1$. Hint: Since $\mathbf{Z}_n$ is finite, f is bijective iff f is an embedding iff $\ker(f) = \{0 + n\mathbf{Z}\}$, by Lemma 9.18.

Exercise 9.17. One day in the middle of an abstract algebra lecture, the instructor pointed out that the definition of the quotient group G/N apparently never relied on the properties of N as a *group*, but only on the fact that N is *normal*, that is, closed under conjugation. Could this be true? And if so, does it give us a whole new world of groups to explore? This exercise investigates these questions.

Let G be a group. For a subset $S \subseteq G$, let us say that S is *closed under conjugation* (by elements of G) if $\forall x \in S \; \forall g \in G, \; gxg^{-1} \in S$. (Note that we do not require S to be a subgroup of G.) Let $\mathfrak{C}$ denote the set of all left cosets of S in G, and suppose that S is closed under conjugation by elements of G.

(a) Prove Lemma 7.25 with S in place of N: that is, prove that for all $a \in G$, we have $aS = Sa$.

(b) Consider the formula $(xS) \cdot (yS) = (xy)S$ for $x, y \in G$. Prove that this formula gives a well-defined group law on $\mathfrak{C}$.

(c) Define a function $\sigma : G \to \mathfrak{C}$ by the formula $\sigma(x) = xS$. Prove that σ is a surjective group homomorphism.

(d) Use part (c) above to show that $\mathfrak{C}$ is isomorphic to an ordinary quotient group of G. Thus our construction does not "really" yield any new groups.

10

Making Groups

10.1 Introduction

One way of producing groups is by starting with a group, taking a subset of the group, and generating a subgroup from this subset. We have seen (Theorem 4.12) that there is always a unique smallest subgroup containing a given subset. In general, looking for subgroups within a given group is a promising way to find "new" groups.

Another way to produce groups is to start with a group, find a normal subgroup of the group, and then form the quotient group.

In this chapter, we explore each of these two ideas in turn.

10.2 A Quotient Engine

In order to produce normal subgroups on demand, it will be helpful to have some results on generating normal subgroups similar to those we obtained for ordinary subgroups. We start with the counterpart for normal subgroups of Lemma 4.20.

Lemma 10.1. *Let G be a group and let $\mathfrak{C}$ be a non-empty collection of normal subgroups of G. Then $\cap_{H \in \mathfrak{C}} H \trianglelefteq G$.*

Proof. Set $I = \cap_{H \in \mathfrak{C}} H$. By Lemma 4.20, we have $I \leq G$. Let $x \in G$ and let $y \in I$. [Show $xyx^{-1} \in I$] [This means $\forall H \in \mathfrak{C}, xyx^{-1} \in H$] Let $H \in \mathfrak{C}$. Then $y \in H$, by definition of I. Since $H \trianglelefteq G$ and $x \in G$, we have $xyx^{-1} \in H$. But since H was an arbitrary member of $\mathfrak{C}$, we have $xyx^{-1} \in \cap_{H \in \mathfrak{C}} H = I$. This shows that $I \trianglelefteq G$. $\qquad\square$

We can use the preceding result to construct normal subgroups with a "top-down" approach:

DOI: 10.1201/9781003252139-10

Definition 10.2. Let G be a group and let $S \subseteq G$. The *normal subgroup of* G *generated by* S is

$$\langle\langle S \rangle\rangle = \bigcap_{\substack{N \trianglelefteq G \\ S \subseteq N}} N.$$

Remark 10.3. Since we may take $N = G$, we are intersecting a non-empty collection here. We therefore have $\langle\langle S \rangle\rangle \trianglelefteq G$ by Lemma 10.1. By construction, every normal subgroup of G which contains S also contains $\langle\langle S \rangle\rangle$. Thus, $\langle\langle S \rangle\rangle$ is the smallest normal subgroup of G which contains S.

Remark 10.4. There is also a "bottom-up" construction of $\langle\langle S \rangle\rangle$; see Exercise 10.1.

Before we put these notions to work, we establish a useful multiplicative formula for the index of a "nested" subgroup: that is, a subgroup of a subgroup.

Lemma 10.5. *Suppose that $H_1 \leq H_2 \leq G$ and that $[G : H_1] < \infty$. Then we have*

$$[G : H_1] = [G : H_2] \cdot [H_2 : H_1].$$

Proof. We first note that in case G is finite, the result follows immediately from Lagrange's Theorem; for in that case, we have $[G : H_1] = |G| / |H_1|$, $[G : H_2] = |G| / |H_2|$, and $[H_2 : H_1] = |H_2| / |H_1|$. However, this argument is of no use in the general case. Instead, we apply the result we used to prove Lagrange's Theorem in the first place.

So suppose that $[G : H_1] < \infty$. Since $H_1 \leq H_2$, Theorem 8.7 lets us write

$$H_2 = \bigcup_{a \in R_1} \cdot\, aH_1, \tag{10.1}$$

where $R_1 \subseteq H_2$ is a complete set of left coset representatives of H_1 in H_2. Let us also choose a complete set of left coset representatives R_2 of H_2 in G, so that

$$G = \bigcup_{b \in R_2} \cdot\, bH_2. \tag{10.2}$$

By definition of the index of a subgroup, we have $[H_2 : H_1] = |R_1|$ and $[G : H_2] = |R_2|$. Note that we do not officially know yet whether these indexes are finite.

$$\left[\text{Idea: } G = \bigcup_{b \in R_2} \cdot\, bH_2 = \bigcup_{b \in R_2} \cdot\, b \cdot \left[\bigcup_{a \in R_1} \cdot\, aH_1 \right] = \bigcup_{b \in R_2} \left[\bigcup_{a \in R_1} \cdot\, baH_1 \right]. \right]$$

Let $\mathfrak{C}$ be the set of all left cosets of H_1 in G. Define a function $f : R_1 \times R_2 \to \mathfrak{C}$ by the formula $(a, b) \mapsto baH_1$. We claim that f is bijective. For surjectivity, let $C \in \mathfrak{C}$. By definition of left coset, we can write $C = xH_1$ for some $x \in G$. By Equation 10.2, we can write $x \in bH_2$ for some $b \in R_2$. So $x = by$ for some $y \in H_2$. By Equation 10.1, we can write $y \in aH_1$ for some $a \in R_1$. So $x = by \in baH_1$. Now the cosets xH_1 and baH_1 have the element

x in common, so the two cosets must be equal: $C = xH_1 = baH_1 = f(a, b)$. This establishes that f is surjective. For injectivity, we refer to Exercise 10.3.

It follows that the number of left cosets of H_1 in G is $[G : H_1] = |\mathcal{C}| = |R_1 \times R_2| = |R_1| \cdot |R_2| = [G : H_2] \cdot [H_2 : H_1]$. In particular, all these indexes are finite. $\qquad\qquad\square$

Now we are ready to unleash the power of normal subgroups. We have seen in Section 7.7.2 that the kernel of a group homomorphism measures the amount of new relations we add in passing from the domain to the image of the homomorphism. In Section 7.7.4 (see especially Theorem 7.30) and later in Section 9.9.1 (especially Theorem 9.7), we saw that the image of a group homomorphism looks like the *quotient* of the domain by the kernel, and that any normal subgroup can serve as a kernel. Combining these ideas, we realize that when we take the quotient of a group G by a normal subgroup $N \trianglelefteq G$, we are merely adding new relations to G by forcing every element of N to be the identity: that is, the quotient group G/N is what we get by starting with the group G and adding all relations of the form $n = e$ for $n \in N$.

The usefulness of the construction $\langle\langle S \rangle\rangle$ is that it allows us to use *any* subset S of G as a set of relations to be forced upon G. Now of course S itself need not be a normal subgroup, or even a subgroup, of G; but $\langle\langle S \rangle\rangle$ will be, and we think of the quotient group $G/\langle\langle S \rangle\rangle$ as the group G modified by adding exactly those relations forced on us by requiring $s = e$ for every $s \in S$.

Example 10.6. We return to the dihedral group D_{2n} in this example. We saw in Section 5.5.2 that D_{2n} is generated by two elements F and R satisfying the relations $F^2 = e$, $R^n = e$, and $FR = R^{-1}F$ (as well as infinitely many other relations). Let $\omega : \mathrm{Fr}(\{x, y\}) \to D_{2n}$ be the group homomorphism which maps x to F and y to R, and let $K = \ker(\omega)$. As we have seen, relations in D_{2n} correspond to elements of K. We have $x^2 \in K$ and $y^n \in K$. To get an element of K from the third relation, we put it in standard form, namely $F^{-1}RFR = e$. This leads to the realization that $x^{-1}yxy \in K$. Set $\mathcal{F} = \mathrm{Fr}(\{x, y\})$, $S = \{x^2, y^n, x^{-1}yxy\} \subseteq \mathcal{F}$, and $N = \langle\langle S \rangle\rangle$. Then we have $S \subseteq K$ and $K \trianglelefteq \mathcal{F}$, so $N \leq K$, as N is the smallest normal subgroup of $\mathcal{F}$ which contains S.

Let $Q = \mathcal{F}/N$, and let $\bar{x} = xN$, $\bar{y} = yN$. Then $\bar{x}, \bar{y} \in Q$, and we have $\bar{x}^2 = xN \cdot xN = x^2N = N$ (since $x^2 \in N$) $= e_Q$. Similarly, we have $\bar{y}^n = e_Q$ and $\bar{x}\bar{y} = \bar{y}^{-1}\bar{x}$. These relations in Q illustrate what we have managed to do: by forming a normal subgroup N of $\mathcal{F}$ containing the three dihedral relations listed above, we forced corresponding relations to hold true in the resulting quotient group Q. We also note that $Q = \langle \bar{x}, \bar{y} \rangle$; for example, the element $(xyy)N$ of Q can be written as $xN \cdot yN \cdot yN = \bar{x}\bar{y}\bar{y} \in \langle \bar{x}, \bar{y} \rangle$. So Proposition 5.10 applies, with $\bar{x}$ and $\bar{y}$ playing the roles of F and R, and we get $|Q| \leq 2n$.

Since ω is surjective, then $D_{2n} \cong \mathcal{F}/K$ by Theorem 9.7. In particular, we have $2n = |D_{2n}| = |\mathcal{F}/K| = [\mathcal{F} : K]$. Now we also have $[\mathcal{F} : N] = |\mathcal{F}/N| = |Q| \leq 2n$. Since $N \leq K \leq \mathcal{F}$ and $[\mathcal{F} : N] < \infty$, we can use Lemma 10.5 to conclude that $[\mathcal{F} : N] = [\mathcal{F} : K] \cdot [K : N]$. So we have $2n \geq [\mathcal{F} : N] =$

$[\mathcal{F} : K] \cdot [K : N] = 2n \cdot [K : N]$, which forces $[K : N] \leq 1$. Since $[K : N]$ is a positive integer, we must have $[K : N] = 1$, and from Exercise 8.1, we conclude that $K = N$.

The upshot of all this is that $\mathrm{Fr}(\{x, y\}) / \langle\langle \{x^2, y^n, x^{-1}yxy\} \rangle\rangle \cong D_{2n}$. In other words, the three relations on which we have been focusing are enough to completely determine the structure of the dihedral group D_{2n}. When we consider the enormous complexity of the set of relations corresponding to K, this is an amazing result.

Since *every* group is the quotient of a free group—as we see by putting together Example 7.12 with Theorem 9.7—this example can (in principle) be generalized: given a group G, we can try to find a small set of generators for G, together with a small set of relations involving these generators, such that G is isomorphic to the corresponding quotient of the free group on these generators.

Example 10.7. Let G be a group. Let's force G to be abelian. More precisely, let us add just enough relations to G so that the resulting quotient group is abelian. For each pair of elements $a, b \in G$, we need to include the relation $ab = ba$. Putting this relation into standard form, we get $aba^{-1}b^{-1} = e$. So let

$$C = \langle\langle \{aba^{-1}b^{-1} \; : \; a, b \in G\} \rangle\rangle.$$

Let $G^{\mathrm{ab}} = G/C$. Then G^{ab} is the largest abelian quotient of G. More precisely, we claim that G^{ab} is an abelian group, and that if $\sigma : G \to H$ is a surjective group homomorphism where H is abelian, then we must have $C \subseteq \ker(\sigma)$; this last assertion follows from Exercise 7.12. In fact, σ factors through G^{ab} (see Exercise 9.15): we have $\sigma = \tilde{\sigma} \circ \nu$, where ν is the natural map, and $\tilde{\sigma}$ is the map on G^{ab} induced by σ. That is, the following diagram commutes (see Definition 7.5):

$$G \xrightarrow{\;\nu\;} G^{\mathrm{ab}} \xrightarrow{\;\tilde{\sigma}\;} H$$
$$\underbrace{\phantom{G \xrightarrow{\;\nu\;} G^{\mathrm{ab}} \xrightarrow{\;\tilde{\sigma}\;} H}}_{\sigma}$$

Thus, H must be isomorphic to a quotient of G^{ab}.

Definition 10.8. An element of the form $aba^{-1}b^{-1}$ is called a *commutator*, and the subgroup C of Example 10.7 is called the *commutator subgroup* of G. The group G^{ab} is known as the *abelianization* of G.

10.3   Room for Everyone Inside

To end this chapter, let us return to the subject of group tables. Consider once again the group table for the group $G = \mathbf{Z}_3$:

$$
\begin{array}{c|ccc}
+ & \bar{0} & \bar{1} & \bar{2} \\
\hline
\bar{0} & \bar{0} & \bar{1} & \bar{2} \\
\bar{1} & \bar{1} & \bar{2} & \bar{0} \\
\bar{2} & \bar{2} & \bar{0} & \bar{1}
\end{array}
$$

We will wring another insight from this little table. Notice that each row of the group table seems to contain *every* element of our group *exactly once*, in some scrambled order. A scrambling of the elements of G is, more formally, what we called a *permutation* of G. Thus, it seems that every row of the group table for G gives us a permutation of G.

But each row of the group table is labeled (on the left) by an element of G. Thus, we can associate to every element of G a permutation of the set of elements of G. In other words, we have a function

$$
\sigma \ : \ G \to \mathrm{Sym}(G)
$$

for which $\sigma(g)$ is the permutation corresponding to the row labeled by g. In the present case, for example, $\sigma(\bar{1})$ is the permutation $\pi_{\bar{1}}$ such that $\pi_{\bar{1}}(\bar{0}) = \bar{1} + \bar{0} = \bar{1}$, $\pi_{\bar{1}}(\bar{1}) = \bar{1} + \bar{1} = \bar{2}$, and $\pi_{\bar{1}}(\bar{2}) = \bar{1} + \bar{2} = \bar{0}$.

Since σ is a very natural-seeming function between two groups, then, according to our principles, σ ought to be a group homomorphism. This claim is justified in the following result, which applies to arbitrary groups. The reader is invited to check that the function π_x defined below is the general version of the function $\pi_{\bar{1}}$. The function π_x is the "left multiplication by x" function, which produces the row of the group table indexed by the element x.

Theorem 10.9 (Cayley's Theorem). *Let G be a group. Define a function*

$$
\sigma \ : \ G \to Sym(G)
$$

by the formula $\sigma(x) = \pi_x$, where $\pi_x(g) = x \cdot g$ for each g in G. Then σ is an embedding of G into $Sym(G)$.

The proof amounts to several formal verifications; no big ideas are needed, but rather we merely unravel definitions.

Proof. Let G be a group. We first show that σ really does map G to $\mathrm{Sym}(G)$.

[Show $\forall a \in G, \pi_a \in \mathrm{Sym}(G)$; i.e., that π_a is a bijection from G to G]

Let $a \in G$. Suppose $x, y \in G$ and $\pi_a(x) = \pi_a(y)$. Then $a \cdot x = a \cdot y$ (by definition of π_a). So $x = y$ by Lemma 3.46. Thus π_a is injective.

Let $z \in G$. [Solve the equation $\pi_a(x) = z$ for x: $a \cdot x = z$, $x = a^{-1}z$]

Let $x = a^{-1}z$. Then $x \in G$, and we have $\pi_a(x) = a \cdot x = a \cdot (a^{-1}z) = z$. So π_a is surjective.

Therefore, π_a is bijective, so $\pi_a \in \text{Sym}(G)$.

[Show that σ is an embedding: an injective group homomorphism]

Let $a, b \in G$. [Show $\sigma(a \cdot b) = \sigma(a) \circ \sigma(b)$]

Then $\sigma(a \cdot b) = \pi_{a \cdot b}$, $\sigma(b) = \pi_b$, and $\sigma(a) = \pi_a$.

[Show $\pi_{a \cdot b} = \pi_a \circ \pi_b$]

[Show $\forall x \in G, \pi_{a \cdot b}(x) = (\pi_a \circ \pi_b)(x)$]

Let $x \in G$. Then $\pi_{a \cdot b}(x) = (a \cdot b)x$ and $(\pi_a \circ \pi_b)(x) = \pi_a(\pi_b(x)) = \pi_a(bx)$ $= a(bx)$. So $\pi_{a \cdot b}(x) = (\pi_a \circ \pi_b)(x)$ by Axiom G1, associativity in G. Thus, σ is a group homomorphism.

[Show σ is injective]

Let $a, b \in G$, and suppose that $\sigma(a) = \sigma(b)$. [Show $a = b$]

Then $\pi_a = \pi_b$. So $\forall x \in G, \pi_a(x) = \pi_b(x)$. Since $e \in G$, we have $\pi_a(e) = \pi_b(e)$. But this means that $a \cdot e = b \cdot e$, and so $a = b$. Thus, σ is injective. This completes the proof. $\qquad \square$

Corollary 10.10. *Every finite group is isomorphic to a subgroup of S_n for some n.*

Proof. Let G be a finite group, and set $n = |G|$. By Cayley's Theorem, we know that $G \hookrightarrow \text{Sym}(G)$; from Exercise 9.6, we have $\text{Sym}(G) \cong S_n$. Composing these two maps, we get an embedding $G \hookrightarrow S_n$. $\qquad \square$

Note that $\text{Sym}(G)$ does not "know" anything about the group operation of G, only about the underlying set of this group; yet Cayley's Theorem tells us that G embeds in $\text{Sym}(G)$ as a group! Cayley's Theorem tells us that the family of symmetric groups is "universal" in a certain sense: every group can be found inside of one of the symmetric groups. Thus, by constructing subgroups of symmetric groups, we can make any group.

10.4 Exercises

Exercise 10.1. Let G be a group and let $S \subseteq G$. Set $T = \{xyx^{-1} : x \in G, y \in S\}$. Prove that $\langle T \rangle = \langle\langle S \rangle\rangle$.

Exercise 10.2. Let G be a group, and let $S = \{aba^{-1}b^{-1} : a, b \in G\}$ be the set of all commutators in G. Prove that $\langle S \rangle = \langle\langle S \rangle\rangle$.

Exercise 10.3. Let f be the function that was defined in the proof of Lemma 10.5. Prove that f is injective.

Exercise 10.4 (Cayley's Theorem is sharp for primes). Let p be a prime integer, and suppose that G is a group of order p. By Corollary 10.10, we know that $G \hookrightarrow S_p$. Prove that if $G \hookrightarrow S_n$ for some integer n, then we must have $n \geq p$. Hint: the order of S_n is $n! = 1 \cdot 2 \cdot 3 \cdot \cdots \cdot n$.

Exercise 10.5. Let $S = \{x_1, \ldots, x_n\}$ be a set of order $n < \infty$, and let $G = \text{Fr}(S)^{\text{ab}}$. (Note: the reader may use any non-empty set S, not just a finite set, with no real increase in difficulty.) We call G the *free abelian group on S*.

(a) Let H be an abelian group, and let $h_1, \ldots, h_n \in H$. Let y_i be the image of x_i under the natural map from $\text{Fr}(S)$ to G. Prove that there is a unique group homomorphism $\sigma : G \to H$ such that for all i we have $\sigma(y_i) = h_i$.

(b) Suppose that H is an abelian group which can be generated by a set of n elements. Prove that H is isomorphic to a quotient of G.

(c) Let $A = \mathbf{Z}^n = \{(k_1, \ldots, k_n) : k_i \in \mathbf{Z}\}$, with binary operation $+$ given by $(k_1, \ldots, k_n) + (\ell_1, \ldots, \ell_n) = (k_1 + \ell_1, \ldots, k_n + \ell_n)$. Prove that $G \cong A$. (For more details about the close relationship between $\mathbf{Z}$ and finitely generated abelian groups, see Chapter 18.)

Exercise 10.6. Let n be an integer with $n \geq 3$. Let G be a group of order $2n$ generated by two elements F and G which satisfy the three dihedral relations $F^2 = e$, $R^n = e$, and $FR = R^{-1}F$. Prove that $G \cong D_{2n}$.

Exercise 10.7. Let n be an integer with $n \geq 3$. Prove that we have

$$\text{Fr}(\{x, y\}) \, / \langle\langle \{x^2, y^n, (xy)^2\} \rangle\rangle \cong D_{2n}.$$

(Compare to Example 10.6.)

11

Rings

11.1 A New Type of Structure

A group is a set with a single binary operation satisfying certain axioms which we abstracted from the properties of ordinary number systems. Motivated by our desire to gain further perspective on ordinary number systems, and generalize them, we next introduce a new kind of algebraic object called a *ring*. Rings are richer than groups, and even more like ordinary number systems, because they have not one but two binary operations.

Definition 11.1. A *ring* is a triple $(R, +, \cdot)$, satisfying the following 3 axioms:
  R1: $(R, +)$ is an abelian group with identity element 0;
  R2: $\cdot$ is an associative binary operation on R;
  R3: $\cdot$ distributes over $+$: that is, for all $x, y, z \in R$, we have $(x + y) \cdot z = x \cdot z + y \cdot z$ and $z \cdot (x + y) = z \cdot x + z \cdot y$.

Remark 11.2. The operation $\cdot$ may not be commutative; if it is, we say that R is a commutative ring. Furthermore, $\cdot$ may not have an identity element; if it does, then we denote this element by 1_R or simply 1, and say that "R has 1." We often speak of the operation $\cdot$ as "multiplication," and call 1_R a multiplicative identity element. Note that by Lemma 3.33, a multiplicative identity element must be unique if it exists.

An element of R may or may not have an inverse under multiplication; we single out those elements which do:

Definition 11.3. Let R be a ring with 1. Then a *unit* of R is an element of R which has an inverse under multiplication.

Notation 11.4. The set of all units of R is denoted $R^\times$ (pronounced "R cross"). That is,
$$R^\times = \{x \in R \ : \ \exists y \in R \text{ s.t. } x \cdot y = y \cdot x = 1\}.$$

Theorem 11.5. *If R is a ring with 1, then $(R^\times, \cdot)$ is a group.*

Proof. Since $(R, \cdot)$ satisfies the group axioms **G1** and **G2**, Exercise 4.8 applies to give the desired result. $\qquad\square$

DOI: 10.1201/9781003252139-11

Example 11.6. Each of the following familiar number systems is a ring under ordinary addition and multiplication: $\mathbf{Z}$, $\mathbf{R}$, $\mathbf{Q}$. In fact, these are commutative rings with 1. The group of units of $\mathbf{Z}$ is the set $\mathbf{Z}^{\times}$ of all integers which have multiplicative inverses (reciprocals) *which are also integers*. Thus, we have $\mathbf{Z}^{\times} = \{1, -1\}$. By contrast, almost every element of $\mathbf{R}$ has a multiplicative inverse in $\mathbf{R}$: we have $\mathbf{R}^{\times} = \mathbf{R} - \{0\}$. Similarly, $\mathbf{Q}^{\times} = \mathbf{Q} - \{0\}$.

Example 11.7. The set of all 2×2 (two-by-two) matrices with real entries is

$$M_2(\mathbf{R}) := \left\{ \begin{bmatrix} a & b \\ c & d \end{bmatrix} : a, b, c, d \in \mathbf{R} \right\}.$$

We claim that $M_2(\mathbf{R})$ is a ring under the operations of "matrix addition" and "matrix multiplication" given by

$$\begin{bmatrix} a & b \\ c & d \end{bmatrix} + \begin{bmatrix} u & v \\ w & x \end{bmatrix} = \begin{bmatrix} a+u & b+v \\ c+w & d+x \end{bmatrix}$$

and

$$\begin{bmatrix} a & b \\ c & d \end{bmatrix} \cdot \begin{bmatrix} u & v \\ w & x \end{bmatrix} = \begin{bmatrix} au+bw & av+bx \\ cu+dw & cv+dx \end{bmatrix}.$$

R1: It is straightforward to verify that $M_2(\mathbf{R})$ is an abelian group under matrix addition.

R2: Matrix multiplication is associative by Exercise 3.4.

R3: The distributive property of matrix multiplication over matrix addition is assigned to the reader as Exercise 11.1.

Exercise 4.10 shows that the group of units of $M_2(\mathbf{R})$ is

$$M_2(\mathbf{R})^{\times} = \{A \in M_2(\mathbf{R}) : |A| \neq 0\},$$

where $|A|$ is defined as in that exercise. $|A|$ is called the *determinant* of A. The group $M_2(\mathbf{R})^{\times}$ is called the *general linear group* (of two-by-two real matrices), and is denoted $GL_2(\mathbf{R})$.

Notation 11.8. Let R be a ring and let $x, y \in R$. As is customary in an abelian group, we denote the inverse of x under addition by $-x$. Also, we write $y - x$ as a shorthand for $y + (-x)$. We use exponential notation for repeated multiplication: x^n denotes the product of n factors of x, if n is a positive integer.

Warning 11.9. We do not attempt to define x^n for a general ring element x when n is negative or zero. The trouble is that in a ring, an element x may not have an inverse under multiplication, and in fact the ring may not even have 1. The reader is invited to verify, however, that the usual law of exponents $x^{m+n} = x^m \cdot x^n$ is true when m and n are positive integers. Furthermore, if x is a *unit*, then we can raise x to any integer power, for in this case, x belongs to the *group* of units, with the operation of ring multiplication.

11.2 Ring Fundamentals

The reader may wonder, in case R has 1, whether we can also write $(-1)\cdot x$ as $-x$. Fortunately, these two expressions are always equal. The following lemma establishes this identity as well as several others which are familiar from the algebra of number systems.

Lemma 11.10 (Basic Properties of Rings). *Let R be a ring. Let $x, y \in R$. Then:*

(i) $0 \cdot x = 0 = x \cdot 0$
(ii) $(-x) \cdot y = -(x \cdot y) = x \cdot (-y)$
(iii) $(-x) \cdot (-y) = x \cdot y$
Further, if R has 1, then:
(iv) $(-1) \cdot x = -x$
(v) $(-1)^2 = 1$

Proof. (i) We have $0+0 = 0$ by Exercise 3.3. So we can say $0 \cdot x = (0+0) \cdot x$ (by substitution) $= 0 \cdot x + 0 \cdot x$ (by the distributive property). By adding $-(0 \cdot x)$ to both sides, we find that $0 = 0 \cdot x$. The identity $x \cdot 0 = 0$ can be proved similarly.

(ii) The trick here is to write $(x + (-x)) \cdot y = x \cdot y + (-x) \cdot y$, and realize that the left-hand side is also equal to $0 \cdot y$, which is just 0 by part (i). Thus $x \cdot y + (-x) \cdot y = 0$, and so $(-x) \cdot y = -(x \cdot y)$. The other identity may be proved similarly by considering $x \cdot (y + (-y))$.

(iii) We have $(-x) \cdot (-y) = -(x \cdot (-y)) = -(-(x \cdot y))$, applying part (ii) twice. By the Laws of Exponents, we have $-(-(x \cdot y)) = x \cdot y$. (Specifically, we can apply Theorem 3.41 part (ii), with $a = x \cdot y$ and $m = n = -1$, interpreting this result in additive notation.)

(iv) Supposing R has 1, then $(-1) \cdot x = -(1 \cdot x)$ (by part (ii)) $= -x$.

(v) We have $(-1)^2 = (-1) \cdot (-1)$ (by definition of exponential notation) $= 1 \cdot 1$ (by part (iii)) $= 1$. □

Example 11.11 (Calculations in a ring). Let R be a ring, and let $a, b, c \in R$. We can take advantage of the commutativity of R under $+$ to simplify the expression

$$a + b - a - c$$

down to

$$b - c.$$

The reader should try to perform this simplification one step at a time, using only one definition or result in each step. However, the familiar expression

$$(a + b) \cdot (a - b)$$

does *not* equal $a^2 - b^2$ in general. Attempting to expand this last expression, we find

$$(a + b) \cdot (a - b) = a \cdot (a - b) + b \cdot (a - b) \qquad \text{(by Axiom } \mathbf{R3})$$
$$= a \cdot a + a \cdot (-b) + b \cdot a + b \cdot (-b) \qquad \text{(by } \mathbf{R3} \text{ again)}$$
$$= a^2 - a \cdot b + b \cdot a - b^2. \qquad \text{(using Lemma 11.10)}$$

But because $\cdot$ need not be commutative, this expression cannot be further simplified.

We next identify some especially nice types of ring, by imposing additional requirements above and beyond the ring axioms. In an arbitrary ring, there is nothing to prevent the product of two non-zero elements from being zero; the following definition ensures that this cannot happen.

The condition "$1 \neq 0$" may seem odd in the definitions below. But in the *trivial ring* (or *zero ring*) $R = \{0\}$ whose only element is 0, it is formally true that 0 is a multiplicative identity as well as an additive identity, so in this ring at least, we do have $1 = 0$; see Exercise 11.2.

Definition 11.12. A ring R is called an *integral domain*, or simply a *domain*, if R is a commutative ring with 1, we have $1 \neq 0$, and

$$\forall x, y \in R, \ x \cdot y = 0 \implies x = 0 \text{ or } y = 0.$$

Another way in which a ring can be well-behaved is that most of its elements can have inverses under multiplication; however, we might be surprised if the element 0 were to have a multiplicative inverse; see Exercise 11.3. The following definition serves up as nice a scenario as we can hope for in this regard:

Definition 11.13. A *field* is a commutative ring R with $1 \neq 0$ such that $R^\times = R - \{0\}$.

Perhaps surprisingly, there is a relationship between the two types of ring just defined:

Proposition 11.14. *Every field is a domain.*

Proof. Let F be a field. Then F is a commutative ring with 1, and $1 \neq 0$. [Show that $\forall x, y \in F, x \cdot y = 0 \implies x = 0$ or $y = 0$] Let $x, y \in F$, and suppose that $x \cdot y = 0$. [Show $x = 0$ or $y = 0$] [This is equivalent to $x \neq 0 \longrightarrow y = 0$] Suppose that $x \neq 0$. [Show $y = 0$] Then $x \in F^\times$ (by definition of *field*). So we can write $x^{-1} \cdot (x \cdot y) = x^{-1} \cdot 0$; so $(x^{-1} \cdot x) \cdot y = 0$ (by Axiom $\mathbf{R2}$ and Lemma 11.10); $1 \cdot y = 0$; and thus $y = 0$. $\qquad \square$

In Chapter 4, we found a natural way to express the relationship between $\mathbf{Z}$ and $\mathbf{R}$ when considered as groups under addition, by introducing the notion of a *subgroup*. Now that we recognize the extra structure on these sets which makes them into *rings*, we would like to have a way to express this new relationship. The reader may wish at this point to review the discussion and definition of subgroup in Chapter 4 before reading further.

Definition 11.15. Let $(R, +, \cdot)$ be a ring. A *subring* of R is a ring $(S, +_S, \cdot_S)$ such that $S \subseteq R$, $+_S = +|_{S \times S}$, and $\cdot_S = \cdot|_{S \times S}$.

Thus, a subring is a subset of a ring which is a ring in its own right, under the "same" operations as its parent ring (restricted, of course, to this subset). In particular, just as with groups and subgroups, there is only one way in which a subset S of a ring R can be a subring of R: the binary operations of addition and multiplication on S are inherited from those on R. As with groups, we seldom use the clumsy notation of restriction of functions when we describe the ring operations of a subring; instead, we use the same symbols $+$ and $\cdot$ for the subring as we do for the parent ring.

Notation 11.16. We write $S \leq R$ to mean S is a subring of R. We write $S < R$ to mean that S is a proper subring of R; that is, $S \leq R$ and $S \neq R$.

Proposition 11.17 (Subring Test). *Let $S \subseteq R$, where R is a ring. Then $S \leq R$ iff both*

    (SRT1): $(S, +) \leq (R, +)$

    and

    (SRT2): *S is closed under multiplication: i.e., $\forall x, y \in S, x \cdot y \in S$.*

Proof. Exercise 11.10. $\qquad\qquad\qquad\qquad\qquad\qquad\qquad\qquad\qquad\qquad\qquad$ $\square$

11.3   Ring Homomorphisms, Ideals, and Quotient Rings

We would like to define a notion of "homomorphism" that applies to rings. Taking our direction from our work with groups, it seems reasonable to define a ring homomorphism to be a function between two rings which commutes with *both* ring operations. This is what we shall do.

Definition 11.18. Let R and S be rings. A *ring homomorphism* from R to S is a function $\sigma : R \to S$ such that for all $x, y \in R$, we have

    (RH1): $\sigma(x + y) = \sigma(x) + \sigma(y)$

    and

    (RH2): $\sigma(x \cdot y) = \sigma(x) \cdot \sigma(y)$.

We also define the same special types of homomorphism for rings that we defined for groups:

Definition 11.19. (i) An *isomorphism* of rings, or *ring isomorphism*, is a bijective ring homomorphism.

    (ii) An *embedding* is an injective homomorphism.

    (iii) An *automorphism* of a ring is a ring isomorphism from a ring to itself.

Notation 11.20. We use the notation $\mathrm{Aut}(R)$ to denote the set of all automorphisms of the ring R. As in the case of groups, the set $\mathrm{Aut}(R)$ forms a group under composition of functions (Exercise 11.13).

Remark 11.21. We are now finally in a position to give a (partial) answer, in the language of abstract algebra, to a general question left over from Chapter 2: for which number systems is the number game winnable? First we need to say what we mean by "number system." Let us say for now that a number system means a subring R of $\mathbf{C}$ with $1 \in R$. Then the number game on R is winnable only if $\mathrm{Aut}(R) = \{\mathrm{id}_R\}$, the trivial group. As with groups, ring automorphisms represent symmetries; and a symmetry of R represents a way to re-arrange the elements of R so that a player cannot tell that a change has been made! Therefore, if there is more than one element in $\mathrm{Aut}(R)$, then we cannot possibly tell the correct labeling of the numbers. We note, however, that even if $\mathrm{Aut}(R)$ is trivial, there is still a question of finding a *finite procedure* for identifying numbers, which is not possible when R is uncountable. We are by no means done with ring automorphisms, however; we have merely given them a name. Ring automorphisms play a central role in Galois Theory (Chapter 16), for example, which will allow us to give a more complete answer to the number game question (see Exercise 16.26), among many other applications.

Remark 11.22. Every ring homomorphism $\sigma : R \to S$ is also a group homomorphism on the additive groups of the rings, because of (**RH1**). Thus, a natural point of interest is the *kernel* of σ viewed in this light, namely, the set of all elements of R which map to 0_S. We shall define the kernel of a ring homomorphism to be exactly this set.

Definition 11.23. The *kernel* of a ring homomorphism $\sigma : R \to S$ is the set

$$\ker(\sigma) = \{x \in R : \sigma(x) = 0_S\}.$$

That is, the kernel of σ is its kernel as a group homomorphism from $(R, +)$ to $(S, +)$.

As in the case of group homomorphisms, the kernel of a ring homomorphism measures how much information is lost in passing from the domain to the image. The case when the kernel is $\{0\}$, which is as small as it is possible to be, corresponds to the situation of an embedding, where we lose no information, and a faithful copy of the domain is transported into the codomain (see Exercise 11.23). In this case, we have $R \cong \sigma(R)$.

When we asked which subsets of a group could be the kernel of a group homomorphism, we arrived at the notion of a normal subgroup. Let us now ask, Which subsets of a ring can be the kernel of a ring homomorphism?

Suppose that $\sigma : R \to S$ is a ring homomorphism, and let $K = \ker(\sigma)$. To start with, since K is the kernel of the group homomorphism $\sigma : (R, +) \to (S, +)$, we must have $K \trianglelefteq (R, +)$, by Theorem 7.19. Now since $(R, +)$ is abelian, *every* subgroup of $(R, +)$ is normal (by Exercise 11.4), so we may capture this condition by writing simply $K \leq (R, +)$.

So far, we have only related the fact that K is a kernel to the *additive* structure of the ring R. There is more to see when we consider the multiplicative structure. For suppose that $x \in K$ and $a \in R$. Then we have

$\sigma(a \cdot x) = \sigma(a) \cdot \sigma(x)$ (by (**RH2**)) $= \sigma(a) \cdot 0$ (since $x \in \ker(\sigma)$) $= 0$ (by Lemma 11.10). But this tells us that $a \cdot x$ is also in the kernel of σ, namely, K. Likewise, we can see that $x \cdot a \in K$.

It turns out that these properties of a kernel of a ring homomorphism are enough to characterize a subset of a ring as a kernel, and so we record them in a definition:

Definition 11.24. Let R be a ring. An *ideal* of R is a subset I of R such that

(**Id1**): $(I, +) \le (R, +)$

and

(**Id2**): $\forall x \in I \; \forall a \in R, a \cdot x \in I$ and $x \cdot a \in I$.

Remark 11.25. We may say that an ideal is an additive subgroup of a ring (**Id1**) which "absorbs" arbitrary ring elements under multiplication (**Id2**).

Remark 11.26. What we have just defined is sometimes referred to as a *two-sided* ideal, since it must absorb elements of R from both sides, the right and the left. We shall not consider one-sided ideals in this text.

Remark 11.27. The name *ideal* is rooted in the history of algebraic number theory. In brief, nineteenth-century mathematicians struggling to prove Fermat's Last Theorem, the statement

$$\forall x, y, z, n \in \mathbf{Z}^+, n > 2 \implies x^n + y^n \ne z^n,$$

encountered difficulties because the familiar uniqueness of factorization of integers into primes fails in the case of more general number systems. The concept of an ideal was invented to salvage unique factorization; while numbers do not always factor uniquely, ideals ("ideal numbers") do, in the appropriate context.

Example 11.28 (Ideals of **Z**). If I is an ideal of **Z**, then $(I, +) \le (\mathbf{Z}, +)$, so I must by cyclic, by Exercise 5.6. So the only candidates for ideals of **Z** are the sets $\langle n \rangle$ for $n \in \mathbf{Z}$. Recall from Example 4.18 that $\langle n \rangle$ is simply the set $n\mathbf{Z}$ of all integer multiples of n. But this set has the absorption property (**Id2**): a multiple of an element of $n\mathbf{Z}$ is another element of $n\mathbf{Z}$, since a multiple of a multiple of n is another multiple of n. Therefore, every such set is an ideal of **Z**. To summarize: the ideals of **Z** are the sets $n\mathbf{Z}$ for each n in **Z**.

Example 11.29 (Ideals of **R**). Suppose that I is an ideal of **R** and that $\alpha \in I - \{0\}$. Then $\alpha \in \mathbf{R}^\times$, so $\alpha^{-1} \in \mathbf{R}$, and we have $\alpha^{-1} \cdot \alpha \in I$, by (**Id2**). Thus $1 \in I$. Using (**Id2**) again, we find that any real number times 1 must be in I, and therefore, all of **R** is in I; so $I = \mathbf{R}$. We conclude that the only possible ideals of **R** are $\{0\}$ and **R**; it is easy to check that both of these sets are in fact ideals of **R**. The same argument generalizes to any field F (Exercise 11.7).

Roughly speaking, ideals of a ring are like normal subgroups of a group: they are the type of object which can be the kernel of a homomorphism, as we shall see. But before we reach this result, we come to another parallel between normal subgroups and ideals: just as we can form the quotient of a group by a normal subgroup, we can form the quotient of a ring by an ideal:

Theorem 11.30 (Existence of a Quotient Ring). *Let R be a ring, and let I be an ideal of R. Then the set Q of all cosets of I in R under addition,*

$$Q = \{a + I \ : \ a \in R\},$$

forms a ring under the operations

$$(a + I) + (b + I) = (a + b) + I$$

and

$$(a + I) \cdot (b + I) = (a \cdot b) + I.$$

We call Q the quotient ring of R by I.

Proof. Let I be an ideal of R, and set $Q = \{a + I \ : \ a \in R\}$. Then $(I, +) \trianglelefteq (R, +)$ by (**Id1**), so Q forms a group under addition, by Theorem 7.26. Note that $(Q, +)$ is just the quotient group R/I where R is considered as a group under addition. Now, $(R/I, +)$ is an abelian group, by Exercise 11.5. (The identity element of $(R/I, +)$ is $0_{R/I} = 0_R + I = I$.)

The main point of the proof is to show that multiplication is well-defined on R/I; this means that the product of two cosets should not depend on the particular coset representatives we choose. So let $C_1, C_2 \in R/I$, and suppose that we can write $C_1 = a_1 + I = b_1 + I$ and $C_2 = a_2 + I = b_2 + I$ with a_1, a_2, b_1, b_2 in R. Then we have $d_1 := a_1 - b_1 \in I$ and $d_2 := a_2 - b_2 \in I$, by Exercise 8.2. Now we have

$$a_1 \cdot a_2 = (b_1 + d_1) \cdot (b_2 + d_2)$$
$$= b_1 \cdot b_2 + b_1 \cdot d_2 + d_1 \cdot b_2 + d_1 \cdot d_2.$$

Since $d_1, d_2 \in I$, we have $b_1 \cdot d_2$, $d_1 \cdot b_2$, $d_1 \cdot d_2 \in I$, by (**Id2**). Therefore, $a_1 \cdot a_2 - b_1 \cdot b_2 \in I$, and so $a_1 \cdot a_2 + I = b_1 \cdot b_2 + I$, by Exercise 8.2. This shows that indeed our definition of $C_1 \cdot C_2$ does not depend on the choice of coset representatives.

Ring axioms **R2** and **R3** are now straightforward to verify in R/I. For **R2**, we have

$$\begin{aligned}
((a + I) \cdot (b + I)) \cdot (c + I) &= ((a \cdot b) + I) \cdot (c + I) \\
&= ((a \cdot b) \cdot c) + I \\
&= (a \cdot (b \cdot c)) + I \qquad \text{(by Axiom **R2** for R)} \\
&= (a + I) \cdot ((b \cdot c) + I) \\
&= (a + I) \cdot ((b + I) \cdot (c + I)).
\end{aligned}$$

Axiom **R3** is left to the reader as Exercise 11.9.   □

Notation 11.31. We write R/I for the quotient ring of R by I. We sometimes read this expression as "R modulo I" or "R mod I."

Remark 11.32. Suppose that R is a ring and I is an ideal of R; then I is a subgroup of the abelian group $(R, +)$, so we can form the quotient group R/I *or* the quotient ring, also denoted R/I. However, no confusion is likely to arise, because these two quotient objects are identical as sets, and also as additive groups; the only difference is that the quotient *ring* has the extra operation of multiplication.

Example 11.33 (Quotients of $\mathbf{Z}$). From Example 11.28, we know that for each n in $\mathbf{Z}$, the set $n\mathbf{Z}$ is an ideal of $\mathbf{Z}$. Therefore, we can form the quotient ring $\mathbf{Z}/n\mathbf{Z}$. Since this quotient ring is the same as the quotient *group* $\mathbf{Z}/n\mathbf{Z}$ with the extra operation of multiplication, then in particular, we have $|\mathbf{Z}/n\mathbf{Z}| = n$ when $n \geq 1$ (see Example 7.29). Again, the elements of the quotient ring $\mathbf{Z}/n\mathbf{Z}$ are the cosets of $n\mathbf{Z}$ in $\mathbf{Z}$, which are just the congruence classes modulo n, namely $a + n\mathbf{Z}$ for $a \in \mathbf{Z}$.

These quotients of $\mathbf{Z}$ provide our first examples of rings which are not domains. For example, in $\mathbf{Z}/6\mathbf{Z}$, we have $(2 + 6\mathbf{Z}) \cdot (3 + 6\mathbf{Z}) = 6 + 6\mathbf{Z} = 0 + 6\mathbf{Z} = 0_{\mathbf{Z}/6\mathbf{Z}}$, yet neither $2 + 6\mathbf{Z}$ nor $3 + 6\mathbf{Z}$ is 0 in $\mathbf{Z}/6\mathbf{Z}$.

For which positive integers n is $\mathbf{Z}/n\mathbf{Z}$ a domain?

To start with, we must have $n \geq 2$, since if $n = 1$, then we are dealing with the trivial or "zero ring" $\mathbf{Z}/\mathbf{Z}$ with only one element, namely $0_{\mathbf{Z}/\mathbf{Z}} = 0 + \mathbf{Z} = \mathbf{Z}$ (see Exercise 11.6). In this trivial ring, we have $1 = 0$, so it is not a domain.

So assume $n \geq 2$. The zero element of $\mathbf{Z}/n\mathbf{Z}$ is $n\mathbf{Z}$. So the condition we need is

$$(a + n\mathbf{Z}) \cdot (b + n\mathbf{Z}) = n\mathbf{Z} \implies a + n\mathbf{Z} = n\mathbf{Z} \text{ or } b + n\mathbf{Z} = n\mathbf{Z}.$$

Simplifying on the left and using the criterion of Exercise 8.2 for when two cosets are equal, we need

$$a \cdot b \in n\mathbf{Z} \implies a \in n\mathbf{Z} \text{ or } b \in n\mathbf{Z}. \tag{11.1}$$

We can rewrite this implication in the language of divisibility, as

$$n \mid (a \cdot b) \implies n \mid a \text{ or } n \mid b. \tag{11.2}$$

But this is just the property that n is *prime*. So we have found: If n is a positive integer, then $\mathbf{Z}/n\mathbf{Z}$ is a domain iff n is prime.

More generally, we can ask, For which ideals I of $\mathbf{Z}$ is $\mathbf{Z}/I$ a domain? Since every ideal of $\mathbf{Z}$ is of the form $n\mathbf{Z}$ for some integer n, and since $n\mathbf{Z} = (-n)\mathbf{Z}$ for any n, the only ideal we haven't yet checked is the zero ideal $0\mathbf{Z} = \{0\}$. We have $\mathbf{Z}/\{0\} \cong \mathbf{Z}$ by Exercise 11.6, and $\mathbf{Z}$ is certainly a domain.

The rings $\mathbf{Z}/n\mathbf{Z}$ are more or less ubiquitous in mathematics because of their usefulness. Indeed, beginning students of mathematics often study these rings before encountering the general concept of a ring itself. The "ring of integers modulo n," as we sometimes refer to $\mathbf{Z}/n\mathbf{Z}$, is often represented using the numbers $0, 1, \ldots, n - 1$ to stand for the congruence classes $0 + n\mathbf{Z}, 1 + n\mathbf{Z}, \ldots (n - 1) + n\mathbf{Z}$. Then the operations of addition and multiplication on $\mathbf{Z}/n\mathbf{Z}$ are defined to be simply "addition and multiplication modulo n": after performing the ordinary addition or multiplication, we reduce the result modulo n to get it into the appropriate range from 0 to $n-1$. The reader should undertake to verify that this description of $\mathbf{Z}/n\mathbf{Z}$ really does give the right ring.

If n, a, and b are integers such that $a \mid b$ and $b \mid n$, then n/b is a quotient of n/a. Conversely, the quotients of n/a are all of the form n/c where $a \mid c$. Remarkably, the same phenomenon occurs in general for quotient rings, if we replace n by a ring R, replace a and b by ideals I and J, and replace divisibility by containment. That is the content of the following result.

Lemma 11.34. *Let R be a ring, and let I be an ideal of R. Then there is a bijection*

$$\omega \ : \ \{\mathfrak{J} \ : \ \mathfrak{J} \text{ is an ideal of } R/I\} \to \{J \ : \ J \text{ is an ideal of } R \text{ s.t. } J \supseteq I\}$$

given by

$$\omega(\mathfrak{J}) = \{a \in R \ : \ a + I \in \mathfrak{J}\}.$$

The proof is requested in Exercise 11.21. Also see Exercise 7.17 for the analogous result for groups.

The significance of Lemma 11.34 is that quotient rings are *simpler* than their parent rings: the ideals of R/I come from the ideals of R which contain I. Thus, in passing from R to R/I, we "lose" all ideals which do not contain I, but keep the rest—modulo I.

In case J is an ideal of R such that $I \subseteq J$, we indicated above that the ring R/J should be a quotient of R/I, corresponding to the situation with ordinary integers. What is actually true is that $R/J \cong (R/I)/\mathfrak{J}$, where $\mathfrak{J} = \omega^{-1}(J)$. If we use the suggestive notation $\mathfrak{J} = J/I$, then we can rewrite this assertion in the following form, which suggests cancellation of fractions:

Lemma 11.35. *Let R be a ring, and let I and J be ideals of R such that $I \subseteq J$. Then*

$$(R/I)/(J/I) \cong R/J,$$

where by J/I we mean

$$J/I = \omega^{-1}(J) = \{a + I \in R/I \ : \ a \in J\}$$

and ω is the function defined in Lemma 11.34. (See Project 23.14 for a justification of this notation in the proper context, where J/I will be given its own structure as a type of algebraic object called a module.*)*

Proof. Exercise 11.22. $\square$

We will now fulfill our mission to complete the analogy between ideals of a ring and normal subgroups of a group, by proving the analog for rings of a combination of Theorems 7.19, 7.30, and 9.7.

Theorem 11.36 (Fundamental Theorem of Ring Homomorphisms). *(i) Let R be a ring, and let I be an ideal of R. Then the natural map*

$$\sigma : R \to R/I$$

given by

$$\sigma(r) = r + I$$

is a surjective ring homomorphism, and $\ker(\sigma) = I$.

(ii) If $\tau : R \to S$ *is any ring homomorphism, then* $\tau(R) \leq S$, $\ker(\tau)$ *is an ideal of R, and we have* $\tau(R) \cong R/\ker(\tau)$.

Proof. (i) Let R be a ring, and let I be an ideal of R. Define $\sigma : R \to R/I$ by $\sigma(r) = r + I$ for $r \in R$. Note first that $(I, +) \trianglelefteq (R, +)$, by **(Id1)**. So by Theorem 7.30, σ is a surjective *group* homomorphism from $(R, +)$ to $(R/I, +)$ with $\ker(\sigma) = I$. Since the kernel of σ as a *ring* homomorphism is the same as its kernel as an additive group homomorphism, it only remains to show that σ respects multiplication. So let $x, y \in R$. Then $\sigma(x \cdot y) = (x \cdot y) + I$ $= (x + I) \cdot (y + I) = \sigma(x) \cdot \sigma(y)$, as required.

(ii) Let $\tau : R \to S$ be a ring homomorphism. Then τ is also a group homomorphism, $\tau : (R, +) \to (S, +)$. By Theorem 7.13, we have $(\tau(R), +) \leq$ $(S, +)$. Further, $\tau(R)$ is closed under multiplication, since if $a, b \in \tau(R)$, then we can write $a = \tau(\alpha), b = \tau(\beta)$ for some $\alpha, \beta \subset R$, and then $a \cdot b = \tau(\alpha) \cdot \tau(\beta) =$ $\tau(\alpha \cdot \beta) \in \tau(R)$. Therefore, we have $\tau(R) \leq S$ by the Subring Test.

Let $K = \ker(\tau)$. By Theorem 7.19, we have $(K, +) \trianglelefteq (R, +)$, which establishes property **(Id1)** for K. Let $r \in R$ and $x \in K$. Then we have

$$
\begin{aligned}
\tau(r \cdot x) &= \tau(r) \cdot \tau(x) &&\text{(by } (\mathbf{RH2}) \text{)} \\
&= \tau(r) \cdot 0_S &&\text{(since } x \in K) \\
&= 0_S &&\text{(by Lemma 11.10).}
\end{aligned}
$$

Thus, $r \cdot x \in K$. Similarly, we can see that $x \cdot r \in K$. This establishes **(Id2)**, so K is an ideal of R.

Now by Theorem 9.7, the function $f : R/K \to \tau(R)$ given by the formula $f(r + K) = \tau(r)$ is a well-defined group isomorphism. Since K is an ideal of R, we can regard R/K as a *ring*, whose elements and addition operation are the same as those of the group $(R/K, +)$. To check whether f respects multiplication, let $\alpha, \beta \in R/K$. Then we can write $\alpha = r + K$ and $\beta = s + K$ for some $r, s \in R$. We have $f(\alpha \cdot \beta) = f((r + K) \cdot (s + K)) = f((r \cdot s) + K)$ $= \tau(r \cdot s) = \tau(r) \cdot \tau(s)$ (since τ is a ring homomorphism) $= f(r + K) \cdot f(s + K)$ $= f(\alpha) \cdot f(\beta)$. Now we know that f is a ring homomorphism. Since f is an isomorphism of additive groups, then f is a bijective function. This tells us that f is also a ring isomorphism. $\square$

11.4   Exercises

Exercise 11.1. Prove that matrix multiplication distributes over matrix addition in $M_2(\mathbf{R})$: that is, verify Axiom **R3** for $M_2(\mathbf{R})$.

Exercise 11.2. (a) Let $R = \{a\}$ be a set of size one. Show that R forms a ring with $1 = 0$. (There is only one way to form a binary operation on R!)

(b) Suppose that R is a ring with $1 = 0$. Prove that R has only one element, so that $R = \{0\}$, the zero ring.

Exercise 11.3. Prove that if R is a ring with 1, and $0 \in R^{\times}$, then $R = \{0\}$. Moral: you *can* divide by 0, but only if $0 = 1$.

Exercise 11.4. Prove that if G is an abelian group and $H \leq G$, then $H \trianglelefteq G$.

Exercise 11.5. Prove that if G is an abelian group and $H \trianglelefteq G$, then G/H is also abelian.

Exercise 11.6. Let R be a ring.

(a) Prove that the set $\{0\}$ and the set R are ideals of R.

(b) Prove that $R/\{0\} \cong R$.

(c) Prove that $R/R \cong \{0\}$ (the zero ring).

Exercise 11.7. Prove that if F is a field, then the only ideals of F are $\{0\}$ and F.

Exercise 11.8. Prove that if R is a ring and I, J are ideals of R, then $I \cap J$ is also an ideal of R.

Exercise 11.9. Complete the proof of Theorem 11.30 by verifying Axiom **R3** in R/I.

Exercise 11.10. Prove the Subring Test, Proposition 11.17.

Exercise 11.11. Let R be a domain, and let $S \leq R$.

(a) Prove that if $1_R \in S$, then S is also a domain.

(b) Prove that if S is a domain, then $1_S = 1_R$ (and in particular, $1_R \in S$).

(c) Find an example of a commutative ring T with 1 which has a subring V such that V is a domain, but $1_V \neq 1_T$. Hint: look in the ring of 2×2 matrices over $\mathbf{R}$.

Exercise 11.12. Prove the analog of Lemma 9.5 with groups and group isomorphisms replaced by rings and ring isomorphisms.

Exercise 11.13. Prove that if R is a ring, then $\mathrm{Aut}(R)$ is a group under composition of functions. (Exercise 11.12 may help.)

Exercise 11.14. Let $R = \{a + b\sqrt{2} \ : \ a, b \in \mathbf{Z}\}$. Prove that R is a subring of $\mathbf{R}$, the ring of real numbers. (Compare Exercise 2.2.)

Exercise 11.15. Let R be a domain, and suppose that $\mathbf{Z} \leq R$. Let σ be an automorphism of R.

(a) Why must $\sigma(0) = 0$?

(b) Prove that $\sigma(1) = 1$.

(c) Prove by induction that $\sigma(n) = n$ for every $n \in \mathbf{Z}$ such that $n \geq 1$.

(d) Prove that we also have $\sigma(n) = n$ for all $n \in \mathbf{Z}$ such that $n < 0$. Thus, σ maps every integer to itself.

(e) Describe the group $\mathrm{Aut}(\mathbf{Z})$ explicitly. If you have read Chapter 2, then explain how your answer relates to the number game on $\mathbf{Z}$.

Exercise 11.16. Let the ring R be as in Exercise 11.14. Suppose that σ is an automorphism of R.

(a) Prove that $\sigma(n) = n$ for every $n \in \mathbf{Z}$. (The results of the previous two exercises may be used.)

(b) Prove that $\sigma(\sqrt{2}) \in \{\sqrt{2}, -\sqrt{2}\}$.

(c) Use the previous parts of this exercise to prove that there are at most two different automorphisms of R, and give explicit formulas for them.

(d) Check that each of your two candidate automorphisms from part (c) is actually an automorphism of R. If you have read Chapter 2, then explain how your result relates to the number game on R.

Exercise 11.17. This exercise continues the notation of Exercise 11.16. Write $\mathrm{Aut}(R) = \{e, \sigma\}$, where e is the identity map on R, and σ sends $\sqrt{2}$ to $-\sqrt{2}$.

(a) Explain why, for any $\alpha \in R$, we have $\alpha \in \mathbf{Z}$ iff $\sigma(\alpha) = \alpha$.

(b) Prove that for all $\alpha \in R$, we have $\alpha \cdot \sigma(\alpha) \in \mathbf{Z}$. Try to accomplish this by using part (a), without writing α explicitly in terms of $\sqrt{2}$.

(c) Prove that $R^{\times} \cap \mathbf{Z} \subseteq \mathbf{Z}^{\times}$.

(d) Prove that $R^{\times} = \{a + b\sqrt{2} \ : \ a, b \in \mathbf{Z} \text{ and } a^2 - 2b^2 \in \{1, -1\}\}$.

(e) Find some units of R. Can you find all of them?

Exercise 11.18. For a ring R, define the *center* of R to be the set $C(R) :- \{a \in R \ : \ \forall b \in R, ab = ba\}$. Prove that we have $C(R) \leq R$.

Exercise 11.19. Find an example of a ring R with a subset S such that $(S, +) \leq (R, +)$ but S is not an ideal of R.

Exercise 11.20. (a) Write out addition and multiplication tables for the rings $\mathbf{Z}/n\mathbf{Z}$ for each integer n between 1 and 6, inclusive.

(b) Which of the rings in part (a) are domains? Which are fields?

(c) Find all of the ideals of each ring in part (a).

Exercise 11.21. Prove Lemma 11.34.

Exercise 11.22. Prove Lemma 11.35.

Exercise 11.23. Prove that a ring homomorphism is injective if and only if its kernel is $\{0\}$. Hint: Lemma 9.18 may be useful.

Exercise 11.24. This exercise translates Exercise 9.15 to rings. Suppose that $\sigma : R \to S$ is a ring homomorphism with kernel I, and that J is an ideal of R with $J \subseteq I$. Prove that σ factors through R/J, by showing that the map $\tilde{\sigma} : R/J \to S$ given by the formula $a + J \mapsto \sigma(a)$ (for $a \in R$) is a well-defined ring homomorphism.

12

Results on Commutative Rings

12.1 Introduction

In this chapter, we only consider rings which are commutative. Commutative rings include all fields and domains, as well as many other, not-so-nice rings which have "zerodivisors": non-zero elements a and b for which $a \cdot b = 0$.

Our major theme in this chapter will be to relate the properties of quotient rings to the properties of the ideals we "divided" by. In particular, we will discover for which ideals I the quotient ring R/I is a domain, and for which ideals the quotient is a field. We will also discover how to produce ideals more or less at will (Section 12.3).

12.2 Primes and Domains

Revisiting Example 11.33, we notice that $\mathbf{Z}/n\mathbf{Z}$ is a domain iff n is a prime element of $\mathbf{Z}$ (at least, for $n \geq 2$). This suggests that there is a relationship between *primality* ("primeness") and quotient rings being domains. To explore this relationship in general, we first need to come up with a suitable definition of "prime" in more generality, which will apply to a wide class of rings instead of just to the ring of integers $\mathbf{Z}$. We will actually define *two* notions of primeness: one for a ring element, and another for an ideal.

The discussion in Example 11.33 above, and, in particular, Equation 11.1, motivates the following general definition for when an ideal is to be called "prime."

Definition 12.1 (Prime Ideal). Let R be a commutative ring with 1, and let I be an ideal of R. Then I is called *prime* if $I \neq R$ (I is "proper"), and

$$\forall a, b \in R, a \cdot b \in I \implies a \in I \text{ or } b \in I.$$

Notice that in Example 11.33, we characterized prime integers as those integers greater than one which, if they divide a product of two factors, must divide one of the factors. Guided by this idea, we will now define the notion

DOI: 10.1201/9781003252139-12

of prime element in an arbitrary domain. First, we extend the definition of divisibility to arbitrary commutative rings with 1.

Definition 12.2. Let R be a commutative ring with 1, and let $a, b \in R$. To say $b \mid a$ (read "b divides a") means $\exists c \in R$ s.t. $a = b \cdot c$.

Warning 12.3. When we write $b \mid a$, it is important to know which ring R the factors of a are allowed to come from. Notice that no description of R is built in to the notation for divisibility. Thus, for example, while we have $2 \nmid 3$ in **Z**, yet it is true that $2 \mid 3$ in **R**.

Definition 12.4 (Prime Element). Let R be a commutative ring with 1, and let $a \in R - \{0\}$. Then a is *prime* in R if $a \notin R^{\times}$ and

$$\forall b, c \in R, a \mid (b \cdot c) \implies a \mid b \text{ or } a \mid c.$$

Remark 12.5. In our general definition of a prime element, we could not capture the condition that a prime should be greater than 1, because there is no notion of *ordering* in an arbitrary ring. As a result, there are some integers which are prime according to Definition 12.4 but are not prime according to the usual definition of prime for integers; see Exercise 12.3. This is not a bad thing: our general definition of "prime" is actually more natural, even for ordinary integers.

Although our general definition of a prime element does what we want, the reader may be more familiar with the definition of a prime number as a number which has "no" factors: more precisely, no positive factors except for one and itself.

If we try to define primality of an integer in terms of factorization, without referring to any notion of ordering (e.g. positive or negative), then both the "trivial" factors of 1 and -1 should not count as factors, since any integer n can be factored using these numbers. For example, we can write $n = (-n) \cdot (-1) = n \cdot 1$.

The astute reader will recognize 1 and -1 as exactly the *units* of **Z**. In general, any ring element may be "factored" using units. This observation gives rise to the following definition, which describes those elements of a domain which have no non-trivial factors, and so cannot be "reduced" into a product of simpler terms. Informally, the definition simply says that a unit does not count as a factor.

Definition 12.6. Let R be a domain, and let $a \in R - \{0\}$. Then a is *irreducible* in R if $a \notin R^{\times}$ and

$$\forall b, c \in R, a = b \cdot c \implies b \in R^{\times} \text{ or } c \in R^{\times}.$$

Remark 12.7. Although the notions of *prime* and *irreducible* are equivalent for ordinary positive integers, they are *not* equivalent for a general domain R. However, it is true in any domain that every prime element is irreducible (Exercise 12.11). It is in a sense the discrepancy between these two notions which measures the extent to which unique factorization fails in R.

Next we come to the first result which relates a nice property of a quotient ring to a property of the corresponding ideal.

Theorem 12.8. *Let R be a commutative ring with 1, and let I be an ideal of R. Then R/I is a domain iff I is prime.*

Proof. ($\Rightarrow$): Suppose that R/I is a domain. [Show I is prime] [Show I is proper and $\forall a, b \in R, a \cdot b \in I \implies a \in I$ or $b \in I$]

By definition of *domain*, we have $0_{R/I} \neq 1_{R/I}$. Now $0_{R/I} = I = 0 + I$ and $1_{R/I} = 1 + I$, so we can conclude from Exercise 8.2 that $1 - 0 \notin I$, i.e., $1 \notin I$. In particular, $I \neq R$, which means I is proper.

Let $a, b \in R$, and suppose that $a \cdot b \in I$. [Idea: we must somehow work with properties of R/I to conclude that $a \in I$ or $b \in I$. How can we get into R/I given a and b? Answer: form the cosets $a + I$ and $b + I$.]

Then $a + I$, $b + I \in R/I$, and we have $(a \cdot b) + I = 0 + I$ (since $a \cdot b \in I$). We can rewrite this as $(a + I) \cdot (b + I) = 0_{R/I}$. Since R/I is a domain, this forces $a + I = 0_{R/I}$ or $b + I = 0_{R/I}$, which in turn means $a \in I$ or $b \in I$, as desired.

($\Leftarrow$): Suppose that I is prime. [Show R/I is a domain] [Show $0_{R/I} \neq 1_{R/I}$ and $\forall x, y \in R/I, x \cdot y = 0_{R/I} \implies x = 0_{R/I}$ or $y = 0_{R/I}$]

First, assume for a contradiction that $0_{R/I} = 1_{R/I}$. Writing this equation using coset notation, it says $0 + I = 1 + I$. By Exercise 8.2, we have $1 \in I$. Since $1 \in R^\times$, we conclude from Exercise 12.6 that $I = R$, so I is not proper. This contradicts the definition of prime ideal.

Let $x, y \in R/I$, and suppose that $x \cdot y = 0_{R/I}$. Then $x = a + I$ and $y = b + I$ for some $a, b \in R$ (by definition of R/I). So we have $(a + I) \cdot (b + I) = 0_{R/I} = 0 + I$. Thus $(a \cdot b) + I = 0 + I$, and so $a \cdot b \in I$, by Exercise 8.2. Since I is prime, we know that $a \in I$ or $b \in I$. This in turn gives $a + I = 0 + I$ or $b + I = 0 + I$, which means $x = 0_{R/I}$ or $y = 0_{R/I}$, as desired. $\quad\square$

12.3 The Ideal Generated by a Set

Let R be a commutative ring with 1, and let $S \subseteq R$ be a non-empty subset of R. What is the *smallest* ideal I of R which contains S?

First of all, if $s \in S$ and $r \in R$, then we must have $rs \in I$, by (**Id2**). More generally, if $s_1, s_2, \ldots, s_n \in S$ and $r_1, r_2, \ldots, r_n \in R$, then (**Id2**) requires that we have $r_1 s_1, r_2 s_2, r_n s_n \in I$, and then (**Id1**) forces I to be closed under addition, so we must also have $r_1 s_1 + r_2 s_2 + \cdots + r_n s_n \in I$. It turns out that we can stop here: the following result says that the set of all elements of this form is already an ideal! Because of its importance, this kind of set deserves its own notation:

Notation 12.9. Let R be a commutative ring with 1. Let $S \subseteq R$. Then

$$(S) := \left\{ \sum_{i=1}^{k} r_i s_i \ : \ k \in \mathbf{N}, r_i \in R, s_i \in S \right\}. \tag{12.1}$$

We refer to (S) as the *ideal generated by S*; this terminology is justified by Theorem 12.12 below. If $S = \{s_1, s_2, \ldots, s_n\}$ is a *finite* set, then we also write $(s_1, s_2, \ldots, s_n)$ for $(\{s_1, s_2, \ldots, s_n\})$, omitting the set braces inside of the parentheses.

Definition 12.10. The expression $\sum_{i=1}^{k} r_i s_i$ is called an *R-linear combination of elements of S*. Thus the set (S) is the set of all R-linear combinations of elements of S.

Remark 12.11. We allow 0 as a natural number, and so we may get a sum with $k = 0$ terms in the expression above. We interpret a sum with no terms (or "empty sum") to be equal to 0. Conveniently, this takes care of the case when $S = \emptyset$; for the smallest ideal containing the empty set as a subset is simply the smallest ideal of all, which is just the ideal $\{0\}$. In practice, we usually write (0) for this zero ideal, in preference to $\{0\}$ or $(\{0\})$ or even $(\emptyset)$.

Theorem 12.12. *The set (S) given by Equation 12.1 is an ideal of R. It is the smallest ideal of R which contains S.*

Proof. Suppose that R is a commutative ring with 1 and $S \subseteq R$. Set $I = (S)$. The main point of the proof is to show that sums and negations of R-linear combinations of elements of S are again R-linear combinations of elements of S, as is the product of an R-linear combination with an element of R.

[Show (**Id1**): $(I, +) \leq (R, +)$] [Use the Subgroup Test]

[**ST0**] By Remark 12.11, we know that $0 \in I$, so I is non-empty.

[**ST1**] Let $\alpha \in I$. Then we can write $\alpha = \sum_{i=1}^{k} a_i x_i$ for some $k \in \mathbf{N}$, $a_i \in R$, and $x_i \in S$, by definition of (S). Therefore, we have

$$-\alpha = -\left(\sum_{i=1}^{k} a_i x_i \right)$$

$$= \sum_{i=1}^{k} (-a_i x_i)$$

$$= \sum_{i=1}^{k} (-a_i) x_i. \qquad \text{(by Lemma 11.10)}$$

Since $-a_i \in R$ for each i, the element $-\alpha$ is again an R-linear combination of elements of S, and so $-\alpha \in I$.

[**ST2**] Let $\alpha, \beta \in I$, and write $\alpha = \sum_{i=1}^{k} a_i x_i$, $\beta = \sum_{i=1}^{\ell} b_i y_i$ with $k, \ell \in \mathbf{N}$, $a_i, b_i \in R$, and $x_i, y_i \in S$. Then $\alpha + \beta = a_1 x_1 + \cdots + a_k x_k + b_1 y_1 + \cdots + b_\ell y_\ell \in (S)$, since it is an R-linear combination of elements of S.

Therefore, we have $(I, +) \leq (R, +)$.

[Show (**Id2**): I absorbs elements of R under multiplication from either side]

Let $\alpha = \sum_{i=1}^{k} a_i x_i \in I$, and let $r \in R$. Then we have

$$r\alpha = r \left(\sum_{i=1}^{k} a_i x_i \right)$$

$$= \sum_{i=1}^{k} r(a_i x_i) \qquad \text{(by distributing repeatedly using Axiom **R3**)}$$

$$= \sum_{i=1}^{k} (ra_i) x_i \qquad \text{(by Axiom **R2**, associativity of multiplication)}$$

which is again an R-linear combination of elements of S. Since R is assumed to be a commutative ring, we have $\alpha r = r\alpha \in I$ as well.

We have proved that (S) is an ideal of R. That (S) is the *smallest* ideal of R containing S is justified by the discussion immediately preceding this theorem, together with Exercise 12.4. □

Remark 12.13. Only in the final step of the proof above did we use the fact that R was commutative. However, if R is not assumed to be commutative, then our construction of (S) yields only the smallest *left* ideal of R containing S. The reader is invited to find a simple expression for the smallest two-sided ideal containing S in a general (non-commutative) ring.

When we studied the subgroup generated by a set, the simplest case, that of a subgroup generated by a single element, received special attention: we called such groups *cyclic*. The corresponding notion here is the ideal generated by a single element of a ring:

Definition 12.14. An ideal which is generated by a single element is called *principal*. That is, if R is a commutative ring with 1, then an ideal I of R is called principal if $I = (a)$ for some $a \in R$.

A principal ideal (a) in a commutative ring R (with 1) is the set of all R-linear combinations of elements of the singleton set $\{a\}$. Technically, in our definition of the ideal (S), we did not require the elements s_i in the sum on the right-hand side of Equation 12.1 to be distinct. Thus, we have

$$(a) = \left\{ \sum_{i=1}^{k} r_i s_i \ : \ k \in \mathbf{N}, r_i \in R, s_i \in \{a\} \right\}$$

$$= \left\{ \sum_{i=1}^{k} r_i a \ : \ k \in \mathbf{N}, r_i \in R \right\}$$

$$= \left\{ \left(\sum_{i=1}^{k} r_i \right) \cdot a \ : \ k \in \mathbf{N}, r_i \in R \right\}$$

$$= \{ r \cdot a \ : \ r \in R \},$$

where the last equality holds because a finite sum of elements of R is a single element of R and vice versa. Thus, the principal ideal generated by a is the set of all "multiples" of a by elements of R. We may therefore use coset notation for this set, using the multiplication operation of the ring R, and write

$$(a) = Ra = aR.$$

More generally, there is no reason to include repeated elements of S among the s_i's when we build the ideal (S), since we could factor out the common s_i terms. Thus, we have the following simplified description of (S) in case S is a finite set:

Lemma 12.15. *Let $S = \{s_1, \ldots, s_n\}$ be a finite subset of a commutative ring R with 1. Then we have*

$$(S) = \left\{ \sum_{i=1}^{n} r_i s_i \ : \ r_i \in R \right\} = Rs_1 + \cdots + Rs_n.$$

$\square$

While units get in the way of factoring *elements* of a ring, they quietly go away when we introduce them into *ideals*, as the following result shows. This is part of what makes the theory of ideals the "right" place to talk about factorization.

Lemma 12.16. *Let R be a domain, and let $a, b \in R$. Then $(a) = (b)$ iff $b = u \cdot a$ for some $u \in R^\times$. Moreover, if $a \neq 0$ and $(a) = (ca)$ for some $c \in R$, then $c \in R^\times$.*

Proof. Exercise 12.5. $\square$

Example 12.17. We saw in Example 11.28 that in the ring $\mathbf{Z}$, every ideal has the form $n\mathbf{Z}$ for some n in $\mathbf{Z}$. We can recognize now that such ideals are *principal*: we have $n\mathbf{Z} = (n)$, using the notation introduced in this section. Thus, every ideal of $\mathbf{Z}$ is principal.

Rings with this property are important enough to get their own designation, although the term we use to describe them is not very imaginative:

Definition 12.18. A domain in which every ideal is principal is called a *principal ideal domain*, or PID.

We now have quite an impressive hierarchy of types of ring. By Exercise 12.7, every field is a PID, so we can say

$$\text{field} \implies \text{PID} \implies \text{domain} \implies \text{commutative}.$$

To conclude this section, we will ask another question related to building ideals in a commutative ring with 1: What is the smallest ideal which contains both a given ideal I and a given element x? The reader should attempt to prove the following result (Exercise 12.8):

Lemma 12.19. *Let R be a commutative ring with 1, let I be an ideal of R, and let $x \in R$. Then the set*

$$J = I + Rx = \{c + y \cdot x \ : \ c \in I \text{ and } y \in R\}$$

is an ideal of R, and J is the smallest ideal of R which contains both I and x.

12.4   Fields and Maximal Ideals

A *field*, a commutative ring with $1 \neq 0$ in which every non-zero element has a multiplicative inverse (Definition 11.13), is in a sense the nicest type of ring. Let us now explore the question, For which ideals I is R/I a field?

Suppose that R is a commutative ring with 1, I is an ideal of R, and R/I is a field.

Then we have $(R/I)^\times = R/I - \{0_{R/I}\}$. Recall that $0_{R/I} = I = 0 + I$. So for a typical element $x + I$ of R/I, to say $x + I \neq 0_{R/I}$ means exactly that $x \notin I$, i.e., $x \in R - I$.

Now suppose that $x \in R - I$. Then $x + I \in (R/I)^\times$, so there exists $y \in R$ such that $(y + I) \cdot (x + I) = 1_{R/I} = 1 + I$. This equation says that $(y \cdot x) + I = 1 + I$, which in turn tells us that $1 - y \cdot x \in I$.

Set $c = 1 - y \cdot x$. Then $c \in I$, and we have $y \cdot x + c = 1$. This statement tells us that any ideal J of R which contains both I and x must also contain 1: for if $I \subseteq J$ and $x \in J$, then we have $y \cdot x \in J$ by (**Id2**), so $y \cdot x + c \in J$ by (**Id1**).

Another way to make this argument is that since $c \in I$ and $y \cdot x \in Rx$, then we have $1 \in I + Rx$. But $I + Rx$ is the smallest ideal of R containing both I and x, so J must contain 1.

An ideal of R which contains 1 must contain $(1) = R \cdot 1 = R$, and therefore must be equal to R itself. Therefore, our argument shows that for *any* element x outside of I, there is *no proper ideal* of R which contains both I and x. More informally speaking, we can't make I any bigger as an ideal without capturing the entire ring R; there is no ideal of R bigger than I except for R itself. This tells us that I is *maximal* in the following sense:

Definition 12.20. Let R be a commutative ring with 1, and let I be an ideal of R. Then I is called *maximal* if I is proper, and for any ideal J of R,

$$I \subset J \subseteq R \implies J = R.$$

This discussion motivates the following beautiful result:

Theorem 12.21. *Let R be a commutative ring with 1, and let I be an ideal of R. Then R/I is a field iff I is maximal.*

Proof. ($\Rightarrow$): Suppose that R/I is a field. [Show I is maximal: use the definition of *maximal*]

[Show I is proper]

Assume for a contradiction that $I = R$. Then $1 \in I$, so $0 + I = 1 + I$, which says that $0_{R/I} = 1_{R/I}$, contradicting the definition of *field*. Therefore, I must be proper.

[Show for all ideals J, $I \subset J \subseteq R \implies J = R$]

Suppose that J is an ideal of R and $I \subset J \subseteq R$. [Show $J = R$]

Then $\exists x \in J - I$ (since $I \subset J$). Now since $x \notin I$, we have $x + I \neq 0 + I$, so $x + I \in R/I - \{0_{R/I}\}$. Therefore, $x + I \in (R/I)^{\times}$ (by definition of *field*). So there is some element $y + I$ of R/I (with $y \in R$) such that $(y + I) \cdot (x + I) = 1_{R/I} = 1 + I$. Thus, $(y \cdot x) + I = 1 + I$, and $1 - y \cdot x \in I$. Set $c = 1 - y \cdot x$. Then $c \in I \subset J$, and we have $y \cdot x + c = 1$. Now $c \in J$, and also $y \cdot x \in J$ (because $x \in J$ and $y \in R$), and so $y \cdot x + c \in J$, since J is a group under addition **(Id1)**. Therefore, $1 \in J$, and so $J \supseteq (1) = R$. Since we also have $J \subseteq R$, we conclude that $J = R$, as desired.

($\Leftarrow$): Suppose that I is maximal. [Show R/I is a field]

[Show R/I is a commutative ring with $1_{R/I} \neq 0_{R/I}$]

First, R/I is a commutative ring, by Exercise 12.1. Furthermore, we have $1_{R/I} \neq 0_{R/I}$ because I is proper: if $1 + I = 0 + I$, then $1 \in I$, so $I = R$, and I would not be proper.

[Show $(R/I)^{\times} = R/I - \{0_{R/I}\}$]

Let $\alpha \in R/I - \{0_{R/I}\}$. Then $\alpha = x + I$ for some $x \in R - I$. Let $J = I + Rx$. Then J is an ideal of R containing both I and x, so we have $I \subset J \subseteq R$. By definition of *maximal*, we must have $J = R$. In particular, since $1 \in R$, we have $1 \in J$, so we can write $1 = c + y \cdot x$ for some $c \in I$ and $y \in R$. Therefore, $1 - y \cdot x \in I$, and so we have $y \cdot x + I = 1 + I$ in R/I. We can rewrite this last equation as $(y + I) \cdot (x + I) = 1_{R/I}$, which shows that $x + I \in (R/I)^{\times}$.

We have shown that $R/I - \{0_{R/I}\} \subseteq (R/I)^{\times}$. Now we certainly can't have $0_{R/I} \in (R/I)^{\times}$, because then, by Exercise 11.3, we would have $R/I = \{0_{R/I}\}$, whereas we know that R/I has at least two distinct elements, since we already proved that $1_{R/I} \neq 0_{R/I}$. Thus, we must have $(R/I)^\times = R/I - \{0_{R/I}\}$. We conclude that R/I is a field. $\qquad\square$

As a corollary to Theorem 12.21, we get a relationship between the notions of *prime* and *maximal* for ideals in a commutative ring with 1.

Corollary 12.22. *In a commutative ring with 1, every maximal ideal is prime.*

Proof. Let R be a commutative ring with 1, and let M be a maximal ideal of R. Then R/M is a field (by Theorem 12.21). Therefore, R/M is a domain (by Proposition 11.14). Now by Theorem 12.8, we conclude that M is prime. $\qquad\square$

Remark 12.23. The relationship between the notions of *prime* and *maximal* may not be obvious from the original definitions of these two terms (the reader is invited to scrutinize both definitions). Now, however, we can say that if an ideal is *not* prime, then it is *not* maximal. So it is tempting to ask whether there is a simple way to enlarge a non-prime ideal to a bigger, but still proper, ideal. This is accomplished in Exercise 12.9.

Example 12.24. What are the maximal ideals of $\mathbf{Z}$? By Corollary 12.22, the only candidates for maximal ideals are among the prime ideals of $\mathbf{Z}$. We found in Example 11.33 that the only prime ideals of $\mathbf{Z}$ are (0) and the ideals (p), where p is a positive prime number.

Recall that an ideal is maximal if it is proper and it is not contained in any other proper ideal. When does one ideal of $\mathbf{Z}$ contain another? Let us take two ideals $n\mathbf{Z}$ and $m\mathbf{Z}$. If $n\mathbf{Z} \subseteq m\mathbf{Z}$, then we must have in particular that $n \in m\mathbf{Z}$, and so n must be an integer multiple of m; in other words, we must have $m \mid n$. Conversely, if $m \mid n$, then every multiple of n is also a multiple of m, so we have $n\mathbf{Z} \subseteq m\mathbf{Z}$. This argument shows that $n\mathbf{Z} \subseteq m\mathbf{Z}$ iff $m \mid n$: "To divide is to contain." See Exercise 12.10 for a general version of this principle.

Now, no positive prime integer is a multiple of another, so there are no proper containment relationships among prime ideals of $\mathbf{Z}$ of the form (p) for $p > 0$. Therefore, the only proper containment relationships among prime ideals of $\mathbf{Z}$ must involve the zero ideal, (0). And in fact, we have

$$(0) \subset (p)$$

for every prime $p \in \mathbf{Z}$. We conclude that the maximal ideals of $\mathbf{Z}$ are precisely the ideals of the form (p), where p is a positive prime.

As a consequence of Example 12.24 and Theorem 12.21, we are entitled to state the following corollary:

Corollary 12.25. $\mathbf{Z}/p\mathbf{Z}$ *is a field whenever p is a prime element of $\mathbf{Z}$.* $\qquad\square$

Notation 12.26. When p is a prime element of $\mathbf{Z}$, then the field $\mathbf{Z}/p\mathbf{Z}$ is sometimes written $\mathrm{GF}(p)$, where "GF" stands for *Galois field*. Evariste Galois (pronounced "gal WAH", rhymes with *François* and *Ah!*) was a brilliant mathematician who is responsible for many of the crucial insights in the origins of abstract algebra.

Remark 12.27. We saw in Example 12.24 that the only proper containment relationships among prime ideals of $\mathbf{Z}$ are of the form

$$(0) \subset (p)$$

where p is an ordinary (positive) prime of $\mathbf{Z}$. In general, in a commutative ring R with 1, a collection of prime ideals $P_0, P_1, \ldots, P_k$ such that

$$P_0 \subset P_1 \subset \cdots \subset P_k$$

is called a *chain* of prime ideals of *length k*. The *dimension* of R (also called the *Krull dimension*) is defined to be

$\dim(R) = $ the maximum length of any chain of prime ideals of R.

Thus, we have $\dim(\mathbf{Z}) = 1$. The reader may wonder why we start numbering at zero instead of one here: the answer is that *dimension* in this sense is intimately related to "dimension" in the classic sense of geometry, and starting at zero is the correct choice to make the two notions of dimension match.

12.5 Exercises

Exercise 12.1. Prove that if R is a commutative ring and I is an ideal of R, then R/I is also commutative.

Exercise 12.2. Let R be a commutative ring with 1. Prove that R is a domain iff (0) is a prime ideal of R.

Exercise 12.3. Let R be a commutative ring with 1.

(a) Prove that if a is a prime element of R, and u is a unit of R, then au is also a prime element of R.

(b) Prove that if R is a domain, a is an irreducible element of R, and u is a unit of R, then au is also an irreducible element of R.

(c) Find all of the prime elements of $\mathbf{Z}$ (using the *general* definition of "prime," Definition 12.4). Part (a) may help!

Exercise 12.4. Prove that if R is a commutative ring with 1 and $S \subseteq R$, then $S \subseteq (S)$.

Exercise 12.5. Prove Lemma 12.16.

Exercise 12.6. Let R be a commutative ring with 1, and let $a \in R$. Prove that $(a) = R$ iff $a \in R^\times$.

Exercise 12.7. Prove that every field is a PID. (Exercise 11.7 will be useful here.)

Exercise 12.8. Prove Lemma 12.19. Can you generalize this result to find the smallest ideal containing a given ideal I together with a finite collection of elements $x_1, \ldots, x_n$ of R?

Exercise 12.9. Let R be a commutative ring with 1, and let I be an ideal of R. Suppose that $a, b \in R$ and $a \cdot b \in I$ but $a \notin I$ and $b \notin I$. Prove that $I + Ra \neq R$ and $I + Rb \neq R$. This shows "directly" (well, by contrapositive, but without using quotient rings) that a maximal ideal must be prime.

Exercise 12.10 (To Divide Is To Contain). Let R be a commutative ring with 1, and let $a, b \in R$. Prove that $(a) \supseteq (b)$ iff $a \mid b$ in R.

Exercise 12.11. Let R be a domain, and let $\alpha \in R$. Prove that if α is a prime element of R, then α is irreducible in R.

Exercise 12.12. Let R be a commutative ring with 1, and let $\alpha \in R - \{0\}$. Prove that (α) is a prime ideal of R iff α is a prime element of R.

Exercise 12.13. Let p be a positive prime integer, and let n be any integer.
 (a) Why must $|(\mathbf{Z}/p\mathbf{Z})^\times| = p - 1$?
 (b) Prove that if $p \nmid n$, then $n + p\mathbf{Z} \in (\mathbf{Z}/p\mathbf{Z})^\times$. Conclude that $n^{p-1} \equiv 1 \pmod{p}$ if $p \nmid n$.
 (c) Prove that $n^p \equiv n \pmod{p}$, whether or not $p \mid n$. This is *Fermat's Little Theorem.*

Exercise 12.14. Fermat's Little Theorem (see Exercise 12.13) can often be used to quickly show that a given integer m is *not* prime, without the need to factor m. If we can factor $m - 1$, then a converse can also be approached. As an illustration in a very special case, let r be a positive integer, and let $m = 2^r + 1$.
 (a) Prove that if $n^{m-1} \not\equiv 1 \pmod{m}$ for some positive integer $n < m$, then m is not prime.
 (b) Show how we can compute n^{m-1} modulo m in only r steps.
 (c) Prove that if $n^{(m-1)/2} \equiv -1 \pmod{m}$ for some integer n, then m is prime. Hint: Show that if p is a prime which divides m, then for such an n, the order of $n + p\mathbf{Z}$ in the group $(\mathbf{Z}/p\mathbf{Z})^\times$ is $m - 1$.

Exercise 12.15. Let p be a positive prime of $\mathbf{Z}$ such that $p \equiv 3 \pmod{4}$. Prove that $(\mathbf{Z}/p\mathbf{Z})^\times$ has no element of order 4. Conclude that the congruence

$$n^2 \equiv -1 \pmod{p}$$

has no integer solution n.

Exercise 12.16. Let n be a positive integer. Let $R = \mathbf{Z}/n\mathbf{Z}$, and let $a \in \mathbf{Z}$. Prove that if $\gcd(a, n) > 1$ then $a + n\mathbf{Z} \notin R^\times$.

Exercise 12.17. This exercise continues Exercise 12.16. Let $U = \{b + n\mathbf{Z} : b \in \mathbf{Z} \text{ and } \gcd(b, n) = 1\} \subseteq R$. Let $a \in \mathbf{Z}$ with $\gcd(a, n) = 1$, and set $\alpha = a + n\mathbf{Z} \in U$. Define a function $f : R \to R$ by $f(\beta) = \alpha\beta$ for $\beta \in R$.
 (a) Prove that f is a group homomorphism from G to G, where $G = (R, +)$.

(b) Prove that f is injective. Since R is finite, conclude that f is bijective, hence $f \in \text{Aut}(G)$.

(c) Deduce that $\alpha \in R^{\times}$.

(d) Conclude that $R^{\times} = U$.

Exercise 12.18. Use Exercise 9.16 together with Exercise 12.17 to prove that for any positive integer n, we have $\text{Aut}(\mathbf{Z}/n\mathbf{Z}, +) \cong (\mathbf{Z}/n\mathbf{Z})^{\times}$ via the map $\sigma \mapsto \sigma(1 + n\mathbf{Z})$.

13

Vector Spaces

13.1 Introduction

In this chapter, we introduce our third type of algebraic structure, the *vector space*. Some readers may have encountered vectors, or vector spaces, in previous mathematics courses, or even in other areas of study, such as physics. To motivate our definition, we first give an example.

Example 13.1 (Informal Physics Vectors). Let us consider the set of all "arrows" which can be drawn inside the ordinary Cartesian plane, $\mathbf{R}^2$. An arrow may have its "tail" at any point, and its "head" (where we draw the arrowhead, or tip) at any other point—or possibly at the same point as the tail. We consider two arrows to be the same if they point in the same direction and have the same length, regardless of *where* the arrows are placed within the plane; see Figure 13.1.

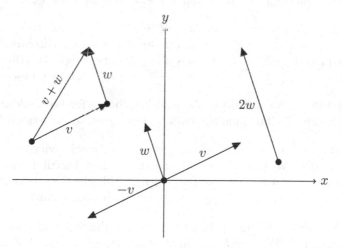

FIGURE 13.1: Vectors in $\mathbf{R}^2$ as arrows

We define a way to "add" two arrows v and w by drawing them so that the tail of w is located at the head of v, and take the sum to be the arrow whose head is located at w's head and whose tail is located at v's tail.

DOI: 10.1201/9781003252139-13

We also define a way to "multiply" an arrow by a real number. To multiply the arrow v by the number c, we stretch out v by a factor of c, keeping its direction the same; only, if $c < 0$, we flip the direction by 180 degrees, and then stretch by a factor of $|c|$.

This example follows the style which some scientists use to describe vectors; an arrow in this example is a vector, and the collection of *all* arrows will be a vector space.

The example above used real numbers in an essential way, as was common in the classical treatment of vectors. It turns out, however, that the theory of vector spaces works just as well when we replace **R** by an arbitrary field.

13.2   Abstract Vector Spaces

To capture the conditions necessary for a general, or *abstract*, vector space, we need a general description of the addition and multiplication operations.

Definition 13.2. Let $(F, +_F, \cdot_F)$ be a field. A *vector space* over F (or *F-vector space*) is a triple $(V, +_V, \cdot_V)$, where $(V, +_V)$ is an abelian group and $\cdot_V$ is a function

$$\cdot_V \; : \; F \times V \to V$$

satisfying the following axioms for all $x, y \in V$ and all $a, b \in F$:

(VS1) :	$(a \cdot_F b) \cdot_V x$	$=$	$a \cdot_V (b \cdot_V x)$	[Associativity]
(VS2(a)) :	$a \cdot_V (x +_V y)$	$=$	$a \cdot_V x +_V a \cdot_V y$	[Left Distributivity]
(VS2(b)) :	$(a +_F b) \cdot_V x$	$=$	$a \cdot_V x +_V b \cdot_V x$	[Right Distributivity]
(VS3) :	$1_F \cdot_V x$	$=$	x	[Unitary Law]

We usually are less formal, using the same symbol $+$ for both addition operations and using $\cdot$ for both multiplication operations, so the axioms become

(VS1) :	$(a \cdot b) \cdot x$	$=$	$a \cdot (b \cdot x)$	[Associativity]
(VS2(a)) :	$a \cdot (x + y)$	$=$	$a \cdot x + a \cdot y$	[Left Distributivity]
(VS2(b)) :	$(a + b) \cdot x$	$=$	$a \cdot x + b \cdot x$	[Right Distributivity]
(VS3) :	$1_F \cdot x$	$=$	x	[Unitary Law]

Note that by (VS1), we can safely write $a \cdot b \cdot x$ without parentheses. A *vector* is an element of V. We call F the *field of scalars* or *base field* of the vector space, and we refer to an element of F as a *scalar*. The operation $\cdot_V$ is called *scalar multiplication*.

Remark 13.3. As with groups and rings, we often suppress the operations when we talk about vector spaces, and refer to V by itself as a vector space.

Example 13.4. The Cartesian plane $\mathbf{R}^2$ is a vector space over $\mathbf{R}$ via the operations
$$(x, y) + (w, z) = (x + w, y + z)$$
and
$$c \cdot (x, y) = (cx, cy)$$
for $x, y, w, z, c \in \mathbf{R}$. This vector space is equivalent to the one in Example 13.1, if we identify an arrow whose tail is located at (a, b) and head at (c, d) with the vector $(c - a, d - b)$, which gives the displacement from the tail to the head.

In this example, we say that addition and scalar multiplication are performed *componentwise*, for evident reasons (look at the formulas above). This example generalizes to any Cartesian power of any field: if F is a field and n is a positive integer, then the set F^n is a vector space under the operations of componentwise addition and scalar multiplication.

Example 13.5. We claim that $\mathbf{C}$ is "naturally" a vector space over $\mathbf{R}$. Why? Well, $\mathbf{R}$ and $\mathbf{C}$ are fields, and $\mathbf{R} \leq \mathbf{C}$. So $\mathbf{C}$ is an abelian group under addition, and there is a natural way to multiply a real number by a complex number to produce a complex number. We know that the associative and distributive laws are true in these familiar number systems, and that 1 is a multiplicative identity element.

This example generalizes: if F is a field and R is a ring with 1 such that $F \leq R$ (as rings), then R is naturally a vector space over F: we use the addition and multiplication operations which R possesses as a ring to make R into an F-vector space (see Exercise 13.1). In this situation, we can even multiply a vector by another vector, since the multiplication operation is defined for any two elements of R, and not just for an element of F with an element of R. However, we shall not discuss the idea of "multiplying two vectors" in this chapter (such a notion is useful and important, but we shall not need it here; see e.g. Definition 23.62 for one such possibility).

Lemma 13.6 (Basic Properties of Vector Spaces). *Let V be a vector space over a field F. Then for all $a \in F$ and all $x \in V$, we have:*
  (1) $0_F \cdot x = 0_V$
  (2) $a \cdot 0_V = 0_V$
  (3) $(-1_F) \cdot x = -x$

Proof. (1): Let $x \in V$. Then $0_F \cdot x = (0_F + 0_F) \cdot x = 0_F \cdot x + 0_F \cdot x$. Since $0_F \cdot x \in V$, and V is an abelian group under addition, we can use the Cancellation Laws for groups to conclude that $0_V = 0_F \cdot x$, as desired.

(2): Let $a \in F$. Then $a \cdot 0_V = a \cdot (0_V + 0_V) = a \cdot 0_V + a \cdot 0_V$. Cancellation in $(V, +)$ again lets us conclude that $a \cdot 0_V = 0_V$.

(3): Let $x \in V$. We have $0_V = 0_F \cdot x$ (by (1) above) $= (1_F + (-1_F)) \cdot x = 1_F \cdot x + (-1_F) \cdot x = x + (-1_F) \cdot x$. Adding $-x$ to both sides completes the proof. $\square$

Now that we have some experience defining subgroups and subrings, it should not be difficult to define a "substructure" for a new kind of algebraic object: in this case, for a vector space. The idea again is that one vector space W naturally lives inside of another vector space V if $W \subseteq V$ and W "inherits" its operations from V. Instead of calling W a sub-vectorspace, we follow common practice and call it a *subspace*.

Definition 13.7. Let $(V, +_V, \cdot_V)$ be a vector space over a field F. A *subspace* of V is a vector space $(W, +_W, \cdot_W)$, also over F, such that $W \subseteq V$, $+_W = +_V|_{W \times W}$, and $\cdot_W = \cdot_V|_{F \times W}$.

Notation 13.8. We write $(W, +_W, \cdot_W) \leq (V, +_V, \cdot_V)$ to indicate the subspace relationship. Just as with groups and subgroups, if we are given a vector space $(V, +, \cdot)$ over a field F together with a subset $W \subseteq V$, then there is at most one way to define operations on W to make W a subspace of V: namely, by restricting the operations of V. Accordingly, we write $W \leq V$ to mean $(W, +|_{W \times W}, \cdot|_{F \times W}) \leq (V, +, \cdot)$. We also write $W < V$ to indicate that W is a proper subspace of V.

As usual, we will use the same symbols (most often just $+$ and $\cdot$) for addition and scalar multiplication on both a vector space and a subspace. This will cause no confusion, since the operations only differ in their domains, not in the way they produce output.

If V is a vector space over F, then every subspace W of V is also a subgroup of V under addition, as we can see by looking at the definitions. The only extra condition we need for W to be a subspace of V is that W be closed under scalar multiplication. This is the content of the following result:

Proposition 13.9 (Subspace Test). *Let V be a vector space over a field F, and let $W \subseteq V$. Then $W \leq V$ iff*
[SST1]: W forms a subgroup of V under addition, and
[SST2]: W is closed under scalar multiplication: that is,

$$\forall x \in W \ \forall a \in F, \ a \cdot x \in W.$$

Proof. Left to the reader as Exercise 13.2. $\qquad\qquad\qquad\qquad\qquad\qquad \square$

Let V be a vector space over a field F, and let $S \subseteq V$. What is the smallest subspace W of V which contains S?

Well, if $s \in S$ and $a \in F$, then we must have $a \cdot s \in W$. If $s_1, \ldots, s_k \in S$ and $a_1, \ldots, a_k \in F$, then we must have $a_i \cdot s_i \in W$ for each i; and since W must be closed under addition, this gives $a_1 \cdot s_1 + a_2 \cdot s_2 + \cdots + a_k \cdot s_k \in W$.

Notation 13.10. Let V be a vector space over a field F, and let $S \subseteq V$. Then the set

$$\langle S \rangle := \{ a_1 s_1 + \cdots + a_k s_k \ : \ k \in \mathbf{N}, a_i \in F, s_i \in S \}$$

is called the *space spanned by S (over F)*. Here as elsewhere, a sum with zero terms is taken to be the additive identity element, in this case 0_V.

Lemma 13.11. *Let V be a vector space over a field F, and let $S \subseteq V$. Then $\langle S \rangle$ is a subspace of V; it is the smallest subspace of V which contains S.*

Proof. The proof is so similar to that of Theorem 12.12 that we leave it to the reader as Exercise 13.3. $\square$

The reader may have expected to see $\langle S \rangle$ referred to as "the subspace generated by S," but by tradition, we use the term *spanned* here instead. The element $a_1 s_1 + \cdots + a_k s_k$ (with $a_i \in F$ and $s_i \in S$) is referred to as an *F-linear combination of $s_1, \ldots, s_k$*; sometimes we refer to the elements a_i as *coefficients*. We may also describe $\langle S \rangle$ by saying that $\langle S \rangle$ is the set of all F-linear combinations of elements of S.

Often, we take the opposite point of view to that in Lemma 13.11. Namely, we start with a vector space W, and ask for a set S with the property that $\langle S \rangle = W$. From this point of view, we say that S *spans* W.

There is an analogue for vector spaces of Lemma 12.15:

Lemma 13.12. *Let V be a vector space over a field F, and let $S = \{s_1, \ldots, s_n\}$ be a finite subset of V. Then we have*

$$\langle S \rangle = \left\{ \sum_{i=1}^{n} a_i s_i \; : \; a_i \in F \right\}.$$

Proof. Exercise 13.4. $\square$

Remark 13.13. As a special case of Lemma 13.12, we have $\langle \emptyset \rangle = \{0_V\}$; compare Remark 12.11.

13.3 Bases: Generalized Coordinate Systems

Let us consider once more the vector space $\mathbf{R}^2$ endowed with componentwise addition and scalar multiplication, as in Example 13.4. We are used to thinking of a point in $\mathbf{R}^2$ in terms of its x and y coordinates; indeed, if $P = (x, y)$ is any point in $\mathbf{R}^2$, then the Cartesian coordinates x and y specify P uniquely; this is part of what makes the Cartesian coordinate system in the plane so useful.

Let us attempt to isolate the coordinates x and y *as real numbers*, as opposed to pieces inside of a vector, starting with the equation $P = (x, y)$. We have $P = (x, y) = (x, 0) + (0, y) = x \cdot (1, 0) + y \cdot (0, 1)$. Every vector in $\mathbf{R}^2$ can be written as an $\mathbf{R}$-linear combination of the two vectors $(1, 0)$ and $(0, 1)$. Furthermore, the real numbers x and y in the expression for P are *unique*: we can say that every vector in $\mathbf{R}^2$ can be written uniquely as an $\mathbf{R}$-linear combination of $(1, 0)$ and $(0, 1)$.

The following definition captures this important property of Cartesian coordinates by simply replacing the special vectors $(1,0)$ and $(0,1)$ by arbitrary vectors, and replacing the vector space $\mathbf{R}^2$ by an arbitrary vector space.

Definition 13.14. Let V be a vector space over a field F. A *basis* of V over F is a set $B \subseteq V$ such that every vector in V can be written uniquely as an F-linear combination of elements of B.

Remark 13.15. A basis can be thought of as a coordinate system for a vector space, with coordinates which are elements of the field F. More specifically, if B is a basis of V over F, then every vector $v \in V$ has unique "coordinates" with respect to B: namely, the coefficients we need in order to write v as an F-linear combination of elements of B.

Remark 13.16. The plural of *basis* is *bases*.

Example 13.17. The set $B := \{(1,0),(0,1)\}$ is a basis of the vector space $\mathbf{R}^2$ under componentwise operations. Indeed, this was our motivating example for the definition of *basis*. The coordinates of a vector $v \in \mathbf{R}^2$ with respect to this basis are the usual Cartesian coordinates of v.

Example 13.18. It is also instructive to try to find a basis of $\mathbf{C}$ over $\mathbf{R}$. We claim that the set $\{1, i\}$ is such a basis. Indeed, every complex number z has a unique representation as $z = x + iy = x \cdot 1 + y \cdot i$ with $x, y \in \mathbf{R}$. This is exactly what it means to say that the set $\{1, i\}$ is a basis of $\mathbf{C}$ over $\mathbf{R}$!

Remark 13.19. The bases we examined in the previous two examples are by no means the only bases for $\mathbf{R}^2$ or $\mathbf{C}$ (over $\mathbf{R}$), although they are perhaps the most natural. In fact, it turns out that any two points P and Q which do not lie on a straight line containing the origin will form a basis of $\mathbf{R}^2$ over $\mathbf{R}$. In a sense, "most" sets of 2 elements of $\mathbf{R}^2$ form a basis of $\mathbf{R}^2$ over $\mathbf{R}$. Later, in Theorem 13.24, we shall prove that every basis of a given vector space has the same size; so the number 2 cannot be altered here!

Example 13.20. Since every subspace is a vector space in its own right, we can look for a basis of a given subspace W of a vector space V over a field F. The smallest subspace of any vector space V is the zero subspace, $W = \{0_V\}$. We don't have many choices when it comes to finding a basis B for $\{0_V\}$, since we must have $B \subseteq \{0_V\}$. We may be tempted to try $B = \{0_V\} = W$, but this is no good: for the vector 0_V has more than one representation as an F-linear combination of 0_V, for instance, $0_V = 0_F \cdot 0_V = 1_F \cdot 0_V$. In fact, the empty set $\emptyset$ is the unique basis of the zero subspace. (Compare Remark 13.13.)

It is often helpful to split up the definition of *basis* into two parts: *existence* and *uniqueness*. Let us illustrate what we mean.

Let V be a vector space over a field F, and let B be a basis of V over F.

Then every element of V can be written as an F-linear combination of elements of B (temporarily ignoring the stronger condition of *uniquely* in the definition of basis). This is exactly what it means to say that B *spans* V; in our notation, that $\langle B \rangle = V$. This is the "existence" part of the definition of basis: it says that, given any $v \in V$, there exist basis elements $b_1, \ldots, b_k$ and field elements $a_1, \ldots, a_k$ such that $v = a_1 \cdot b_1 + \cdots + a_k \cdot b_k$.

Next, the *uniqueness* part of the definition of basis says that, if we can write

$$v = \sum_{i=1}^{k} a_i \cdot b_i = \sum_{i=1}^{k} \tilde{a}_i \cdot b_i \tag{13.1}$$

with $v \in V$, coefficients $a_i, \tilde{a}_i \in F$, and distinct basis vectors $b_i \in B$, then we must have $a_i = \tilde{a}_i$ for all i from 1 to k. In other words, the coefficients of the basis elements in the expression for v as an F-linear combination of elements of B are uniquely determined by v.

The very careful reader may at this point have something to complain about. Namely, in Equation 13.1, we implicitly assumed that our two ways of writing v both used the same set $b_1, \ldots, b_k$ of basis elements. But what if we have

$$v = a_1 b_1 + a_2 b_2 = \tilde{a}_2 b_2 + \tilde{a}_3 b_3$$

for some $a_1, a_2, \tilde{a}_1, \tilde{a}_2 \in F$ and distinct $b_1, b_2, b_3 \in B$? Surely, the uniqueness part of the basis definition has something to say about this situation, too?

Yes, it does. The simple insight here is that we can write both representations of v in an expanded form, so that they both use all three of the basis vectors b_1, b_2, and b_3:

$$v = a_1 b_1 + a_2 b_2 + 0_F \cdot b_3 = 0_F \cdot b_1 + \tilde{a}_2 b_2 + \tilde{a}_3 b_3.$$

This is in the form of Equation 13.1, so we conclude that $a_1 = 0_F$, $a_2 = \tilde{a}_2$, and $b_3 = 0_F$. In general, whenever we have two representations of a vector as a linear combination of basis elements, we can always use the same finite set of basis elements in both representations, by adding extra terms with zero coefficients, as needed. And of course, in case our basis is a finite set, we can always use *all* elements of the basis in representing any vector.

It turns out that the question of uniqueness in the definition of a basis can be reduced to the question of whether the vector 0_V has a unique representation; the only idea involved in the proof is to move everything to the same side of the equation:

Lemma 13.21. *Let V be a vector space over a field F, and let $B \subseteq V$. Then B is a basis of V over F iff $\langle B \rangle = V$ and 0_V has a unique representation as an F-linear combination of elements of B (namely, where all coefficients are 0_F).*

Proof. ($\Rightarrow$): Suppose that B is a basis of V over F. Then *every* vector in V has a *unique* representation as an F-linear combination of elements of B. So B spans V, and in particular, the vector $0_V \in V$ has a unique representation as an F-linear combination of elements of B.

($\Leftarrow$): Suppose that B spans V and that 0_V has a unique representation as an F-linear combination of elements of B. Let $\alpha \in V$. Since $\langle B \rangle = V$, we can write $\alpha = \sum_{i=1}^{k} a_i \cdot b_i$ for some positive integer k and some coefficients $a_i \in F$ and distinct basis vectors $b_i \in B$. Suppose that we also have $\alpha = \sum_{i=1}^{\ell} \tilde{a}_i \cdot \tilde{b}_i$ with $\tilde{a}_i \in F$ and $\tilde{b}_i \in B$. First, by adding terms with zero coefficients as needed, we may assume that $\ell = k$ and $\tilde{b}_i = b_i$ for all i from 1 to k. Therefore, we have

$$\alpha = \sum_{i=1}^{k} a_i \cdot b_i = \sum_{i=1}^{k} \tilde{a}_i \cdot b_i.$$

From this, we get

$$\sum_{i=1}^{k} (a_i - \tilde{a}_i) b_i = 0_V.$$

But we can also write

$$\sum_{i=1}^{k} 0_F \cdot b_i = \sum_{i=1}^{k} 0_V = 0_V.$$

Now the uniqueness of the representation of 0_V as an F-linear combination of basis elements tells us that $a_i - \tilde{a}_i = 0_F$ for each i. Therefore, $a_i = \tilde{a}_i$ for each i, as required. $\square$

Next we isolate and give a name to the condition that the zero vector has a unique representation as an F-linear combination of elements of a set B.

Definition 13.22. Let V be a vector space over a field F. Let $S \subseteq V$. We say that S is *linearly independent over F* if whenever $s_1, \ldots, s_n$ are distinct elements of S and $a_1, \ldots, a_n$ are arbitrary (not necessarily distinct) elements of F, then

$$\sum_{j=1}^{n} a_j s_j = 0_V \implies \forall j, a_j = 0_F. \tag{13.2}$$

Otherwise, we say that S is *linearly dependent over F*.

Using Definition 13.22, we can restate Lemma 13.21 as follows:
Lemma 13.21'. *Let V be a vector space over a field F, and let $B \subseteq V$. Then B is a basis of V over F iff B spans V over F and B is linearly independent over F.*

We have noted that every basis of a vector space V must span V. The next result says that a basis of V spans V "efficiently": we cannot remove any elements of a basis without losing this spanning property.

Theorem 13.23. *A basis is a minimal spanning set, and conversely. That is: Let V be a vector space over a field F, and let $S \subseteq V$. Then S is a basis of V over F iff S spans V and no proper subset of S spans V.*

Proof. ($\Rightarrow$): Suppose that S is a basis of V over F. Then S spans V, by Lemma 13.21. Assume for a contradiction that S is not a *minimal* spanning set for V. Then there is some vector $s \in S$ such that $S - \{s\}$ also spans V. Let $\tilde{S} = S - \{s\}$. Now since $s \in V = \langle \tilde{S} \rangle$, we can write

$$s = a_1 \cdot \tilde{s}_1 + \cdots + a_k \cdot \tilde{s}_k \qquad (13.3)$$

for some $a_i \in F$ and $\tilde{s}_i \in \tilde{S}$. More simply, we can also write

$$s = 1_F \cdot s. \qquad (13.4)$$

But Equations 13.3 and 13.4 are each representations of s as an F-linear combination of elements of S (since $\tilde{S} \subseteq S$). The coefficient of s in Equation 13.4 is 1_F, whereas in Equation 13.3 the coefficient of s is 0_F (since s cannot be among the $\tilde{s}_i$'s). Since $1_F \neq 0_F$ (by the definition of a field), the representation of s as an F-linear combination of elements of S is not unique. This contradicts the definition of a basis, and completes the proof of the forward implication.

($\Leftarrow$): Suppose that S is a minimal spanning set for V over F.
[Show: S is a basis of V over F]
[Strategy: Use Lemma 13.21]
[Show: $\langle S \rangle = V$ and 0_V has a unique representation]
Then in particular, S spans V. Suppose that $0_V = \sum_{i=1}^{k} a_i s_i$ where $k \in \mathbf{N}$, $a_i \in F$, and $s_1, \ldots, s_k$ are distinct elements of S. [Show that $a_i = 0_F$ for all i]

Assume for a contradiction that $a_i \neq 0_F$ for some i; without loss of generality, say $a_1 \neq 0_F$. [Idea: Solve for s_1]

Since F is a field and $a_1 \in F - \{0_F\}$, we have $a_1 \subset F^\times$, so we can multiply both sides by a_1^{-1} and solve for s_1:

$$0_V = a_1 s_1 + a_2 s_2 + \cdots + a_k s_k$$
$$a_1^{-1} \cdot 0_V = a_1^{-1} \cdot (a_1 s_1 + a_2 s_2 + \cdots + a_k s_k)$$
$$0_V = s_1 + \left(a_1^{-1} a_2\right) s_2 + \cdots + \left(a_1^{-1} a_k\right) s_k$$
$$s_1 = (-a_1^{-1} a_2) s_2 + \cdots + (-a_1^{-1} a_k) s_k \qquad (13.5)$$

[Idea: since we can write s_1 in terms of the other s_i's, we don't really need s_1 in S] Let $\tilde{S} = S - \{s_1\}$. Equation 13.5 shows that we have $s_1 \in \langle \tilde{S} \rangle$. We certainly also have $\tilde{S} \subseteq \langle \tilde{S} \rangle$. Thus, $\tilde{S} \cup \{s_1\} \subseteq \langle \tilde{S} \rangle$. But this just says that $S \subseteq \langle \tilde{S} \rangle$. Now, $\langle \tilde{S} \rangle$ is a subspace of V which contains S, and $\langle S \rangle$ is the *smallest* subspace of V which contains S; therefore, $\langle S \rangle \subseteq \langle \tilde{S} \rangle$. We know that S spans V, so $\langle S \rangle = V$, and $V \subseteq \langle \tilde{S} \rangle$. But $\langle \tilde{S} \rangle$ is also a subspace of V, which gives the containment $\langle \tilde{S} \rangle \subseteq V$. Therefore, $\langle \tilde{S} \rangle = V$. So $\tilde{S}$ is a proper subset of S which spans V, contradicting the fact that S is a *minimal* spanning set for V.

This contradiction shows that $a_i = 0_F$ for all i, and thus 0_V has a unique representation as an F-linear combination of elements of S. By Lemma 13.21, S is a basis of V over F. $\qquad \square$

Theorem 13.24 (Invariance of Basis Size). *Suppose that V is a vector space over a field F with a finite basis B of size n. Then every basis of V over F has size n.*

Proof. Let V be a vector space over a field F. Let B be a basis of V over F. Suppose that B is finite, and let $n = |B|$. Suppose that C is any basis of V over F. We proceed by induction on $|B - C|$.

Base Case: $|B - C| = 0$.

In this case, we have $B \subseteq C$. Now we cannot have $B \subset C$, because C is a minimal spanning set for V (by Theorem 13.23), and B also spans V. Therefore, $B = C$, so $|C| = n$, as required.

Inductive Step: Let $k = |B - C| \geq 1$, and suppose that for any basis $\tilde{C}$ of V over F with $|B - \tilde{C}| < k$, we have $|\tilde{C}| = n$.

[Idea: Replace an element of $C - B$ in C by an element of $B - C$]

We complete the proof using a series of claims, which we outline now:

Claim 1: $\exists x \in C - B$.

Claim 2: $\exists y \in B - C$ such that $y \notin \langle C - \{x\} \rangle$.

Claim 3: The set $T := (C - \{x\}) \cup \{y\}$ is a basis of V over F.

Proof of Claim 1: If $C \subseteq B$, then, arguing as we did in the base case above, we would have $C = B$, contradicting $|B - C| \geq 1$. Therefore, $\exists x \in C - B$.

Proof of Claim 2: Let $C' = C - \{x\}$. Since C is a minimal spanning set for V, and $C' \subset C$, we must have $\langle C' \rangle \neq V$. But if $B \subseteq \langle C' \rangle$, then we would have $V = \langle B \rangle \subseteq \langle C' \rangle$, since $\langle B \rangle$ is the smallest subspace of V containing B. Therefore, $\exists y \in B - \langle C' \rangle$.

Now as $y \notin \langle C' \rangle$, we have $y \notin C' = C - \{x\}$. Since $x \notin B$ but $y \in B$, we also have $y \neq x$. Therefore, $y \notin C$.

Proof of Claim 3: Set $T = C' \cup \{y\}$. Since $y \in V$ and C is a basis for V over F, we can write $y = \sum_{i=1}^{m} a_i c_i$ for unique coefficients $a_i \in F$, with distinct $c_i \in C$ (for some $m \in \mathbf{N}$). We contend that x must be one of the c_i's with a non-zero coefficient: for otherwise, we would have $y \in \langle C' \rangle$, contradicting the choice of y in the proof of Claim 2 above. Without loss of generality, $x = c_1$, and $a_1 \neq 0_F$. So we have

$$y = a_1 x + \sum_{i=2}^{m} a_i c_i. \tag{13.6}$$

Since $a_1 \neq 0_F$, we can solve Equation 13.6 for x:

$$x = a_1^{-1} y + \sum_{i=2}^{m} (-a_1^{-1} a_i) c_i.$$

Since we have $c_i \in C'$ for all $i \geq 2$, this shows that $x \in \langle \{y\} \cup C' \rangle = \langle T \rangle$. Now we know that $x \in \langle T \rangle$, and certainly $C' \subseteq \langle T \rangle$, so $C = C' \cup \{x\} \subseteq \langle T \rangle$. Therefore $\langle C \rangle \subseteq \langle T \rangle$; and since C is a basis of V, this gives $\langle T \rangle = V$.

Consider a representation of 0_V as an F-linear combination of elements of T:

$$0_V = wy + \sum_{i=1}^{q} w_i c_i' \tag{13.7}$$

with $w, w_i \in F$ and distinct elements $c'_i \in C'$. First, if $w \neq 0_F$, then we could solve Equation 13.7 for y in terms of the c'_i's and find that $y \in \langle C' \rangle$, contradicting the choice of y. Therefore, we have $w = 0$. Now, since C is a basis for V over F and $C' \subseteq C$, the only way to write 0_V as an F-linear combination of elements of C' is to make all the coefficients equal to 0_F. Therefore, we have $w_i = 0_F$ for all i.

We have shown that T spans V and that 0_V has a unique representation as an F-linear combination of elements of T. By Lemma 13.21, T is a basis of V over F.

Now we complete the proof of Theorem 13.24. Note that $|B - T| = |B - C| - 1$, since we formed T out of C by replacing an element of $C - B$ with an element of $B - C$. Specifically, we have $B - T = (B - C) - \{y\}$. By inductive hypothesis, we conclude that $|T| = |B| = n$. In particular, T is finite, and we have $C = (T - \{y\}) \cup \{x\}$, so $|C| = |T| = n$, as required. $\square$

Definition 13.25. Let V be a vector space over a field F. The *dimension of V over F* is the size of any basis of V over F. This size, if finite, is independent of the particular basis chosen, by Theorem 13.24.

Notation 13.26. We write $\dim_F(V)$ for the dimension of V over F.

Remark 13.27. The reader may also be curious about infinite-dimensional vector spaces. Such vector spaces certainly exist and are important, although we are avoiding them in our results (such as Theorem 13.24). It turns out that even in the infinite-dimensional case, two bases of a given vector space will have the same infinite size (i.e., there exists a bijection from one to the other), and so Definition 13.25 still makes sense.

Remark 13.28. A natural question is, Does every vector space have a basis? In practice, the vector spaces we shall study will all have bases, but it is important to justify any assertion that a particular vector space has a basis. In general, the Axiom of Choice implies that any vector space has a basis—if we are willing to accept that axiom. At any rate, when we write "$\dim_F(V) = n < \infty$," we are asserting that V has a finite basis over F—and thus, justification is required.

Example 13.29. Looking back at Example 13.17, we found that the set $\{(1,0), (0,1)\}$ is a basis for $\mathbf{R}^2$ over $\mathbf{R}$ (under componentwise operations). This basis has size 2, so we have $\dim_{\mathbf{R}}(\mathbf{R}^2) = 2$. No matter what other basis of $\mathbf{R}^2$ we come up with, that basis must also have size 2. We say that $\mathbf{R}^2$ is a *two-dimensional* vector space over $\mathbf{R}$. This agrees with our geometric understanding of "dimension" as the number of independent coordinates we need in order to describe an element of a space. In general, the dimension of a vector space is the number of basis elements, and each basis element acts like a coordinate.

The following result says that a basis of a subspace can always be enlarged to form a basis of the whole space; this ensures that smaller vector spaces have smaller dimensions. See Exercise 13.8 for a strengthening of this result where we drop the hypothesis that $\dim_F(W) < \infty$.

Lemma 13.30. *Let $W \leq V$ be vector spaces over a field F. Suppose that $k = \dim_F(W) < \infty$ and that $B = \{b_1, \ldots, b_k\}$ is a basis for W over F. Also suppose $\dim_F(V) < \infty$. Then B is contained in some basis of V over F, and consequently we have $\dim_F(W) \leq \dim_F(V)$.*

Proof. Let $n = \dim_F(V)$, and let $\tilde{B} = \{\beta_1, \ldots, \beta_n\}$ be a basis for V over F. We induct on the quantity $d := |\tilde{B} - W|$.

Base Case: $d = 0$. Then $\tilde{B} \subseteq W$, so $\langle \tilde{B} \rangle \leq W$; but also, $\langle \tilde{B} \rangle = V \supseteq W$; so $W = V$, and B is already a basis for V over F.

Inductive Step: Suppose that $d \geq 1$. Then $\exists i$ such that $\beta_i \notin W$. Set $B' = B \cup \{\beta_i\}$, and let $W' = \langle B' \rangle \leq V$. By construction, B' spans W'. Suppose that $\left(\sum_{j=1}^{k} a_j b_j \right) + a\beta_i = 0$ with $a_j, a \in F$. If $a \neq 0$, then we could solve for β_i and find that $\beta_i \in \langle B \rangle = W$, which is false. Therefore, $a = 0$. Since B is linearly independent over F, we can conclude that $a_j = 0$ for each j. This shows that B' is linearly independent over F. By Lemma 13.21, B' is a basis of W' over F. Now we certainly have $|\tilde{B} - W'| < d$, so by inductive hypothesis, B' is contained in some basis of V over F. Since $B \subseteq B'$, we are done. $\qquad\square$

13.4 Exercises

Exercise 13.1. Suppose that $(R, +, \cdot)$ is a ring with 1 and F is a field which is a subring of R. Prove that $(R, +, \cdot|_{F \times R})$ is a vector space over F.

Exercise 13.2. Supply a proof of the Subspace Test, Proposition 13.9.

Exercise 13.3. Prove Lemma 13.11.

Exercise 13.4. Prove Lemma 13.12.

Exercise 13.5. Let V be a vector space over F.

(a) Prove that any subset of a linearly independent set is linearly independent: that is, if $S \subseteq V$ is linearly independent over F and $T \subseteq S$, then T is also linearly independent over F.

(b) Prove that any superset of a linearly dependent set is linearly dependent: that is, if $S \subseteq V$ is linearly dependent over F and $T \supseteq S$, then T is also linearly dependent over F.

Exercise 13.6. Let V be a vector space over a field F, and let $S \subseteq V$. Prove that if S is linearly independent over F, then S is a basis of $\langle S \rangle$ over F.

Exercise 13.7. Prove that a basis (of a given vector space V) is the same thing as a maximal linearly independent subset of V. (Compare this to Theorem 13.23.)

Exercise 13.8 (A Subspace of a Finite-Dimensional Vector Space is Finite-Dimensional). Let V be a vector space over a field F, with $\dim_F(V) = n < \infty$.

(a) Prove that if S is a subset of V such that $|S| > n$, then S is linearly dependent over F. (Hint: Exercises 13.5 and 13.6 may be useful.)

(b) Let $W \leq V$. Prove that W has a finite basis over F, and that $\dim_F(W) \leq \dim_F(V)$.

Exercise 13.9. This exercise concerns a variation on Lemma 13.30. Prove or disprove that if $W \leq V$ are finite-dimensional vector spaces over a field F, then any basis B of V over F must contain a subset B' which is a basis of W over F.

Exercise 13.10. Let V be a vector space over a field F, and suppose that F is infinite. Use the following steps to prove that V cannot be written as a finite union of proper subspaces.

(a) Assume that $V = \cup_{j=1}^{n} V_j$, where $V_j < V$, and n is minimal. Show that for all j, we have $V_j \not\subseteq \cup_{i \neq j} V_i$.

(b) Note that we must have $n \geq 2$, and for $j \in \{1, 2\}$ choose $\alpha_j \in V_j - \cup_{i \neq j} V_i$. Prove that $\exists c_1, c_2 \in F$ and $i \in \{1, \dots, n\}$ such that $c_1 \neq c_2$ and both $\alpha_1 + c_1 \alpha_2$ and $\alpha_1 + c_2 \alpha_2$ belong to V_i.

(c) Prove that we have both $\alpha_1 \in V_i$ and $\alpha_2 \in V_i$, contradicting the choice of α_j in part (b).

In the remaining exercises, we make use of the following definitions.

Definition 13.31. The notion corresponding to a "homomorphism" in the category of vector spaces is called a *linear transformation*. If V and W are vector spaces over the same field F, then we define a linear transformation from V to W to be a function $t : V \to W$ such that
(LT1) $\forall a, b \in V$, $t(a + b) = t(a) + t(b)$, and
(LT2) $\forall a \in V \ \forall c \in F$, $t(c \cdot a) = c \cdot t(a)$.

Definition 13.32. The *kernel* of a linear transformation $t : V \to W$ is

$$\ker(t) := \{a \in V \ : \ t(a) = 0_W\}.$$

Definition 13.33. A linear transformation is called an *isomorphism* (of vector spaces) if it is bijective, and an *embedding* if it is injective. We write $V \cong W$ to mean that there is an isomorphism from V to W.

Exercise 13.11. Prove that
(i) The kernel of a linear transformation is a subspace of its domain, and
(ii) The image of a linear transformation is a subspace of its codomain.

Exercise 13.12. Let V be a vector space of finite dimension n over a field F. Prove that $V \cong F^n$, where, as usual, F^n denotes the vector space with underlying set equal to the nth Cartesian power of F under componentwise addition and scalar multiplication (see Example 13.4). Thus, a finite-dimensional vector space over any given field is uniquely determined up to isomorphism by its dimension.

Exercise 13.13. Let $t : V \to W$ be a linear transformation, where V and W are vector spaces over a field F, and suppose that $\dim_F(V) < \infty$. Prove that $\dim_F(\ker(t)) + \dim_F(t(V)) = \dim_F(V)$.

Exercise 13.14. Let U, V, and W be vector spaces over the same field F, and suppose that $\sigma : U \to V$ and $\tau : V \to W$ are linear transformations. Prove that $\tau \circ \sigma$ is also a linear transformation.

Exercise 13.15. [Matrices and Linear Transformations, part 1] Let V and W be finite-dimensional vector spaces over a field F, with $\dim_F(V) = n$ and $\dim_F(W) = m$. Suppose that $B = \{b_1, \ldots, b_n\}$ and $C = \{c_1, \ldots, c_m\}$ are bases of V and W, respectively, over F. Let $\mathcal{T}$ denote the set of all linear transformations from V to W, and let $M_{m,n}(F)$ denote the set of all m by n matrices with entries in F; here, m is the number of rows, and n is the number of columns. Let $t : V \to W$ be a linear transformation.

(a) Explain why for each integer j between 1 and n, we can write $t(b_j) = \sum_{i=1}^{m} T_{i,j} c_i$ for uniquely determined elements $T_{i,j} \in F$.

(b) Let T be the m by n matrix whose entry in row i and column j is $T_{i,j}$. Prove that for all $x \in V$, we have $t(x) = \sum_{i=1}^{m} y_i c_i$, where $x = \sum_{j=1}^{n} x_j b_j$ with $x_j \in F$, and $y_i = \sum_{j=1}^{n} T_{i,j} x_j$. Note: we say that T is the matrix of t with respect to the bases B and C. Also note that we are assuming that we know the ordering of the basis elements: thus, it would be more accurate to say that $B = (b_1, \ldots, b_n)$ is an *ordered basis* of V over F, and similarly for C.

(c) Prove that the map $t \mapsto T$ is a bijection from $\mathcal{T}$ to $M_{m,n}(F)$.

Exercise 13.16. [Matrices and Linear Transformations, part 2] Suppose that U, V, and W are finite-dimensional vector spaces over a field F with dimensions n, m, and ℓ, respectively. Choose ordered bases of U, V, and W over F. Let $t_1 : U \to V$ and $t_2 : V \to W$ be linear transformations.

(a) Set $t = t_2 \circ t_1$. Let M, N, and T be the matrices of t_1, t_2, and t respectively, with respect to the given bases. Prove that we have

$$T_{i,j} = \sum_{k=1}^{m} N_{i,k} \cdot M_{k,j} \tag{13.8}$$

for all i, j with $1 \le i \le \ell$ and $1 \le j \le n$.

(b) Use Equation 13.8 to define matrix multiplication for arbitrary matrices $M \in M_{m,n}(F)$ and $N \in M_{\ell,m}(F)$: namely, define $N \cdot M = T$. Prove that matrix multiplication is associative whenever the product of three matrices is defined, that is, when the number of columns of the first matrix is equal to the number of rows of the second matrix in each product.

Exercise 13.17. [Matrices and Linear Transformations, part 3] Let V be a vector space over a field F, with $\dim_F(V) = n < \infty$. Suppose that B is a basis of V over F. Let R be the set of all linear transformations from V to V. When we represent a linear transformation in R as a matrix, we will use the same basis B for both the domain and the codomain.

(a) For $t, u \in R$, define $t + u$ to be the function $s : V \to V$ given by the formula $s(a) = t(a) + u(a)$ for $a \in V$. Prove that $s \in R$.

(b) Prove that $(R, +, \circ)$ is a ring with 1, where $\circ$, as usual, denotes composition of functions.

(c) Let $t, u \in R$, and let $s = t + u$. Let S, T, and U be the matrices of s, t, and u, respectively, with respect to B. Prove that we have $S_{i,j} = T_{i,j} + U_{i,j}$ for all i and j, where, as usual, for a matrix M, the notation $M_{i,j}$ denotes the entry of M in row i and column j.

Remark 13.34. We usually denote $M_{n,n}(F)$ more simply as $M_n(F)$. Because of the bijective correspondence between linear transformations in R and matrices in $M_n(F)$, the set of all $n \times n$ matrices over F forms a ring under matrix addition and matrix multiplication. By construction, we have $M_n(F) \cong R$. The multiplicative identity element of $M_n(F)$ is denoted I_n or simply I, and is called the $n \times n$ *identity matrix*.

The group of units of $M_n(F)$ is denoted $\mathrm{GL}_n(F)$ and is called the *general linear group* of $n \times n$ matrices over F (compare Example 11.7).

Exercise 13.18. [Matrices and Linear Transformations, part 4] Let V be a vector space over a field F, with $\dim_F(V) = n < \infty$. Let R be the ring of all linear transformations from V to V over F. For an element $a \in F$, define a function $\mu_a : V \to V$ by the formula $\mu_a(t) = a \cdot t$ for $t \in V$.

(a) Show that μ_a is a linear transformation.

(b) Show that the matrix of μ_a is the $n \times n$ matrix A with

$$A_{i,j} = \begin{cases} a, & \text{if } i = j; \\ 0, & \text{if } i \neq j, \end{cases}$$

no matter what basis we choose for V. Note: such a matrix A is called a *scalar matrix*.

(c) Prove that the center $C(R)$ of R is the set $S := \{\mu_a : a \in F\}$. (See Exercise 11.18.)

(d) Prove that the map $a \mapsto \mu_a$ is a ring isomorphism from F to $C(R)$.

(e) Define a function $\triangle : F \times R \to R$ by the formula $\triangle(a, t) = \mu_a \circ t$. Prove that $(R, +, \triangle)$ is an F-vector space, where $+$ is the addition operation from Exercise 13.17.

(f) Use the correspondence between matrices and linear transformations together with part (e) above to make $M_n(F)$ into an F-vector space. Prove that for any matrix $T \in M_n(F)$ and any $a \in F$, the matrix aT can be obtained by multiplying every entry of T by a.

14

Polynomial Rings

14.1 Polynomials Over a Commutative Ring

Polynomials form a central topic of study from the very first courses of school algebra, immediately after the basics of arithmetic. What *is* a polynomial, though?

Informally, we might say that a polynomial is a thing made out of powers of a variable (often called x) multiplied by various coefficients (which are numbers), and added together, as in

$$14x^2 - 24x + 7.$$

Even more important is the question: What can we *do* with polynomials? At the least, we can add and multiply polynomials. This suggests that the set of all polynomials may form a *ring*; and this is true.

In our treatment of polynomials, we will not restrict the coefficients in polynomials to be numbers; instead, we will allow the coefficients to come from *any* fixed commutative ring with 1. In a sense, polynomials are the most "generic" things that can be added and multiplied commutatively; as a result, the subject of commutative rings is *all about* polynomials.

Definition 14.1. Let R be a commutative ring with 1. The *polynomial ring in the variable x with coefficients in R* is the ring

$$R[x] := \{a_0 + a_1 x + a_2 x^2 + \cdots + a_n x^n \ : \ n \in \mathbf{N}, a_i \in R\},$$

with addition and multiplication given by

$$\left(\sum_{k=0}^{n} a_k x^k\right) + \left(\sum_{k=0}^{m} b_k x^k\right) = \sum_{k=0}^{\max\{m,n\}} (a_k + b_k) x^k \tag{14.1}$$

and

$$\left(\sum_{k=0}^{n} a_k x^k\right) \cdot \left(\sum_{k=0}^{m} b_k x^k\right) = \sum_{k=0}^{m+n} \left(\sum_{j=0}^{k} a_j b_{k-j}\right) x^k. \tag{14.2}$$

In these formulas, we treat missing coefficients as zero: that is, as 0_R. In other

DOI: 10.1201/9781003252139-14

words, we identify $0_R x^k$ with $0_{R[x]}$. More simply, we will write $0x^k = 0$. We also follow the convention of writing $1_R x^k$ as x^k and ax^0 as a for $a \in R$.

Two polynomials are considered to be equal iff all their corresponding coefficients are equal: that is,

$$\sum_{k=0}^{n} a_k x^k = \sum_{k=0}^{m} b_k x^k \text{ iff } a_k = b_k \text{ for all } k \text{ s.t. } 0 \leq k \leq \max\{m, n\}. \quad (14.3)$$

Definition 14.2. By a *term* of the polynomial $\sum_{k=0}^{n} a_k x^k$, we mean any of the individual expressions $a_k x^k$.

Remark 14.3. The reader should verify that the formulas defining polynomial addition and multiplication give the same results as the usual process for performing these operations on polynomials as taught in school algebra. In fact, the definitions of polynomial addition and multiplication are designed exactly so that polynomials follow the same rules of arithmetic as ordinary numbers: for example, $x^2 \cdot x^3 = x^5$, $x + x = 2x$, and so on. These formulas are true whether x is a variable in a polynomial ring or a number in **R**.

Remark 14.4. We have $R \leq R[x]$. We can view R as the set of all "constant polynomials" in $R[x]$.

Remark 14.5. As usual when we choose a variable name, the name x is not special here; we may speak of polynomials in other variables, and form polynomials rings such as $R[t]$ or $R[y]$. The important thing is that our variable not already be in use (especially that it not be an element of R!).

Remark 14.6. The polynomial ring $R[x]$ can be thought of as the result of taking the ring R and adding a new element, x, which behaves in the most "free" way possible (while commuting with elements of R). Here, as in our discussion of free groups in Chapter 6, the meaning of *free* is *relation-free*. The way that this freeness manifests itself in the case of polynomials is packed into the definition of when two polynomials are equal, Equation 14.3. Specifically, when the right side of Equation 14.3 is 0, we have

$$\sum_{k=0}^{n} a_k x^k = 0 \text{ iff } a_k = 0 \text{ for all } k \text{ s.t. } 0 \leq k \leq n. \quad (14.4)$$

This equation may be viewed as saying that there are no non-trivial algebraic relations among x and the elements of R.

Remark 14.7. We have not managed to provide a very satisfactory notion of what kind of thing a polynomial really is, in terms of standard mathematical objects. Formally, we should define a polynomial with coefficients in R to be a *function* $f : \mathbf{N} \to R$ such that $f(j) = 0$ for all large enough values of $j \in \mathbf{N}$. Then f corresponds to the polynomial $\sum_j f(j) x^j$. In practice, however, this formal treatment of polynomials is more clumsy than the familiar presentation recalled above. The reader is invited to verify all statements about polynomials from this formal point of view, now and in the future, while sticking with the familiar polynomial notation in actual work.

Lemma 14.8. *If R is a commutative ring with 1, then so is $R[x]$.*

Proof. Let R be a commutative ring with 1. So far, we have not even established that $R[x]$ is a ring. We outline the proof below.

Let $f, g, h \in R[x]$. Let $n \in \mathbf{N}$ be the highest power of x which appears in any of the polynomials f, g, and h. By adding extra terms with a coefficient of 0 as needed, we can write all three polynomials in terms of the powers of x up to x^n. That is, we can write $f = \sum_{k=0}^{n} a_k x^k$, $g = \sum_{k=0}^{n} b_k x^k$, and $h = \sum_{k=0}^{n} c_k x^k$ with $a_k, b_k, c_k \in R$. First, we see from Equation 14.1 that $R[x]$ is closed under addition, since for $a_k, b_k \in R$ we have $a_k + b_k \in R$. Associativity of $R[x]$ under addition likewise follows from associativitiy of R under addition: we have $(f+g)+h = \sum_{k=0}^{n}[(a_k+b_k)+c_k]x^k$ and $f+(g+h) = \sum_{k=0}^{n}[a_k + (b_k + c_k)]x^k$. The element 0_R is an additive identity element for $R[x]$, and the element $\sum_{k=0}^{n}(-a_k)x^k$ is an additive inverse for $\sum_{k=0}^{n} a_k x^k$. Commutativity of addition in $R[x]$ follows from the same property in R. Thus, $(R[x], +)$ is an abelian group.

We can see from Equation 14.2 that $R[x]$ is closed under multiplication, since R is closed under multiplication and addition. To prove associativity and commutativity of multiplication in $R[x]$, it will be helpful to rewrite the polynomial multiplication formula (Equation 14.2) in a more symmetric way:

$$\left(\sum_{k=0}^{n} a_k x^k\right) \cdot \left(\sum_{k=0}^{m} b_k x^k\right) = \sum_{k=0}^{m+n}\left(\sum_{i+j=k} a_i b_j\right) x^k. \tag{14.5}$$

Since we have $\sum_{i+j=k} a_i b_j = \sum_{i+j=k} b_j a_i = \sum_{j+i=k} b_j a_i$, it follows that multiplication in $R[x]$ is commutative. For associativity, we compute

$$(f \cdot g) \cdot h = \left(\sum_{k=0}^{2n}\left(\sum_{i+j=k} a_i b_j\right) x^k\right) \cdot \left(\sum_{k=0}^{n} c_k x^k\right) \tag{14.6}$$

$$= \sum_{\ell=0}^{3n}\left(\sum_{m+k=\ell}\left(\sum_{i+j=m} a_i b_j\right) c_k\right) x^\ell \tag{14.7}$$

$$= \sum_{\ell=0}^{3n}\left(\sum_{i+j+k=\ell} a_i b_j c_k\right) x^\ell. \tag{14.8}$$

The computation of $f \cdot (g \cdot h)$ yields the same expression. The proof of the distributive property in $R[x]$ is left to the reader in Exercise 14.1. Finally, note that x^0 is a multiplicative identity for $R[x]$. □

Recall that the *degree* of a polynomial in x is the highest power of x which appears in the polynomial. We next define "degree" formally, for a polynomial with coefficients in any commutative ring with 1.

Definition 14.9. Let R be a commutative ring with 1, and let $f \in R[x] - \{0\}$. Write $f = \sum_{k=0}^{n} a_k x^k$ with $n \in \mathbf{N}$ and $a_k \in R$. The *degree of f in x* is

$$\deg(f) = \deg_x(f) = \max\{k \ : \ a_k \neq 0\}.$$

We also define $\deg(0) = -\infty$.

The usefulness of the degree concept as a measure of the "size" of a polynomial is illustrated in the following result.

Lemma 14.10. *Let R be a domain, and let $f, g \in R[x] - \{0\}$. Then $\deg(f \cdot g) = \deg(f) + \deg(g)$. In particular, $R[x]$ is also a domain.*

Proof. We can write $f = a_m x^m + a_{m-1} x^{m-1} + \cdots + a_1 x + a_0$ and $g = b_n x^n + b_{n-1} x^{n-1} + \cdots + b_1 x + b_0$ for some $m, n \in \mathbf{N}$ and $a_i, b_j \in R$ with $a_m \neq 0$ and $b_n \neq 0$. Thus we have $f \cdot g = a_m b_n x^{m+n} + $ (terms of smaller degree) . Since R is a domain and $a_m, b_n \in R - \{0\}$, we have $a_m b_n \neq 0$. Therefore, $\deg(f \cdot g) = m + n = \deg(f) + \deg(g)$, which proves the first statement. In particular, $f \cdot g \neq 0$. Together with the fact that $1_R \in R[x]$ is a multiplicative identity element for $R[x]$, and $1_R \neq 0_R = 0_{R[x]}$, this establishes that $R[x]$ is a domain. □

Another, even easier, result tells us how degrees behave with respect to addition:

Lemma 14.11. *Let R be a commutative ring with 1, and let $f, g \in R[x]$. Then $\deg(f+g) \leq \max\{\deg(f), \deg(g)\}$. Further, $\deg(f+g) = \max\{\deg(f), \deg(g)\}$ if $\deg(f) \neq \deg(g)$.*

Proof. Left to the reader as Exercise 14.3. □

One thing that we can't do with polynomials is divide them by each other: at least, if we try to divide two polynomials, the result will in general not be a polynomial. The following result tells us exactly which polynomials have multiplicative inverses, over a domain. The answer is, only the constant polynomials which already had inverses in R. Again, the notion of *degree* is a key in the proof.

Lemma 14.12. *If R is a domain, then $(R[x])^{\times} = R^{\times}$.*

Proof. Suppose that R is a domain.

[Show $(R[x])^{\times} \subseteq R^{\times}$] Let $f \in (R[x])^{\times}$. Then $\exists g \in R[x]$ such that $f \cdot g = 1$. By Lemma 14.10, we have $\deg(f) + \deg(g) = \deg(1)$. But $\deg(1) = 0$; and since $f, g \neq 0$, we have $\deg(f), \deg(g) \geq 0$. The only possibility now is that $\deg(f) = \deg(g) = 0$, and so $f, g \in R$. Since $f \cdot g = 1 = 1_R$, we can say $f \in R^{\times}$.

[Show $R^{\times} \subseteq (R[x])^{\times}$] Let $a \in R^{\times}$. Then $\exists b \in R$ such that $a \cdot b = 1_R$. Since $R \leq R[x]$ and $1_R = 1_{R[x]}$, this means that $a \in (R[x])^{\times}$. □

Besides doing addition and multiplication with polynomials, we can also *evaluate* a polynomial. For example, if $f(x) = x^2 + 3x - 1 \in \mathbf{R}[x]$, then we can evaluate f at $x = 5$ to get $f(5) = 5^2 + 3 \cdot 5 - 1 = 39$. We would like to generalize this idea to polynomials over an arbitrary commutative ring with 1.

Definition 14.13. Let R and S be commutative rings with 1 such that $R \leq S$, and let $\alpha \in S$. If $f \in R[x]$ with

$$f = c_n x^n + c_{n-1} x^{n-1} + \cdots + c_0,$$

then we define

$$f(\alpha) := c_n \alpha^n + c_{n-1} \alpha^{n-1} + \cdots + c_0 \in S.$$

We read $f(\alpha)$ as "f of α," or "f evaluated at $x = \alpha$." In particular, since $R \leq R[x] \ni x$, we can consider $f(x)$, and we see that $f(x) = f$.

The reader may have noticed that we began using "function notation" in Definition 14.13, writing $f(x)$ instead of just f. This allows us to use a consistent notation to denote both f and $f(\alpha)$. Stepping back a bit, we realize that evaluation of a polynomial is a way to turn a polynomial into a ring element. That is, evaluation at an element of S gives us a function from $R[x]$ to S:

Definition 14.14. Let R and S be commutative rings with 1 such that $R \leq S$, and let $\alpha \in S$. The *evaluation-at-α map* is the function $\varepsilon_\alpha : R[x] \to S$ given by

$$\varepsilon_\alpha(f) = f(\alpha)$$

for $f \in R[x]$.

Remark 14.15. In school algebra, the ability to evaluate a polynomial at a number allows us to produce a function from a polynomial. In fact, polynomials are often viewed in this light, and even confused with the corresponding functions. For example, if $f(x) = x^2 \in \mathbf{R}[x]$, then f gives a function $\sigma_f : \mathbf{R} \to \mathbf{R}$ via the formula $\sigma_f(\alpha) = \alpha^2$ for $\alpha \in \mathbf{R}$. However, we should be careful to distinguish the polynomial x^2 from the squaring function σ_f. In general, we wish to make a distinction between a polynomial and the corresponding evaluation function. Not only are they technically different kinds of object, but the correspondence between them may not be one-to-one: see Exercise 14.4.

There is also a difference in point of view between our definition of the evaluation-at-α map and the school algebra view of a polynomial as a function. Namely, in school algebra, a polynomial f is fixed, while the element α (where f is to be evaluated) varies over the set of all real numbers. In Definition 14.14, it is α which is fixed, while the polynomial f varies over the ring $R[x]$.

Since ε_α is a function from one ring to another, our guiding principles tell us that if it is to be worth studying, ε_α should be a ring homomorphism. This is indeed the case:

Proposition 14.16. *Let R and S be commutative rings with 1 such that $R \leq S$, and let $\alpha \in S$. Then the evaluation-at-α map ε_α is a ring homomorphism from $R[x]$ to S.*

Proof. [Show $\forall f, g \in R[x]$, $\varepsilon_\alpha(f + g) = \varepsilon_\alpha(f) + \varepsilon_\alpha(g)$ and $\varepsilon_\alpha(f \cdot g) = \varepsilon_\alpha(f) \cdot \varepsilon_\alpha(g)$; this means $(f + g)(\alpha) = f(\alpha) + g(\alpha)$ and $(f \cdot g)(\alpha) = f(\alpha) \cdot g(\alpha)$]

Let $f, g \in R[x]$. Write $f = \sum_{j=0}^{n} a_j x^j$ and $g = \sum_{j=0}^{n} b_j x^j$ with $a_j, b_j \in R$ (as previously noted, we can always do this even if f and g have different degrees, by adding terms to either f or g whose coefficients are 0). We have $f(\alpha) = \sum_{j=0}^{n} a_j \alpha^j$ and $g(\alpha) = \sum_{j=0}^{n} b_j \alpha^j$. Now $f + g = \sum_{j=0}^{n}(a_j + b_j)x^j$, so $(f + g)(\alpha) = \sum_{j=0}^{n}(a_j + b_j)\alpha^j = \sum_{j=0}^{n} a_j \alpha^j + \sum_{j=0}^{n} b_j \alpha^j = f(\alpha) + g(\alpha)$.

The proof that $(f \cdot g)(\alpha) = f(\alpha) \cdot g(\alpha)$ is likewise purely formal. The key point is that the formula defining the product of f and g, namely

$$\left(\sum_{j=0}^{n} a_j x^j \right) \cdot \left(\sum_{j=0}^{n} b_j x^j \right) = \sum_{i=0}^{2n} \left(\sum_{j+k=i} a_j b_k \right) x^i,$$

is the same as the formula for the corresponding product of ring elements:

$$\left(\sum_{j=0}^{n} a_j \alpha^j \right) \cdot \left(\sum_{j=0}^{n} b_j \alpha^j \right) = \sum_{i=0}^{2n} \left(\sum_{j+k=i} a_j b_k \right) \alpha^i.$$

$\square$

Notation 14.17. Let $\varepsilon_\alpha : R[x] \to S$ be the evaluation-at-α map. We denote the image of ε_α by $R[\alpha]$. That is,

$$R[\alpha] := \varepsilon_\alpha(R[x]) = \{\varepsilon_\alpha(f) : f \in R[x]\} = \{f(\alpha) : f \in R[x]\}. \qquad (14.9)$$

Remark 14.18. We can also write

$$R[\alpha] = \{c_0 + c_1 \alpha + c_2 \alpha^2 + \cdots + c_n \alpha^n : n \in \mathbf{N}, c_i \in R\}.$$

Thus, $R[\alpha]$ is the "set of all polynomials in α with coefficients in R." Even though α is an element of a ring S, and not necessarily a "variable," we use the same square-bracket notation in both cases. A unified point of view will be presented in Chapter 15.

Example 14.19. Let $R = S = \mathbf{R}$, and let $\alpha = 3 \in \mathbf{R}$. Consider the evaluation-at-3 map $\varepsilon_3 : \mathbf{R}[x] \to \mathbf{R}$, $f(x) \mapsto f(3)$. For example, we have $\varepsilon_3 : x^2 + 7 \mapsto 3^2 + 7 = 16$.

We claim that ε_3 is surjective. An easy way to see this is by considering the constant polynomials $c \in \mathbf{R} \leq \mathbf{R}[x]$. Evaluating a constant polynomial at $x = 3$ has no effect on the constant, so $\varepsilon_3(c) = c$ for all $c \in \mathbf{R}$.

What is $\ker(\varepsilon_3)$? We have $f \in \ker(\varepsilon_3)$ iff $f(3) = 0$. Recall that if $f(3) = 0$, we say that 3 is a *root* of f. From school algebra, the reader may also recall the

Root-Factor Theorem, which tells us that 3 is a root of f iff $x-3$ is a factor of f. (Later, we shall prove a version of the Root-Factor Theorem for ourselves, Theorem 14.24.) Thus, we are led to the conclusion that $\ker(\varepsilon_3)$ is the set of all multiples of $x-3$ in $\mathbf{R}[x]$, which is just the principal ideal $(x-3)$. Now by the Fundamental Theorem of Ring Homomorphisms, Theorem 11.36, we can say $\mathbf{R}[x]/(x-3) \cong \mathbf{R}$. We interpret this result as follows. By forcing $x-3$ to be zero in $\mathbf{R}[x]$, we collapse the ring $\mathbf{R}[x]$ into $\mathbf{R}$ by, in effect, substituting $x=3$ in each polynomial in $\mathbf{R}[x]$.

14.2   Polynomials Over a Field

As we might expect, since fields are the nicest type of ring, polynomial rings with coefficients from a field, or polynomial rings "over" a field (as we may call them) are the nicest type of polynomial rings.

We are familiar with the notion of quotients and remainders from ordinary arithmetic. Namely, suppose that $f, g \in \mathbf{N}$ with $g \neq 0$. When we divide f by g, we can express the result as an integer plus a remainder term: more precisely, we can write

$$f/g = q + r/g, \tag{14.10}$$

where $q, r \in \mathbf{N}$ and

$$0 \le r < g. \tag{14.11}$$

The situation with polynomials is not so nice in general. Lemma 14.12 tells us that (over a domain) the only polynomials we can "divide" by are those that are not "really" polynomials, but only constants: actually, only units of the coefficient ring. Nevertheless, when we are working with coefficients from a *field*, it is possible to effectively simulate the results of division, as the next theorem shows. By replacing integers with polynomials over a field, clearing away denominators in Equation 14.10, and replacing the bound on the size of the remainder in Equation 14.11 by a bound on the *degree* of the remainder, we arrive at the following (true!) result.

Theorem 14.20 (Quotient-Remainder Theorem). *Let F be a field. Let $f \in F[x]$ and let $g \in F[x] - \{0\}$. Then there exist unique polynomials $q, r \in F[x]$ such that $f = q \cdot g + r$ and $\deg(r) < \deg(g)$. (It may be that $r = 0$, in which case we agree that $\deg(r) = -\infty < 0 \le \deg(g)$, and our result is still true in this case.)*

Proof. Let F be a field, $f, g \in F[x]$, $g \neq 0$. [Show existence of q and r] Let $S = \{f - a \cdot g \; : \; a \in F[x]\}$. Then we have $S \subseteq F[x]$, and S is non-empty. Let $m = \min\{\deg(s) \; : \; s \in S\}$. Assume for a contradiction that $m \ge d := \deg(g)$. Then $\exists s \in S$ s.t. $\deg(s) = m$, and since $s \in S$ we can write $s = f - a \cdot g$ for some $a \in F[x]$. Now we have

$$s = c_m x^m + c_{m-1} x^{m-1} + \cdots + c_0$$

and

$$g = b_d x^d + b_{d-1} x^{d-1} + \cdots + b_0$$

for some coefficients $c_i, b_i \in F$ with c_m and b_d non-zero. [Idea: Perform one step of a "polynomial division" of s/g. The leading coefficient would be $c_m x^m / (b_d x^d) = b_d^{-1} c_m x^{m-d}$.] Since F is a field and $b_d \neq 0_F$, we have $b_d^{-1} \in F$, and we set $h = b_d^{-1} \cdot c_m \cdot x^{m-d}$. Note that $m - d \geq 0$ by assumption, so $h \in F[x]$. Set $\tilde{s} = s - h \cdot g$. Then we have

$$\tilde{s} = s - h \cdot g = (c_m - b_d^{-1} c_m \cdot b_d) x^m + \text{ (terms of lower degree)},$$

from which we see that $\deg(\tilde{s}) < m$. But $\tilde{s} = s - h \cdot g = f - a \cdot g - h \cdot g = f - (a + h) \cdot g \in S$ and $\deg(\tilde{s}) < \deg(s)$, contradicting that s has the smallest degree of any element of S. This contradiction shows that $m < d$. Therefore, there exist $q, r \in F[x]$ such that $f = q \cdot g + r$ and $\deg(r) < \deg(g)$.

[Show uniqueness of q and r] Now suppose that $q, r, \tilde{q}, \tilde{r} \in F[x]$ are such that $f = q \cdot g + r = \tilde{q} \cdot g + \tilde{r}$ and $\deg(r), \deg(\tilde{r}) < d$. Then we have $(q - \tilde{q}) \cdot g = \tilde{r} - r$. Assume for a contradiction that $q \neq \tilde{q}$. Then $q - \tilde{q} \neq 0$, and since also $g \neq 0$ by hypothesis, we can apply Lemma 14.10 to get $\deg((q - \tilde{q}) \cdot g) = \deg(q - \tilde{q}) + \deg(g)$. By Lemma 14.11, we have $\deg(\tilde{r} - r) < d$. Combining these facts, we have $d > \deg(\tilde{r} - r) = \deg((q - \tilde{q}) \cdot g) = \deg(q - \tilde{q}) + \deg(g) \geq \deg(g) = d$. This contradiction shows that $q = \tilde{q}$.

Now since $q \cdot g + r = \tilde{q} \cdot g + \tilde{r}$ and $q = \tilde{q}$, it follows that $r = \tilde{r}$. This proves uniqueness. $\square$

Our hunch that polynomial rings over a field ought to be well-behaved is made precise in the next result, which says that such rings are principal ideal domains. Before stating this result, we need one definition.

Definition 14.21. A polynomial is *monic* if its leading coefficient is 1. That is, let R be a commutative ring with 1, and let $f \in R[x] - \{0\}$, with $\deg(f) = d$. Write $f = c_d x^d + c_{d-1} x^{d-1} + \cdots + c_1 x + c_0$, with $c_i \in R$. Then f is called monic if $c_d = 1$.

Theorem 14.22. *Let F be a field, and let I be an ideal of $F[x]$. If $I \neq (0)$, then there is a unique monic polynomial $f \in F[x]$ such that $I = (f)$. In particular, $F[x]$ is a principal ideal domain.*

Proof. First, since every field is a domain (Proposition 11.14), $F[x]$ is a domain by Lemma 14.10.

Suppose that I is an ideal of $F[x]$ with $I \neq (0)$. [Idea: If I is really principal, generated by a polynomial f, then every polynomial in I is a multiple of f. Thus, f should have the smallest degree of any (non-zero) element of I.] Let $m = \min\{\deg(a) : a \in I - \{0\}\}$. Then $\exists g \in I - \{0\}$ such that $\deg(g) = m$.

So we can write $g = \sum_{j=0}^{m} c_j x^j$ with $c_j \in F$ and $c_m \neq 0$. Since F is a field, we have $c_m \in F^\times$, and we set $f = c_m^{-1} \cdot g \in F[x]$. Observe that f is monic and $\deg(f) = \deg(g)$.

[Show that $I = (f)$] Let $h \in I$. By the Quotient-Remainder Theorem (Theorem 14.20), there exist $q, r \in F[x]$ such that $\deg(r) < \deg(f)$ and $h = q \cdot f + r$. So $r = h - q \cdot f$. Since $f, h \in I$ and $q \in F[x]$, we have $r \in I$. Now $\deg(r) < \deg(f)$, but there is no non-zero polynomial in I whose degree is less than the degree of f. Therefore, $r = 0$. It follows that $h = q \cdot f \in (f)$. Thus, $I \subseteq (f)$. Conversely, since $f \in I$, we have $(f) \subseteq I$. This shows that $I = (f)$.

[Show uniqueness of f] Suppose that $\tilde{f} \in F[x]$ is another monic polynomial such that $I = (\tilde{f})$. Then $(f) = (\tilde{f})$, so by Lemma 12.16, we have $\tilde{f} = u \cdot f$ for some $u \in (F[x])^\times$. Now by Lemma 14.12, we can say $u \in F^\times$. In particular, we have $\deg(u) = 0$, so by Lemma 14.10, we have $\deg(\tilde{f}) = \deg(f)$. Since f is monic of degree m, we can write $f = x^m +$ (terms of smaller degree). Since $\tilde{f} = u \cdot f$ and $u \in F^\times$, we have $\tilde{f} = u \cdot x^m +$ (terms of smaller degree). But $\tilde{f}$ is also monic, which forces $u = 1$. Therefore, $\tilde{f} = f$, as desired.

Now we know that every non-zero ideal of $F[x]$ is principal; since the zero ideal (0) is also principal, we have proved that $F[x]$ is a PID. $\qquad\square$

After defining the term "root" in a suitably general manner, we will be ready to state and prove the Root-Factor Theorem for polynomials over an arbitrary field.

Definition 14.23. Let R be a commutative ring with 1, and let $f \in R[x]$. An element $\alpha \in R$ is called a *root* of f if $f(\alpha) = 0$.

Theorem 14.24 (Root-Factor Theorem). *Let F be a field. Let $f \in F[x]$ and $\alpha \in F$. Then α is a root of f iff $(x - \alpha) \mid f$ in $F[x]$.*

Proof. ($\Rightarrow$): Suppose that α is a root of f. By the Quotient-Remainder Theorem (Theorem 14.20), there exist unique polynomials $q, r \in F[x]$ such that

$$f = (x - \alpha) \cdot q + r \qquad (14.12)$$

and $\deg(r) < 1$. Since r has degree less than 1, we must have $r \in F$. Evaluating both sides of Equation 14.12 at $x = \alpha$, and using the fact that evaluation at α is a ring homomorphism (by Proposition 14.16), we get

$$f(\alpha) = (\alpha - \alpha) \cdot q(\alpha) + r(\alpha). \qquad (14.13)$$

Now since $r \in F$, we have $r(\alpha) = r$. Since α is a root of f, we have $f(\alpha) = 0$. Therefore, $r = 0$, so $f = (x - \alpha) \cdot q$, and $(x - \alpha) \mid f$.

($\Leftarrow$): Suppose that $(x - \alpha) \mid f$ in $F[x]$. This means that $f = (x - \alpha) \cdot q$ for some $q \in F[x]$. Evaluating at α, we find $f(\alpha) = 0$, so α is a root of f. $\qquad\square$

Next we state and prove a result familiar from school algebra: a polynomial of degree n can have at most n roots—if we are working over a *field*. But see Exercise 14.5 for a counterexample to this result over a ring which is not a domain.

Theorem 14.25 (Root Bound on Polynomials). *If $f \in F[x]$ is a non-zero polynomial over a field F, then the number of roots of f in F is at most the degree of f.*

Proof. Let F be a field, and let $f \in F[x] - \{0\}$. Set $R_f = \{\alpha \in F : f(\alpha) = 0\}$, the set of all roots of f in F. We proceed by induction on $d := \deg(f)$.

Base Case: $d = 0$.

Then $f \in F - \{0\}$, so f is a non-zero constant polynomial, which has no roots in F. So $R_f = \emptyset$ and $|R_f| = 0 = \deg(f)$.

Inductive Step: Suppose that $d \geq 1$, and assume inductively that if $g \in F[x] - \{0\}$ and $\deg(g) < d$, then $|R_g| \leq \deg(g)$.

If f has no roots in F, then we have $|R_f| = 0 < d$, so we are done. Therefore, we may assume that there exists a root α of f in F.

By the Root-Factor Theorem (Theorem 14.24), we can write

$$f = (x - \alpha) \cdot g \tag{14.14}$$

for some $g \in F[x]$. Since $f \neq 0$, we must also have $g \neq 0$. Thus, Lemma 14.10 applies to give $\deg(f) = \deg(x - \alpha) + \deg(g)$. So $d = 1 + \deg(g)$ and $\deg(g) = d - 1$. By the inductive hypothesis, we have

$$|R_g| \leq \deg(g) = d - 1. \tag{14.15}$$

We claim that

$$R_f = R_g \cup \{\alpha\}. \tag{14.16}$$

To prove this claim, let $\beta \in R_f$. Then we have $f(\beta) = 0 = (\beta - \alpha) \cdot g(\beta)$. Since F is a domain (by Proposition 11.14), we have $\beta - \alpha = 0$ or $g(\beta) = 0$. Thus, either $\beta = \alpha$ or $\beta \in R_g$. This shows that $R_f \subseteq R_g \cup \{\alpha\}$. Conversely, Equation 14.14 shows immediately that $R_g \cup \{\alpha\} \subseteq R_f$.

Equations 14.15 and 14.16 together allow us to write $|R_f| \leq |R_g| + 1 \leq (d - 1) + 1 = d$, as desired. $\square$

14.3 Exercises

Exercise 14.1. (a) Prove the distributive property for polynomials over a commutative ring with 1 by using the formulas for the ring operations given in Definition 14.1.

(b) Find an example of an abelian group $(S, +)$ and an associative binary operation $\cdot$ on S with a subset $T \subseteq S$ such that $\langle T \rangle = S$, and such that for all $t, u, v \in T$, we have $t \cdot (u + v) = t \cdot u + t \cdot v$ and $(u + v) \cdot t = u \cdot t + v \cdot t$, but such that $(S, +, \cdot)$ is *not* a ring. That is, to check the distributive property for S, it is not enough to check it for a set which generates S under addition.

Exercise 14.2. Let R be a commutative ring with 1. Prove the converse of the second part of Lemma 14.10: If $R[x]$ is a domain, then R is a domain.

Exercise 14.3. Prove Lemma 14.11.

Exercise 14.4. Let $R = \mathbf{Z}/3\mathbf{Z}$, and let $f(x) = x^3 - x \in R[x]$. Let $\sigma_f : R \to R$ be the function which corresponds to f via evaluation, namely $\sigma_f(a) = f(a)$ for $a \in R$.

  (a) Find the values $\sigma_f(a)$ for each a in R.

  (b) Find another polynomial $g \in R[x]$ such that $g \neq f$ but $\sigma_g = \sigma_f$.

Exercise 14.5. Find a commutative ring R with 1 such that the polynomial $x^2 \in R[x]$ has more than 2 distinct roots in R. Hint: Choose R to be the ring of integers modulo n, for an appropriate choice of n.

Exercise 14.6. Let F be a field. Prove that every polynomial of degree 1 in $F[x]$ is irreducible in $F[x]$.

Exercise 14.7. Let F be a field and let $f \in F[x]$.

  (a) Prove that if f is irreducible in $F[x]$ *and* f has a root in F, then $\deg(f) = 1$.

  (b) Suppose that $\deg(f) \in \{2, 3\}$. Prove that f is irreducible in $F[x]$ iff f does not have any roots in F.

  (c) Find a counterexample to the biconditional in part (b) using a field F of your choice and a polynomial f of degree 4.

Exercise 14.8. Let F be a field and let $V = F[x]$.

  (a) Prove that V is a vector space over F under the operations $+$ and $\cdot$ of $F[x]$ (after restricting the latter operation to $F \times V$).

  (b) Prove that the set $B := \{x^j : j \in \mathbf{N}\}$ is a basis of V over F; thus, V is an infinite-dimensional F-vector space.

Exercise 14.9. Let F be a field and let $V = F[x]$. You may use the result of Exercise 14.8 here.

  (a) Let V_n be the set of all polynomials in V of degree at most n. Prove that $V_n \leq V$ and $\dim_F(V_n) = n + 1$.

  (b) Suppose that $f \in V$ and that f factors into linear terms as $f = a \cdot \prod_{j=1}^{n}(x - r_j)$ with $a, r_1, \ldots, r_n \in F$, $a \neq 0$, and $r_1, \ldots, r_n$ distinct. For each $i \in \{1, \ldots, n\}$, let $f_i = a \cdot \prod_{j \neq i}(x - r_j)$. Prove that $\{f_1, \ldots, f_n\}$ is a basis of V_{n-1} over F. Hint: start with a linear dependence relation involving the f_i, and then apply carefully chosen evaluation maps. Note: this result justifies the method of partial fractions in calculus to integrate a rational expression in the case when the denominator factors into distinct linear terms.

Exercise 14.10. (a) Let $\sigma : R \to S$ be a homomorphism between two commutative rings with 1 such that $\sigma(1_R) = 1_S$. Let a be an element of S. Prove that there is a unique ring homomorphism $\overline{\sigma} : R[x] \to S$ such that $\overline{\sigma}|_R = \sigma$ and $\overline{\sigma}(x) = a$. (Notice that this construction generalizes the evaluation mappings.)

  (b) Prove that if T shares the property of $R[x]$ expressed in part (a) above, then $T \cong R[x]$. More precisely:

Suppose that T is a commutative ring with 1 such that $R \leq T$. Suppose that $\alpha \in T$. Finally, suppose that for every commutative ring S with 1, for every element $a \in S$, and for every ring homomorphism $\sigma : R \to S$ which maps 1_R to 1_S, there is a unique ring homomorphism $\overline{\sigma} : T \to S$ which extends σ and maps α to a. Prove that $T \cong R[x]$ via an isomorphism sending α to x.

Exercise 14.11. Let R be a commutative ring with 1. The *polynomial ring over R in the variables* $x_1, \ldots, x_n$ is defined recursively as $R[x_1, \ldots, x_n] := (R[x_1, \ldots, x_{n-1}])[x_n]$. Prove that if S is a commutative ring with 1, $\sigma : R \to S$ is a ring homomorphism with $\sigma(1_R) = 1_S$, and $a_1, \ldots, a_n$ are elements of S, then there is a unique ring homomorphism $\overline{\sigma} : R[x_1, \ldots, x_n] \to S$ such that $\overline{\sigma}|_R = \sigma$ and $\overline{\sigma}(x_j) = a_j$ for all $j \in \{1, \ldots, n\}$. Note: for a polynomial $f \in R[x_1, \ldots, x_n]$ it is natural to denote the image $\overline{\sigma}(f)$ by $f(a_1, \ldots, a_n)$, and to refer to $f(a_1, \ldots, a_n)$ as "f evaluated at $(a_1, \ldots, a_n)$." We define $R[a_1, \ldots, a_n]$ to be the image of σ. We define $\{a_1, \ldots, a_n\}$ to be *algebraically independent over R* (with respect to σ) if $\overline{\sigma}$ is injective; this is a generalization of the notion of a transcendental element, and is especially important when $R \leq S$ and σ is just the inclusion map.

Exercise 14.12. Let R be a ring with 1 (not necessarily commutative), and let $C = C(R)$ be the center of R (see Exercise 11.18). Let $a \in R$. Prove that the set $S := \{c_0 + c_1 a + c_2 a^2 + \cdots + c_n a^n : n \in \mathbf{N}, c_j \in C\}$ is a commutative ring with $S \leq R$. Thus we have $S = C[a]$, the image of the evaluation-at-a map from $C[x]$ to S.

Exercise 14.13. Find a commutative ring R with 1 such that $(R[x])^{\times} \neq R^{\times}$. (Compare to Lemma 14.12.)

Exercise 14.14 (Two-Thirds Rule for Factoring a Polynomial over a Field). Let $F \leq K$ be fields, and let $f, g, h \in K[x] - (0)$ be such that $f = g \cdot h$. Prove that if any 2 of these 3 polynomials are in $F[x]$, then so is the third.

Exercise 14.15 (Composition of Polynomials). Let R be a commutative ring with 1, and let $S = R[x]$. Observe that since $R \leq S$, then for any $f \in S$ we may consider the evaluation-at-f map $\varepsilon_f : S \to S$ which sends x to f. Define a binary operation $\circ$ on S by the formula $g \circ f = \varepsilon_f(g)$ for any $f, g \in S$. We may refer to $\circ$ as *composition*, since the standard notation for evaluation allows us to write $g \circ f = g(f)$.

(a) Prove that $\circ$ is associative.

(b) Prove that x is an identity element for $\circ$.

(c) Now suppose that R is a domain. Prove that if $f, g \in S - (0)$, then we have $\deg(g \circ f) = \deg(g) \cdot \deg(f)$. Use this formula to help show that the elements of S which have inverses with respect to $\circ$ are precisely the polynomials of degree 1.

15

Field Theory

15.1 Extension Fields

The theory of polynomial rings over a field is closely connected to the general theory of fields. We illustrate this connection in the following example.

Example 15.1. Consider the evaluation-at-i map

$$\varepsilon_i \; : \; \mathbf{R}[x] \to \mathbf{C}$$

which maps a polynomial $f(x)$ with real coefficients to the complex number $f(i)$.

We claim that ε_i is surjective. To see this, consider the image under ε_i of a typical linear polynomial $a + bx$ with $a, b \in \mathbf{R}$. We have $\varepsilon_i(a + bx) = a + bi$, which is the general form of a complex number.

Set $K = \ker(\varepsilon_i)$. Since ε_i is a ring homomorphism (by Proposition 14.16), we know that K is an ideal of $\mathbf{R}[x]$. By Theorem 14.22, either $K = (0)$ or there is a unique monic polynomial $f(x) \in \mathbf{R}[x]$ such that $K = (f)$; further, from the proof of that theorem, f would be a non-zero polynomial of smallest degree in K.

We have $f \in K$ iff $\varepsilon_i(f) = 0$ iff $f(i) = 0$. In other words, to find elements of K, we must seek polynomial relations satisfied by i with coefficients in $\mathbf{R}$. The basic formula

$$i^2 = -1$$

gives us such a polynomial relation. It can be rewritten as $i^2 + 1 = 0$, so we realize that $g(i) = 0$ where $g(x) = x^2 + 1$. Thus we have $g \in K$; in particular, $K \neq (0)$. Since $\deg(g) = 2$, we should ask whether there are any non-zero polynomials of degree 1 in K. Well, if $f(x) = a + bx$ with $a, b \in \mathbf{R}$, then $f(i) = a + bi$, which is not zero unless $a = b = 0$. Therefore, g has the smallest degree of any non-zero polynomial in K. Note that g is monic, so in fact g is the generator of K described in Theorem 14.22.

Now we know that $\ker(\varepsilon_i) = (g) = (x^2 + 1)$. Since ε_i is surjective (as shown earlier in this example), the Fundamental Theorem of Ring Homomorphisms (Theorem 11.36) tells us that

$$\mathbf{R}[x]/(x^2 + 1) \cong \mathbf{C}. \tag{15.1}$$

DOI: 10.1201/9781003252139-15

Notice that everything on the left-hand side of (15.1) involves only *real* quantities. By starting with the field $\mathbf{R}$, we were able to produce the field $\mathbf{C}$, as follows: First, add in a new element, x, to $\mathbf{R}$, to get $\mathbf{R}[x]$. This x is a blank slate for our handiwork, because it does not have any algebraic relations with elements of $\mathbf{R}$. Next, we force the equation $x^2 = -1$ to have a solution by forming the quotient ring $\mathbf{R}[x]/(x^2 + 1)$. In this quotient ring, let $\bar{x}$ denote the image of x under the natural map

$$\mathbf{R}[x] \to \mathbf{R}[x]/(x^2 + 1), \quad f \mapsto f + (x^2 + 1).$$

In other words, set $\bar{x} = x + (x^2 + 1) \in S := \mathbf{R}[x]/(x^2 + 1)$. Then we have $\bar{x}^2 = -1_S$.

If we were familiar with $\mathbf{R}$ but did not know about $\mathbf{C}$, then we could actively seek to "extend" $\mathbf{R}$ by forming such a quotient ring. Desiring a new "number" i satisfying $i^2 = -1$, we would simply form the quotient ring $\mathbf{C} := \mathbf{R}[x]/(x^2+1)$, check that $\mathbf{R} \hookrightarrow \mathbf{C}$, and agree to use the notation $i := x+(x^2+ 1) \in \mathbf{C}$. From this point of view, we need not fear i or label i dubiously as an "imaginary" number: instead, i is simply an element of a quotient of $\mathbf{R}[x]$.

Definition 15.2. Let F and K be fields. If F is a subring of K,

$$F \leq K,$$

then we also say that F is a *subfield* of K, or that K is an *extension field* of F. We may also say that K is a *superfield* of F.

Remark 15.3. Whether we use the term *subfield* or *extension field* depends on our point of view. If we start out knowing the bigger field K, then we usually refer to F as a subfield of K; if we start out only knowing the smaller field F, and then discover or construct K, then we usually speak of K as being an extension field of F instead of referring to F as a subfield of K.

If K is an extension field of F and $\alpha \in K$, then one important question we can ask is, what algebraic relations, if any, does α satisfy with respect to F? For example, if $\alpha = i \in \mathbf{C}$, then what algebraic relations does i satisfy with respect to $\mathbf{R}$? From Example 15.1, we see that the answer to this question is tied up with the kernel of the evaluation-at-i map from $\mathbf{R}[x]$ to $\mathbf{C}$. We next explore this relationship in general.

Proposition 15.4. *Let F and K be fields. Suppose that F is a subfield of K, and let $\alpha \in K$. Let $\varepsilon_\alpha : F[x] \to K$ be the evaluation-at-α map. Then one of the following must be the case:*

(i) $\ker(\varepsilon_\alpha) = (f)$ for a unique non-zero monic polynomial $f \in F[x]$. In this case, f is irreducible in $F[x]$, we have $f(\alpha) = 0$, and $F[\alpha] \cong F[x]/(f)$.

(ii) $\ker(\varepsilon_\alpha) = (0)$. In this case, α is not a root of any non-zero polynomial in $F[x]$, and we have $F[\alpha] \cong F[x]$.

Proof. Let $I = \ker(\varepsilon_\alpha)$. First suppose that $I \neq (0)$. By Proposition 14.16, ε_α is a ring homomorphism. So by Theorem 11.36, I is an ideal of $F[x]$. We know from Theorem 14.22 that there is a unique monic polynomial $f \in F[x]$ such that $I = (f)$. We have $f \neq 0$ since $I \neq (0)$. Also by Theorem 11.36, we know that the image of ε_α is a subring of K; that is, $F[\alpha] \leq K$. Now K is a field, hence also a domain (by Proposition 11.14); and $F[\alpha]$, since it is a subring of K containing 1, must also be a domain (by Exercise 11.11). The Fundamental Theorem of Ring Homomorphisms (Theorem 11.36 again—what a useful theorem!) now tells us that $F[\alpha] \cong F[x]/(f)$. By Theorem 12.8, we can say that (f) is a prime ideal of $F[x]$. We can use Exercise 12.12 to assert that f is a prime element of $F[x]$. Since $F[x]$ is a domain (for example, by Theorem 14.22), then by Exercise 12.11, we conclude that f is irreducible in $F[x]$. We have $f(\alpha) = 0$ because $f \in \ker(\varepsilon_\alpha)$.

Now suppose that $I = (0)$. This means that for every polynomial $f \in F[x] - (0)$, we have $f(\alpha) \neq 0$, as claimed. Theorem 11.36 gives $F[\alpha] \cong F[x]/(0)$. But $F[x]/(0) \cong F[x]$ (by Exercise 11.6). Therefore, $F[\alpha] \cong F[x]$ (by Exercise 11.12). $\square$

The following definition attaches some names to the wealth of information contained in the outcomes of the two cases in Proposition 15.4.

Definition 15.5. Let F and K be fields with $F \leq K$, and let $\alpha \in K$.

Case (i): If α is a root of some non-zero polynomial in $F[x]$, then we say that α is *algebraic over* F. Let f be the unique non-zero monic polynomial which generates the kernel of the evaluation map $\varepsilon_\alpha : F[x] \to K$ in this case according to Proposition 15.4. Then f is called the *irreducible polynomial* for α over F in the variable x. We use the notation $f = \text{Irr}(\alpha, F, x)$.

Case (ii): If α is not a root of any non-zero polynomial in $F[x]$, then we say that α is *transcendental over* F.

Definition 15.6. Let F and K be fields with $F \leq K$. If α is algebraic over F for all α in K, then we say that K is an *algebraic extension* of F, or that K is *algebraic over* F.

Remark 15.7. Now we can see how the square-brackets notation $F[\alpha]$, when α is an element of a ring containing F, meshes with the notation $F[x]$ for a polynomial ring in the variable x. They actually amount to the same thing *when α is transcendental over F*. In other words, a transcendental element behaves like an independent variable!

Example 15.8. Consider the element $\sqrt{2} \in \mathbf{R}$. Since $\mathbf{Q} \leq \mathbf{R}$, we can ask whether $\sqrt{2}$ is algebraic over $\mathbf{Q}$. The answer is yes, because (for instance) $\sqrt{2}$ is a root of the non-zero polynomial $x^2 - 2 \in \mathbf{Q}[x]$. Notice that many other polynomials in $\mathbf{Q}[x]$ also have $\sqrt{2}$ as a root, such as $x^4 - 4$ and $x^3 + x^2 - 2x - 2$, but it is not hard to see that in fact $x^2 - 2 = \text{Irr}(\sqrt{2}, \mathbf{Q}, x)$.

Example 15.9. We are not in a position yet to display fully supported examples of transcendental elements, but some attempt should be made. It can be shown (but is not at all obvious) that the real number π is transcendental over $\mathbf{Q}$. Thus, by Proposition 15.4, we have $\mathbf{Q}[\pi] \cong \mathbf{Q}[x]$ via a map sending π to x. This is much stronger than saying that π is irrational. In fact, π "transcends" any description by a finite amount of rational numbers: π not only cannot be expressed as a ratio of integers, but cannot be realized as the root of any non-zero polynomial with rational coefficients.

As another example, we could take a variable x and try to show that x is transcendental over $\mathbf{Q}$—as indeed it should be. The trouble here is that x would first have to live in some field containing $\mathbf{Q}$. It turns out that the polynomial ring $\mathbf{Q}[x]$ naturally embeds in a field of "rational functions" $\mathbf{Q}(x) := \{f/g : f,g \in \mathbf{Q}[x] \text{ and } g \neq 0\}$; see Chapter 20. With the field extension $\mathbf{Q} \leq \mathbf{Q}(x)$ in hand, we can say that indeed x is transcendental over $\mathbf{Q}$.

For an explicit construction of a real number which is transcendental over $\mathbf{Q}$, see Exercise 15.17.

In Proposition 15.4, we found that the kernel of any evaluation map ε_α from $F[x]$ to K, where F is a subfield of K and α is algebraic over F, is generated by an irreducible polynomial $f \in F[x]$ such that α is a root of f. We also saw that $F[\alpha] \cong F[x]/(f)$ in this situation.

Next, we take the opposite point of view: we start with a field F and a polynomial $f \in F[x]$ such that f is irreducible in $F[x]$. We would like to find an extension field K of F which contains a root of f.

The idea here is to form the quotient ring $Q = F[x]/(f)$. In this ring, we have forced f to equal zero; more precisely, we would like to say that $f(\bar{x}) = 0$, where $\bar{x} = x + (f) \in Q$. Thus, $\bar{x}$ would be a root of f in Q. For the evaluation of f at $\bar{x}$ to make sense, Q should be a ring which contains F, and if we were lucky enough to find that Q is actually a *field* containing F, this would allow us to achieve our goal by setting $K = Q$. This turns out to work, as the following result indicates.

Theorem 15.10. *Let F be a field, and let $f \in F[x]$. Suppose that f is irreducible in $F[x]$. Then:*

  (1) (f) is a maximal ideal of $F[x]$.
  (2) $F[x]/(f) =: K$ is a field which naturally contains F.
  (3) The element $\alpha := x + (f) \in K$ is a root of f in K.
  (4) $\dim_F(K) = \deg(f)$.

Proof. Set $R = F[x]$.

(1) First, we have $f \notin R^\times$ by definition of *irreducible*, so (f) is a proper ideal of R by Exercise 12.6. Suppose that $(f) \subset I \subseteq R$, for some ideal I of R. [Show $I = R$] Since R is a PID (by Theorem 14.22), $\exists g \in R$ such that $I = (g) = Rg$. Now $f \in (f) \subseteq Rg$, so $\exists h \in R$ such that $f = hg$. Since f is irreducible in R, either $h \in R^\times$ or $g \in R^\times$. If $h \in R^\times$, then $(f) = (g) = I$ by Lemma 12.16, a contradiction. Therefore, $g \in R^\times$, so $I = (g) = R$ (by Exercise 12.6 again). This shows that (f) is a maximal ideal of R.

(2) Since (f) is maximal, Theorem 12.21 tells us that $R/(f)$ is a field. Now consider the map

$$\omega \; : \; F \to K, \quad c \mapsto c + (f).$$

This is just the natural map $F[x] \to K$ restricted to F, so ω is a ring homomorphism. Now $\ker(\omega)$ is an ideal of F (by Theorem 11.36), hence is either (0) or F (by Exercise 11.7). In case $\ker(\omega) = F$, we would have $\omega(1_F) = 0_K$, so $1_F + (f) = 0_F + (f)$, hence $1_F = 1_R \in (f)$; so $(f) = R$, which is impossible since maximal ideals are proper. Therefore, $\ker(\omega) = (0)$. By Exercise 11.23, ω is an embedding. Since $F \hookrightarrow K$ via the natural map ω, we regard F as a subfield of K, and write $F \leq K$. (This is a slight abuse of notation, since actually $\omega(F) \leq K$; but such slurring of distinctions becomes one of the most important coping mechanisms for the advanced algebraist who must deal with many technically distinct but naturally isomorphic structures at the same time!)

(3) Again, we regard F as a subfield of K, by identifying $c \in F$ with $c + (f) \in K$. Thus the evaluation-at-α map $\varepsilon_\alpha \; : \; F[x] \to K$ makes sense. We have, for any $g \in F[x]$, the equality

$$g(x + (f)) = g(x) + (f). \tag{15.2}$$

(See Exercise 15.3 for a detailed justification of this equation.) We have $f(x) \in F[x]$, so we compute $\varepsilon_\alpha(f) = f(\alpha) = f(x + (f)) = f(x) + (f) = 0 + (f) = 0_K$, since certainly $f(x) \in (f)$.

(4) Let $d = \deg(f)$. From the proof of part (2) above, we know that $d \geq 1$. We claim that the set

$$B := \{1 + (f), x + (f), x^2 + (f), \ldots, x^{d-1} + (f)\}$$

is a basis for K over F.

[Show that B spans K over F] Let $a \in K$. Then $\exists g \in R$ such that $a = g + (f)$. By Theorem 14.20, there are unique elements $q, r \in R$ such that $g = q{\cdot}f + r$ and $\deg(r) < d$. Now $q \cdot f \in (f)$, so we have $g + (f) = r + (f)$ by Exercise 8.2. Since $\deg(r) < d$, we can write $r = \sum_{j=0}^{d-1} c_j x^j$ for unique elements $c_j \in F$. Now we have $a = r + (f) = \left(\sum_{j=0}^{d-1} c_j x^j\right) + (f) = \sum_{j=0}^{d-1} \left(c_j x^j + (f)\right)$ $= \sum_{j=0}^{d-1} c_j \cdot \left(x^j + (f)\right) \in \langle B \rangle$, the vector space spanned by B over F. Since a was an arbitrary element of K, we have shown that $\langle B \rangle = K$.

[Show that 0_K has a unique representation as an F-linear combination of elements of B] Finally, suppose that $c_j \in F$ for j between 0 and $d-1$, and that $\sum_{j=0}^{d-1} c_j \cdot (x^j + (f)) = 0_K$. Then we have $r + (f) = 0_K = 0_R + (f)$, where $r = \sum_{j=0}^{d-1} c_j x^j$. Therefore, $r \in (f)$. So $r = fg$ for some $g \in R$. But $\deg(r) < \deg(f)$. It follows that $g = 0$ and $r = 0$. So $c_j = 0$ for all j. This argument also shows that the elements $b_j := x^j + (f)$ for $j \in \{0, \ldots, d-1\}$ are distinct, since if $b_i = b_j$ with $i \neq j$, then choosing $c_i = 1$, $c_j = -1$, and all other c_k to be 0 would result in the contradiction $c_i = c_j = 0$.

By Lemma 13.21, B is a basis for K over F. Thus, $\dim_F(K) = |B| = d$. $\square$

Although Theorem 15.10 may appear highly technical, it has a corollary which gets right to the heart of field theory: every non-constant polynomial over a field has a root in some extension field. Before proving this corollary, we need a lemma on factoring polynomials over a field.

Lemma 15.11. *Let F be a field, and let $f \in F[x] - F$ be a non-constant polynomial. Then f factors into irreducible polynomials: that is, we can write*

$$f = \prod_{j=1}^{n} f_i \qquad (15.3)$$

for some $n \in \mathbf{Z}^+$ where each f_i is irreducible in $F[x]$.

Proof. We proceed by induction on $d := \deg(f)$.

Base Case: $d = 1$. In this case, f itself is irreducible, by Exercise 14.6. So we may take $n = 1$ and $f_1 = f$.

Inductive Step: Suppose that the assertion of the lemma is true for all polynomials of degree less than $\deg(f)$. If f is irreducible, then we are done. So suppose that f is not irreducible in $F[x]$. This means that we can write $f = g \cdot h$ for some $g, h \in F[x]$ where neither g nor h is a unit in $F[x]$. Thus, $g, h \notin F$, so $\deg(g) \geq 1$ and $\deg(h) \geq 1$. Since $\deg(f) = \deg(g) + \deg(h)$ (by Lemma 14.10), we have $\deg(g) < \deg(f)$ and $\deg(h) < \deg(f)$. Thus our inductive hypothesis applies to g and h, so both g and h may be factored into products of irreducible polynomials in $F[x]$. Multiplying two such factorizations yields the desired factorization of f. □

Corollary 15.12 (Existence of Roots). *Let F be a field, and let $f \in F[x] - F$ be a non-constant polynomial. Then there exists an extension field K of F such that f has a root in K.*

Proof. By Lemma 15.11, there exists a polynomial $g \in F[x]$ such that g is irreducible in $F[x]$ and $g \mid f$. By Theorem 15.10, the field $K = F[x]/(g)$ is an extension field of F in which g has a root. But since g is a factor of f, every root of g is also a root of f; so f has a root in K. □

Example 15.13 (Finite Fields). Suppose that F is a finite field; that is, F is a field and $|F| < \infty$. We have already seen how to get our hands on such things: namely, take any prime $p \in \mathbf{Z}$; then by Corollary 12.25, $\mathbf{Z}/p\mathbf{Z}$ is a finite field, which has order p. Are there any finite fields whose order is not prime?

It turns out that there are. Our idea is to find an irreducible polynomial $f \in F[x]$ of some degree d. Then we can find an extension field K of F in which f has a root: namely, we can let $K = F[x]/(f)$. By Theorem 15.10, we will have $\dim_F(K) = d$. Suppose that $B = \{b_1, \ldots, b_d\}$ is a basis for K over F. Then every element $\alpha \in K$ has a unique representation as $\alpha = a_1 b_1 + \cdots + a_d b_d$ with $a_i \in F$. Since the number of distinct d-tuples $(a_1, \ldots, a_d) \in F^d$ is $|F^d| = |F|^d$, it follows that $|K| = |F|^d$.

The upshot of this discussion is that if we can find a polynomial $f \in F[x]$ which is irreducible in $F[x]$ and has degree d, then we can find an extension field K of F such that $|K| = |F|^d$. In this way, we may hope to find a field whose order is a power of a prime.

Let us compute a particular example. Let $F = \mathbf{Z}/(2)$, a finite field of order 2. Let us look for an irreducible polynomial in $F[x]$ of degree 2. There are exactly 4 polynomials of degree 2 in $F[x]$; three of these factor non-trivially: $x^2 = x \cdot x$; $x^2 + 1 = (x + 1) \cdot (x + 1)$ (remember that the coefficients of these polynomials are integers modulo two!); $x^2 + x = x \cdot (x + 1)$. This leaves $x^2 + x + 1 =: f$, which is irreducible in $F[x]$ by Exercise 14.7, since $f(0_F) = 1_F = f(1_F)$. So let $K = F[x]/(f)$. Then K is a field of order $2^2 = 4$.

By the proof of Theorem 15.10, the set $B = \{1_F + (f), x + (f)\}$ is a basis for K over F. Set $\rho = x + (f)$ and write $1 = 1_F + (f)$ (since $1_F + (f)$ is, after all, the multiplicative identity element of K). Then we have $B = \{1, \rho\}$. The four elements of K are the elements $a_1 \cdot 1 + a_2 \cdot \rho$ where $a_i \in F$. Since $F = \{0_F, 1_F\}$, we find $K = \{0, 1, \rho, 1 + \rho\}$. By Theorem 15.10, we know that ρ is a root of f, which means that

$$\rho^2 + \rho + 1 = 0. \tag{15.4}$$

We claim that all arithmetic in K is "modulo two." For example, we have $\rho + \rho = 1_F \cdot \rho + 1_F \cdot \rho = (1_F + 1_F) \cdot \rho = 0_F \cdot \rho = 0$. Even more to the point, we have $-1 = -1_K = -1_F + (f) = 1_F + (f) = 1_K = 1$ in K. This fact together with Equation 15.4 determines the structure of K completely. For example, we compute $\rho \cdot \rho = \rho^2 = -(\rho + 1) = \rho + 1$ (since $-1 = 1$ in K). In Exercise 15.4, the reader is asked to complete the addition and multiplication tables for the field K.

15.2   Splitting Fields

We can take the idea of Corollary 15.12 one step farther, by showing that every non-constant polynomial over a field F has *all* its roots in some appropriate extension field of F. First we make a related definition.

Definition 15.14. Let F be a subfield of K, and let $f \in F[x]$. We say that f *factors completely over* K or *splits (completely) over* K if we can write

$$f = c \cdot \prod_{i=1}^{n} (x - \rho_i) \tag{15.5}$$

for some $c \in F$ and some elements $\rho_i \in K$.

We note that in Equation 15.5, if $f \neq 0$, then we must have $n = \deg(f)$. Also, each ρ_i is a root of f in K; and the leading term of f must be cx^n.

Corollary 15.15. *Let F be a field, and let $f \in F[x]$. Then there is an extension field K of F such that f factors completely over K.*

Proof. We proceed by induction on $d := \deg(f)$.

Base Case: $d \leq 0$. Then $f \in F$, so we may take $K = F$ and write $f = c$, which has the form of Equation 15.5 with $n = 0$.

Inductive step: Suppose that $f \in F[x]$ and $\deg(f) \geq 1$. Suppose inductively that for every field $\tilde{F}$ and every polynomial in $\tilde{F}[x]$ of degree less than $\deg(f)$, there is an extension field of $\tilde{F}$ over which that polynomial factors completely.

By Corollary 15.12, there is an extension field L of F such that f has a root $\rho \in L$. By the Root-Factor Theorem (Theorem 14.24), there exists $g \in L[x]$ such that $f = (x - \rho) \cdot g$. Now we must have $\deg(g) = \deg(f) - 1$ by Lemma 14.10, so by inductive hypothesis, there is an extension field K of L such that g factors completely over K. It follows that f factors completely over K. $\square$

Now that we know that every polynomial over a field factors completely in some extension of that field, it is natural to ask for a "smallest" extension where a given polynomial factors completely. That is the aim of the following definition.

Definition 15.16. Let F be a field, and let $f \in F[x]$. An extension field K of F is called a *splitting field* for f over F if

(SF1) f factors completely over K, and

(SF2) K is minimal with respect to property (SF1): that is, there is no field L such that f factors completely over L and $F \leq L < K$.

We do not yet have any guarantee that a splitting field exists in general. Let us try to see how we could construct a splitting field.

Suppose that F is a field and $f \in F[x] - F$. Let E be an extension field of F such that f factors completely in E. Let $\rho_1, \ldots, \rho_n$ be the roots of f in E. Suppose that K is a splitting field for f over F, with $K \leq E$. Then f must factor completely in K. So it may seem reasonable that K should at least contain all of the roots ρ_i of f. Thus, our idea is to start with F and form the ring of all "polynomials in $\rho_1, \ldots, \rho_n$ with coefficients from F," which we will denote $F[\rho_1, \ldots, \rho_n]$.

In order to make this notation precise, we introduce two definitions. First, as with groups and ideals which can be generated by one element, we have a term to describe extension fields which can be generated by throwing in a single "new" element:

Definition 15.17. Let $F \leq K$ be fields. We say that K is a *simple* extension of F if $\exists \alpha \in K$ such that $K = F[\alpha]$.

Then we proceed to the case of several elements:

Definition 15.18. Let $F \leq K$ be fields, and let $\rho_1, \ldots, \rho_n$ be elements of K, wtih $n \in \mathbf{N}$. Then we recursively define $F[\rho_1, \ldots, \rho_n] = (F[\rho_1, \ldots, \rho_{n-1}])[\rho_n]$. We read $F[\rho_1, \ldots, \rho_n]$ as "F *adjoin* $\rho_1, \ldots, \rho_n$." When $n = 0$, we define $F[\rho_1, \ldots, \rho_n] = F$.

One nice feature of this construction is given by:

Lemma 15.19. *Let $F \leq K$ be fields, and let $\rho_1, \ldots, \rho_n$ be elements of K. Let $R = F[\rho_1, \ldots, \rho_n]$. Then R is the smallest subring of K which contains both the field F and all of the ρ_i.*

Proof. R certainly contains F and the ρ_i. On the other hand, any ring which contains F and the ρ_i must contain all polynomials in the ρ_i with coefficients in F, by closure of a ring under addition and multiplication. □

In Lemma 15.19, suppose that the ρ_i are all of the roots of a polynomial $f \in F[x]$, as before. Then we may expect that the ring $R = F[\rho_1, \ldots, \rho_n]$ is at least a "splitting ring" for f over F, if not a splitting *field*. But latent in our previous results is the amazing fact that, in this situation, R *must already be a field!*

Proposition 15.20. *Let $F \leq K$ be fields, and let $\rho_1, \ldots, \rho_n \in K$. Suppose that the ρ_i are algebraic over F. Then $F[\rho_1, \ldots, \rho_n]$ is a field; it is the smallest extension field of F in K which contains all the ρ_i.*

Proof. In light of Lemma 15.19, we need only show that $F[\rho_1, \ldots, \rho_n]$ is a field. We proceed by induction on n. We take our base case to be $n = 0$. In the base case, we have $F[\rho_1, \ldots, \rho_n] = F$, which is a field by hypothesis.

For the inductive step, suppose that $L := F[\rho_1, \ldots, \rho_n]$ is a field, and that $\rho_{n+1} =: \alpha$ is an element of K which is algebraic over F. By Exercise 15.1, α must also be algebraic over L. Set $f = \mathrm{Irr}(\alpha, L, x)$.

By Proposition 15.4, we have $L[\alpha] \cong L[x]/(f)$. Since f is irreducible over $L[x]$, Theorem 15.10 says that $L[x]/(f)$ is a field, as desired. □

Theorem 15.21. *Let $F \leq K$ be fields, and let $f \in F[x]$ be a non-zero polynomial which factors completely over K as*

$$f = c \cdot \prod_{i=1}^{n} (x - \rho_i) \qquad (15.6)$$

with $c \in F$ and $\rho_i \in K$. Then the ring $L := F[\rho_1, \ldots, \rho_n]$ is the unique splitting field for f in K.

Proof. We observe that the ρ_i are algebraic over F, since they are roots of the non-zero polynomial $f \in F[x]$. Therefore, by Proposition 15.20, L is a field. Now f factors completely over L, since $\rho_1, \ldots, \rho_n \in L$. Next, suppose that $\tilde{L}$ is another field such that f factors completely over $\tilde{L}$ and $F \leq \tilde{L} \leq K$. It will suffice to show that $\tilde{L} \supseteq L$. Since f factors completely over $\tilde{L}$, we can write

$$f = a \cdot \prod_{i=1}^{n} (x - \omega_i) \qquad (15.7)$$

with $a \in F$ and $\omega_i \in \tilde{L}$. Pick one of the ρ_j, and observe that, on the one hand, we have $f(\rho_j) = 0$ (from Equation 15.6), while on the other hand, we have $f(\rho_j) = a \cdot \prod_{i=1}^{n}(\rho_j - \omega_i)$ (from Equation 15.7). Therefore,

$$a \cdot \prod_{i=1}^{n}(\rho_j - \omega_i) = 0. \qquad (15.8)$$

Now all the terms in Equation 15.8 belong to K, which is a field, and thus a domain. Since $f \notin F$, we have $a \neq 0$. Therefore, we must have $\rho_j = \omega_i$ for some i. This shows that $\rho_j \in \tilde{L}$. Since j was arbitrary, we must have $\rho_1, \ldots, \rho_n \in \tilde{L}$. By Proposition 15.20, we can say that $\tilde{L} \supseteq L$. This completes the proof. $\qquad \square$

Corollary 15.22. *Let $f \in F[x]$, where F is a field. Then there exists a splitting field K for f over F.*

Proof. If $f \in F$, then we may take $K = F$. Otherwise, we combine Corollary 15.15 and Theorem 15.21 to get our result. $\qquad \square$

Example 15.23. Here is a tiny example to illustrate the preceding notions. Set $F = \mathbf{Q}$, $f(x) = x^2 - 2 \in F[x]$, and $K = \mathbf{R}$. Then f factors in $\mathbf{R}[x]$ as $f = (x - \sqrt{2})(x + \sqrt{2})$. Therefore, $\mathbf{Q}[\sqrt{2}, -\sqrt{2}]$ is a splitting field for f over $\mathbf{Q}$. Notice that we don't need to include $-\sqrt{2}$ here, since $-\sqrt{2} \in \mathbf{Q}[\sqrt{2}]$. Thus, $\mathbf{Q}[\sqrt{2}]$ is already a splitting field for f over $\mathbf{Q}$.

Suppose that F is a field and $\rho_1, \ldots, \rho_n$ are the roots of an irreducible polynomial $f \in F[x]$ in some extension field K of F. We can form the splitting field $F[\rho_1, \ldots, \rho_n]$ of f over F in a series of steps (as it was originally defined):

$$F[\rho_1, \ldots, \rho_n]$$

$$\vert$$

$$\vdots$$

$$\vert$$

$$F[\rho_1, \rho_2] \qquad\qquad (15.9)$$

$$\vert$$

$$F[\rho_1]$$

$$\vert$$

$$F$$

We refer to Diagram 15.9, naturally enough, as a *tower of fields*. (Sometimes, we may speak of a series of field extensions $F_1 \leq F_2 \leq \cdots \leq F_n$ as a "tower" even when it is displayed horizontally.)

Recall that whenever we have an extension of fields $L \leq K$, we may view K as a vector space over L, and thus speak of the dimension $\dim_L(K)$. This dimension is one of the most important invariants of a field extension. Befitting its importance, this invariant has more than one name and notation:

Definition 15.24. If $L \leq K$ are fields, then the *index of L in K*, also called the *degree of K over L*, is the integer $[K : L] := \dim_L(K)$. We say that K is a *finite* extension of L if $[K : L] < \infty$.

Now set $F_i = F[\rho_1, \ldots, \rho_i]$ for $i \in \{0, 1, \ldots, n\}$. (We take $F_0 = F$.) Note that ρ_i is algebraic over F_{i-1}, since ρ_i is algebraic over F and $F \leq F_{i-1}$. Set $f_i = \mathrm{Irr}(\rho_i, F_{i-1}, x) \in F_{i-1}[x]$. By Proposition 15.4, we have $F_{i-1}[\rho_i] \cong F_{i-1}[x]/(f_i)$. By Theorem 15.10, we then have $[F_i : F_{i-1}] = \deg(f_i)$ if $1 \leq i \leq n$. This last equation is useful enough to single out as a general lemma:

Lemma 15.25. *Let $L \leq K$ be fields, let $\alpha \in K$, and suppose that α is algebraic over L. Let $d = \deg(\mathrm{Irr}(\alpha, L, x))$. Then $\{1, \alpha, \alpha^2, \ldots, \alpha^{d-1}\}$ is a basis of $L[\alpha]$ over L, and $[L[\alpha] : L] = d$.*

Proof. Exercise 15.8. $\qquad\square$

How is the overall degree $[F_n : F]$ related to the intermediate degrees $[F_i : F_{i-1}]$? The answer is given by the following lemma, which says that degrees of field extensions multiply.

Lemma 15.26. *If $F \leq L \leq K$ are fields and the degrees $[K : L]$ and $[L : F]$ are finite, then $[K : F]$ is also finite, and we have*

$$[K : F] = [K : L] \cdot [L : F]. \tag{15.10}$$

Proof. Let $m = [L : F]$ and $n = [K : L]$. Then there are bases $A = \{a_1, \ldots, a_m\}$ of L over F and $B = \{b_1, \ldots, b_n\}$ of K over L. The idea is to show that the set $C := \{a_i b_j : 1 \leq i \leq m, 1 \leq j \leq n\}$ is a basis of K over F.

Let $\alpha \in K$. Since B is a basis of K over L, there exist unique elements $x_1, \ldots, x_n \in L$ such that $\alpha = \sum_{j=1}^n x_j b_j$. Since A is a basis of L over F, for each j there exist unique elements $w_{1,j}, \ldots, w_{m,j} \in F$ such that $x_j = \sum_{i=1}^m w_{i,j} a_i$. Thus, we can write $\alpha = \sum_{i=1}^m \sum_{j=1}^n w_{i,j} a_i b_j$. This shows that C spans K over F.

Next, suppose that we have $0 = \sum_{i,j} w_{i,j} a_i b_j$ with $w_{i,j} \in F$. We can rewrite this as $0 = \sum_{i=1}^m \left(\sum_{j=1}^n w_{i,j} b_j \right) a_i$. Since A is a basis of L over F, Lemma 13.21 implies that $\sum_{j=1}^n w_{i,j} b_j = 0$ for each i. But since B is a basis of K over L, the same lemma gives $w_{i,j} = 0$ for all i and j.

Our final point in this proof is that $|C| = mn$; that is, the elements $a_i b_j$ are distinct for distinct values of i and j. For if we had $a_i b_j = a_q b_r$ with $(i, j) \neq (q, r)$, then we would have $0 = 1_F \cdot a_i b_j + (-1_F) \cdot a_q b_r$, which contradicts the result obtained in the previous paragraph regarding the uniqueness of representation of 0. Using Lemma 13.21 in the backwards direction, we deduce that C is a basis of K over F. We can now say that $[K : F] = |C| = mn$, as desired. $\qquad\square$

Example 15.27. Let $F = \mathbf{Q}$ be our "ground field," as we sometimes refer to the smallest field in a tower. Let $f(x) = x^4 - 2 \in \mathbf{Q}[x]$. Then we can factor f in $\mathbf{C}[x]$ as follows: $f(x) = (x^2 - \sqrt{2}) \cdot (x^2 + \sqrt{2}) = (x - \sqrt[4]{2})(x + \sqrt[4]{2})(x - i\sqrt[4]{2})(x + i\sqrt[4]{2})$. Thus, f factors completely in $\mathbf{C}[x]$. The set of all roots of f in $\mathbf{C}$ is $R_f = \{\pm \sqrt[4]{2}, \pm i\sqrt[4]{2}\} = \{i^k \cdot \sqrt[4]{2} : k = 0, 1, 2, 3\}$. Let $K = \mathbf{Q}[\sqrt[4]{2}, -\sqrt[4]{2}, i\sqrt[4]{2}, -i\sqrt[4]{2}]$. By Theorem 15.21, K is the unique splitting field for f in $\mathbf{C}$.

Set $L = \mathbf{Q}[\sqrt[4]{2}]$. By Proposition 15.20, L is a field, since $\sqrt[4]{2}$ is algebraic over $\mathbf{Q}$; moreover, this proposition tells us that L is the smallest field which contains both $\mathbf{Q}$ and $\sqrt[4]{2}$. Since the real field $\mathbf{R}$ contains both $\mathbf{Q}$ and $\sqrt[4]{2}$, we must have $L \subseteq \mathbf{R}$.

We note that $-\sqrt[4]{2} \in \mathbf{Q}[\sqrt[4]{2}]$ since $-1 \in \mathbf{Q}$. Similarly, $-i\sqrt[4]{2} \in \mathbf{Q}[\sqrt[4]{2}, i\sqrt[4]{2}]$. Therefore, $\mathbf{Q}[\sqrt[4]{2}, i\sqrt[4]{2}] = K$. Consider the tower of field extensions

$$\mathbf{Q}\left[\sqrt[4]{2}, i\sqrt[4]{2}\right] = K$$

$$|$$

$$\mathbf{Q}\left[\sqrt[4]{2}\right] = L$$

$$|$$

$$\mathbf{Q} = F$$

We claim that $f = \mathrm{Irr}(\sqrt[4]{2}, \mathbf{Q}, x)$. We have $f \in \mathbf{Q}[x]$, f is monic, and $f(\sqrt[4]{2}) = 0$. Assume for a contradiction that f factors in $\mathbf{Q}[x]$ as $f = g \cdot h$, where $g, h \in \mathbf{Q}[x] - \mathbf{Q}$. We may assume (after multiplying g and h by suitable non-zero rational numbers) that g and h are monic. By the Root-Factor Theorem (Theorem 14.24), neither g nor h can be linear, since f has no roots in $\cdot\mathbf{Q}$. Therefore, $\deg(g) = \deg(h) = 2$. The four distinct roots of f in $\mathbf{C}$ must be roots of g or h. Since g and h can have at most two distinct roots each (by Theorem 14.25), we must have $g(\rho_1) = g(\rho_2) = 0$ for two distinct roots ρ_1 and ρ_2 of f in $\mathbf{C}$. This forces $g = (x - \rho_1)(x - \rho_2)$ (see Exercise 15.11). Since $g \in \mathbf{Q}[x]$, we must have $\rho_1\rho_2 \in \mathbf{Q}$. Looking at the roots of f in $\mathbf{C}$, we see that the product $\rho_1\rho_2$ of two of them must be $\pm\sqrt{2}$ or $\pm i\sqrt{2}$, none of which are in $\mathbf{Q}$. This contradiction shows that f is irreducible in $\mathbf{Q}[x]$. Thus, by Exercise 15.2, $f = \mathrm{Irr}(\sqrt[4]{2}, \mathbf{Q}, x)$, as claimed. From Lemma 15.25 it now follows that $[L : \mathbf{Q}] = [\mathbf{Q}[\sqrt[4]{2}] : \mathbf{Q}] = \deg(f) = 4$.

Since $K \not\subseteq \mathbf{R}$ but $L \subseteq \mathbf{R}$, we must have $K \neq L$. By Exercise 15.5, this forces $[K : L] > 1$. On the other hand, $i\sqrt[4]{2}$ is a root of the polynomial $x^2 + \sqrt{2} \in L[x]$. Therefore, $\deg(\mathrm{Irr}(i\sqrt[4]{2}, L, x)) \leq 2$, and so $[K : L] \leq 2$. It follows that $[K : L] = 2$. By Lemma 15.26, we have $[K : \mathbf{Q}] = 8$.

Remark 15.28. Exercise 15.2 tells us that f is the irreducible polynomial over $\mathbf{Q}$ for *any* of the roots of f, not just for $\sqrt[4]{2}$. Thus, for instance, we have $f = \mathrm{Irr}(i\sqrt[4]{2}, \mathbf{Q}, x)$. It follows that $[\mathbf{Q}[i\sqrt[4]{2}] : \mathbf{Q}] = 4$. Yet notice that $[L[i\sqrt[4]{2}] : L] = [K : L] = 2$. In general, adding an algebraic quantity to an extension field carries less of a penalty in terms of degree than adding that same quantity to the ground field; this is because a polynomial which is irreducible over the ground field may factor non-trivially over an extension field. Further, this phenomenon always occurs when we add successive roots of the same irreducible polynomial to a given ground field; for by the Root-Factor Theorem, we are peeling off at least one linear factor every time we add a root to an extension field. For details, see Exercise 15.10.

Remark 15.29. The hardest part of Example 15.27 was showing that the polynomial f is irreducible in $\mathbf{Q}[x]$. In general, it can be difficult to decide whether a given polynomial is irreducible over a given field; there is, however, a test for irreducibility called *Eisenstein's criterion*, which applies in this example (see Lemma 21.5).

15.3 Exercises

Exercise 15.1. Suppose that $F \leq L \leq K$ is a tower of field extensions, $\alpha \in K$, and α is algebraic over F. Prove that α is also algebraic over L.

Exercise 15.2. Let $F \leq K$ be fields and let $\alpha \in K$. Suppose that $f(x) \in F[x]$ is a non-zero monic polynomial, irreducible in $F[x]$, and that $f(\alpha) = 0$. Prove that $f = \text{Irr}(\alpha, F, x)$.

Exercise 15.3. Assume the hypotheses and notation of Theorem 15.10. Justify the following statements involving computations in K.

  (a) $(x + (f))^2 = x^2 + (f)$.
  (b) For all positive integers j, we have $(x + (f))^j = x^j + (f)$.
  (c) For all $a \in F$ and all positive integers j, we have $a \cdot (x^j + (f)) = ax^j + (f)$. (Hint: Recall how we are identifying F with a certain subfield of K.)
  (d) By now you should find it straightforward to justify Equation 15.2; do this.

Exercise 15.4. Compute the addition and multiplication tables for the finite field K of order 4 described in Example 15.13.

Exercise 15.5. Let $F \leq K$ be fields. Prove that $[K : F] = 1$ iff $K = F$.

Exercise 15.6 (Finite Implies Algebraic). Let $F \leq K$ be fields with $[K : F] < \infty$. Prove that K is algebraic over F.

Exercise 15.7. Let K be a finite extension field of F. Prove that there exist finitely many elements $\alpha_1, \ldots, \alpha_n$ of K such that $K = F[\alpha_1, \ldots, \alpha_n]$.

Exercise 15.8. Prove Lemma 15.25.

Exercise 15.9. Let $F \leq L \leq K$ be fields such that $[K : F]$ is finite. Prove that both $[K : L]$ and $[L : F]$ are finite. (This allows us, for example, to use Lemma 15.26 under more general hypotheses than those stated in that Lemma.)

Exercise 15.10. Let F be a field, and let $f \in F[x]$, with $\deg(f) = n \geq 1$. Write $f = c \cdot \prod_{i=1}^{n}(x - \rho_i)$ with $c \in F$ and $\rho_1, \ldots, \rho_n$ in some splitting field K for f over F, and set $F_i = F[\rho_1, \ldots, \rho_i]$ for $i : 0 \leq i \leq n$. Note that $F_0 = F$ (since we are adjoining no extra elements in that case) and $F_n = K$ (by Theorem 15.21). Prove that we have $[F_i : F_{i-1}] \leq n - i + 1$ for all $i : 1 \leq i \leq n$. Conclude that $[K : F] \leq n!$.

Exercise 15.11. Let F be a field and $f \in F[x]$.

(a) Suppose that α and β are distinct roots of f in F. Prove that $(x - \alpha)(x - \beta)$ divides f in $F[x]$.

(b) Generalize part (a) to the situation where $\alpha_1, \ldots, \alpha_r$ are r distinct roots of f in F. (This result can be seen as a step towards proving a unique factorization theorem in the ring $F[x]$; we will examine such things more thoroughly in Section 20.3.)

Exercise 15.12. Prove that in Example 15.27, the splitting field K in $\mathbf{C}$ for $x^4 - 2$ over $\mathbf{Q}$ can also be written as $K = \mathbf{Q}[\sqrt[4]{2}, i]$.

Exercise 15.13. Let $F \leq K$ be fields with $[K : F] < \infty$, and suppose that F is infinite and that there are only finitely many fields L between F and K. (This hypothesis may seem strange, but it occurs in many important cases; for an example, see the proof of Theorem 16.32, the Fundamental Theorem of Galois Theory.) Prove that K is a simple extension of F. Hint: Exercise 13.10 could be useful here.

Exercise 15.14. Let $F \leq K$ be fields with $[K : F] = r < \infty$, and suppose that F is finite, of order n.

(a) Prove that $|K| = n^r$.

(b) Prove that every non-zero element of K is a root of the polynomial $f_r(x) := x^{n^r - 1} - 1 \in F[x]$.

(c) Let $\alpha \in K - (0)$ and suppose that $F[\alpha] \neq K$. Prove that there is a proper factor m of r such that α is a root of the polynomial $f_m(x)$.

(d) Using the crude estimate that the number of proper factors of r is at most $r/2$, prove that $\exists \alpha \in K - (0)$ such that $F[\alpha] = K$; that is, every finite extension of a finite field is simple. Hint: The inequality $t(n^t - 1) < n^{2t} - 1$ is true for real numbers $t \geq 1$ (show this!).

Exercise 15.15. In this exercise, you will show that the field of complex numbers $\mathbf{C}$ has no quadratic extensions (that is, extension fields of degree 2 over $\mathbf{C}$). Essentially, the reason is that $\mathbf{C}$ already contains square roots of all its elements.

(a) Let $\alpha \in \mathbf{C}$. Let $r = |\alpha|$, the absolute value of α. Show that we can write $\alpha = r(\cos(\theta) + i\sin(\theta))$ for some real number $\theta \in [0, 2\pi)$.

(b) Let $\beta = \sqrt{r}(\cos(\theta/2) + i\sin(\theta/2))$. Show that $\beta^2 = \alpha$.

(c) Prove that every monic quadratic polynomial $f \in \mathbf{C}[x]$ factors non-trivially in $\mathbf{C}[x]$ as a product of linear polynomials.

(d) Conclude that there is no field F such that $[F : \mathbf{C}] = 2$.

Exercise 15.16. We have seen (Example 15.8 combined with Lemma 15.25) that the index $\lfloor \mathbf{Q}[\sqrt{2}] : \mathbf{Q} \rfloor$ is 2 as a *field extension*. But what is this index when $\mathbf{Q}$ and $\mathbf{Q}[\sqrt{2}]$ are considered as groups under addition?

Exercise 15.17. In this exercise (which feels more like analysis than algebra), you will establish the existence of real numbers which are transcendental over $\mathbf{Q}$. Let $f \in \mathbf{Q}[x]$ be a polynomial of degree $r > 0$, and suppose that α is a root of f in $\mathbf{R}$.

(a) Let $D \in \mathbf{Z}^+$ be a common denominator for f : that is, $D \cdot f \in \mathbf{Z}[x]$. Show that if $q = n/d \in \mathbf{Q}$ with $n, d \in \mathbf{Z}$, $d \neq 0$, and $f(q) \neq 0$, then we have $|f(q)| \geq 1/(d^r D)$.

(b) Show that there is a constant M which depends only on f such that $|f(\alpha + \delta)| \leq M \cdot |\delta|$ for every $\delta \in [-1, 1]$.

The idea now is to exploit the tension between the inequalities from parts (a) and (b) above, which go in opposite directions, by constructing a real number α that is absurdly close to—but not *equal* to—an entire sequence of rational numbers. To this end:

(c) Set $\alpha = \sum_{j=1}^{\infty} 2^{-j!}$. Show (e.g. by comparison with a geometric series) that the series for α is convergent, and that $|\alpha - \alpha_k| < 2^{1-(k+1)!}$, where α_k is the kth partial sum for α, i.e., $\alpha_k = \sum_{j=1}^{k} 2^{-j!}$.

(d) Assume for a contradiction that α is algebraic over $\mathbf{Q}$, and set $f = \text{Irr}(\alpha, \mathbf{Q}, x)$. Note that $\alpha_k \in \mathbf{Q}$, and reach a contradiction by substituting α_k for q in part (a) and $\alpha_k - \alpha$ for δ in part (b).

Exercise 15.18. Let $F \leq K$ be a field extension.

(a) Prove that if $\alpha \in K$ and α is algebraic over F, then $F[\alpha]$ is a field which is algebraic over F.

(b) Prove that if $\alpha_1, \alpha_2 \in K$ and each α_i is algebraic over F, then $F[\alpha_1, \alpha_2]$ is a field and is algebraic over F.

(c) Set $L = \{\alpha \in K \mid \alpha \text{ is algebraic over } F\}$. Use part (b) above to prove that L is a field and $F \leq L \leq K$.

16

Galois Theory

16.1   Field Embeddings

Evariste Galois is credited with inventing the concept of a group, as well as the notions of normal subgroup, symmetric groups, and a host of other ideas central to abstract algebra. Yet even among these epochal inventions, the contribution of Galois to the theory of automorphisms of field extensions is so astonishingly beautiful that it was singled out to bear his name: Galois Theory. In this wonderful branch of mathematics, many of the key ideas of algebra come together to create a symphony of incomparable splendor.

We shall see that automorphisms of fields are intimately related to roots of polynomials. Before we study automorphisms, however, we will step back to the more general case of field isomorphisms. Since we know that the only ideals of a field L are (0) and the field L itself (by Exercise 11.7), then the only possibilities for a ring homomorphism $L \to R$ are embeddings and the zero map; thus, we do not really lose anything by speaking of field embeddings. The first result in this chapter says that any field isomorphism naturally extends to an isomorphism of the corresponding polynomial rings.

Lemma 16.1. *Let $\sigma : L \to M$ be an isomorphism of fields. Then σ extends to an isomorphism $\tilde{\sigma} : L[x] \to M[x]$, where*

$$\tilde{\sigma}\left(\sum_{j=0}^{n} c_j x^j\right) := \sum_{j=0}^{n} \sigma(c_j) x^j. \tag{16.1}$$

Proof. Since $M \leq M[x]$, we can consider σ to be an embedding $\sigma : L \hookrightarrow M[x]$. By Exercise 14.10, there is a unique extension of σ to a ring homomorphism $\tilde{\sigma} : L[x] \to M[x]$ such that $\tilde{\sigma}(x) = x$. Exercise 16.1 asks the reader to complete the proof. $\qquad\square$

Our approach to the study of automorphisms of fields is somewhat indirect. Given a field K, we wish to study $\mathrm{Aut}(K)$, the group of automorphisms of K. More generally, given a field extension $F \leq K$, we wish to study those automorphisms of K which restrict to the identity map on F; we call such things *automorphisms of K over F*. In even more generality, we define:

DOI: 10.1201/9781003252139-16

Definition 16.2. Let R, S, and T be commutative rings with 1 such that $R \leq S$ and $R \leq T$. A ring homomorphism $\sigma : S \to T$ is said to be *over R* if $\sigma(a) = a$ for every $a \in R$.

The set of all automorphisms of K over F has a special notation:

Notation 16.3. Let F and K be fields with $F \leq K$. Then

$$\text{Aut}(K/F) := \{\sigma \in \text{Aut}(K) : \forall a \in F, \ \sigma(a) = a\}, \tag{16.2}$$

the *automorphism group of K over F* (see Exercise 16.2).

Remark 16.4. The notation K/F here does *not* denote any sort of quotient. When we say that an automorphism is "over F," we are using the term "over F" as an *adjective* describing a property of the automorphism, not performing division. Here, the sense of "over F" is that F is our ground field, which we consider to be inviolate, and not subject to the whims of an automorphism; F is also called the *fixed field*, for evident reasons.

Our primary case of interest is when the extension is finite: that is, when $[K : F] < \infty$. Our strategy will be to start at the bottom, with the identity automorphism $\text{id}_F : F \to F$. Since $[K : F] < \infty$, we can construct K by adjoining a finite number of elements of K to F; that is, we can write

$$K = F[\alpha_1, \dots, \alpha_r] \tag{16.3}$$

for some $\alpha_1, \dots, \alpha_r \in K$ (by Exercise 15.7). We successively ask, for increasing values of i, what are the embeddings of $F[\alpha_1, \dots, \alpha_i]$ into K over F? When we reach $i = r$, we will know all the embeddings of K into K over F, which are simply the automorphisms of K over F (according to Exercise 16.22).

At a typical intermediate stage, we will know all the embeddings of $L := F[\alpha_1, \dots, \alpha_i]$ into K over F, and we will want to understand the embeddings of $L[\alpha_{i+1}]$ into K. Now, an embedding $\sigma : L \hookrightarrow K$ has an image $M = \sigma(L)$ with the property that $\sigma : L \to M$ is an isomorphism. Writing α for α_{i+1}, we note that α is algebraic over L by Exercise 15.6. Let $f = \text{Irr}(\alpha, L, x)$. An extension of σ to an embedding $\tau : L[\alpha] \hookrightarrow K$ is completely determined by what it does to α. Since α is a root of the polynomial $f \in L[x]$, then we might expect that $\tau(\alpha)$ is a root of $\tilde{\sigma}(f) \in M[x]$. The following result confirms our suspicions in a quite satisfying way.

Theorem 16.5 (Extending Field Isomorphisms). *Suppose that L and M are subfields of the fields $\tilde{L}$ and $\tilde{M}$, respectively. Suppose that $\sigma : L \to M$ is an isomorphism. Let $\alpha \in \tilde{L}$ and suppose that α is algebraic over L. Let $\tilde{\sigma} : L[x] \to M[x]$ be as in Lemma 16.1. Set $f = \text{Irr}(\alpha, L, x)$ and $g = \tilde{\sigma}(f) \in M[x]$. Then g is irreducible in $M[x]$, and the extensions of σ to an embedding of $L[\alpha]$ into $\tilde{M}$ correspond bijectively with the roots of g in $\tilde{M}$. Namely, for every root β of g in $\tilde{M}$, there is a unique extension of σ to an isomorphism from $L[\alpha]$ to $M[\beta]$ which maps α to β; and every extension of σ to an embedding of $L[\alpha]$ into $\tilde{M}$ is of this form.*

Proof. By Lemma 16.1, we can extend σ to an isomorphism $\tilde{\sigma} : L[x] \to M[x]$ sending x to itself. Set $g = \tilde{\sigma}(f)$. Let $\beta \in \tilde{M}$ be a root of g. The idea of the proof is to show that the unique candidate for the desired map $\tau : L[\alpha] \to M[\beta]$, namely

$$\sum_{j=0}^{n} c_j \alpha^j \mapsto \sum_{j=0}^{n} \sigma(c_j)\beta^j, \tag{16.4}$$

is a well-defined isomorphism. To accomplish this, it is convenient to first step back to the level of polynomial rings. Now let $\psi : M[x] \to M[\beta]$ be the evaluation-at-β map. Set $\tilde{\tau} = \psi \circ \tilde{\sigma}$. It is not hard to verify (Exercise 16.3) that $\mathrm{Irr}(\beta, M, x) = g$. Therefore, we have $\ker(\psi) = (g)$. It follows that $\ker(\tilde{\tau}) = \tilde{\sigma}^{-1}((g)) = (f)$. By the Fundamental Theorem of Ring Homomorphisms (Theorem 11.36), $\tilde{\tau}$ induces an isomorphism $L[x]/(f) \cong \tilde{\tau}(L[x]) = M[\beta]$. But $L[x]/(f) \cong L[\alpha]$ via the map sending x to α (see Proposition 15.4). It follows that there is an isomorphism $\tau : L[\alpha] \to M[\beta]$ which extends σ and which sends α to β. The map τ must be given by Formula 16.4, in order to be a ring homomorphism with the properties just stated. This shows that every root β of g in $\tilde{M}$ gives a distinct embedding of $L[\alpha]$ into $\tilde{M}$ which extends σ.

To complete the proof, let $\alpha_1, \ldots, \alpha_r$ be the distinct roots of g in $\tilde{M}$. Suppose that $\tau : L[\alpha] \hookrightarrow \tilde{M}$ is an embedding which extends σ; that is, $\tau|_L = \sigma$. Let $\beta = \tau(\alpha) \in \tilde{M}$. Write $f = \sum_{j=0}^{d} b_j x^j$ with $b_j \in L$. Then $g = \tilde{\sigma}(f) = \sum_{j=0}^{d} \sigma(b_j)x^j$. Thus, we have $\tau(f(\alpha)) = \tau\left(\sum_{j=0}^{d} b_j \alpha^j\right) = \sum_{j=0}^{d} \tau(b_j)\tau(\alpha^j) = \sum_{j=0}^{d} \tau(b_j)(\tau(\alpha))^j = \sum_{j=0}^{d} \tau(b_j)\beta^j = \sum_{j=0}^{d} \sigma(b_j)\beta^j = g(\beta)$. But also, we have $f(\alpha) = 0$ since $f = \mathrm{Irr}(\alpha, L, x)$, and so $\tau(f(\alpha)) = \tau(0) = 0$. It follows that $g(\beta) = 0$. Thus $\beta = \alpha_i$ for some i, and τ is the map given by Formula 16.4. $\square$

To get some idea of what Theorem 16.5 is saying, let us consider the special case when $L = M$ and $\sigma = \mathrm{id}_L$ is the identity map. So let L be any field, let $f \in L[x]$ be irreducible in $L[x]$, and let α, β be roots of f in some extension field $\tilde{L}$ of L. Then Theorem 16.5 asserts the existence of an isomorphism $\tau : L[\alpha] \to L[\beta]$ making the following diagram commute (the vertical maps are the natural embeddings):

$$
\begin{array}{ccc}
L[\alpha] & \xrightarrow{\ \tau\ } & L[\beta] \\
\uparrow & & \uparrow \\
L & \xrightarrow{\ \mathrm{id}_L\ } & L
\end{array}
\tag{16.5}
$$

The big idea here is that any two roots α and β of the irreducible polynomial f "look alike over L": there is an isomorphism $\tau : L[\alpha] \to L[\beta]$ which fixes L pointwise and sends α to β.

Example 16.6. The complex numbers i and $-i$ "look alike" over the real field $\mathbf{R}$. Both quantities are roots of the polynomial $x^2 + 1$, which is irreducible in $\mathbf{R}[x]$. Consequently, there is an isomorphism $\tau : \mathbf{R}[i] \to \mathbf{R}[-i]$ which fixes

$\mathbf{R}$ pointwise and sends i to $-i$. Now $\mathbf{R}[i] = \mathbf{R}[-i] = \mathbf{C}$, and the map τ is just complex conjugation.

Motivated by Example 16.6, we make the following definition.

Definition 16.7. Let F be a subfield of K and let $\alpha, \beta \in K$. We say that α and β are *conjugate over* F if there exists an isomorphism $\tau : F[\alpha] \to F[\beta]$ such that τ is the identity map on F and $\tau(\alpha) = \beta$.

Remark 16.8. The definition of conjugate above is distinct from the definition of conjugate in group theory, Definition 7.15. It should be clear from context which definition is meant. In general, a "conjugate" is something closely related to—conjoined or coupled with—some original object. In both groups and fields, we note that a conjugate is the image of the original object under an isomorphism.

Example 16.9. The numbers $\sqrt{2}$ and $-\sqrt{2}$ are conjugate over $\mathbf{Q}$. For both numbers are roots of the polynomial $x^2 - 2$, which is irreducible in $\mathbf{Q}[x]$. Compare this discussion to the results we obtained earlier about the number game on $\mathbf{Z}[\sqrt{2}]$, or, equivalently, about the group $\mathrm{Aut}(\mathbf{Z}[\sqrt{2}])$.

Next we utilize Theorem 16.5 to prove a lemma which will be used later in this chapter.

Lemma 16.10. *Suppose that* $\sigma : L \hookrightarrow K$ *is an embedding of fields, and that* $\tilde{L}$ *is a finite extension field of* L. *Then there exists an extension field* $\tilde{K}$ *of* K *and an embedding* $\tau : \tilde{L} \hookrightarrow \tilde{K}$ *such that* τ *extends* σ.

Proof. We proceed by induction on $[\tilde{L} : L]$.

Base Case: $[\tilde{L} : L] = 1$. Then by Exercise 15.5, we have $\tilde{L} = L$, so we may take $\tilde{K} = K$ and $\tau = \sigma$.

Inductive Step: Suppose $[\tilde{L} : L] > 1$. Then $\exists \alpha \in \tilde{L} - L$. By Exercise 15.6, α is algebraic over L. Set $M = \sigma(L)$, so that σ is an isomorphism from L to M. Let $\tilde{\sigma} : L[x] \to M[x]$ be the isomorphism of Lemma 16.1. Set $f = \mathrm{Irr}(\alpha, L, x)$ and $g = \tilde{\sigma}(f) \in M[x] \subseteq K[x]$. Note that $g \notin K$, so by Corollary 15.12, there exists an extension field K' of K such that g has a root β in K'. By Theorem 16.5, there is an isomorphism $\sigma' : L[\alpha] \to M[\beta]$ which extends σ. Now set $L' = L[\alpha]$, and (for consistency of notation) $\tilde{L}' = \tilde{L}$. Then we have $[\tilde{L}' : L'] = [\tilde{L} : L[\alpha]] = [\tilde{L} : L] / [L[\alpha] : L] < [\tilde{L} : L]$. Thus, by inductive hypothesis applied to the primed entities, there exists an extension field $\tilde{K}$ of K' and an embedding $\tau : \tilde{L}' \hookrightarrow \tilde{K}$ such that τ extends σ'. Finally, $\tilde{K}$ is an extension field of K, τ is an embedding of $\tilde{L}$ into $\tilde{K}$, and τ extends σ, as required. $\qquad\square$

16.2 Separable Extensions

In order to understand Theorem 16.5 better, if we are given an irreducible polynomial $g \in M[x]$, then we would like to know how many roots g has in a given extension field $\tilde{M}$ of M.

The number of roots of g in $\tilde{M}$ is a function of both $\tilde{M}$ and g. First, $\tilde{M}$ may simply not contain some of the roots of g; this problem can be overcome by extending M until g factors completely, and we will consider this aspect of the situation in the following section.

Second, we would like to know how many distinct roots g has when we do manage to factor g completely. For this quantity provides a fundamental limit on the number of roots of g in any field. Most of the time, we expect that a polynomial of degree n should have n distinct roots, but of course this does not always happen.

For example, let $f(x) = (x-2)^2 \cdot (x+3) \cdot (x-9) \in \mathbf{R}[x]$. Then $\deg(f) = 4$, but f has only 3 distinct roots, since the number 2 is a double root of f. Ideas from calculus turn out to be useful in identifying this type of situation. Taking the derivative using the product rule, we find

$$f'(x) = 2(x-2) \cdot (x+3) \cdot (x-9) + (x-2)^2 \cdot 1 \cdot (x-9) + (x-2)^2 \cdot (x+3) \cdot 1. \quad (16.6)$$

Notice that $x-2$ is a factor of f' as well as of f, so we have $f(2) = f'(2) = 0$.

In order to apply this idea more generally, we want to define the notion of derivative for polynomials over arbitrary commutative rings with 1. But we cannot use the standard definition of a derivative as a limit, since we have no notion of "limit" in general rings. Instead, we simply define the derivative of a polynomial by using the usual formula from calculus.

Definition 16.11. Let R be a commutative ring with 1, and let $f \in R[x]$. Write $f = \sum_{j=0}^{n} a_j x^j$ with $a_j \in R$. The *derivative of f with respect to x* is the polynomial

$$f' = \sum_{j=1}^{n} j \cdot a_j x^{j-1} \in R[x]. \quad (16.7)$$

Remark 16.12. In Equation 16.7, the notation $j \cdot a_j$ stands for $a_j + a_j + \cdots + a_j$ (j terms). Recall that we use this multiplicative notation in place of exponential notation in a commutative group whose operation is denoted $+$ (addition).

Using the above definition of derivative, the product rule for derivatives of polynomials is still true in our general setting:

Lemma 16.13 (Product Rule). *Let R be a commutative ring with 1, and let $f, g \in R[x]$. Then we have $(fg)' = f'g + fg'$.*

Proof. Write $f = \sum_{j=0}^{n} a_j x^j$, $g = \sum_{j=0}^{m} b_j x^j$ with $a_j, b_j \in R$. Then we have $f' = \sum_{j=0}^{n-1} \hat{a}_j x^j$ and $g' = \sum_{j=0}^{m-1} \hat{b}_j x^j$, where $\hat{a}_j = (j+1)a_{j+1}$ and

$\hat{b}_j = (j+1)b_{j+1}$. By Definition 14.1, we have $fg = \sum_{k=0}^{m+n} c_k x^k$, where $c_k = \sum_{j=0}^{k} a_j b_{k-j}$. So $(fg)' = \sum_{k=1}^{m+n} k c_k x^{k-1}$, and the coefficient of x^k in $(fg)'$ is $(k+1)c_{k+1} = \sum_{j=0}^{k+1}(k+1)a_j b_{k+1-j}$. Similarly, the coefficient of x^k in $f'g$ is $\sum_{j=0}^{k} \hat{a}_j b_{k-j} = \sum_{j=0}^{k}(j+1)a_{j+1}b_{k-j} = \sum_{j=1}^{k+1} ja_j b_{k-j+1}$, and the coefficient of x^k in fg' is $\sum_{j=0}^{k} a_j \hat{b}_{k-j} = \sum_{j=0}^{k} a_j(k-j+1)b_{k-j+1}$. Thus, the coefficient of x^k in $f'g + fg'$ is $\sum_{j=0}^{k+1}(k+1)a_j b_{k+1-j}$, which matches the coefficient of x^k in $(fg)'$. $\qquad\square$

Definition 16.14. Let F be a subfield of K, and let $\alpha \in K$. Let $f \in F[x]$. We say that α is a *repeated root* or *multiple root* of f if $(x-\alpha)^2$ divides f in $K[x]$. The *multiplicity* of α as a root of f is the greatest integer m such that $(x-\alpha)^m$ divides f in $K[x]$.

Proposition 16.15. *Let F be a subfield of K, and let $\alpha \in K$. Let $f \in F[x]$. Then*

(i) α is a repeated root of f iff $f(\alpha) = f'(\alpha) = 0$.

(ii) If f is irreducible in $F[x]$ and α is a repeated root of f, then $f'(x) = 0$.

Proof. (i) ($\Rightarrow$): Suppose that α is a repeated root of f. This means that $f = (x-\alpha)^2 \cdot g$ for some $g \in K[x]$. So we have $f(\alpha) = (\alpha-\alpha)^2 \cdot g(\alpha) = 0$. Moreover, using the Product Rule, we have $f'(x) = ((x-\alpha)^2)' \cdot g(x) + (x-\alpha)^2 \cdot g'(x) = 2(x-\alpha)g(x) + (x-\alpha)^2 \cdot g'(x)$, and so $f'(\alpha) = 0$ also.

($\Leftarrow$): Suppose that $f(\alpha) = f'(\alpha) = 0$. Then by Theorem 14.24, we can say that $x - \alpha$ divides f in $K[x]$. So $f = (x-\alpha) \cdot h$ for some $h \in K[x]$. Taking derivatives, we find $f' = 1 \cdot h + (x-\alpha) \cdot h'$. Therefore, $0 = f'(\alpha) = h(\alpha)$. By Theorem 14.24 again, we have $h = (x-\alpha) \cdot g$ for some $g \in K[x]$. Thus $f = (x-\alpha)^2 \cdot g$, so α is a repeated root of f.

(ii) Suppose that f is irreducible in $F[x]$ and α is a repeated root of f. Then by part (i), we have $f(\alpha) = f'(\alpha) = 0$. Consider the evaluation-at-α map $\varepsilon_\alpha : F[x] \to K$. By Proposition 15.4 and Definition 15.5, we know that $\ker(\varepsilon_\alpha) = (g)$ where $g = \mathrm{Irr}(\alpha, F, x)$ is the irreducible polynomial of α over F. Now both f and f' are in $\ker(\varepsilon_\alpha)$, so we have $g \mid f$ and $g \mid f'$ in $F[x]$. Therefore, $f = g \cdot h$ for some polynomial $h \in F[x]$. Since f and g are irreducible in $F[x]$, we must have $h \in (F[x])^\times = F^\times$. Thus $\deg(f) = \deg(g)$. But since $g \mid f'$, we can write $f' = g \cdot q$ for some $q \in F[x]$. Now $\deg(f') < \deg(f)$, while if $q \neq 0$ then $\deg(g \cdot q) = \deg(g) + \deg(q) = \deg(f) + \deg(q) \geq \deg(f)$, and thus $\deg(g \cdot q) > \deg(f')$, a contradiction. It follows that $q = 0$, and so $f' = 0$, as desired. $\qquad\square$

Example 16.16. Proposition 16.15 (ii) leads us to ask, when does $f'(x) = 0$ in $F[x]$? Certainly if $f \in F$, that is, if f is constant, then $f' = 0$. In classical calculus—if $F \leq \mathbf{C}$—this is the only way to get $f'(x) = 0$.

But consider $f(x) = x^p - 1 \in \mathbf{F}_p[x]$, where $\mathbf{F}_p = \mathbf{Z}/p\mathbf{Z}$ and p is a prime of $\mathbf{Z}$. Here we have $f'(x) = px^{p-1} - 0 = 0$, yet $f \notin \mathbf{F}_p$. For example, when

$p = 2$, we have $f(x) = x^2 - 1 = (x - 1)^2$. More generally, it turns out that $x^p - 1 = (x - 1)^p$ in $\mathbf{F}_p[x]$ (see Corollary 21.21).

We note that none of the polynomials f given in this example is actually irreducible. To find an example of an irreducible polynomial f such that $f'(x) = 0$, we must look further afield—to so-called *function fields*; see Exercise 22.14.

To capture the condition that a given field extension is well-behaved with respect to the number of roots of its associated irreducible polynomials, namely, that the number of roots is equal to the degree, we make the following definition.

Definition 16.17. Let K be an algebraic extension field of F. For an individual element $\alpha \in K$, we say that α is *separable over F* if $\mathrm{Irr}(\alpha, F, x)$ has no repeated roots. We say that K is separable over F if for all $\alpha \in K$, α is separable over F.

Remark 16.18. In Definition 16.17, we did not specify *where* to look for repeated roots. It turns out that this does not matter. By Exercise 16.6, if an irreducible polynomial $f = \mathrm{Irr}(\alpha, F, x)$ has a repeated root in any field whatsoever, then all roots of f are repeated roots, including α itself. This accords with our principle that all roots of an irreducible polynomial look the same over the ground field.

Example 16.19. Suppose that F and K are fields between $\mathbf{Q}$ and $\mathbf{C}$, so that $\mathbf{Q} \le F \le K \le \mathbf{C}$, and that K is algebraic over F. Let $\alpha \in K$, and let $f = \mathrm{Irr}(\alpha, F, x)$. Since f is irreducible in $F[x]$, we must have $f \notin F$, because the definition of *irreducible* excludes 0 and units. Now since $F \le \mathbf{C}$, we have $f \in \mathbf{C}[x] - \mathbf{C}$, so $f'(x) \ne 0$. Thus by Proposition 16.15, f has no repeated roots. We conclude that K is separable over F.

Looking at Example 16.16, the reader may notice that what allowed f' to be 0 without f being constant was that, in $\mathbf{F}_p[x]$, multiplying something non-zero by a non-zero integer could turn out to give zero. This does not contradict the fact that $\mathbf{F}_p[x]$ is a domain, as guaranteed by Theorem 14.10; for the ring of ordinary integers $\mathbf{Z}$ is not a subring of $\mathbf{F}_p[x]$. For convenience, set $R = \mathbf{F}_p[x]$. With a little creativity, we realize that although $\mathbf{Z} \not\le R$, there *is* a natural-seeming map from $\mathbf{Z}$ to R which sends $1_{\mathbf{Z}}$ to 1_R and in general sends n to $n \cdot 1_R$. What went wrong in allowing f' to be 0 without f being constant was that the kernel of this map from $\mathbf{Z}$ to R was non-trivial. This observation generalizes as follows.

Lemma 16.20. *Let R be a commutative ring with 1. Then the natural map* $\chi : \mathbf{Z} \to R$ *given by the formula* $n \mapsto n \cdot 1_R$ *is a ring homomorphism.*

Proof. Exercise 16.16. □

Definition 16.21. Let R be a commutative ring with 1. Let $\chi : \mathbf{Z} \to R$, $n \mapsto n \cdot 1_R$ be the natural map. The *characteristic* of R is the unique non-negative integer c such that $\ker(\chi) = (c)$. We write $\mathrm{char}(R) = c$.

Example 16.22. In the case of an ordinary number system R, the natural map from $\mathbf{Z}$ to R is an embedding, corresponding to the subring relation $\mathbf{Z} \leq R$. Consequently, the ordinary number systems, such as $\mathbf{Q}$, $\mathbf{R}$, and $\mathbf{C}$, have characteristic 0. On the other hand, the "modular" ring $\mathbf{Z}/n\mathbf{Z}$ has characteristic n; the term *natural map* is actually defined two different ways in this case, but they coincide to give the same mapping from $\mathbf{Z}$ to $\mathbf{Z}/n\mathbf{Z}$, which has kernel $n\mathbf{Z}$.

Finally we can give a relatively simple condition on a field which is sufficient (but not necessary!) to guarantee separability.

Proposition 16.23. *If F is a field of characteristic 0, then every algebraic extension field of F is separable over F.*

Proof. Suppose that $\operatorname{char}(F) = 0$ and that K is an algebraic extension field of F. Let $\alpha \in K$, and set $f = \operatorname{Irr}(\alpha, F, x) \in F[x]$. By the definition of irreducible polynomial, f is monic and non-constant, so the leading term of f is $1_F \cdot x^n$ where $n = \deg(f) \geq 1$. Thus, the term of f' of degree $n - 1$ is $n \cdot 1_F \cdot x^{n-1}$. Because $\operatorname{char}(F) = 0$, we have $n \cdot 1_F \neq 0_F$, and thus $f' \neq 0$ in $F[x]$. By Proposition 16.15, f cannot have repeated roots. $\qquad\square$

16.3   Normal Extensions

We turn again to Theorem 16.5 and ask, how can we guarantee the maximum possible number of embeddings of $L[\alpha]$ into $\tilde{M}$? In the previous section, we defined the notion of *separability*, which ensures that the number of roots of an irreducible polynomial is equal to its degree. A non-zero polynomial over a field can never have more roots than its degree, by Theorem 14.25. The separability of the field extension $\tilde{M}/M$ thus guarantees the maximum possible number of embeddings of $L[\alpha]$ into $\tilde{M}$—provided in addition that $\tilde{M}$ contains all the roots of our irreducible polynomial g. This motivates the following definition.

Definition 16.24. Let K be an algebraic extension field of F. We say that K is *normal over* F if for all α in K, the polynomial $\operatorname{Irr}(\alpha, F, x)$ factors completely over K.

Let us imagine how we might construct a normal extension of a given field F. We could start with a monic polynomial f which is irreducible in $F[x]$, and then form a splitting field K for f over F. At this point, we are at least assured that f factors completely over K. So if α is a root of f in K, then $\operatorname{Irr}(\alpha, F, x)$ (which is f) factors completely over K. But what of the irreducible polynomials of the other elements of K? In general, K could be infinite, while f has only finitely many roots. Must we continue to extend K by adjoining the remaining roots of the irreducible polynomials of all these other elements? Fortunately, the answer is no.

Proposition 16.25. *Let K be a finite extension field of F. Then the following conditions are equivalent:*

(1) K is a splitting field over F for some non-zero polynomial $f \in F[x]$.

(2) For all extension fields $\tilde{K}$ of K, and all embeddings $\sigma : K \hookrightarrow \tilde{K}$ over F, we have $\sigma(K) \subseteq K$.

(3) K is normal over F.

Proof. [(1)$\Rightarrow$(2)] Suppose that K is a splitting field for $f \in F[x] - (0)$. Then we can write $f(x) = c \cdot \prod_{j=1}^{n}(x - \rho_j)$ for some $c \in F$, $\rho_j \in K$; and by Theorem 15.21 we have $K = F[\rho_1, \ldots, \rho_n]$. Let $\tilde{K}$ be an extension field of K, and suppose that $\sigma : K \hookrightarrow \tilde{K}$ is an embedding of K into $\tilde{K}$ over F. We note that f need not be irreducible in $F[x]$, but that for each j, we must have $\mathrm{Irr}(\rho_j, F, x)$ divides f in $F[x]$, since $f(\rho_j) = 0$. By considering the restriction of σ to $F[\rho_j]$ and applying Theorem 16.5, we see that for each j, $\sigma(\rho_j)$ must be a root of f; because K is a domain, this forces $\sigma(\rho_j)$ to be one of the ρ_i's. Since $K = F[\rho_1, \ldots, \rho_n]$, and σ acts as the identity map on F, it follows that $\sigma(K) \subseteq K$, as required.

[(2)$\Rightarrow$(3)] Suppose that (2) holds. Let $\alpha \in K$. By Exercise 15.6, we know that α is algebraic over F; set $f = \mathrm{Irr}(\alpha, F, x)$. Then we have $f \in F[x] \subseteq K[x]$. By Exercise 15.10, there is a finite extension field K' of K such that f factors completely over K'. So we can write $f = c \cdot \prod_{j=1}^{n}(x - \alpha_j)$ for some $c \in F$ and $\alpha_j \subset K'$. Now fix a value j such that $1 \leq j \leq n$. By Theorem 16.5, there is an isomorphism $\sigma : F[\alpha] \to F[\alpha_j]$ which extends the identity map on F and sends α to α_j. Using Lemma 16.10 with $L := F[\alpha]$, $\tilde{L} := K$, and $K := K'$, we can find an extension field $\tilde{K}$ of K' and an embedding $\tau : K \hookrightarrow \tilde{K}$ such that τ extends σ. By (2), we must have $\tau(K) \subseteq K$, hence in particular $\tau(\alpha) \in K$. Since τ extends σ, this says $\alpha_j \in K$. Since j was arbitrary, we can say that f factors completely over K. Thus K is normal over F.

[(3)$\Rightarrow$(1)] Suppose that K is normal over F. Since $[K : F] < \infty$, we can write $K = F[\alpha_1, \ldots, \alpha_n]$ for some elements $\alpha_1, \ldots, \alpha_n \in K$ (by Exercise 15.7). Set $f_i = \mathrm{Irr}(\alpha_i, F, x)$ for each i between 1 and n, and set $f = f_1 \cdot f_2 \cdots f_n \in F[x]$. By definition of *normal*, each f_i factors completely over K, say as

$$f_i = c_i \cdot \prod_{j=1}^{D_i}(x - \rho_{i,j}) \tag{16.8}$$

with $c_i \in F$ and $\rho_{i,j} \in K$. It follows that f also factors completely over K. By Theorem 15.21, the field $L := F[\{\rho_{i,j}\}] \subseteq K$ is the unique splitting field for f in K. But since $f_i = \mathrm{Irr}(\alpha_i, F, x)$, we have $f_i(\alpha_i) = 0$ for each i, and thus by Equation 16.8 we must have $\alpha_i = \rho_{i,j}$ for some j. It follows that $F[\alpha_1, \ldots, \alpha_n] \subseteq L$. Hence $L = K$, so K is a splitting field for f over F. $\square$

Example 16.26. Let $F = \mathbf{Q}[\sqrt{2}]$, $K = \mathbf{Q}[\sqrt[4]{2}]$, and $L = \mathbf{Q}[\sqrt[4]{2}, i]$. Then F

is normal over $\mathbf{Q}$, since F is the splitting field over $\mathbf{Q}$ of the polynomial $x^2 - 2 \in \mathbf{Q}[x]$, whose roots are $\sqrt{2}$ and $-\sqrt{2}$. Also, K is the splitting field over F of the polynomial $g := x^2 - \sqrt{2} \in F[x]$, since $K = F[\sqrt[4]{2}]$ and the roots of g are $\sqrt[4]{2}$ and $-\sqrt[4]{2}$. Therefore, K is normal over F. But K is *not* normal over $\mathbf{Q}$. To see this, we observe that $\mathrm{Irr}(\sqrt[4]{2}, \mathbf{Q}, x) = x^4 - 2 =: f$ by Example 15.27; but f does not factor completely over K, since (by Exercise 15.12) L is a splitting field for f over $\mathbf{Q}$, and $K \subset L$. A factorization of f into irreducibles in $K[x]$ is $x^4 - 2 = (x^2 + \sqrt{2})(x - \sqrt[4]{2})(x + \sqrt[4]{2})$. In fact, L is the smallest extension field of K in $\mathbf{C}$ which is normal over $\mathbf{Q}$.

16.4   Galois Extensions

When we combine the two conditions defined in the previous two sections, namely, separability and normality, then we get the "best" kind of field extensions: that is, we get field extensions with the greatest number of automorphisms, in a precise sense (see Exercise 16.14). Many other benefits also follow.

Definition 16.27. Let K be an extension field of F. We say that K is *Galois over* F if K is both normal and separable over F.

Notation 16.28. If K is a finite Galois extension of F, then we denote $\mathrm{Aut}(K/F)$ by $\mathrm{Gal}(K/F)$, and call this the *Galois group of* K *over* F.

One nice feature of Galois extensions is that the top field in a finite Galois extension is Galois over any intermediate field, all the way down to the bottom field in the extension. This is the content of the following lemma.

Lemma 16.29. *Let* K *be a finite Galois extension of* F. *Let* L *be a field such that* $F \leq L \leq K$. *Then* K *is Galois over* L, *and we have* $\mathrm{Gal}(K/L) \leq \mathrm{Gal}(K/F)$.

Proof. [Show: K is normal and separable over L] Let $\alpha \in K$. Since K is algebraic over F (e.g., by definition of *normal*), we can set $f = \mathrm{Irr}(\alpha, F, x)$. Now $f \in F[x] \subseteq L[x]$ and $f \neq 0$, but $f(\alpha) = 0$. Thus, α is algebraic over L, and we set $g = \mathrm{Irr}(\alpha, L, x)$. Since $f \in L[x]$ and $f(\alpha) = 0$, we have $g \mid f$ in $L[x]$. Because K is normal over F, we can factor f completely in $K[x]$. It follows (by Exercise 16.20) that g also factors completely in $K[x]$, using a subset of the linear factors of f; thus, K is normal over L. Since K is separable over F, f has no repeated roots in K, and so g also has no repeated roots. Thus, K is Galois over L. If $\sigma \in \mathrm{Gal}(K/L)$, then we have $\sigma \in \mathrm{Aut}(K)$, and $\sigma(a) = a$ for all $a \in L$; hence $\sigma(a) = a$ for all $a \in F$, so $\sigma \in \mathrm{Gal}(K/F)$. $\square$

We are almost ready for the main result of this chapter, the Fundamental Theorem of Galois Theory. The setting is a finite Galois extension K/F, with

Galois group G. The result tells us that the intermediate fields between F and K exactly correspond to the subgroups of the finite group G, even so far as to say that *normal* extensions of F in K correspond to (has the reader dared to guess that such an amazing thing could be true?) *normal subgroups* of G.

The correspondence between intermediate fields and subgroups requires a little explanation. Given an intermediate field L with $F \leq L \leq K$, we can form the group $\mathrm{Gal}(K/L)$, using Lemma 16.29. From the other side, given a subgroup $H \leq G$, we can form the set of all elements of K which are fixed by every element of H:

Definition 16.30. Let K/F be a finite Galois extension, and let $H \leq \mathrm{Gal}(K/F)$. The *fixed field of H* is the set

$$K^H := \{a \in K \ : \ \forall \sigma \in H, \ \sigma(a) = a\}. \tag{16.9}$$

Lemma 16.31. *Let K/F be a finite Galois extension, and let $H \leq \mathrm{Gal}(K/F)$. Then K^H is a field, and $F \leq K^H \leq K$.*

Proof. Exercise 16.21. □

One final remark is in order before the main theorem. We say that the correspondence just described is *order-reversing*. This is because if $H \leq M \leq G$, then $K^H \supseteq K^M$; and if $F \leq L \leq M \leq K$, then $\mathrm{Gal}(K/L) \supseteq \mathrm{Gal}(K/M)$. These assertions are straightforward to check, and should be verified by the reader.

Theorem 16.32 (Fundamental Theorem of Galois Theory). *Let K/F be a finite Galois extension. Let $G = \mathrm{Gal}(K/F)$. Then:*

(1) $|G| = [K : F]$.

(2) Define a function γ from the set of all fields between F and K to the set of all subgroups of G by

$$\gamma \ : \ \{L \ : \ F \leq L \leq K\} \to \{H \ : \ H \leq G\},$$
$$\gamma(L) = \mathrm{Gal}(K/L).$$

Then γ is a bijection, and for $H \leq G$, we have $\gamma^{-1}(H) = K^H$.

(3) For any subgroup H of G, we have $H \trianglelefteq G$ iff K^H is normal over F, in which case K^H is Galois over F and we have $\mathrm{Gal}(K^H/F) \cong G/H$.

Proof. (1) We will prove that for every intermediate field L with $F \leq L \leq K$, the number of embeddings of L into K over F is exactly $[L : F]$. We proceed by induction on $k := [L : F]$. When $k = 1$, then $L = F$ (by Exercise 15.5), and the unique embedding of L into K over F is the identity map on F. Inductively, suppose that $F < \tilde{L} \leq K$ and that our hypothesis holds for every field L with $[L : F] < \tilde{k} = [\tilde{L} : F]$. We can realize $\tilde{L}$ as the top field in a tower of simple extensions starting at F, by Exercise 15.7. So in particular, we can write $\tilde{L} = L[\alpha]$ for some field L and some $\alpha \in K$, where

$F \leq L < \tilde{L} \leq K$. By Lemma 16.29, K is Galois over L; hence α is algebraic over L, and we set $f = \mathrm{Irr}(\alpha, L, x)$. Let $d = \deg(f)$. By Lemma 15.25, we have $[\tilde{L} : L] = d$. Suppose that σ is an embedding of L into K over F. Set $M = \sigma(L)$. Then M is a field, $F \leq M \leq K$, and $\sigma : L \to M$ is an isomorphism. Set $g = \tilde{\sigma}(f) \in M[x]$, where $\tilde{\sigma}$ is the isomorphism of Lemma 16.1. Then $\deg(g) = \deg(f) =: d$. By Lemma 16.10, we can extend σ to an embedding $\tau : K \hookrightarrow \tilde{K}$ for some extension field $\tilde{K}$ of K; and by Proposition 16.25, we can take $\tilde{K} = K$. By Exercise 16.22, τ is an automorphism of K; since τ extends σ, then $\tau \in \mathrm{Aut}(K/F)$. Let $\beta = \tau(\alpha) \in K$. Then $\tau(\alpha)$ is a root of g by Exercise 16.4, so $g = \mathrm{Irr}(\beta, M, x)$ by Exercise 16.3. Now K is Galois over M (by Lemma 16.29), so g factors completely over K with distinct roots. So by Theorem 16.5, the number of extensions of σ to an embedding of $\tilde{L}$ into K is equal to the number of roots of g in K, which is equal to the degree d of g (note how both normality and separability were used here). Since our inductive hypothesis holds for L, the number of distinct embeddings of L into K over F is equal to $k := [L : F]$. Therefore, the total number of embeddings of $\tilde{L}$ into K over F is equal to $d \cdot k$. But by Lemma 15.26, we also have $[\tilde{L} : F] = [\tilde{L} : L] \cdot [L : F] = d \cdot k$. This completes the inductive step. To complete the proof of (1), we note that an embedding of K into K over F is the same thing as an automorphism of K over F, by Exercise 16.22.

(2) We first show that γ is injective. So suppose that L_1 and L_2 are fields between F and K such that $\gamma(L_1) = \gamma(L_2)$; that is, $\mathrm{Gal}(K/L_1) = \mathrm{Gal}(K/L_2)$. Assume for a contradiction that $L_1 \neq L_2$. Then without loss of generality, we may suppose $\exists \alpha \in L_1 - L_2$. Set $L = L_2[\alpha]$. Then we have $F \leq L_2 < L \leq K$. Now we have $\mathrm{Gal}(K/L) \subseteq \mathrm{Gal}(K/L_2)$; and since $[K : L_2] > [K : L]$, then part (1) tells us that in fact $\mathrm{Gal}(K/L) \subset \mathrm{Gal}(K/L_2)$. But let $\sigma \in \mathrm{Gal}(K/L_2)$. Then we have $\sigma \in \mathrm{Gal}(K/L_1)$ as well, and so $\sigma(\alpha) = \alpha$. It follows that σ must fix every element of $L_2[\alpha] = L$, and so $\sigma \in \mathrm{Gal}(K/L)$. We have deduced that $\mathrm{Gal}(K/L_2) \subseteq \mathrm{Gal}(K/L)$, a contradiction. Thus, γ is injective.

At this point, we note that the injectivity of γ implies that there are only finitely many fields L between F and K; this is because the finite group G can have only finitely many subgroups. From Exercises 15.13 and 15.14, it now follows that K is a simple extension of F; that is, $\exists \alpha \in K$ such that $K = F[\alpha]$.

To complete the proof of (2), we will show that for every $H \leq G$, we have $\mathrm{Gal}(K/K^H) = H$. Note that for each $\sigma \in H$, σ acts as the identity map on K^H, by definition of K^H; and thus, $\sigma \in \mathrm{Gal}(K/K^H)$. Therefore, we have $H \leq \mathrm{Gal}(K/K^H)$. We will now apply the Kaleidoscope Principle (see Exercise 5.9) to construct what "should" be the irreducible polynomial of α over K^H. Namely, set $f = \prod_{\sigma \in H}(x - \sigma(\alpha)) \in K[x]$. Let $\tau \in H$; then (identifying τ with its extension $\tilde{\tau} : K[x] \to K[x]$ of Lemma 16.1) we have
$$\tau(f) = \tau\left(\prod_{\sigma \in H}(x - \sigma(\alpha))\right) = \prod_{\sigma \in H} \tau(x - \sigma(\alpha)) = \prod_{\sigma \in H}(\tau(x) - \tau(\sigma(\alpha)))$$
$$= \prod_{\sigma \in H}(x - (\tau\sigma)(\alpha)).$$
As σ varies over the elements of H, the quantity $\tau\sigma$ also varies over the elements of H (although these elements are permuted according to the map π_τ of Cayley's Theorem, Theorem 10.9). This observation lets us conclude that $\tau(f) = f$ (Wow!). Since τ was an arbitrary element of H, we

have $f \in K^H[x]$, by definition of K^H. Let $g = \mathrm{Irr}(\alpha, K^H, x) \in K^H[x]$. Since $e_G \in H$, then α is a root of f in K. It follows that g divides f in $K^H[x]$. Hence (using Lemma 15.25) we have $[K^H[\alpha] : K^H] = \deg(g) \leq \deg(f) = |H|$. But since $F[\alpha] = K$ and $F \leq K^H \leq K$, we have $K^H[\alpha] = K$. Therefore, $|H| \geq [K : K^H], = |\mathrm{Gal}(K/K^H)|$ by part (1). Since $H \leq \mathrm{Gal}(K/K^H)$, this forces $H = \mathrm{Gal}(K/K^H)$, as desired. We note incidentally that this gives $\deg(g) = |H| = \deg(f)$, so in fact we do have $g = f$.

(3) ($\Rightarrow$) Suppose that $H \trianglelefteq G$. Let $\beta \in K^H$. Then $\beta \in K$, so β is algebraic over F, and we can set $q = \mathrm{Irr}(\beta, F, x)$. Since K is normal over F, we know that q factors completely over K; thus, to show that K^H is normal over F, it is enough to show that any root of q in K actually lies in K^H. So let $\tilde{\beta}$ be a root of q in K. Let $h = \prod_{\sigma \in G}(x - \sigma(\beta))$. Note that $\gamma(K^G) = \mathrm{Gal}(K/K^G) = G = \mathrm{Gal}(K/F)$, so $K^G = F$ by part (2). Furthermore, we have $\tau(h) = h$ for any $\tau \in G$ (as in the proof of (2)), and so $h \in K^G[x] = F[x]$. Since $\mathrm{id}_K \in G$, then β is a root of h, so $h(\beta) = 0$. It follows that q divides h in $F[x]$. Since $\tilde{\beta}$ is a root of q, then $\tilde{\beta}$ must also be a root of h, and so $\tilde{\beta} = \sigma(\beta)$ for some $\sigma \in G$. [Show $\tilde{\beta} \in K^H$] Let $\tau \in H$. [Show $\tau(\tilde{\beta}) = \tilde{\beta}$] Then $\tau(\tilde{\beta}) = \tau(\sigma(\beta)) = (\tau\sigma)(\beta)$. Since $H \trianglelefteq G$, we have $H\sigma = \sigma H$ (by Lemma 7.25), so we can write $\tau\sigma = \sigma\tilde{\tau}$ for some $\tilde{\tau} \in H$. Therefore, $(\tau\sigma)(\beta) = (\sigma\tilde{\tau})(\beta) = \sigma(\tilde{\tau}(\beta)) = \sigma(\beta)$ (since $\beta \in K^H$ and $\tilde{\tau} \in H$) $= \tilde{\beta}$. We have shown that $\tau(\tilde{\beta}) = \tilde{\beta}$ for any $\tau \in H$, and thus $\tilde{\beta} \in K^H$. But $\tilde{\beta}$ was an arbitrary root of q in K. Thus, q factors completely over K^H, and so K^H is indeed normal over F.

($\Leftarrow$) Suppose that K^H is normal over F. Let $\sigma \in G$ and $\tau \in H$. [Show $\sigma\tau\sigma^{-1} \in H$] Set $\psi = \sigma\tau\sigma^{-1} \in G$. [Strategy: Show that ψ fixes every element of K^H; part (2) will then show that $\psi \in H$] Let $\alpha \in K^H$. Since K^H is normal over F, then $\sigma^{-1}(K^H) \subseteq K^H$ by Proposition 16.25 (2). Therefore, $\sigma^{-1}(\alpha) \in K^H$. Since $\tau \in H$, we can say $\tau(\sigma^{-1}(\alpha)) = \sigma^{-1}(\alpha)$. Thus, we have $\psi(\alpha) = (\sigma\tau\sigma^{-1})(\alpha) = \sigma(\sigma^{-1}(\alpha)) = \alpha$. Since α was an arbitrary element of K^H, we have shown that $\psi \in \mathrm{Gal}(K/K^H)$. By part (2), this gives $\psi \in H$, which was our goal.

We have established the "iff" part of (3). Now we suppose that K^H is normal over F, and first prove that K^H is Galois over F. Let $\alpha \in K^H$. Then $\alpha \in K$, so $\mathrm{Irr}(\alpha, F, x)$ has no repeated roots (since K is separable over F). This shows that K^H is separable over F, and hence Galois over F. Set $Q = \mathrm{Gal}(K^H/F)$. Define a function $\omega : G \to Q$ by $\sigma \mapsto \sigma|_{K^H}$. We are guaranteed by Proposition 16.25 (2) that this restriction really does map K^H into K^H, and hence defines an element of $\mathrm{Gal}(K^H/F)$ by Exercise 16.22. It is immediate that ω is a group homomorphism. To show that ω is surjective, let $\tau \in Q$. By Lemma 16.10, there is an extension of τ to an embedding $\tilde{\tau} : K \to \tilde{K}$, for some extension field $\tilde{K}$ of K. But by Proposition 16.25 (2), we must have $\tilde{\tau}(K) \subseteq K$, so we have $\tilde{\tau} \in G$ (by Exercise 16.22) and $\omega(\tilde{\tau}) = \tau$; thus, ω is surjective. We have $\ker(\omega) = \{\sigma \in G : \sigma|_{K^H} = \mathrm{id}_{K^H}\}$. Certainly, if $\sigma \in H$, then σ restricts to the identity map on K^H, by definition of K^H; thus, $H \leq \ker(\omega)$. By the Fundamental Theorem of Group Homomorphisms (Theorem 9.7), we have $Q \cong G/\ker(\omega)$. Since all groups in question are finite, we have $|Q| = |G/\ker(\omega)| = |G|/|\ker(\omega)|$. But also, by parts (1) and (2), we have $|Q| = [K^H : F] = [K : F]/[K : K^H] = |G|/|\mathrm{Gal}(K/K^H)| = |G|/|H|$. This forces $|\ker(\omega)| = |H|$, and thus $\ker(\omega) = H$, which gives $Q \cong G/H$, as desired. □

16.5 Exercises

Exercise 16.1. Complete the proof of Lemma 16.1 by showing that $\tilde{\sigma}$ satisfies Equation 16.1 and that $\tilde{\sigma}$ is an isomorphism.

Exercise 16.2. Let $R \leq S$ be commutative rings with 1, and define

$$\mathrm{Aut}(S/R) := \{\sigma \in \mathrm{Aut}(S) : \sigma(a) = a \text{ for all } a \in R\}.$$

(a) Prove that $\mathrm{Aut}(S/R) \leq \mathrm{Aut}(S)$.

(b) Let T be the image of the natural map from $\mathbf{Z}$ to S, so $T \leq S$ (see Definition 16.21). Prove that $\mathrm{Aut}(S/T) = \mathrm{Aut}(S)$.

Exercise 16.3. Let $\sigma : L \to M$ be an isomorphism of fields. Let $\tilde{\sigma} : L[x] \to M[x]$ be the extension of σ given by Lemma 16.1. Let $\tilde{L}$ be an extension field of L, let $\alpha \in \tilde{L}$ be algebraic over L, and set $f = \mathrm{Irr}(\alpha, L, x)$ and $g = \tilde{\sigma}(f)$. Let β be a root of g in some extension field $\tilde{M}$ of M. Prove that $\mathrm{Irr}(\beta, M, x) = g$. Note: This exercise is used in the proof of Theorem 16.5, so you should avoid using that theorem here!

Exercise 16.4. Let K be a field, and let $\tau \in \mathrm{Aut}(K)$. Let $\alpha \in K$, and suppose that $f \in K[x]$ and $f(\alpha) = 0$. Let $\beta = \tau(\alpha)$. Let $g = \tilde{\tau}(f)$, where $\tilde{\tau} : K[x] \to K[x]$ is the isomorphism of Lemma 16.1. Prove that $g(\beta) = 0$.

Exercise 16.5. Let $\sigma : \mathbf{Q}[\sqrt{2}] \to \mathbf{Q}[\sqrt{2}]$ be the embedding which sends $\sqrt{2}$ to $-\sqrt{2}$. What are the extensions of σ to an embedding $\tau : \mathbf{Q}[\sqrt[4]{2}] \to \mathbf{Q}[\sqrt[4]{2}]$? What is $\mathrm{Aut}(\mathbf{Q}[\sqrt[4]{2}]/\mathbf{Q})$?

Exercise 16.6. Let F be a field and suppose that $f \in F[x]$ is monic and irreducible in $F[x]$. Let α, β be roots of f in some extension field K of F.

(a) Prove that $f = \mathrm{Irr}(\alpha, F, x) = \mathrm{Irr}(\beta, F, x)$.

(b) Prove that α is a repeated root of f iff β is a repeated root of f.

Exercise 16.7. Let K/F be a finite Galois extension. Let $\alpha, \beta \in K$. Prove that $\beta = \sigma(\alpha)$ for some $\sigma \in \mathrm{Gal}(K/F)$ iff $\mathrm{Irr}(\alpha, F, x) = \mathrm{Irr}(\beta, F, x)$. In the language of Definition 16.7, this says that two elements of K are conjugate over F iff they have the same irreducible polynomial over F.

Exercise 16.8. Prove that if K is separable over F, and L is any field such that $F \leq L \leq K$, then

(a) K is separable over L, and

(b) L is separable over F.

(c) Conclude that for any two fields L_1 and L_2 such that $F \leq L_1 \leq L_2 \leq K$, we have that L_2 is separable over L_1.

Exercise 16.9. Let F be a field and let $f \in F[x] - (0)$. Prove that any two splitting fields for f over F are isomorphic over F. That is, prove that if K and L are splitting fields for f over F, then there is an isomorphism $\sigma : K \to L$ such that σ is the identity map on F.

Exercise 16.10. Let $F \leq K$ be fields with $[K : F] = n < \infty$. Prove that for every embedding σ of F into a field L, there are at most n extensions of σ to an embedding of K into L.

Exercise 16.11. Let K be a finite extension field of F. Prove that the following conditions are equivalent:

(1) $K = F[\alpha_1, \ldots, \alpha_r]$ for some $\alpha_1, \ldots, \alpha_r \in K$ where each α_i is separable over F.

(2) For every embedding σ of F into a field M, there exists a superfield L of M such that the number of extensions of σ to an embedding $\tau : K \hookrightarrow L$ is equal to $[K : F]$.

(3) There exists an extension field L of K such that the number of distinct embeddings of K into L over F is equal to $[K : F]$.

(4) K is separable over F.

Exercise 16.12. Suppose that K is a finite extension field of F.

(a) Prove that there is a finite extension field L of K with the properties that (1) L is normal over F and (2) no smaller extension field of K is normal over F. (Note: L is called a *normal closure* of K over F.)

(b) Prove that if K is separable over F, then so is L.

(c) Prove that L is unique up to isomorphism over K: that is, if M is another normal closure of K over F, then there is an isomorphism from M to L over K.

Exercise 16.13. Let K be a finite extension field of F. Prove that K is Galois over F if and only if K is a splitting field over F for some non-constant polynomial $f \in F[x]$ such that f has no repeated roots.

Exercise 16.14. Let K be a finite extension field of F. Prove that $|\text{Aut}(K/F)| \leq [K : F]$, with equality iff K is Galois over F.

Exercise 16.15. Decide, with proof, whether each of the following field extensions is (1) separable, and (2) normal.

(a) $\mathbf{Q}[\sqrt{10}]$ over $\mathbf{Q}$

(b) $\mathbf{Q}[\sqrt[3]{2}]$ over $\mathbf{Q}$

(c) $\mathbf{F}_2[x]/(x^2 + x + 1)$ over $\mathbf{F}_2$, where $\mathbf{F}_2 = \mathbf{Z}/2\mathbf{Z} = \{0, 1\}$.

Exercise 16.16. Prove Lemma 16.20.

Exercise 16.17. Prove that the characteristic of a field must be 0 or a prime number. Hint: Exercise 11.11 may be a useful ingredient in the proof.

Exercise 16.18. Prove that if K is an extension field of F, then $\text{char}(K) = \text{char}(F)$.

Exercise 16.19. [Binomial Theorem] Let R be a commutative ring with 1, let $a, b \in R$, and let $n \in \mathbf{N}$. Prove by induction on n that we have

$$(a + b)^n = \sum_{k=0}^{n} \binom{n}{k} \cdot a^k b^{n-k}, \tag{16.10}$$

where, as usual, $\binom{n}{k}$ denotes the binomial coefficient $\binom{n}{k} := \frac{n!}{k!(n-k)!}$. In particular, your proof should show that $\binom{n}{k} \in \mathbf{N}$.

Exercise 16.20. Let F be a field, and let $f \in F[x] - (0)$. Suppose that K is an extension field of F such that f factors completely over K. Also suppose that $g \in F[x]$ and g divides f in $F[x]$. Prove that g factors completely in $K[x]$, and that if α is a root of g in K with multiplicity m, then α is a root of f with multiplicity $M \geq m$.

Exercise 16.21. Prove Lemma 16.31.

Exercise 16.22. Let K be a finite extension field of F, let L be an intermediate field, and let $\sigma : L \hookrightarrow K$ be an embedding over F.

(a) Prove that σ is a linear transformation of F-vector spaces.

(b) Prove that $\dim_F(\sigma(L)) = \dim_F(L)$. (Exercise 13.13 may be of use here.) Conclude that if $L = K$, then σ is an automorphism of K.

Exercise 16.23. Let K/F be a finite Galois extension, and let $G = \mathrm{Gal}(K/F)$. For $S \subseteq G$, define $K^S := \{a \in K : \forall \sigma \in S, \sigma(a) = a\}$. Prove that $K^S = K^{\langle S \rangle}$.

Exercise 16.24. (The reader may find Exercise 15.14 relevant.) Let K be a finite field.

(a) Prove that $\mathrm{char}(K) = p$ for some prime p, and hence that $\mathbf{Z}/p\mathbf{Z} \hookrightarrow K$ via the natural map from $\mathbf{Z}$ to K; let F be the image of this map, so that $\mathbf{Z}/p\mathbf{Z} \cong F \leq K$. ($F$ is called the *prime subfield* of K.)

(b) Let $r = [K : F]$. Prove that K is a splitting field for the polynomial $x^{p^r} - x$ over F.

(c) Prove that K is Galois over F.

Exercise 16.25. Let K be the splitting field for $x^4 - 2$ over $\mathbf{Q}$ in $\mathbf{C}$ (see Example 15.27).

(a) Why is K Galois over $\mathbf{Q}$?

(b) Why does $G := \mathrm{Gal}(K/\mathbf{Q})$ have order 8?

(c) Let $F = \mathbf{Q}[\sqrt{2}]$ and $L = \mathbf{Q}[\sqrt[4]{2}]$, so that $\mathbf{Q} \leq F \leq L \leq K$. Show that L is Galois over F but not over $\mathbf{Q}$.

(d) Use the previous parts of this exercise together with the Fundamental Theorem of Galois Theory to show that there is a tower of groups $H_1 \leq H_2 \leq G$ such that $H_1 \trianglelefteq H_2$ and $H_2 \trianglelefteq G$ but $H_1 \ntrianglelefteq G$. Thus, the property of being a normal subgroup is not transitive.

(e) Find concise descriptions of all 8 elements of G. Hint: this is easy if you start by asking where $\sqrt[4]{2}$ can be mapped, and then asking where i can be mapped.

Exercise 16.26. Let K be a finite extension field of $\mathbf{Q}$. Show that the number game on K is winnable iff $\mathrm{Aut}(K)$ is the trivial group.

17

Direct Sums and Direct Products

17.1 Introduction

In this chapter, we present two constructions for combining several objects from the same category into a single object of that category. These notions apply to all of the categories we have studied—groups, rings, and vector spaces. We model our constructions on a construction we already know for combining two or more sets into a single set: the Cartesian product.

17.2 Direct Products

We could start by defining a construction without giving any background, but we prefer to motivate it instead by setting it in context. So, in the spirit of abstract algebra, we start by asking: What properties does the Cartesian product have?

Consider two sets A_1 and A_2, and let $C = A_1 \times A_2$. First, notice that there are "natural" maps (that is, functions) from C to A_i. Namely, we can define the "projection" maps $\pi_i : C \to A_i$ by $\pi_1(x, y) = x$ and $\pi_2(x, y) = y$ for $(x, y) \in C$.

Now, the existence of maps to both A_1 and A_2 is not enough to characterize the Cartesian product of A_1 and A_2; lots of other sets besides C also have maps to both A_i's. But what makes C special is that it captures all the information in both A_1 and A_2 while being *minimal* in a certain sense. We want to understand these properties in a purely functional way.

So suppose that we have a set D and maps $\alpha_i : D \to A_i$. We claim that we can interpose C between D and the A_i's. More precisely, there is a map $f : D \to C$ through which the maps α_i factor. The reader is invited to construct f and verify the previous statement before reading on.

The unique choice for the function f is given by the formula $f(x) = (\alpha_1(x), \alpha_2(x)) \in C$ for $x \in D$. With this choice of f, we have $\alpha_i = \pi_i \circ f$ for $i \in \{1, 2\}$. The following diagram illustrates the situation. The dashed line labeled "$\exists! f$" indicates that there is a unique function f from D to C which

DOI: 10.1201/9781003252139-17

makes the diagram commute. We say that C satisfies the *universal property* illustrated by Diagram 17.1.

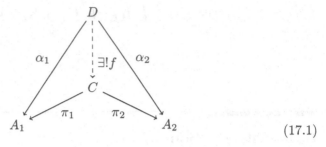

$$(17.1)$$

Suppose that $\tilde{C}$ is another set with the same universal property as C. That is, $\tilde{C}$ comes with maps $\tilde{\pi}_i : \tilde{C} \to A_i$, and Diagram 17.1 works (for any D and α_i) with $\tilde{C}$ and $\tilde{\pi}_i$ in place of C and π_i. Then by the universal property of C, setting $D = \tilde{C}$ and $\alpha_i = \tilde{\pi}_i$ in Diagram 17.1, we get a unique map $f : \tilde{C} \to C$ such that $\tilde{\pi}_i = \pi_i \circ f$. Likewise, using the universal property of $\tilde{C}$, we get a unique map $g : C \to \tilde{C}$ such that $\pi_i = \tilde{\pi}_i \circ g$. It follows that $\pi_i = \pi_i \circ f \circ g$. Set $h = f \circ g$, so that $\pi_i = \pi_i \circ h$.

Let $(x, y) \in C$, and write $(a, b) = h(x, y)$. Then we have $x = \pi_1(x, y) = (\pi_1 \circ f \circ g)(x, y) = (\pi_1 \circ h)(x, y) = \pi_1(h(x, y)) = \pi_1(a, b) = a$. Similarly, we find $y = b$. Therefore, $h = f \circ g = \mathrm{id}_C$.

We can arrive at the same result, $f \circ g = \mathrm{id}_C$, by using the universal property of C alone, without taking advantage of the special form of C and the π_i maps as we did in the previous paragraph. To see this, put $D = C$ and $\alpha_i = \pi_i$ in Diagram 17.1. Then by the universal property of C, there is a unique map $\omega : C \to C$ such that $\pi_i \circ \omega = \pi_i$ for $i \in \{1, 2\}$. But we know that, on the one hand, $\pi_i \circ h = \pi_i$, while on the other hand, certainly $\pi_i \circ \mathrm{id}_C = \pi_i$. Therefore, the uniqueness of ω tells us that $\omega = h = \mathrm{id}_C$.

The advantage of this last argument is that we can use it with the roles of C and $\tilde{C}$ reversed, to find that $g \circ f = \mathrm{id}_{\tilde{C}}$. Thus, f and g are inverse to each other; so f is a bijection from C to $\tilde{C}$.

This result says that any set C satisfying the universal property of Diagram 17.1 is essentially unique: there is a bijection from C to any other such set, which furthermore is "compatible" with the projection maps. This situation is typical of objects with universal properties defined by commutative diagrams: they are typically unique up to a unique isomorphism. Thinking in terms of commutative diagrams and universal properties is the mainstay of a branch of mathematics known as *category theory*.

The beauty of the category-theoretic approach is that we can use the same diagram, Diagram 17.1, to define the notion of a "product" for *any* type of algebraic objects, whether groups, rings, or vector spaces; the only modification we must make is to require that all the maps be "morphisms" of the appropriate type: group homomorphisms, ring homomorphisms, or linear transformations, respectively! The following definition illustrates this idea for rings.

Definition 17.1. Let R_1 and R_2 be rings. Then a ring P is called a *direct product* of R_1 with R_2 if there are ring homomorphisms $\pi_i : P \to R_i$ such that for any ring R and any pair of ring homomorphisms $\alpha_i : R \to R_i$, there exists a unique ring homomorphism $f : R \to P$ which makes the following diagram commute:

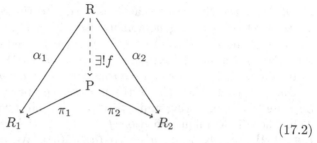

$$(17.2)$$

Remark 17.2. Technically, it is the ring P together with the maps π_i which constitute the direct product of R_1 with R_2.

Remark 17.3. It can be shown that, if it exists, a direct product of rings is unique up to isomorphism (Exercise 17.1); the argument is essentially the same as that we used above for Cartesian products of sets. What we do not yet know is whether a direct product of two given rings must always exist. We shall show next that in fact it does.

Suppose that R_1 and R_2 are rings. We would like to invent a ring P which will be a direct product of R_1 with R_2. First, we try to identify the underlying set for our product ring P. Perhaps we are being simplistic, but let us try to set $P = R_1 \times R_2$; that is, make the underlying set P equal to the Cartesian product of the set R_1 with the set R_2. We'll also take the map π_i to be the ordinary projection map from $R_1 \times R_2$ to R_i. This way, we are at least guaranteed the existence of a unique *function* f in Diagram 17.2, although we do not know whether we can force f to be a ring homomorphism, as we require. Now our task is to come up with binary operations $+$ and $\cdot$ on P to make this happen.

Let $x, y \in P$. We will try to determine what the sum $x + y =: s$ must look like. Since we require the maps π_i to be ring homomorphisms we must have $\pi_i(s) = \pi_i(x + y) = \pi_i(x) + \pi_i(y)$. Remembering that $P = R_1 \times R_2$, we can write $x = (x_1, x_2)$ and $y = (y_1, y_2)$ for some $x_1, y_1 \in R_1$ and $x_2, y_2 \in R_2$. Then $\pi_i(x) = x_i$ and $\pi_i(y) = y_i$. It follows that we have $\pi_i(s) = x_i + y_i$. It is a general fact for an element s of a Cartesian product that $s = (\pi_1(s), \pi_2(s))$. Therefore, we have $s = (x_1 + y_1, x_2 + y_2)$.

We have shown that we must have $(x_1, y_1) + (x_2, y_2) = (x_1 + y_1, x_2 + y_2)$. The same argument will show that $(x_1, y_1) \cdot (x_2, y_2) = (x_1 \cdot y_1, x_2 \cdot y_2)$. These are the only choices for the operations $+$ and $\cdot$ on $R_1 \times R_2 = P$ which could possibly make P a direct product of R_1 with R_2 using the standard projection maps π_i. The following result shows that our efforts have not been in vain.

Lemma 17.4. *Let R_1 and R_2 be rings. Set $P = R_1 \times R_2$, and define two binary operations on P by the formulas $(x_1, x_2) + (y_1, y_2) = (x_1 + y_1, x_2 + y_2)$ and $(x_1, x_2) \cdot (y_1, y_2) = (x_1 \cdot y_1, x_2 \cdot y_2)$ for $x_i, y_i \in R_i$. Then P is a ring, the standard projection maps $\pi_i : P \to R_i$ are ring homomorphisms, and P is a direct product of R_1 with R_2.*

Proof. We leave the proof that P is a ring to the reader as Exercise 17.2. To see that π_i is a ring homomorphism, let $a, b \in P$. Then we can write $a = (a_1, a_2)$ and $b = (b_1, b_2)$ for some $a_1, b_1 \in R_1$ and $a_2, b_2 \in R_2$. So we have $\pi_i(a+b) = \pi_i((a_1, a_2)+(b_1, b_2)) = \pi_i((a_1+b_1, a_2+b_2)) = a_i+b_i = \pi_i(a)+\pi_i(b)$, as required. Similarly, we can see that $\pi_i(a \cdot b) = a_i \cdot b_i = \pi_i(a) \cdot \pi_i(b)$.

Now we verify that P has the universal property of the direct product. Suppose that R is a ring and $\alpha_i : R \to R_i$ are two ring homomorphisms. Then there is a unique function $f : R \to P$ such that $\pi_i \circ f = \alpha_i$ for $i \in \{1, 2\}$, namely $f : a \mapsto (\alpha_1(a), \alpha_2(a))$. We must verify that f is a ring homomorphism. So let $a, b \in R$. Then we have $f(a + b) = (\alpha_1(a + b), \alpha_2(a+b)) = (\alpha_1(a)+\alpha_1(b), \alpha_2(a)+\alpha_2(b))$ (since α_i is a ring homomorphism) $= (\alpha_1(a), \alpha_2(a)) + (\alpha_1(b), \alpha_2(b))$ (by definition of $+$ on P) $= f(a) + f(b)$ (by definition of f). Similarly, we find that $f(a \cdot b) = f(a) \cdot f(b)$. This completes the proof. $\qquad\square$

Remark 17.5. We refer to the operations $+$ and $\cdot$ defined on P in Lemma 17.4 as *componentwise* operations, just as we did in Example 13.4. For the categories of rings, groups, and vector spaces, it turns out that taking the Cartesian product of the underlying sets and defining componentwise operations will produce a direct product.

Notation 17.6. If R_i are rings for $i \in \{1, 2\}$, then we will understand $R_1 \times R_2$ to denote the ring formed from the Cartesian product of R_1 with R_2 using componentwise operations.

Next we generalize the idea of direct product by allowing more than two "factors."

Definition 17.7. Let $\mathcal{C} = (R_i)_{i \in \mathcal{I}}$ be an indexed collection of rings. Then a ring P is called a *direct product* of $\mathcal{C}$ if there are ring homomorphisms $\pi_i : P \to R_i$ for $i \in \mathcal{I}$ such that for any ring R and any collection of ring homomorphisms $\alpha_i : R \to R_i$, there exists a unique ring homomorphism $f : R \to P$ with the property that, for each $i \in \mathcal{I}$, the following diagram commutes:

$$
\begin{array}{c}
R \\
\Big\downarrow {\scriptstyle \exists! f} \\
P \\
\Big\downarrow {\scriptstyle \pi_i} \\
R_i
\end{array}
\qquad \alpha_i
$$

$$(17.3)$$

The reader should verify that, when $|\mathcal{I}| = 2$, Definition 17.7 is equivalent to Definition 17.1. We shall mostly consider direct products with only finitely many factors; however, Definition 17.7 applies whether the index set $\mathcal{I}$ is finite or infinite.

Just as in the case with only two factors, it turns out that componentwise operations make the Cartesian product of underlying sets into a direct product of rings. For ease of notation, we only treat this result for a finite direct product (that is, when there are only a finite number of factors):

Lemma 17.8. *Let R_1, $R_2, \ldots,$ R_n be rings. Then the Cartesian product $P = R_1 \times R_2 \times \cdots \times R_n$ forms a ring under componentwise operations; this ring is a direct product of $\{R_i\}_{1 \le i \le n}$.*

Proof. We leave the proof to the reader as Exercise 17.3. $\qquad\Box$

Notation 17.9. Let S_1, S_2, $\ldots$, S_n be sets. We denote the Cartesian product $S_1 \times S_2 \times \cdots \times S_n$ by $\prod_{i=1}^{n} S_i$. In case the S_i are rings, we take $\prod_{i=1}^{n} S_i$ to mean the corresponding ring with componentwise operations.

Remark 17.10. The reader may have noticed that in Definition 17.7, the collection $\mathcal{C}$ is unordered, and we did not require the index set $\mathcal{I}$ to be ordered (although there *can* be repetition among the R_i). In Definition 17.1 as well, the order of the factors R_1 and R_2 is immaterial. This can be seen from the symmetry of Diagram 17.2. The reader is invited in Exercise 17.4 to verify this fact directly.

At this point, we give an application of direct products of rings to elementary number theory.

Theorem 17.11 (The Chinese Remainder Theorem). *Let a_1, a_2, $\ldots$, a_n be positive integers which are pairwise relatively prime: that is, we have $\gcd(a_i, a_j) = 1$ whenever $i \ne j$. Then there is a ring isomorphism*

$$f \; : \; \mathbf{Z}/(a_1 \cdot a_2 \cdots a_n) \longrightarrow \prod_{i=1}^{n} \mathbf{Z}/(a_i) \tag{17.4}$$

given by the formula

$$k + (a_1 \cdot a_2 \cdots a_n) \mapsto (k + (a_1), k + (a_2), \ldots, k + (a_n)) \tag{17.5}$$

for $k \in \mathbf{Z}$.

Proof. For each i between 1 and n, consider the natural map $\nu_i \; : \; \mathbf{Z} \to \mathbf{Z}/(a_i)$, $k \mapsto k + (a_i)$. Set $a = a_1 \cdot a_2 \cdots a_n$. Then for each i, we have $a \in (a_i)$, and therefore $(a) \subseteq (a_i)$. It follows from Exercise 11.24 that ν_i factors through $\mathbf{Z}/(a)$ to give a ring homomorphism $\alpha_i \; : \; \mathbf{Z}/(a) \to \mathbf{Z}/(a_i)$, $k + (a) \mapsto k + (a_i)$. Set $P = \prod_{i=1}^{n} \mathbf{Z}/(a_i)$, and let $\pi_i \; : \; P \to \mathbf{Z}/(a_i)$ be the standard projection map. Then P is a direct product of the collection of rings $\{\mathbf{Z}/(a_i)\}_{1 \le i \le n}$ by Lemma 17.8. By the universal property of the direct product, we have a ring homomorphism $f : \mathbf{Z}/(a) \to P$ such that $\alpha_i = \pi_i \circ f$ for each i. Now we must have $f : k + (a) \mapsto (k + (a_1), \ldots, k + (a_n))$ for $k \in \mathbf{Z}$.

Let $x \in \ker(f)$, and write $x = k + (a)$ with $k \in \mathbf{Z}$. Then we have $f(x) = 0_P = (0 + (a_1), \ldots, 0 + (a_n))$. Since $f(x) = (k + (a_1), \ldots, k + (a_n))$, it follows that $k \in (a_i)$ for each i; that is, a_i divides k for each i. Because the a_i's are pairwise relatively prime, this forces their product to divide k. So $a \mid k$, and thus $k + (a) = 0 + (a)$, which is the zero element of the ring $\mathbf{Z}/(a)$. Therefore, $\ker(f) = (0)$. It follows from Exercise 11.23 that f is injective.

Observe next that we have $|\mathbf{Z}/(a)| = a = \prod_{i=1}^{n} a_i = \prod_{i=1}^{n} |\mathbf{Z}/(a_i)| = |\prod_{i=1}^{n} \mathbf{Z}/(a_i)|$. Thus, the domain and codomain of f have the same finite size. Since f is injective, it follows that f must also be surjective. Thus, f is an isomorphism. This completes the proof. $\square$

The Chinese Remainder Theorem is sometimes described in terms of systems of congruences. We treat this result next.

Corollary 17.12. *Let $a_1, a_2, \ldots, a_n$ be positive integers which are pairwise relatively prime. Let $c_1, c_2, \ldots, c_n$ be arbitrary integers, and set $a = a_1 \cdot a_2 \cdots a_n$. Then the system of congruences*

$$x \equiv c_i \pmod{a_i}, \ 1 \leq i \leq n \tag{17.6}$$

has a unique solution x modulo a. That is, there is an integer x which satisfies the system of congruences (17.6), and any other integer solution y is congruent to x modulo a.

Proof. Set $c = (c_1 + (a_1), \ldots, c_n + (a_n)) \in \prod_{i=1}^{n} \mathbf{Z}/(a_i)$. By Theorem 17.11, there is a unique element $b \in \mathbf{Z}/(a)$ such that $f(b) = c$. Writing $b = x + (a)$, we see that x is the desired solution. $\square$

17.3   Direct Sums

By reversing the arrows in Diagram 17.3, we get the definition of our second way to combine algebraic objects, the *coproduct*. To keep this definition general, we use the terms *object* and *morphism*; these can be replaced by *group*, *ring*, or *vector space* and by *group homomorphism*, *ring homomorphism*, or *linear transformation*, respectively.

Definition 17.13. Let $\mathcal{C} = (X_i)_{i \in \mathcal{I}}$ be an indexed collection of objects. Then an object S is called a *coproduct* of $\mathcal{C}$ if there are morphisms $\tau_i : X_i \to S$ for $i \in \mathcal{I}$ such that for any object X and any collection of morphisms $\alpha_i : X_i \to X$, there exists a unique morphism $f : S \to X$ with the property that, for each $i \in \mathcal{I}$, the following diagram commutes:

$$(17.7)$$

Remark 17.14. The terms *object* and *morphism* are actually technical terms in category theory, where we speak of the *category of groups*, the *category of rings*, and so on. We also speak of the *category of sets*, in which a morphism from a set X to a set Y is simply a function from X to Y. See Project 23.4 for some details.

Remark 17.15. Just as in the case of direct products, the notion of coproduct does not depend on any ordering of the collection of component objects.

Next, we investigate whether finite coproducts exist in the category of groups. We start with a collection of just 2 groups, G_1 and G_2. As with direct products, our first candidate for a coproduct of G_1 with G_2 will be the Cartesian product $S = G_1 \times G_2$, with multiplication defined componentwise. In order to proceed with the proof, we need to find a group homomorphism $\tau_i : G_i \to S$ for $i \in \{1,2\}$. Now if $a \in G_1$, say, then it seems natural for τ_1 to map a to an element of the form $(a, x_2) \in S$, but what should x_2 be? It seems unreasonable for x_2 to depend on a at all, since the groups G_1 and G_2 may have nothing to do with each other. Therefore, we choose x_2 to be a distinguished element of G_2. The only distinguished element in a generic group is its identity element. So we set $\tau_1(a) = (a, e_2)$, where e_i denotes the identity element of G_i. Similarly, we define $\tau_2 : G_2 \to S$ by the formula $\tau_2(a) = (e_1, a)$ for $a \in G_2$.

Next we check whether S satisfies the universal property of the coproduct. So suppose that G is a group and $\alpha_i : G_i \to G$ are group homomorphisms for $i \in \{1,2\}$. We must find a group homomorphism $f : S \to G$ with the property that $f \circ \tau_i = \alpha_i$ for each i. Let $a = (a_1, a_2) \in S$. Set $\hat{a}_i = \tau_i(a_i)$. Then we require $f(\hat{a}_i) = f(\tau_i(a_i)) = \alpha_i(a_i)$ for each i. So we need to have $f(\hat{a}_1 \cdot \hat{a}_2) = f(\hat{a}_1) \cdot f(\hat{a}_2)$ (since f is required to be a group homomorphism) $= \alpha_1(a_1) \cdot \alpha_2(a_2)$. But $\hat{a}_1 \cdot \hat{a}_2 = (a_1, e_2) \cdot (e_1, a_2) = (a_1 \cdot e_1, e_2 \cdot a_2) = (a_1, a_2) = a$, so we must have $f(a) = f(a_1, a_2) = \alpha_1(a_1) \cdot \alpha_2(a_2)$. This formula defines a function from S to G, and it is the only candidate for our group homomorphism f.

Now we ask whether this function f is indeed a group homomorphism. So let $a, b \in S$ with $a = (a_1, a_2)$ and $b = (b_1, b_2)$. Then we have $f(a \cdot b) = f(a_1 \cdot b_1, a_2 \cdot b_2) = \alpha_1(a_1 \cdot b_1) \cdot \alpha_2(a_2 \cdot b_2) = \alpha_1(a_1) \cdot \alpha_1(b_1) \cdot \alpha_2(a_2) \cdot \alpha_2(b_2)$ (since the α_i are group homomorphisms). On the other hand, we have $f(a) \cdot f(b) = \alpha_1(a_1) \cdot \alpha_2(a_2) \cdot \alpha_1(b_1) \cdot \alpha_2(b_2)$. In order for f to be a group homomorphism, then, we must have $\alpha_1(b_1) \cdot \alpha_2(a_2) = \alpha_2(a_2) \cdot \alpha_1(b_1)$ for all $b_1 \in G_1$ and all $a_2 \in G_2$. But this is not automatically true! It is true, however, if all our groups are *abelian*. Thus, in the following result, we prove that finite coproducts exist in the category of abelian groups.

Lemma 17.16. *Let $C = \{G_i\}_{1 \le i \le n}$ be a finite collection of abelian groups (not necessarily distinct). Set $S = G_1 \times G_2 \times \cdots \times G_n$, the Cartesian product of the sets $G_1, \ldots, G_n$, with multiplication defined componentwise. Let e_i denote the identity element of G_i. For $i \in \{1, 2, \ldots, n\}$, define a function $\tau_i : G_i \to S$ by the formula $\tau_i(a) = (e_1, \ldots, a, \ldots, e_n)$ for $a \in G_i$, where the ith component is equal to a. Then S is a group, the functions τ_i are group homomorphisms, and S is a coproduct of the collection C in the category of abelian groups.*

Proof. We leave to the reader the proof that S forms an abelian group under componentwise multiplication (Exercise 17.10). To verify that τ_i is a group homomorphism, let $a, b \in G_i$. Then we have $\tau_i(a \cdot b) = (e_1, \ldots, a \cdot b, \ldots, e_n)$, while $\tau_i(a) \cdot \tau_i(b) = (e_1, \ldots, a, \ldots, e_n) \cdot (e_1, \ldots, b, \ldots, e_n) = (e_1 \cdot e_1, \ldots, a \cdot b, \ldots, e_n \cdot e_n) = (e_1, \ldots, a \cdot b, \ldots, e_n)$, as required.

Next we verify that S satisfies the universal property of the coproduct in the category of abelian groups. So suppose that G is an abelian group and $\alpha_i : G_i \to G$ are group homomorphisms for i between 1 and n. Suppose that $f : S \to G$ is a group homomorphism with the property that $f \circ \tau_i = \alpha_i$ for each i. Let $a = (a_1, \ldots, a_n) \in S$. Set $\hat{a}_i = \tau_i(a_i) = (e_1, \ldots, a_i, \ldots, e_n)$, where the a_i occurs in the ith component. Then we have $f(\hat{a}_i) = f(\tau_i(a_i)) = \alpha_i(a_i)$ for each i. So $f(\hat{a}_1 \cdots \hat{a}_n) = f(\hat{a}_1) \cdots f(\hat{a}_n)$ (since f is a group homomorphism) $= \alpha_1(a_1) \cdots \alpha_n(a_n)$. But $\hat{a}_1 \cdots \hat{a}_n = a$, so $f(a) = \alpha_1(a_1) \cdots \alpha_n(a_n)$. This proves that such an f is unique if it exists.

To finish the proof, we define the function $f : S \to G$ by the formula $f(a) = \alpha_1(a_1) \cdots \alpha_n(a_n)$ for $a = (a_1, \ldots, a_n) \in S$, and show that f is indeed a group homomorphism. So let $a, b \in S$ with $a = (a_1, \ldots, a_n)$ and $b = (b_1, \ldots, b_n)$. Then we have $f(a \cdot b) = f(a_1 \cdot b_1, \ldots, a_n \cdot b_n)$ (by definition of multiplication on S) $= \alpha_1(a_1 \cdot b_1) \cdots \alpha_n(a_n \cdot b_n)$ (by definition of f) $= \alpha_1(a_1) \cdot \alpha_1(b_1) \cdots \alpha_n(a_n) \cdot \alpha_n(b_n)$ (since the α_i are group homomorphisms) $= \alpha_1(a_1) \cdots \alpha_n(a_n) \cdot \alpha_1(b_1) \cdots \alpha_n(b_n)$ (since G is abelian) $= f(a) \cdot f(b)$. $\square$

Notation 17.17. The group S defined in Lemma 17.16 is denoted $G_1 \oplus G_2 \cdots \oplus G_n$ or

$$\oplus_{i=1}^{n} G_i. \tag{17.8}$$

The symbol $\oplus$ is read "direct sum." Even though this is just one particular construction of a coproduct, we often use the term *direct sum* to mean *coproduct* in the category of abelian groups. This should cause no confusion in practice. Some authors distinguish between an "external direct sum" as given by Equation 17.8 above, and an "internal direct sum," by which they mean a group G for which the natural map of Definition 17.19 below is an isomorphism.

Remark 17.18. We have discovered that coproducts exist in the category of abelian groups, where we call them *direct sums*; accordingly, we will use additive notation for our group operations. Indeed, the term "direct sum" and the notation $\oplus$ already suggest that we should be doing this! On the other hand, we have not proved that coproducts do not exist in the category of all groups; see Exercise 17.13.

Next, we turn our point of view around, and ask: under what conditions can we "split" a given abelian group G into a direct sum?

First, we observe that if $G \cong \oplus_{i=1}^{n} G_i$, then each "direct summand" G_i is isomorphic to a subgroup of G, since the maps τ_i defined in Lemma 17.16 are injective. Therefore, we look for direct summands of G inside of G itself.

Suppose that G is an abelian group with subgroups $G_1, \ldots, G_n$. We have natural embeddings $\alpha_i : G_i \to G$ for each i, so the defining property of the direct sum (Definition 17.13) gives us a unique group homomorphism $f : \oplus_{i=1}^{n} G_i \to G$ such that $f \circ \tau_i = \alpha_i$ for each i. Observe that we must have $f(g_1, \ldots, g_n) = g_1 + \cdots + g_n$, since f is a group homomorphism.

The map f described in the preceding paragraph occurs often enough that we give it a name.

Definition 17.19. Let $(G, +)$ be an abelian group, and let $G_1, \ldots, G_n$ be subgroups of G. The *natural map* from $\oplus_{i=1}^{n} G_i$ to G is the group homomorphism sending $(g_1, \ldots, g_n)$ to $g_1 + \cdots + g_n$.

Now that we have a map between G and a direct sum, we look for a condition which guarantees this map to be an isomorphism. The relevant condition is very similar to the condition which defines the notion of a basis of a vector space.

Proposition 17.20. *Let $(G, +)$ be an abelian group, and let $G_1, \ldots, G_n$ be subgroups of G. Then the following are equivalent:*

(i) For each $g \in G$, there exist unique elements $g_1, \ldots, g_n$, with $g_i \in G_i$, such that $g = g_1 + \cdots + g_n$.

(ii) The natural map $\sigma : \oplus_{i=1}^{n} G_i \to G$ is an isomorphism.

Proof. (i) $\implies$ (ii): Let $\sigma : \oplus_{i=1}^{n} G_i \to G$ be the natural map. Now σ must be injective by the uniqueness hypothesis on the g_i's, and σ is surjective by the existence hypothesis on the g_i's. Thus, σ is a bijective group homomorphism, as required.

(ii) $\implies$ (i): This is likewise immediate from the definition of the natural map. □

The conclusion of Proposition 17.20 still holds if we drop the "unique" condition from the preceding hypothesis, and replace it with the condition that 0_G has only the trivial representation as a sum of elements from the G_i's.

Corollary 17.21. *Let* $(G, +)$ *be an abelian group, and let* $G_1, \ldots, G_r$ *be subgroups of* G. *Suppose that*

   (i) for each $g \in G$, *there are elements* $g_1, \ldots, g_r$, *with* $g_i \in G_i$, *such that* $g = g_1 + \cdots + g_r$, *and*

   (ii) if $g_1 + \cdots + g_r = 0_G$, *with* $g_i \in G_i$, *then* $g_i = 0$ *for each* i.

   Then $\oplus_{i=1}^r G_i \cong G$ *via the natural map* $(g_1, \ldots, g_r) \mapsto g_1 + \cdots + g_r$.

Proof. This proof is very similar to the proof of Lemma 13.21 from Definition 13.14; we leave the proof to the reader as Exercise 17.11. □

17.4   Exercises

Exercise 17.1. Let R_1 and R_2 be rings. Prove that a direct product of R_1 with R_2 is unique up to isomorphism: that is, any two direct products of R_1 with R_2 are isomorphic.

Exercise 17.2. Let R_1 and R_2 be rings. Prove that $R_1 \times R_2$ forms a ring under componentwise addition and multiplication.

Exercise 17.3. Prove Lemma 17.8.

Exercise 17.4. Prove by exhibiting an explicit isomorphism that for any two rings R_1 and R_2, we have $R_1 \times R_2 \cong R_2 \times R_1$. Explain why this result also follows from Exercise 17.1.

Exercise 17.5 (Direct Product of Groups). Let G_1, G_2 be two groups.

   (a) What does it mean to say that a group G is a direct product of G_1 with G_2?

   (b) Let $G = G_1 \times G_2$, with multiplication in G defined componentwise. Prove that G is a direct product of G_1 with G_2.

   (c) Suppose that G is a group with $G_1, G_2 \leq G$. Define a function σ : $G_1 \times G_2 \to G$ by $(g_1, g_2) \mapsto g_1 g_2$. Prove that σ is a group isomorphism if and only if all of the following conditions are satisfied: $\langle G_1 \cup G_2 \rangle = G$, $G_1 \cap G_2 = \{e_G\}$, and $G_i \trianglelefteq G$ for $i \in \{1, 2\}$.

   (d) Prove that in case G_1 and G_2 are abelian, then $G_1 \times G_2$ under componentwise multiplication is also abelian, and is a direct sum of G_1 and G_2; in fact, $G_1 \times G_2 = G_1 \oplus G_2$, and this generalizes to n factors.

Exercise 17.6. Let $R_1, R_2, \ldots, R_n$ be rings with 1. Prove that $(\prod_{i=1}^n R_i)^\times = \prod_{i=1}^n R_i^\times$.

Exercise 17.7. Let n be a positive integer, and set $R = \mathbf{Z}/n\mathbf{Z}$. We denote $|R^\times|$ by $\phi(n)$. Note: ϕ is called *Euler's totient function* or the *Euler phi function*.

   (a) Deduce from Exercise 12.17 that we have

$$\phi(n) = |\{a \in \mathbf{Z} \; : \; 1 \leq a \leq n \text{ and } \gcd(a, n) = 1\}|.$$

(b) Show that if p is a positive prime integer and r is a positive integer, then $\phi(p^r) = p^r - p^{r-1} = p^{r-1}(p-1)$.

(c) Use the Chinese Remainder Theorem and Exercises 17.5 and 17.6 to deduce that if $n > 1$ and $n = \prod_{i=1}^{k} p_i^{r_i}$ is the prime factorization of n, then $(\mathbf{Z}/n\mathbf{Z})^\times \cong \oplus_{i=1}^{k}(\mathbf{Z}/p_i^{r_i}\mathbf{Z})^\times$ via the map $a + n\mathbf{Z} \mapsto (a + p_1^{r_1}\mathbf{Z}, \ldots, a + p_k^{r_k}\mathbf{Z})$. Conclude that $\phi(n) = \prod_{i=1}^{k} p_i^{r_i-1}(p_i - 1)$.

(d) Prove that if m is an integer such that $\gcd(m, n) = 1$, then we have $m^{\phi(n)} \equiv 1 \pmod{n}$. This result is known as *Euler's Theorem*; it is a generalization of Fermat's Little Theorem (see Exercise 12.13).

Exercise 17.8. Use Exercise 17.7 to prove that if m and n are positive integers with $\gcd(m, n) = 1$, then we have $\phi(mn) = \phi(m) \cdot \phi(n)$.

Exercise 17.9 (Factoring a Commutative Ring). Let R be a commutative ring with 1, and suppose that $e \in R$ is an element such that $e^2 = e$.

(a) Prove that $(1 - e)^2 = 1 - e$.

(b) Prove that eR and $(1 - e)R$ are commutative subrings of R which possess multiplicative identity elements.

(c) Prove that $R \cong eR \times (1 - e)R$.

Exercise 17.10. Prove that the set S defined in Lemma 17.16 forms an abelian group under componentwise multiplication.

Exercise 17.11. Prove Corollary 17.21.

Exercise 17.12 (Arbitrary Direct Sums). The reader may have noticed that the formula defining the natural map from a direct sum in Definition 17.19 does not seem to bode well for the direct sum of an *infinite* collection, since we would need to sum an infinite series. However, things are not really so bad. Given an arbitrary indexed collection of abelian groups $\mathcal{C} = (G_i)_{i \in \mathcal{I}}$, let $\oplus_{i \in \mathcal{I}} G_i$ denote the subset of $\prod_{i \in \mathcal{I}} G_i$ where all but finitely many of the components are 0. Prove that $\oplus_{i \in \mathcal{I}} G_i$ is a direct sum of the collection $\mathcal{C}$. Note: we use the notation $(g_i)_{i \in \mathcal{I}}$ to denote a typical element of the direct product $\prod_{i \in \mathcal{I}} G_i$ whose i^{th} component is g_i. This is consistent with our notation for an indexed collection (see Example 1.32) with indexing function $g : \mathcal{I} \to \cup_{i \in \mathcal{I}} G_i$ and the additional requirement that $g_i \in G_i$ for each $i \in \mathcal{I}$.

Exercise 17.13 (Coproduct of Two Groups). Let G_1 and G_2 be arbitrary groups (not necessarily abelian). In the text, we could not force Diagram 17.7 to commute when S is the Cartesian product of G_1 with G_2. But it turns out we just didn't try hard enough to find a suitable underlying set S. For $i \in 1, 2$, let $Y_i = \{x_{g,i} : g \in G_i\}$, where $x_{g,i}$ are distinct symbols. Let $F_i = \text{Fr}(Y_i\}$ be the free group on Y_i, and let $F = \text{Fr}(Y_1 \cup Y_2)$. Let $\sigma_i : F_i \to G_i$ be the group homomorphism induced by $x_{g,i} \mapsto g$. Note that F_i is a subgroup of F, and let $N = \langle\langle \ker(\sigma_1) \cup \ker(\sigma_2) \rangle\rangle \trianglelefteq F$ and $S = F/N$.

(a) Prove that the map $\tau_i : G_i \to S$ given by the formula $g \mapsto x_{g,i}N$ is a well-defined group homomorphism.

(b) Prove that the group S together with the maps τ_i satisfy Definition 17.13, and thus S is a coproduct of G_1 and G_2 in the category of groups.

(c) As mentioned in the text, we only use the term "direct sum" when we require all of our groups to be abelian, that is, in the category of abelian groups. The reader may wonder whether the group S constructed above will be isomorphic to the direct sum $G_1 \oplus G_2$ in case both G_1 and G_2 are abelian. Prove that the answer in general is *no*. Hint: Even if both G_i are abelian, we can have group homomorphisms $\alpha_i : G_i \to G$ where G is non-abelian.

Exercise 17.14. Let G_1 and G_2 be abelian groups.

(a) Use the construction of Exercise 17.13, but with free groups replaced by free abelian groups, to produce an abelian group S, and prove that S is a coproduct of G_1 and G_2 in the category of abelian groups. (See Exercise 10.5 for the definition of a free abelian group.)

(b) Prove a uniqueness result analogous to that of Exercise 17.1 in the category of abelian groups, and use this to prove that the group S from part (a) above is isomorphic to $G_1 \oplus G_2$.

18

The Structure of Finite Abelian Groups

18.1 Introduction

In this chapter, we attempt to decompose finite abelian groups into direct sums until we cannot go further. In the end, we shall see that all finite abelian groups are really just things cobbled together from the group $(\mathbf{Z}, +)$ using quotients and direct sums.

18.2 Preliminaries

Looking back at Corollary 17.21, we would like to be able to say that the image of the natural map from $\oplus_{i=1}^{r} G_i$ to G is the "sum" of the subgroups G_1 through G_r. More generally, if we are in a group which may not be abelian, we would like to speak of the "product" of two subsets of the group. Therefore, to start our discussion, we generalize the idea of the product of a set with an element, by considering the product of two sets.

Definition 18.1. Let S be a set with an associative binary operation $\cdot$, and let $A, B \subseteq S$. Then we define the *setwise product* of A with B to be

$$A \cdot B = \{a \cdot b \ : \ a \in A \text{ and } b \in B\}. \tag{18.1}$$

Remark 18.2. We note that in case A or B is a singleton set, this definition agrees with the notion of a coset: we have $\{a\} \cdot B = a \cdot B$ and $A \cdot \{b\} = A \cdot b$.

Remark 18.3. Just as with cosets, it is true that setwise products are associative: that is, we always have $(A \cdot B) \cdot C = A \cdot (B \cdot C)$ for $A, B, C \subseteq S$. The proof is a straightforward consequence of the associativity of the original $\cdot$ operation. As a result, we write expressions such as $A \cdot B \cdot C$ without parentheses.

Remark 18.4. When G is an abelian group, we use additive notation for the group operation, and speak of the *setwise sum* $A + B := \{a + b \ : \ a \in A \text{ and } b \in B\}$.

DOI: 10.1201/9781003252139-18

It is natural to ask whether the setwise product of two subgroups must also be a subgroup. In general, the answer is no (see Exercise 18.1). But the situation is better if one of the subgroups is normal:

Lemma 18.5. *Let G be a group, and suppose that $H \leq G$ and $N \trianglelefteq G$. Then we have $H \cdot N = N \cdot H$ and $H \cdot N \leq G$.*

Proof. To prove the first part of the lemma, let $x \in H \cdot N$. Then $x = a \cdot b$ for some $a \in H$ and some $b \in N$. So $x \in aN$. But $aN = Na$ by Lemma 7.25, so $x = c \cdot a$ for some $c \in N$. Thus, $x \in N \cdot H$. This shows that $H \cdot N \subseteq N \cdot H$. A similar argument shows the opposite inclusion, so we have $H \cdot N = N \cdot H$.

To prove that $H \cdot N \leq G$, we use the subgroup test. First, we have $e \in H$ and $e \in N$, so $e \cdot e \in H \cdot N$. Thus $H \cdot N$ is non-empty. Next, let $x \in H \cdot N$, and write $x = a \cdot b$ with $a \in H$ and $b \in N$. Then we have $x^{-1} = b^{-1} \cdot a^{-1}$ (by Exercise 3.15). Now $b^{-1} \in N$ and $a^{-1} \in H$, so we have $x^{-1} \in N \cdot H$. But $N \cdot H = H \cdot N$ by the first part of our proof. Therefore, $H \cdot N$ is closed under inverses. Finally, choose $y \in H \cdot N$, and write $y = c \cdot d$ with $c \in H$ and $d \in N$. Then we have $x \cdot y = (a \cdot b) \cdot (c \cdot d) = a \cdot (b \cdot c) \cdot d$. Now $b \cdot c \in N \cdot H = H \cdot N$, so we can write $b \cdot c = \tilde{c} \cdot \tilde{b}$ for some $\tilde{b} \in N$ and $\tilde{c} \in H$. Therefore, $x \cdot y = a \cdot (\tilde{c} \cdot \tilde{b}) \cdot d = (a \cdot \tilde{c}) \cdot (\tilde{b} \cdot d)$. Since H and N are subgroups of G, we have $a \cdot \tilde{c} \in H$ and $\tilde{b} \cdot d \in N$. Therefore, $x \cdot y \in H \cdot N$, so $H \cdot N$ is closed under the group operation of G. This completes the proof. $\square$

Corollary 18.6. *Let G be an abelian group, and let H_1, H_2 be subgroups of G. Then the setwise sum $H_1 + H_2$ is a subgroup of G.*

Proof. Since G is abelian, every subgroup of G is normal, by Exercise 11.4. The result then follows immediately from Lemma 18.5. Alternatively, since $H_1 + H_2$ is the image of the natural map from $H_1 \oplus H_2$ to G, the result follows from Theorem 7.13. $\square$

Notation 18.7. Let G be an abelian group, and let $(G_i)_{i \in S}$ be an indexed collection of subgroups of G, where the index set S is finite. We will use the notation $\sum_{i \in S} G_i$ to denote the setwise sum of the G_i's. Note that since G is abelian, the order of the summands does not matter. Also note that this sum is a subgroup of G, by repeated application of Corollary 18.6.

We next state a condition for an abelian group to split as a direct sum, which is often easier to check than that in Proposition 17.20.

Proposition 18.8. *Let G be an abelian group, and let $\{G_i \; : \; i \in S\}$ be a finite collection of subgroups of G. Suppose that*
(i) For each $g \in G$, we can write $g = \sum_{i \in S} g_i$ with $g_i \in G_i$, and
(ii) For each $i \in S$, we have

$$G_i \bigcap \left(\sum_{j \in S - \{i\}} G_j \right) = \{0_G\},$$

the trivial subgroup.
Then the natural map $\oplus_{i \in S} G_i \to G$ is an isomorphism.

Proof. We will show that the g_i's in condition (i) are unique. Suppose that $\sum_{i \in S} g_i = \sum_{i \in S} h_i$ with g_i, $h_i \in G_i$. Let $j \in S$. Then solving for the j components and taking advantage of the fact that G is abelian, we have

$$g_j - h_j = \sum_{i \in S - \{j\}} (g_i - h_i). \tag{18.2}$$

Now the left side of Equation 18.2 belongs to G_j, while the right side belongs to $\sum_{i \in S - \{j\}} G_i$. By hypothesis, this forces both sides to be 0_G, the identity element of G. Therefore, $g_j = h_j$. Since j was an arbitrary element of S, the uniqueness of the g_i's follows. Now the proposition follows from Proposition 17.20. □

In case two subgroups are finite, there is a beautiful and useful formula for the size of their setwise product.

Lemma 18.9. *Let G be a group, and suppose that $H_i \leq G$ and $|H_i| < \infty$ for $i \in \{1, 2\}$. Then we have*

$$|H_1 \cdot H_2| = |H_1| \cdot |H_2| \ / \ |H_1 \cap H_2|. \tag{18.3}$$

Proof. Consider the function

$$\sigma : H_1 \times H_2 \to H_1 \cdot H_2, \quad (a, b) \mapsto a \cdot b. \tag{18.4}$$

From the definition of $H_1 \cdot H_2$, we see that σ is surjective. Let $w \in H_1 \cdot H_2$, and consider the set $P := \sigma^{-1}(w)$. Since σ is surjective, P is non empty; let us fix an element $(a, b) \in P$. Then $w = a \cdot b$.

Define a function $\tau : H_1 \cap H_2 \to P$ by $q \mapsto (a \cdot q, q^{-1} \cdot b)$. Suppose that $(x, y) \in P$. Then we have $x \cdot y = w = a \cdot b$, and so $a^{-1} \cdot x = b \cdot y^{-1}$. Set $z = a^{-1} \cdot x$. Since $a, x \in H_1 \leq G$, we have $z \in H_1$; similarly, $z = b \cdot y^{-1} \in H_2$. Thus $z \in H_1 \cap H_2$. Therefore, we have $\tau(z) = (a \cdot z, z^{-1} \cdot b) = (x, y)$, so τ is surjective. If $\tau(w_1) = \tau(w_2)$, then $a \cdot w_1 = a \cdot w_2$, and so $w_1 = w_2$. Therefore, τ is injective.

We now know that τ is a bijection, and so $|P| = |H_1 \cap H_2|$. Since P is the pre-image under σ of an arbitrary point in $H_1 \cdot H_2$, and since σ is surjective, the desired formula follows. □

18.3   Splitting into p-Subgroups

We now turn to the application of the preceding ideas to finite abelian groups. Our first concern is for subgroups whose order is a power of a given prime number, p.

Definition 18.10. Let $p \in \mathbf{N}$ be a prime number. A *p-group* is a finite group whose order is a power of p. A *p-subgroup* of a group G is a p-group which is a subgroup of G.

Proposition 18.11. *Let G be a finite abelian group, and let $p \in \mathbf{N}$ be a prime number. Then there is a unique maximal p-subgroup G_p of G. That is, there exists a unique subgroup $G_p \leq G$ such that*

(i) *G_p is a p-group, and*

(ii) *If H is any p-subgroup of G, then we have $H \subseteq G_p$.*

Proof. The set of all p-subgroups of G is non-empty, since the trivial subgroup is a p-group $(1 = p^0)$. Since $|G| < \infty$, there is some p-subgroup G_p of G whose size is largest among all p-subgroups of G. By construction, G_p satisfies condition (i) above. Suppose that H is a p-subgroup of G, and consider the setwise sum $K := G_p + H$. Since G is abelian, Corollary 18.6 applies to give $K \leq G$. From Lemma 18.9, we see that $|K|$ divides $|G_p| \cdot |H|$; since both G_p and H are p-groups, it follows that K is also a p-group. Also, since $0_G \in H$, we have $G_p \subseteq K$. Thus, K is a p-subgroup of G which contains G_p; by maximality of $|G_p|$, we must have $K = G_p$. Therefore, $H \subseteq G_p + H = K = G_p$, so G_p satisfies condition (ii).

As for uniqueness, suppose that $\tilde{G}_p$ is any subgroup of G satisfying both (i) and (ii). Then we may substitute G_p for H in condition (ii) for $\tilde{G}_p$ to find $G_p \subseteq \tilde{G}_p$; reversing the roles gives $\tilde{G}_p \subseteq G_p$. Thus, $\tilde{G}_p = G_p$, as desired. $\square$

The groups G_p turn out to be direct summands of G, as the next result indicates.

Proposition 18.12. *Let G be a finite abelian group, and for each prime number p, let G_p denote the maximal p-subgroup of G. Then we have*

$$G \cong \bigoplus_{p \in S} G_p, \tag{18.5}$$

via the inverse of the natural map $\oplus_{p \in S} G_p \to G$, where S is the set of primes which divide the order of G.

Proof. Let $n = |G|$, and for $p \in S$, let e_p be the exact power of p which divides n; that is, $n = \prod_{p \in S} p^{e_p}$. For each prime $p \in S$, we seek an integer n_p which satisfies

$$n_p \equiv 1 \pmod{p^{e_p}} \text{ and} \tag{18.6}$$

$$n_p \equiv 0 \pmod{q^{e_q}}, \text{ if } q \in S \text{ and } q \neq p. \tag{18.7}$$

By the Chinese Remainder Theorem (Corollary 17.12), there is a unique solution to this system of congruences modulo n. For each element $a \in G$, we set $a_p = n_p \cdot a$.

Note that we have $n_p p^{e_p} \equiv 0 \pmod{q^{e_q}}$ for every prime $q \in S$, and therefore $n_p p^{e_p} \equiv 0 \pmod{n}$. It follows that we have $p^{e_p} \cdot a_p = 0_G$, by Lemma 8.24.

Thus, the order of $\langle a_p \rangle$ must divide p^{e_p}, by Exercise 5.7. Therefore, $\langle a_p \rangle$ is a p-subgroup of G, and so we have $\langle a_p \rangle \subseteq G_p$. In particular, $a_p \in G_p$.

Next, set $k = \sum_{q \in S} n_q$. Note that for each prime $p \in S$, we have $k \equiv 1 \pmod{p^{e_p}}$, since k is a sum of terms of which one is congruent to 1 and the rest are congruent to 0 modulo p^{e_p}. By the Chinese Remainder Theorem, it follows that $k \equiv 1 \pmod{n}$. Let $v = |a|$ be the order of a as an element of G. Then $v \mid n$ by Lemma 8.24, so we also have $k \equiv 1 \pmod{v}$. By Exercise 5.7, we have $k \cdot a = 1 \cdot a = a$ since $k \equiv 1 \pmod{v}$. Therefore, $\sum_{p \in S} a_p = \sum_{p \in S} n_p \cdot a = k \cdot a = a$.

Finally, fix $p \in S$, and set $\hat{G}_p = \sum_{q \in S - \{p\}} G_q$. Note that the order of $\hat{G}_p$ is a divisor of $\prod_{q \in S - \{p\}} q^{e_q}$, and hence is relatively prime to p. The intersection $G_p \cap \hat{G}_p$ is a subgroup of both G_p and $\hat{G}_p$, by Lemma 4.20. Therefore, its order, $\left| G_p \cap \hat{G}_p \right|$, divides both $|G_p|$ and $\left| \hat{G}_p \right|$, by Lagrange's Theorem. Since these two numbers are relatively prime, it follows that $G_p \cap \hat{G}_p = \{0_G\}$. Now the theorem follows from Proposition 18.8. $\square$

Corollary 18.13. *Let G be a finite abelian group of order n. Let $n = \prod_{p \in S} p^{e_p}$ be the prime factorization of n. Let G_p be the maximal p-subgroup of G. Then we have $|G_p| = p^{e_p}$.*

Proof. By Theorem 18.12, we have $G \cong \oplus_{p \in S} G_p$, and so $|G| = |\oplus_{p \subset S} G_p| = \prod_{p \in S} |G_p|$. Since G_p is a p-group, our result follows from unique factorization in $\mathbb{Z}$. $\square$

18.4 Structure of Abelian p-Groups

Now we know that every finite abelian group decomposes into a direct sum of p-groups. A natural subject of further inquiry is the structure of abelian p-groups.

Currently, we have a modest set of machinery available to construct abelian p-groups. Namely, we can form cyclic p-groups and take direct sums. The end result of this construction will be a group of the form

$$G = \bigoplus_{i=1}^{r} G_i \tag{18.8}$$

where $G_i = \langle g_i \rangle$ and $|g_i| = p^{e_i}$ for arbitrary positive integers e_i.

Given the group G, how can we recover the elements g_i and obtain the decomposition of Equation 18.8? Let us begin with the simplest case, when $r = 1$ (we ignore the case $r = 0$, which yields the trivial group). In this case, dropping the subscripts, we have $G = \langle g \rangle \cong \mathbb{Z}_{p^e}$.

Now, we cannot quite hope to recover g given G alone, because of symmetry considerations: in general, G will have many different generators, related to each other by automorphisms of G (see Exercise 18.2). So we must be content to find any element which generates G. This is equivalent to finding an element in G of maximal order.

We could simply ask for an element whose order is $|G|$; but this approach does not generalize to the case when G is not cyclic. Instead, we dissect G into equivalence classes according to the order of its elements, and ask how we can isolate elements at the top level of this hierarchy. A fruitful idea is the following: if $|g| = p^k$ where $k > 0$, then $|pg| = p^{k-1}$. This fact can be verified at once: $p^{k-1} \cdot (p \cdot g) = p^k \cdot g = 0$, so $|pg|$ divides p^{k-1}; but on the other hand, if $m \cdot (pg) = 0$ with $m > 0$, then $(pm)g = 0$, so $pm \geq p^k$ and $m \geq p^{k-1}$. (It also follows from Exercise 18.3.)

We claim that the subgroup $\langle pg \rangle =: H$ of G contains all elements of order less than p^k—that is, all elements lower than the top level. Certainly, every element of H has order at most $|H| = p^{k-1}$. On the other hand, if $q \in G - H$, then $q = c \cdot g$ for some $c \in \mathbf{N}$ with $p \nmid c$, so $|q| = p^k$ (by Exercise 18.3).

Finally, to isolate the elements in G of maximal order, we consider the quotient group G/H. A non-trivial element of G/H is a coset $a + H$ where $a \in G - H$, and thus gives an element a of G which has maximal order.

To generalize these ideas, we replace $\langle pg \rangle$ by the set of all p-multiples of elements of G:

Notation 18.14. Let G be an abelian group and let $k \in \mathbf{Z}$. Then

$$kG := \{ka \ : \ a \in G\}.$$

Note that when $G = \langle g \rangle$, we have $pG = \langle pg \rangle$. We also have:

Lemma 18.15. *If G is an abelian group and $k \in \mathbf{Z}$, then $kG \leq G$.*

Proof. This follows from Lemma 3.44 and the Laws of Exponents (Theorem 3.41) for groups. $\square$

Now given an abelian p-group G, to see whether G decomposes as in Equation 18.8, our strategy is to look at the quotient group G/pG. Since by definition the pth multiple of every element of G lies in pG, it follows that every element of G/pG has order at most p. Before stating and proving this result formally, we make the following definitions.

Definition 18.16. Let G be a group. We say that G has *exponent* k (where k is a positive integer) if for every $g \in G$ we have $g^k = e$ (or, in additive notation, $kg = 0$) and k is the minimum such number.

Definition 18.17. Let p be prime in $\mathbf{N}$. An *elementary* abelian p-group is an abelian p-group of exponent at most p.

Remark 18.18. An abelian p-group of exponent at most p must have exponent p or 1. In the latter case, the group must be trivial.

Lemma 18.19. *Let G be an abelian p-group. Then G/pG is an elementary abelian p-group.*

Proof. First note that $pG \leq G$ by Lemma 18.15, so $pG \trianglelefteq G$ by Exercise 11.4. Thus the quotient $G/pG =: Q$ is defined; and this quotient is abelian by Exercise 11.5. Since $|G|$ is a power of p and $|Q| = |G| / |pG|$, then $|Q|$ is also a power of p; so Q is a p-group. Finally, let $\alpha \in Q$. Then we have $\alpha = a + pG$ for some $a \in G$. So $p \cdot \alpha = pa + pG$. But $pa \in pG$, so we have $pa + pG = 0 + pG$ (by Exercise 8.2) $= 0_Q$. Thus Q has exponent at most p, which completes the proof. $\qquad\square$

When we use additive notation for an elementary abelian p-group G, we find that given any two elements $a, b \in G$ and any $m \in \mathbf{Z}$, we can form the sum $a + b$ and the product $m \cdot a$. Further, since G has exponent p, we have $p \cdot a = 0$ (the identity element of G). It follows from Exercise 5.7 that $m \cdot a$ only depends on m modulo p. A flash of insight shows us that it is not so much the integer m that is multiplying a in the expression $m \cdot a$, but rather the coset $m + p\mathbf{Z}$, an element of $\mathbf{Z}/p\mathbf{Z}$. Now $\mathbf{Z}/p\mathbf{Z}$ is a *field*, and so it seems we may be dealing with a vector space. The following result confirms this hunch.

Lemma 18.20. *Let $(G, +)$ be an elementary abelian p-group, where p is prime. Let $\mathbf{F}_p = \mathbf{Z}/p\mathbf{Z}$. Then the operation $\cdot : \mathbf{F}_p \times G \to G$ given by $(a \mid p\mathbf{Z}) \cdot g = a \cdot g$ makes G into an $\mathbf{F}_p$-vector space.*

Proof. The main point is to show that $\cdot$ is well-defined; this follows from Exercise 5.7. The vector space axioms follow from the Laws of Exponents, Theorem 3.41 and Lemma 3.44. We leave the details to the reader in Exercise 18.5 (a). $\qquad\square$

With vector spaces come many useful notions, including the notions of a basis and dimension. Exploiting the natural vector space structure of an elementary abelian p-group, we find below that every elementary abelian p-group is a direct sum of cyclic p-groups. The reader is encouraged to verify that the isomorphism of Corollary 18.22 below can be realized by mapping a basis of the vector space G to a set of generators for the terms $\mathbf{Z}_p$ in the direct sum.

Notation 18.21. For consistency and clarity of notation, when p is a positive prime integer we reserve $\mathbf{Z}_p$ to stand for the additive group $(\mathbf{Z}/p\mathbf{Z}, +)$ and $\mathbf{F}_p$ for the field $(\mathbf{Z}/p\mathbf{Z}, +, \cdot)$. Thus, in the following result, the isomorphism is a group isomorphism.

Corollary 18.22. *Let G be an elementary abelian p-group of order p^r. Then $G \cong \oplus_{i=1}^{r} \mathbf{Z}_p$.*

Proof. Endow G with the natural vector space structure of Lemma 18.20. Since G is finite as a set, and G certainly spans G as an $\mathbf{F}_p$-vector space, there must be a minimal spanning set B for G over $\mathbf{F}_p$. This set B must be a basis of G over $\mathbf{F}_p$, by Theorem 13.23. By Exercise 13.12, we have $G \cong \mathbf{F}_p^d$ as $\mathbf{F}_p$-vector spaces, where $d = |B|$. We must have $d = r$ since $p^r = |G| = \left|\mathbf{F}_p^d\right| = p^d$. Notice that $(\mathbf{F}_p^r, +) = \oplus_{i=1}^r \mathbf{Z}_p$. Since a vector space isomorphism is also an isomorphism of additive groups, our result follows. $\square$

Suppose that we are given a finite direct sum of cyclic p-groups

$$G = \bigoplus_{i=1}^{r} \langle g_i \rangle. \tag{18.9}$$

Results such as Proposition 17.20 and Corollary 17.21 lead us to think that, as in the case of a basis of a vector space, a set of generators $\{g_i\}$ in such a direct sum decomposition must generate the group G "efficiently." One possible measure of such "efficiency" (which worked for vector space bases) is that every element of G should have a unique representation in the form

$$c_1 g_1 + c_2 g_2 + \cdots + c_r g_r \tag{18.10}$$

with integer coefficents c_i. Now, we cannot hope for such a condition to be true in a finite group, because Expression 18.10 only depends on the values c_i modulo the order of g_i. So we should restrict these values to $0 \le c_i < |g_i|$. The total number of such expressions is equal to the product of the $|g_i|$. Therefore, a necessary condition for Equation 18.9 to hold is that the product of the orders of the g_i is equal to the number of elements of G; note that this is a lower bound for the value of the product of the orders of the g_i given that they generate G.

The following result utilizes this approach to prove that every finite abelian p-group must in fact be a direct sum of the groups generated by such elements.

Proposition 18.23. *Let G be an abelian p-group. Then G is a direct sum of cyclic groups. Further, if $g_1, \ldots, g_n$ are elements of G which generate G as a group, then the natural map $\sigma : \oplus_{i=1}^n \langle g_i \rangle \to G$ is an isomorphism iff $\prod_{i=1}^n |g_i| = |G|$.*

Proof. Suppose that G is an abelian p-group. Let $S = \{g_1, \ldots, g_n\}$ be a subset of G such that $\langle S \rangle = G$, and such that the product of the orders of the g_i is minimal. Set $G_i = \langle g_i \rangle$. Assume for a contradiction that there is a relation of the form $x_1 + \cdots + x_n = 0_G$ with $x_i \in G_i$ and x_i not all 0_G.

Since $x_i \in \langle g_i \rangle$, by Lemma 5.6 we can write $x_i = k_i g_i$ for some $k_i \in \{0, 1, 2, \ldots, r_i - 1\}$, where $r_i = |g_i|$. Since not all of the x_i are 0_G, then not all of the k_i can be 0, so there is a maximal power p^t of p which divides all of the k_i. By maximality, there is some i for which $p^t \mid\mid k_i$ (note: this is read "p^t exactly divides k_i," and means that p^t divides k_i but p^{t+1} does not divide k_i); without loss of generality, we have $p^t \mid\mid k_1$.

Now set $c_i = k_i / p^t$ for all i such that $1 \le i \le n$. Set $\tilde{g}_1 = \sum_{i=1}^n c_i g_i$, and set $\tilde{g}_i = g_i$ if $2 \le i \le n$. Finally, set $\tilde{S} = \{\tilde{g}_1, \ldots, \tilde{g}_n\}$. Notice that $p \nmid c_1$, but $|g_1|$ is a power of p by Lagrange's Theorem. Therefore, $\gcd(c_1, |g_1|) = 1$;

and so we have $\langle c_1 g_1 \rangle = \langle g_1 \rangle$ by Exercise 18.3. It follows that $g_1 \in \langle c_1 g_1 \rangle = \langle \tilde{g}_1 - \sum_{i=2}^{n} c_i g_i \rangle = \langle \tilde{g}_1 - \sum_{i=2}^{n} c_i \tilde{g}_i \rangle \subseteq \langle \tilde{S} \rangle$. Since we also have $g_i = \tilde{g}_i \in \langle \tilde{S} \rangle$ for $i \geq 2$, we see that $S \subseteq \langle \tilde{S} \rangle$, so $G = \langle S \rangle \subseteq \langle \tilde{S} \rangle$ and $\langle \tilde{S} \rangle = G$. But in going from S to $\tilde{S}$ we have reduced the product of the orders of the g_i, since $|\tilde{g}_i| = |g_i|$ for $i \geq 2$ and $|\tilde{g}_1| \leq p^t \leq k_1 < |g_1|$.

This contradiction shows that there is no non-trivial relation among the G_i's, and so the natural map $\oplus_{i=1}^{n} G_i \to G$ is an isomorphism by Corollary 17.21. Now we have $|G| = \prod_{i=1}^{n} |G_i| = \prod_{i=1}^{n} |g_i|$, which finishes the proof. $\square$

Example 18.24. Suppose that G is an abelian group of order 7^3. Then by Proposition 18.23, G must be a direct sum of cyclic groups. There aren't very many ways for this to happen, since each non-trivial cyclic direct summand of G must have order 7^m where $m \in \{1, 2, 3\}$, and these powers must total 3. Here are the possibilities: $G \cong \mathbf{Z}/7\mathbf{Z} \oplus \mathbf{Z}/7\mathbf{Z} \oplus \mathbf{Z}/7\mathbf{Z}$ or $G \cong \mathbf{Z}/7\mathbf{Z} \oplus \mathbf{Z}/7^2\mathbf{Z}$ or $G \cong \mathbf{Z}/7^3\mathbf{Z}$. Notice that each of these possibilities really is distinct, because the exponents of these groups are 7, 7^2, and 7^3, respectively, so no two of these groups are isomorphic.

Generalizing this example, we see that an abelian group of order p^n, where p is prime, corresponds to a way of writing n as a sum of positive integers. Indeed, if $n = \sum_{i=1}^{k} m_i$ with $m_i \in \mathbf{Z}^+$, then we can form the group $\oplus_{i=1}^{k} \mathbf{Z}/p^{m_i}\mathbf{Z}$, which is an abelian group of order p^n; and by Proposition 18.23, every abelian group of order p^n has this form. Because changing the order of the direct summands does not change the result (up to isomorphism), we are free to require $m_1 \geq m_2 \geq \cdots \geq m_k$.

It is natural to ask whether two different non-increasing sequences $\ell_1 \geq \cdots \geq \ell_j$ and $m_1 \geq \cdots \geq m_k$ necessarily give rise to non-isomorphic groups under our direct sum construction. The reader is encouraged at this point to experiment with examples to arrive at a conjecture about this question.

Let us attempt to form a strategy to prove that an abelian p-group in fact determines a unique non-increasing exponent sequence. The idea is that this sequence should be somehow "intrinsic" to the group, that is, invariant under isomorphism. We wish to study this notion of invariance under isomorphism more generally and precisely. To do this, it is more convenient to talk about a *subgroup* H which is invariant under isomorphism than about sequences of numbers. But at most, we can hope for H to be invariant only under those isomorphisms which carry the group G to *itself*.

Definition 18.25. Let G be a group, and let $H \leq G$. We say that H is a *characteristic* subgroup of G if for all automorphisms σ of G, we have $\sigma(H) = H$.

Remark 18.26. An automorphism must fix a characteristic subgroup as a set, but not necessarily pointwise.

Remark 18.27. The concept of a characteristic subgroup is completely separate from the notion of the characteristic of a commutative ring (Definition 16.21).

As a rule of thumb, we expect that any subgroup which can be defined "generically"—that is, in an arbitrary group (or arbitrary abelian group)—should be a characteristic subgroup. Examples include the center of a group and the commutator subgroup (Exercise 18.7). Similarly, we have:

Lemma 18.28. *Let G be an abelian group, and let $k \in \mathbf{Z}$. Then kG is a characteristic subgroup of G.*

Proof. Let $\sigma \in \mathrm{Aut}(G)$, and let $\alpha \in kG$. Then $\alpha = kg$ for some $g \in G$. Therefore, $\sigma(\alpha) = \sigma(kg) = k\sigma(g)$ (by Proposition 7.4) $\in kG$. This shows that $\sigma(kG) \subseteq kG$. Since $\sigma^{-1} \in \mathrm{Aut}(G)$, we similarly obtain $\sigma^{-1}(kG) \subseteq kG$, so that $kG \subseteq \sigma(kG)$. Thus $\sigma(kG) = kG$, as desired.   □

We also expect intrinsically defined characteristic subgroups to be preserved under isomorphism:

Lemma 18.29. *Let $k \in \mathbf{Z}$, and suppose that $\sigma : G \to H$ is an isomorphism of abelian groups. Then we have $\sigma(kG) = kH$.*

Proof. The proof is similar to that of Lemma 18.28.   □

What does kG actually look like for an abelian group G? Consider the special case that $G = \mathbf{Z}/p^e\mathbf{Z}$ is a cyclic p-group. Then kG is cyclic, generated by $k + p^e\mathbf{Z}$ (see Exercise 5.6). By Exercise 18.3, we see that $kG \cong \mathbf{Z}/(p^e/\gcd(k,p^e))\mathbf{Z}$. In particular, when $k = p$ and $e \geq 1$, we find that

$$p(\mathbf{Z}/p^e\mathbf{Z}) \cong \mathbf{Z}/p^{e-1}\mathbf{Z}. \tag{18.11}$$

We can handle a general abelian p-group by combining this equation with the observation that

$$k(A \oplus B) = (kA) \oplus (kB) \tag{18.12}$$

for every integer k and all abelian groups A and B.

We use these facts to prove that the exponent sequence of an abelian p-group is in fact isomorphism-invariant.

Lemma 18.30. *Let p be a prime number, and let $\ell_1 \geq \cdots \geq \ell_j$ and $m_1 \geq \cdots \geq m_k$ be two non-increasing sequences of positive integers. Then $\oplus_{i=1}^{j}\mathbf{Z}/p^{\ell_i}\mathbf{Z} \cong \oplus_{i=1}^{k}\mathbf{Z}/p^{m_i}\mathbf{Z}$ iff $j = k$ and $\ell_i = m_i$ for all i.*

Proof. ($\Leftarrow$): This is trivial. ($\Rightarrow$): Set $G = \oplus_{i=1}^{j}\mathbf{Z}/p^{\ell_i}\mathbf{Z}$ and $H = \oplus_{i=1}^{k}\mathbf{Z}/p^{m_i}\mathbf{Z}$. Let $\sigma : G \to H$ be an isomorphism. Let $e = \sum_{i=1}^{j}\ell_i$. We proceed by induction on e. In the base case, $e = 1$, we have $j = 1$, $\ell_1 = 1$, and $G \cong \mathbf{Z}/p\mathbf{Z} \cong H$, so just from comparing cardinalities, we see that $k = 1$ and $m_1 = 1$. Inductively, assume that $e > 1$ and that the result is true for all sequences with sum less than e. By Lemma 18.29 we have $\sigma(pG) = pH$, so the restriction of σ to pG gives an isomorphism $pG \cong pH$. Now $pG \cong \oplus_{i=1}^{j}\mathbf{Z}/p^{\ell_i-1}\mathbf{Z}$ and $pH \cong \oplus_{i=1}^{k}\mathbf{Z}/p^{m_i-1}\mathbf{Z}$, so we would like to use our inductive hypothesis on these smaller p-groups. The only problem is that some of the exponents may be 0 here. To address this issue, let us separate the exponent sequences into $S_0 := \{i : \ell_i = 1\}$, $S_+ := \{i : \ell_i > 1\}$, $T_0 := \{i : m_i = 1\}$, and $T_+ := \{i : m_i > 1\}$. The terms in the direct sums for pG and pH coming from S_0 and T_0 do not contribute anything, by Exercise 18.10. Thus we have $pG \cong \oplus_{i \in S_+} \mathbf{Z}/p^{\ell_i - 1}\mathbf{Z}$ and $pH \cong \oplus_{i \in T_+} \mathbf{Z}/p^{m_i - 1}\mathbf{Z}$. Now our inductive hypothesis applies to give $S_+ = T_+$ and $\ell_i - 1 = m_i - 1$ for all $i \in S_+$. Since $G \cong H$, we have $|G| = |H|$, so $\sum_{i=1}^{j} \ell_i = \sum_{i=1}^{k} m_i$. We can write this as $\sum_{i \in S_0} \ell_i + \sum_{i \in S_+} \ell_i = \sum_{i \in T_0} m_i + \sum_{i \in T_+} m_i$. The sums involving S_+ and T_+ are equal, so we have $\sum_{i \in S_0} \ell_i = \sum_{i \in T_0} m_i$. But since $\ell_i = 1$ for all $i \in S_0$ and $m_i = 1$ for all $i \in T_0$, we have $|S_0| = |T_0|$ and thus $S_0 = T_0$. This completes the proof. $\qquad\square$

18.5 The Fundamental Theorem

Finally, we combine our results on decomposing abelian groups into a summary result.

Theorem 18.31 (Fundamental Theorem of Finite Abelian Groups). *Every finite abelian group is a direct sum of cyclic groups. If G is a finite abelian group and S is the set of all primes which divide the order of G, then we have*

$$G \cong \bigoplus_{p \subseteq S} \left(\oplus_{i=1}^{n_p} G_{p,i} \right), \tag{18.13}$$

where $G_{p,i}$ is cyclic of order $p^{e_{p,i}}$, for uniquely determined positive integers n_p and non-increasing sequences of positive integers $e_{p,1} \geq \cdots \geq e_{p,n_p}$.

Proof. Let G be a finite abelian group. We can decompose G into a direct sum of abelian p-groups by Theorem 18.12, and then decompose each of these factors into a direct sum of cyclic groups by Theorem 18.23, to get a representation of G in the form of Formula 18.13.

Suppose we are given a representation of G as in Formula 18.13. The inner direct sum gives a maximal p-subgroup of the right-hand side, which must be isomorphic to G_p, the unique maximal p-subgroup of G, by Exercise 18.9. The uniqueness of the decomposition of G_p follows from Lemma 18.30. $\qquad\square$

We have now shown that every finite abelian group is a direct sum of cyclic groups of prime-power order. Our effort at understanding finite abelian groups may as well stop here, since by Exercise 18.11, there is no non-trivial way to decompose a cyclic p-group.

18.6   Exercises

Exercise 18.1. Find an example of a group G and two subgroups H_1, H_2 of G such that the setwise product $H_1 \cdot H_2$ is not a subgroup of G. Hint: take $G = S_3$.

Exercise 18.2. Let G be a finite cyclic group. Suppose that $G = \langle g \rangle = \langle \tilde{g} \rangle$ where $g, \tilde{g} \in G$. Prove that there is a unique automorphism $\sigma \in \mathrm{Aut}(G)$ such that $\sigma(g) = \tilde{g}$.

Exercise 18.3. Let G be a group, and suppose that $g \in G$ is an element of finite order n.

(a) Prove that we have $|g^a| = n/\gcd(n, a)$ for all $a \in \mathbf{Z}$.

(b) Conclude that $\langle g^a \rangle = \langle g \rangle$ if and only if $\gcd(a, n) = 1$.

(c) Show finally that the number of generators of $\langle g \rangle$ is equal to $\phi(n)$. (See Exercise 17.7.)

Exercise 18.4. Let G be a finite cyclic group. Prove the following converse of Lagrange's Theorem for G: If m divides the order of G, then G has a subgroup of order m. Hint: Exercise 18.3 may be useful. (Also compare Exercise 8.11.)

Exercise 18.5. Let G be an abelian p-group.

(a) Prove that G/pG is naturally an F_p-vector space, where F_p denotes the field $(\mathbf{Z}/p\mathbf{Z}, +, \cdot)$. (See Lemma 18.20.)

(b) Prove that if $\{g_1 + pG, \dots, g_n + pG\}$ is a basis of G/pG over F_p, then the set $S := \{g_1, \dots, g_n\}$ generates G as a group. (This is a version of a result known as *Nakayama's Lemma*.) Hint: Let $H = \langle S \rangle$ and prove inductively that for each $k \in \mathbf{N}$, every element of G can be written in the form $h_1 + p h_2 + p^2 h_3 + \cdots + p^{k-1} h_k + p^k a$ for some $h_1, \dots, h_k \in H$ and some $a \in G$.

(c) Prove or disprove that for every S constructed as in part (b) above, we must have $G \cong \oplus_{i=1}^n \langle g_i \rangle$.

Exercise 18.6. Let G be an abelian group and $A, B \leq G$. Prove or disprove that if $G \cong A \oplus B$, then the natural map $f : A \oplus B \to G$ must be an isomorphism.

Exercise 18.7. Let G be a group. Prove that the following subgroups of G are characteristic:

(a) The center, $Z(G)$, of G (see Exercise 9.11).

(b) The commutator subgroup of G (see Definition 10.8).

Exercise 18.8. Let G be a finite abelian group with $|G| = mn$ and $\gcd(m, n) = 1$. Prove that nG is the unique subgroup of G of order m.

Exercise 18.9. Let G be a finite abelian group, and let G_p denote the maximal p-subgroup of G.

(a) Prove that G_p is a characteristic subgroup of G.

(b) Prove that if $\sigma : G \to H$ is an isomorphism, then $\sigma(G_p) = H_p$.

Exercise 18.10. Prove that if G is an abelian group and T is a trivial group (recall this means $|T| = 1$), then we have $G \oplus T \cong G$.

Exercise 18.11. Prove that if G is a cyclic group of prime power order and $G \cong A \oplus B$ for two abelian groups A and B, then either A or B is trivial.

Exercise 18.12. Suppose that $G = \oplus_{i=1}^{n} G_i$, where the G_i are finite abelian groups.

(a) Prove that for any element $g = (g_1, \ldots, g_n)$ of G, we have $|g| = \mathrm{lcm}(|g_1|, \ldots, |g_n|)$.

(b) Prove that $\mathrm{exponent}(G) = \mathrm{lcm}(\mathrm{exponent}(G_1), \ldots, \mathrm{exponent}(G_n))$.

Exercise 18.13. Let G be an abelian p-group of exponent p^e, and let $g \in G$ be any element of order p^e. Prove that there exists a subgroup $A \leq G$ such that the natural map $\langle g \rangle \oplus A \to G$ is an isomorphism. Thus, $\langle g \rangle$ is a "natural" direct summand of G.

Exercise 18.14. Prove that if G is a finite abelian group whose order is divisible by the prime number p, then G contains an element of order p.

Exercise 18.15. Suppose that the natural map $\oplus_{i=1}^{n} G_i \to G$ is an isomorphism, where G is an abelian group and $G_i \leq G$ for each i. Let $H_i \leq G_i$ be subgroups of G_i ($1 \leq i \leq n$). Prove that the natural map $\oplus_{i=1}^{n} H_i \to H$ is an isomorphism, where $H := \sum_{i=1}^{n} H_i$.

Exercise 18.16. Prove the following converse of Lagrange's Theorem for finite abelian groups: If G is a finite abelian group and m divides the order of G, then G has a subgroup of order m. Hint: Exercises 18.4 and 18.15 may be useful here.

Exercise 18.17. Let G be a finite abelian group.

(a) Prove that G is cyclic if and only if for every prime p which divides the order of G, the maximal p-subgroup of G is cyclic. (G is "cyclic iff locally cyclic.")

(b) Show that the statement of part (a) is false if we remove the hypothesis that G is abelian, and we replace "the" maximal p-subgroup with *every* maximal p-subgroup.

Exercise 18.18. A positive integer n is called *squarefree* if no perfect square greater than 1 is a factor of n. Prove that every finite abelian group of squarefree order is cyclic.

Exercise 18.19. For a positive integer n, let C_n denote the number of isomorphism classes of abelian groups of order n.

(a) Why must C_n be finite?

(b) Prove that if $\gcd(m, n) = 1$, then $C_{mn} = C_m \cdot C_n$.

(c) Find the number of isomorphism classes of abelian groups of order 1000000. (You do not need to list all of the isomorphism classes explicitly.)

Exercise 18.20. Let G be the free abelian group on a set of n elements (see Exercise 10.5). Prove that $G \cong \oplus_{i=1}^{n} \mathbf{Z}$.

19

Group Actions

19.1 Groups Acting on Sets

We have seen several examples of groups whose elements are functions, and whose group operation is composition of functions. Here are two examples:

(1). $\mathrm{Sym}(S)$ is the group of all permutations of the set S, whose elements are bijective functions from S to S. If $\sigma, \tau \in \mathrm{Sym}(S)$ and $x \in S$, then we have $(\sigma \cdot \tau)(x) = \sigma(\tau(x))$. We also have $e(x) = x$, where e is the identity element of $\mathrm{Sym}(S)$.

(2). $\mathrm{GL}_n(F)$ is the group of all invertible $n \times n$ matrices over the field F (see Exercises 13.15 through 13.17). Since matrices represent linear transformations, we may identify a matrix $M \in \mathrm{GL}_n(F)$ as a linear transformation $M : V \to V$, for a vector space V over F with $\dim_F(V) = n$ and a fixed basis of V. Again, if $M, N \in \mathrm{GL}_n(F)$ and $v \in V$, then we have $(M \cdot N)(v) = M(N(v))$, and we also have $I(v) = v$, where I is the identity element of $\mathrm{GL}_n(F)$.

We will now attempt to generalize these examples. Given a group G and a set S, we want to define what it means for G to "act on" S. We only need to capture the idea that elements of G behave like functions from S to S under composition, and that the identity element of G behaves like the identity function.

Definition 19.1. Let G be a group and let S be a set. A *group action* of G on S is a function $\cdot : G \times S \to S$ such that for all $g, h \in G$ and all $x \in S$, we have

    (i) $(gh) \cdot x = g \cdot (h \cdot x)$ and
    (ii) $e \cdot x = x$.

Example 19.2. As expected, if S is any non-empty set, then the group $G = \mathrm{Sym}(S)$ acts on S via the rule $\sigma \cdot x = \sigma(x)$ for $\sigma \in G$ and $x \in S$. In fact, if $H \leq G$ is any subgroup of $\mathrm{Sym}(S)$, then H acts on S in the same way. For a particular instance, take $S = \{1, 2, 3, 4, 5\}$, $h = \begin{bmatrix} 1 & 2 & 3 & 4 & 5 \\ 4 & 1 & 3 & 2 & 5 \end{bmatrix}$, and $H = \langle h \rangle$. The reader can verify that $|h| = 3$, so that $H = \{e, h, h^2\}$. Now the different elements of S are not treated equally by the action of H: the elements 3 and 5 always get sent to themselves when we apply any element of H, whereas the elements 1, 2, and 4 get sent to each other in various ways.

DOI: 10.1201/9781003252139-19

Example 19.2 leads us to ask in general: given a group G acting on a set S, and an element $x \in S$, what are the possibilities for where x can be sent by elements of G?

Definition 19.3. Given an action $\cdot$ of a group G on a set S, and an element $x \in S$, the *orbit* of x under the action of G is

$$\mathrm{Orb}_G(x) := \{g \cdot x \ : \ g \in G\}.$$

It only makes sense that the more group elements send x to itself, the smaller the orbit of x can be. The next result captures this idea nicely.

Lemma 19.4 (Orbit-Stabilizer Lemma). *Let G be a finite group acting on a set S. For $x \in S$, define the stabilizer of x in G to be*

$$\mathrm{Stab}_G(x) := \{g \in G \ : \ g \cdot x = x\}.$$

Then:

(i) $\mathrm{Stab}_G(x) \leq G$, and
(ii) $|\mathrm{Orb}_G(x)| = [G \ : \ \mathrm{Stab}_G(x)]$.

Proof. Let $x \in S$.

(i). We use the subgroup test. Since $e \cdot x = x$ (by Definition 19.1, part (ii)), we have $e \in \mathrm{Stab}_G(x)$. Let $g, h \in \mathrm{Stab}_G(x)$. Then we have $(g \cdot h) \cdot x = g \cdot (h \cdot x)$ (by Definition 19.1, part (i)) $= g \cdot x = x$. Thus, $g \cdot h \in \mathrm{Stab}_G(x)$. We also have $g^{-1} \cdot x = g^{-1} \cdot (g \cdot x) = (g^{-1} \cdot g) \cdot x = e \cdot x = x$, so $g^{-1} \in \mathrm{Stab}_G(x)$.

(ii). Set $T = \mathrm{Stab}_G(x)$. We claim that for all $a, b \in G$, we have $a \cdot x = b \cdot x$ iff a and b belong to the same left coset of T in G. For if $a, b \in cT$ for some $c \in T$, then we have $a = c \cdot t$ for some $t \in T$, and so $a \cdot x = (c \cdot t) \cdot x = c \cdot (t \cdot x)$ $= c \cdot x$ (since $t \in \mathrm{Stab}_G(x)$), and similarly, $b \cdot x = c \cdot x$. Conversely, if $a \cdot x = b \cdot x$, then we have $a^{-1} \cdot (a \cdot x) = a^{-1}(b \cdot x)$, from which we see that $x = (a^{-1} \cdot b) \cdot x$; hence $a^{-1} \cdot b \in T$, so that both a and b belong to the coset aT. This establishes a bijection from the set of all left cosets of T in G to the orbit of x under the action of G, given by $aT \mapsto a \cdot x$. It follows that $|\mathrm{Orb}_G(x)|$ is the number of left cosets of T in G, which is by definition equal to the index of T in G, as desired.   □

Example 19.5. Let S, h, and H be as in Example 19.2. Then the orbit of 1 under the action of H is the set $\{e \cdot 1, h \cdot 1, h^2 \cdot 1\} = \{1, 4, 2\}$. The orbit of 3 under the action of H is the set $\{e \cdot 3, h \cdot 3, h^2 \cdot 3\} = \{3\}$. The orbit of 5 under the action of H, similarly, is the set $\{5\}$. Notice that the orbit of 2 under this action is $\{e \cdot 2, h \cdot 2, h^2 \cdot 2\} = \{2, 1, 4\}$, which is the same as the orbit of 1. We extrapolate from this example as follows.

Lemma 19.6. *Suppose that $\cdot$ is a group action of G on S. Then the orbits of the elements of S form a partition of S.*

Proof. Given two orbits O and O', we must show that they are either the same or disjoint. So suppose that they are not disjoint, and let $x \in O \cap O'$. Since O is an orbit, we have $O = \mathrm{Orb}_G(y)$ for some $y \in S$. Since $x \in O = \{g \cdot y \; : \; g \in G\}$, we have

$$x = a \cdot y \tag{19.1}$$

for some $a \in G$. Now if $b \in G$, we have $b \cdot x = b \cdot (a \cdot y) = (b \cdot a) \cdot y \in \mathrm{Orb}_G(y)$. Therefore, $\mathrm{Orb}_G(x) \subseteq \mathrm{Orb}_G(y)$. But we can solve Equation 19.1 for y to get $y = a^{-1} \cdot x$, showing that $y \in \mathrm{Orb}_G(x)$. Thus we also get $\mathrm{Orb}_G(y) \subseteq \mathrm{Orb}_G(x)$. We now can say $\mathrm{Orb}_G(x) = O$. But likewise, we must have $\mathrm{Orb}_G(x) = O'$, so $O = O'$, as desired.

Finally, every element of S is in some orbit: namely, if $x \in S$, then $x \in \mathrm{Orb}_G(x)$, since we have $e \cdot x = x$. This finishes the proof. $\square$

Suppose as in Examples 19.2 and 19.5 that h is a permutation of a finite set S, and $x \in S$. Then we can form $\langle h \rangle =: H \leq \mathrm{Sym}(S)$. Now H acts on S via $a \cdot y = a(y)$ for $a \in H$ and $y \in S$. Because H is *cyclic*, we have an especially nice way to represent the orbit $\mathrm{Orb}_H(x)$. Namely, $\mathrm{Orb}_H(x) = \{h^0 \cdot x, h^1 \cdot x, \ldots, h^{n-1} \cdot x\} = \{x, h \cdot x, h^2 \cdot x, \ldots, h^{n-1} \cdot x\}$, where $n = |H| = |h|$. We note that there may be fewer than n elements in the orbit of x under H. But we have $\mathrm{Stab}_H(x) \leq H$, so $\mathrm{Stab}_H(x) = \langle h^r \rangle$ for some $r \in \mathbf{N}$ with $r \mid n$ (by Exercise 8.11). We leave it to the reader to verify that $|\mathrm{Orb}_H(x)| = r$ and $\mathrm{Orb}_H(x) = \{x, h \cdot x, h^2 \cdot x, \ldots, h^{r-1} \cdot x\}$. Supposing as we have that both x and h are given, then the set $\mathrm{Orb}_H(x)$ has a natural ordering given in precisely this fashion.

Notation 19.7 Let S be a finite set. Let $\sigma \in \mathrm{Sym}(S)$, and let $x \in S$. Then the *cycle* of σ corresponding to x is the ordered r-tuple $(x, \sigma x, \sigma^2 x, \ldots, \sigma^{r-1} x)$, where $r = |\mathrm{Orb}_{\langle \sigma \rangle}(x)|$. We interpret such r-tuples as permutations in their own right, as follows: if $x_0, x_1, \ldots, x_{r-1}$ are distinct elements of S, then by $(x_0, x_1, \ldots, x_{r-1})$ we mean the permutation $\tau \in \mathrm{Sym}(S)$ defined by $\tau(x_i) = x_{i+1}$ if $0 \leq i < r-1$, $\tau(x_{r-1}) = x_0$, and $\tau(y) = y$ for $y \notin \{x_0, x_1, \ldots, x_{r-1}\}$. We say that $(x_0, x_1, \ldots, x_{r-1})$ is an *r-cycle*.

Cycle notation is useful for understanding permutations of finite sets; see Exercises 19.5 through 19.7.

The next result strengthens the connection between group actions and symmetric groups.

Lemma 19.8. *Let G be a group and let S be a set. Let $\cdot$ be a group action of G on S. For each $g \in G$, let $\sigma_g : S \to S$ be defined by $\sigma_g(x) = g \cdot x$. Then $\sigma_g \in \mathrm{Sym}(S)$, and the function $f : G \to \mathrm{Sym}(S)$, $g \mapsto \sigma_g$ is a group homomorphism. The collection of all group actions of G on S is in bijective correspondence with the set of all group homomorphisms from G to $\mathrm{Sym}(S)$ via the map $\cdot \mapsto f$.*

Proof. See Exercise 19.1. $\square$

An important example of group actions is given by the following result.

Lemma 19.9. *Let G be a group. The function $\Delta \; : \; G \times G \to G$ given by $g \Delta h = g h g^{-1}$ is a group action of G on G, and for this group action we have $Stab_G(h) = C_G(h)$ (see Exercise 4.12).*

Proof. Let $f, g, h \in G$. We have $e \Delta h = e h e^{-1} = e h e = h$ and $(fg) \Delta h = (fg) h (fg)^{-1} = f g h g^{-1} f^{-1} = f(g \Delta h) f^{-1} = f \Delta (g \Delta h)$, as required. The result about stabilizers is immediate. $\qquad\square$

Definition 19.10. The group action described in Lemma 19.9 is called *conjugation*. An orbit of an element of G under conjugation is called a *conjugacy class*.

Example 19.11. In a similar vein, if G is a group and $X \subseteq G$ is any nonempty subset of G, then we can form the set of all conjugates of X, namely $S := \{g X g^{-1} \; : \; g \in G\}$. G acts on S via conjugation in the obvious way: for $g \in G$ and $Y \in S$, let $g \Delta Y = g Y g^{-1}$. Since Y has the form $Y = h X h^{-1}$ for some $h \in G$, we have $g Y g^{-1} = g h X h^{-1} g^{-1} = (gh) X (gh)^{-1} \in S$, so this action is well-defined.

Example 19.12. Set $F = \mathbf{Q}$ and $K = \mathbf{Q}[\sqrt[4]{2}, i]$, so that K is the unique splitting field for $f := x^4 - 2$ over $\mathbf{Q}$ in $\mathbf{C}$ (see Exercise 16.25). Then K/F is a finite Galois extension; set $G = \mathrm{Gal}(K/F)$. We have $G \leq \mathrm{Aut}(K) \leq \mathrm{Sym}(K)$, so the inclusion map $G \hookrightarrow \mathrm{Sym}(K)$ is a group homomorphism. By Lemma 19.8, we get an action of G on K. This action is given by $\sigma \cdot \alpha = \sigma(\alpha)$ for $\sigma \in G$ and $\alpha \in K$. The set of all roots of f in K is $R = \{i^j \sqrt[4]{2} : j \in \{0, 1, 2, 3\}\}$. Every element of K has the form $\sum c_{j,k} (\sqrt[4]{2})^j i^k$ with $c_{j,k} \in \mathbf{Q}$, where $j \in \{0, 1, 2, 3\}$ and $k \in \{0, 1\}$. Let us find the orbits of various elements of K under the action of G. From Exercise 16.25, we know that for each $j \in \{0, 1, 2, 3\}$ and $k \in \{1, 3\}$, there is an element $\sigma \in G$ such that $\sigma(\sqrt[4]{2}) = i^j \sqrt[4]{2}$ and $\sigma(i) = i^k$; and these 8 possibilities for σ give all the elements of G. So, for example, we have $\mathrm{Orb}_G(i) = \{i, i^3\} = \{i, -i\}$ and $\mathrm{Orb}_G(\sqrt[4]{2}) = \{i^j \sqrt[4]{2} : j \in \{0, 1, 2, 3\}\} = \{\sqrt[4]{2}, i\sqrt[4]{2}, -\sqrt[4]{2}, -i\sqrt[4]{2}\}$. Since $\sqrt{2} = (\sqrt[4]{2})^2 \in K$, we can compute $\mathrm{Orb}_G(\sqrt{2} + 5i \sqrt[4]{2}) = \{\sqrt{2} + 5i \sqrt[4]{2}, -\sqrt{2} - 5 \sqrt[4]{2}, \sqrt{2} - 5i \sqrt[4]{2}, -\sqrt{2} + 5 \sqrt[4]{2}\}$. We note that the orbit of an element $\alpha \in K$ under the action of G is just the set of all conjugates of α over F. Further, if we know $\sigma(r)$ for every $r \in R$, then we know $\sigma(a)$ for every $a \in K$—this is because $K = F[R]$, and σ must fix every element of F. Notice also that the action of G on K restricts nicely to an action of G on R, because $\sigma(R) = R$ (setwise, not necessarily pointwise!) for any $\sigma \in G$. Thus, we want to say that there is a group homomorphism $\omega : G \to \mathrm{Sym}(R)$, and further that ω loses no information about G—i.e., ω is an embedding. The reader is asked to prove these statements, in more generality, in Exercises 19.9 and 19.10.

Our next application of group actions involves "translating" a finite Galois extension by adjoining a single element:

Proposition 19.13. *Let K/F be a finite Galois extension. Suppose that $\tilde{K} = K[\alpha]$ is a finite simple extension field of K. Set $\tilde{F} = F[\alpha]$. Then $\tilde{K}$ is a finite Galois extension of $\tilde{F}$, and we have $\mathrm{Gal}(\tilde{K}/\tilde{F}) \hookrightarrow \mathrm{Gal}(K/F)$.*

Proof. Note that $\tilde{K}$ is finite over F, by Lemma 15.26. We can write $K = F[R]$, where R is the set of all roots of f in K, for some polynomial $f \in F[x]$ such that f splits completely in K and has no repeated roots (by Exercise 16.13 and Theorem 15.21). Thus we have $\tilde{K} = F[R \cup \{\alpha\}] = \tilde{F}[R]$. Certainly we have $f \in \tilde{F}[x]$, and R is also the set of all roots of f in $\tilde{K}$—for $K \le \tilde{K}$, and f already factors completely over K. So we can use Exercise 16.13 again to conclude that $\tilde{K}$ is Galois over $\tilde{F}$. Set $G = \mathrm{Gal}(K/F)$ and $\tilde{G} = \mathrm{Gal}(\tilde{K}/\tilde{F})$. By Exercise 19.9, we have an embedding $G \hookrightarrow \mathrm{Sym}(R)$ given by $\sigma \mapsto \sigma|_R$, and likewise, $\tilde{G} \hookrightarrow \mathrm{Sym}(R)$. Now consider the function $\omega : \tilde{G} \to G$, $\sigma \mapsto \sigma|_K$. By Proposition 16.25 together with Exercise 16.22, we have $\sigma(K) = K$, so ω is well-defined. It is easy to check that ω is a group homomorphism. If $\sigma \in \tilde{G}$ and $\omega(\sigma) = e_G$, then we have $\sigma|_K = \mathrm{id}_K$, so in particular, we have $\sigma|_R = \mathrm{id}_R$; but then $\sigma = e_{\tilde{G}}$. Thus ω is an embedding. $\square$

19.2 Reaping the Consequences

Next we use group actions to obtain several classical results on the structure of finite groups, which give partial converses to Lagrange's Theorem.

Proposition 19.14. *Let G be a non-trivial p-group. Then $|Z(G)| > 1$; that is, the center of G is non-trivial.*

Proof. Assume for a contradiction that $Z(G) = \{e\}$. Consider the action of G on itself by conjugation. Let $\mathcal{C}$ be the set of all conjugacy classes of G, which are (by definition) the orbits of G under this action. One such orbit is $\{e\}$. If $g \in G - \{e\}$, then by our assumption, $g \notin Z(G)$, so $C_G(g) \ne G$; therefore, $|\mathrm{Orb}(g)| = |G| / |C_G(g)| = p^r$ for some integer $r > 1$. So we have $|G| = \sum_{S \in \mathcal{C}} |S| = 1 + \sum_{S \in \mathcal{C} - \{e\}} |S| \equiv 1 \pmod p$. This contradicts the fact that $|G| \equiv 0 \pmod p$. Thus, our original assumption must be false, so $|Z(G)| > 1$, as desired. $\square$

As an application of Proposition 19.14, we prove a partial converse of Lagrange's Theorem for p-groups.

Proposition 19.15. *Let p be prime, and let G be a p-group. Then for all natural numbers m dividing the order of G, there exists a subgroup of G of order m.*

Proof. Fix p, and suppose that $|G| = p^r$ with $r \in \mathbf{N}$. We proceed by induction on r.

In the base case, we have $|G| = 1$, and we only have to check that G contains a subgroup of order $m = 1$, which is true.

Inductively, suppose that the result is true for all p-groups of order less than p^r. Let m divide p^r. The case $m = 1$ being trivial, we suppose that $m > 1$.

By Proposition 19.14, $Z(G) \neq \{e\}$. Now $Z(G)$ is a p-group by Lagrange's Theorem, so p divides the order of $Z(G)$. Certainly $Z(G)$ is abelian; so by Exercise 18.14, $Z(G)$ contains an element x of order p. Set $H = \langle x \rangle$. Since $H \unlhd G$ (by Exercise 9.11), we can form $Q := G/H$. Since Q is a p-group of order p^{r-1} (by Corollary 8.19), we can find a subgroup $T \leq Q$ with $|T| = m/p$ by inductive hypothesis. By Exercise 7.17, we can lift T to a subgroup L of G, where $L = \{ab : a \in G, b \in H, \text{ and } aH \in T\}$. By Exercise 19.3, we have $T \cong L/H$. Therefore, $|T| = |L|/|H|$ (by Corollary 8.19). So $|L| = |T| \cdot |H| = m$, as desired. $\qquad\square$

Theorem 19.16 (Cauchy's Theorem). *Let G be a finite group and let p be a prime number dividing the order of G. Then G has an element of order p.*

Proof. Fix the prime number p, and consider finite groups G whose order is divisible by p. We proceed by induction on $|G|$. In the base case, we have $|G| = p$, and so G certainly has an element of order p, because G is cyclic (by Theorem 8.20).

Inductively, we may suppose that every proper subgroup of G of order divisible by p contains an element of order p. So we assume

$$G \text{ has no proper subgroups whose order is divisible by } p. \qquad (19.2)$$

Consider the action of G on itself by conjugation, and let $\mathcal{C}$ be the set of all conjugacy classes of G. For every $g \in G - Z(G)$, we have $C_G(g) < G$, and so, by hypothesis (19.2), $p \nmid |C_G(g)|$. Therefore, p divides $|\mathrm{Orb}(g)|$ if $g \notin Z(G)$. Note that the orbit of an element of $Z(G)$ is a singleton set, so $Z(G)$ is a union of orbits, and consequently G acts on $G - Z(G)$ (see Exercise 19.8). Let $\mathfrak{O}$ be the set of orbits of G, and let $\mathfrak{P} \subseteq \mathfrak{O}$ be the set of orbits of $G - Z(G)$. Then we have $|G| = \sum_{O \in \mathfrak{O}} |O| = |Z(G)| + \sum_{O \in \mathfrak{P}} |O| \equiv |Z(G)| \pmod{p}$. Since p divides $|G|$, this forces p to divide $|Z(G)|$ as well. By hypothesis (19.2), we must have $Z(G) = G$. But this means that G is abelian, so G has an element of order p by Exercise 18.14. $\qquad\square$

In Corollary 18.13, we saw that every finite abelian group of order n has a subgroup of order p^r, where p is prime and $p^r \mid\mid n$. By tweaking the argument used to prove Cauchy's Theorem (Theorem 19.16), we can generalize this result to all finite groups, including non-abelian groups. Notice that, by Lagrange's Theorem, there can never be a p-subgroup of order more than p^r, so such subgroups must be maximal p-subgroups. Unlike in the abelian case, however, such subgroups will not in general be unique. We note for what follows that Sylow is pronounced "SEE low".

Theorem 19.17 (Sylow's Theorem, Weak Form). *Let G be a group of order $n < \infty$, and suppose that $p^r \mid\mid n$, where p is prime and r is a positive integer. Then G contains a subgroup of order p^r.*

Proof. We fix p and proceed by induction on n. In the base case, $n = p$, we can take our subgroup to be G itself.

Inductively, suppose that the result is true for all finite groups of order less than n.

Case 1: p divides the order of the center $Z = Z(G)$.

Since Z is an abelian group, there is a subgroup $Y \leq Z$ such that $|Y| = p^t$, where $p^t \mid\mid |Z|$ (by Corollary 18.13). Since $Y \trianglelefteq G$ (by Exercise 9.11 (b)), we can form $Q = G/Y$. Now $p^{r-t} \mid\mid Q$ (since $|Z| \cdot |Q| = n$), and $t \geq 1$, so our inductive hypothesis yields a subgroup $T \leq Q$ with $|T| = p^{r-t}$. By Exercise 7.17, we can lift T to a subgroup L of G, where $L = \{ab : a \in G, b \in Y,$ and $aY \in T\}$. By Exercise 19.3, we have $T \cong L/Y$, so $|L| = |T| \cdot |Y| = p^r$. Thus, L satisfies our requirements.

Case 2: The order of Z is prime to p, i.e., p does not divide $|Z|$.

Subcase (i): p^r divides the order of $C_G(g)$ for some $g \in G - Z$.

Then $C_G(g) \neq G$, since $g \notin Z$. So our inductive hypothesis gives a subgroup of $C_G(g)$ of order p^r, which suits our purposes.

Subcase (ii): For all $g \in G - Z$, p^r does not divide $|C_G(g)|$.

Consider the action of G on itself by conjugation, and let $\mathcal{C}$ be the set of all conjugacy classes of G.

Then p divides $|\mathrm{Orb}_G(g)|$ for all $g \in G - Z$. For $g \in Z$, we have $\mathrm{Orb}_G(g) = \{g\}$. Let $\mathfrak{Z}$ be the collection of orbits of elements of Z. Then $n = |G| = \sum_{O \in \mathcal{C}} |O| = \sum_{O \in \mathfrak{Z}} |O| + \sum_{O \in \mathcal{C} - \mathfrak{Z}} |O| = |Z| + \sum_{O \in \mathcal{C} - \mathfrak{Z}} |O| \equiv |Z| \pmod{p}$. But this is a contradiction, since $p \mid n$ while $p \nmid |Z|$. Thus, subcase (ii) is impossible. □

Definition 19.18. Let G be a finite group, and let p be a prime number. Let p^r be the highest power of p which divides $|G|$. A subgroup of G of order p^r is called a *p-Sylow subgroup* of G.

Example 19.19. We demonstrate how a conjugation action can be used in tandem with Sylow's Theorem to understand the structure of a group. Let G be a group of order 6. By Sylow's Theorem, there is a subgroup $H < G$ with $|H| = 2$ and a subgroup $K \leq G$ with $|K| = 3$. Consider the action of G on the set $S = \{gHg^{-1} : g \in G\}$ by conjugation. Now $\mathrm{Stab}_G(H) = N_G(H)$, the normalizer of H in G, by definition of normalizer (see Exercise 7.16). We know from Exercise 7.16 that $H \leq N_G(H)$, so we have $|N_G(H)| \in \{2, 6\}$, by Lagrange's Theorem.

Case 1: $|N_G(H)| = 6$. Then $H \trianglelefteq G$. We also have $K \trianglelefteq G$, by Exercise 8.7. Lagrange's Theorem forces $H \cap K = \{e\}$ and $\langle H \cup K \rangle = G$. Note that H and K have prime order, and are thus cyclic, hence also abelian; we have $H \cong \mathbf{Z}_2$ and $K \cong \mathbf{Z}_3$. So by Exercise 17.5, we have $G \cong H \times K \cong H \oplus K \cong \mathbf{Z}_2 \oplus \mathbf{Z}_3$, and G is abelian. We prefer to write $G \cong \mathbf{Z}_6$ (justified, for example, by Exercise 18.17).

Case 2: $|N_G(H)| = 2$. Then $N_G(H) = H$ and $|\mathrm{Orb}_G(H)| = 6/2 = 3$. Note that $\mathrm{Orb}_G(H) = S$. Let $f : G \to \mathrm{Sym}(S)$ be the group homomorphism of Lemma 19.8. Then $\ker(f) = \{g \in G : gXg^{-1} = X$ for all $X \in S\} \subseteq N_G(H) = H$. Since $|H| = 2$, this forces $\ker(f) = \{e_G\}$ or $\ker(f) = H$. Now $\ker(f) \trianglelefteq G$, while $N_G(H) = H \neq G$, so $H \ntrianglelefteq G$. Therefore $\ker(f) = \{e_G\}$, so f

is an embedding. We have $\text{Sym}(S) \cong S_3$ (by Exercise 9.6), and $|G| = 6 = |S_3|$, so f is an isomorphism.

We conclude that every group of order 6 is isomorphic to either $\mathbf{Z}_6$ or S_3.

As another application of what we've learned up to this point, we consider finite extension fields of the field $\mathbf{C}$ of complex numbers.

Example 19.20. Suppose that F is a finite extension of $\mathbf{C}$. Since $[\mathbf{C} : \mathbf{R}] = 2$, then F is also finite over $\mathbf{R}$. In order to be able to use Galois Theory, we will extend F to a normal closure K of F over $\mathbf{R}$ (see Exercise 16.12). Then K is finite over $\mathbf{R}$ as well.

Since $\mathbf{R}$ has characteristic zero, K is separable over $\mathbf{R}$, by Proposition 16.23. Thus, K is Galois over $\mathbf{R}$. Set $G = \text{Gal}(K/\mathbf{R})$, and let $n = |G|$.

Let's investigate how n might factor, using our newfound Sylow Theory. Since the number 2 seems to be important here, as the degree of $\mathbf{C}$ over $\mathbf{R}$, let's take a Sylow 2-subgroup H of G. Then $h := |H|$ is the highest power of 2 which divides $|G|$, so we can write $n = hr$, where r is odd. By the Fundamental Theorem of Galois Theory, we have $[K : \mathbf{R}] = n$ and $[K : K^H] = h$. This forces $[K^H : \mathbf{R}] = r$.

Let $\alpha \in K^H$. Then α is algebraic over $\mathbf{R}$, and we can let $f = \text{Irr}(\alpha, \mathbf{R}, x)$. Let $d = \deg(f)$. Now we have $[\mathbf{R}[\alpha] : \mathbf{R}] = d$, by Lemma 15.25. Since $\mathbf{R} \leq \mathbf{R}[\alpha] \leq K^H$, we must have $d \mid r$, so d is also odd.

Here we have a polynomial $f \in \mathbf{R}[x]$ of odd degree, which is irreducible in $\mathbf{R}[x]$. Yet we know from (pre)calculus that the values of a polynomial over $\mathbf{R}$ of odd degree must approach infinity and negative infinity as its argument approaches infinity and negative infinity (not necessarily respectively!). By the Intermediate Value Theorem, we can thus be sure that f has a root in $\mathbf{R}$. Since f is irreducible in $\mathbf{R}[x]$, this forces $d = 1$. But this implies that $\alpha \in \mathbf{R}$.

Since α was an arbitrary element of K^H, we must have $K^H = \mathbf{R}$. Thus, $r = [\mathbf{R} : \mathbf{R}] = 1$, and $n = h$. In other words, n is a power of 2.

Let's put F back in the picture now. We have $\mathbf{R} \leq \mathbf{C} \leq F \leq K$, and $n = [K : \mathbf{R}]$ is a power of 2; it follows that the degree $[K : \mathbf{C}]$ is also a power of 2, say $[K : \mathbf{C}] = 2^m$ where $m \geq 0$. Remembering how nicely p-groups behave (2-groups, in this case), and recalling from Exercise 15.15 that $\mathbf{C}$ has no extension fields of degree 2, we sense that we can say more. Indeed, by Lemma 16.29, K is Galois over $\mathbf{C}$, and we set $H = \text{Gal}(K/\mathbf{C}) < G$. Now $|H| = 2^m$, so if $m > 0$, then H has a subgroup of index 2 (by Proposition 19.15), which by Galois Theory corresponds to an extension field of degree 2 over $\mathbf{C}$; but this is impossible by Exercise 15.15. We conclude that $m = 0$, so $K = \mathbf{C}$. Since $\mathbf{C} \leq F \leq K$, this forces $F = \mathbf{C}$. Thus, we have shown that $\mathbf{C}$ has no non-trivial finite extensions!

What does Example 19.20 tell us about the field $\mathbf{C}$? From field theory, we know that finite extension fields correspond to irreducible polynomials: polynomials which cannot be factored non-trivially. Thus, we have shown that there are essentially no such polynomials over $\mathbf{C}$ (that is, none except linear polynomials). To be precise, we make the following definition.

Definition 19.21. A field F is *algebraically closed* if every polynomial in $F[x]$ factors completely over F.

With this definition in hand, we are ready to prove

Theorem 19.22 (Fundamental Theorem of Algebra). $\mathbf{C}$ *is algebraically closed.*

Proof. Let $f \in \mathbf{C}[x]$. If $f \in \mathbf{C}$, then $f = c$ is a product of zero linear factors, so certainly f factors completely over $\mathbf{C}$.

So suppose $f \notin \mathbf{C}$. Then by Lemma 15.11, we can write $f = \prod_{i=1}^{n} f_i$, where each f_i is irreducible in $\mathbf{C}[x]$. Now each f_i must be linear, for otherwise, $\mathbf{C}[x]/(f_i)$ would be a finite extension field of $\mathbf{C}$ of degree greater than 1 (by Theorem 15.10), which is impossible according to Example 19.20. By factoring out the leading coefficients of the f_i's, we can write f in the form required by Definition 15.14. $\qquad\square$

This may be a good time to pause and reflect on what we have done. As its name suggests, Theorem 19.22, the Fundamental Theorem of Algebra, is a rather important classical result. Carl Friedrich Gauss, considered by many people to be one of the greatest mathematicians of all time, devoted his doctoral work to proving this theorem. By toying around with the ideas and results we proved about groups, fields, polynomials, and their relationships to each other (in Example 19.20), what naturally fell out was precisely this theorem. We are reminded of the episode of the animated series *Dragon Ball*® in which the diminutive young Goku, having been trained by the expert Master Roshi, competes in the World Martial Arts Competition for the first time. Goku's opponent not even taking him seriously enough to face towards him in the ring, Goku attempts to merely get his opponent's attention by delicately tapping him with a single outstretched finger; but such is Goku's strength that this single finger-tap throws his opponent completely off-balance and he crashes down, immediately winning the match for Goku by ring-out. This is a good comparison to what our training in abstract algebra has done for us.

And we are just warming up.

19.3   Exercises

Exercise 19.1. Let G be a group and let S be a set. Let $\cdot$ be a group action of G on S. For each $g \in G$, let $\sigma_g : S \to S$ be defined by $\sigma_g(x) = g \cdot x$.

(a) Prove that $\sigma_g \in \mathrm{Sym}(S)$.

(b) Prove that the function $f : G \to \mathrm{Sym}(S)$ given by $g \mapsto \sigma_g$ is a group homomorphism.

(c) Let $\mathcal{A}$ be the set of all group actions of G on S. Let $\mathcal{H}$ be the set of all group homomorphisms from G to $\mathrm{Sym}(S)$. Define a function $\tau : \mathcal{A} \to \mathcal{H}$ by $\cdot \mapsto f$, where f is defined as in part (b) above. Prove that τ is a bijection.

Exercise 19.2. Suppose that the group G acts on the non-empty set S. Define a relation $\sim$ on S by $x \sim y$ iff $\exists g \in G$ s.t. $g \cdot x = y$. Prove that $\sim$ is an equivalence relation on S, whose equivalence classes are just the orbits of G on S.

Exercise 19.3. Let G be a group, $N \trianglelefteq G$, and $Q = G/N$. Let $T \leq Q$, and let L be the lift of T to G (see Exercise 7.17).

(a) Prove that $T = L/N$.

(b) Prove that if $T \trianglelefteq U \leq Q$ and M is the lift of U to G, then $L \trianglelefteq M$ and $U/T \cong M/L$.

Exercise 19.4. Let S be a finite set, and let $G = \mathrm{Sym}(S)$. Let $x_0, x_1, \ldots, x_{r-1}$ be r distinct elements of S (for some $r \geq 1$). Let σ be the corresponding r-cycle, $\sigma = (x_0, x_1, \ldots, x_{r-1}) \in G$. Let τ be the cycle of σ corresponding to x_0. Prove that $\tau = \sigma$.

Exercise 19.5 (Disjoint cycles commute). Let S be a finite set, and let $G = \mathrm{Sym}(S)$. Let $x_0, x_1, \ldots, x_{r-1}, y_0, y_1, \ldots, y_{s-1}$ be $r + s$ distinct elements of S (for some $r, s \geq 1$). Set $\sigma = (x_0, x_1, \ldots, x_{r-1})$ and $\tau = (y_0, y_1, \ldots, y_{s-1})$. Prove that $\sigma\tau = \tau\sigma$. (We say that σ and τ are *disjoint* cycles since for all i, j, we have $x_i \neq y_j$.)

Exercise 19.6 (Every permutation is the product of its disjoint cycles). Let S be a finite set, and let $G = \mathrm{Sym}(S)$. Let $\sigma \in G$, and let $O_1, \ldots, O_r$ be the orbits of S under the action of $\langle \sigma \rangle$. For each orbit O_i, choose an element $x_i \in O_i$. Let $c_i \in G$ be the cycle of σ corresponding to x_i.

(a) Explain why the cycles $c_1, \ldots, c_r$ are mutually disjoint.

(b) Prove that $\sigma = c_1 c_2 \cdots c_r$. We call this the *disjoint cycle decomposition* of σ.

(c) Prove that $|\sigma| = \mathrm{lcm}(|c_1|, \ldots, |c_r|)$, the least common multiple of the orders of the cycles. (Use Exercise 19.5 together with Exercise 3.14.)

Exercise 19.7 (Cycle structure determines conjugacy class). Let S be a finite set, and let $G = \mathrm{Sym}(S)$.

(a) Let $\tau = (x_0, x_1, \ldots, x_{r-1}) \in G$ be an r-cycle, and let $\sigma \in G$. Show that $\sigma\tau\sigma^{-1} = (\sigma(x_0), \sigma(x_1), \ldots, \sigma(x_{r-1}))$, another r-cycle. Observe that this formula extends to products of cycles, since the conjugation-by-σ map is a group homomorphism.

(b) For a permutation $\tau \in G$, define the *cycle structure* of τ to be the ordered tuple $\mathrm{cs}(\tau) := (|c_1|, |c_2|, \ldots, |c_r|)$, where $\tau = c_1 c_2 \cdots c_r$ is the disjoint cycle decomposition of τ, arranged so that $|c_1| \leq |c_2| \leq \cdots |c_r|$. (Note that we can re-arrange the c_i's as we wish, since they commute with each other.) Prove that for every $\tau \in G$, the conjugacy class of τ is

$$\{\alpha\tau\alpha^{-1} : \alpha \in G\} = \{\sigma \in G : \mathrm{cs}(\sigma) = \mathrm{cs}(\tau)\}.$$

Exercise 19.8. Suppose that a group G acts on a set S. Explain why G also acts on any individual orbit of an element of S, via a restriction of the original action. Generalize to explain why G acts on any union of orbits.

Exercise 19.9. Let K be a finite Galois extension of F, and let $G = \mathrm{Gal}(K/F)$.

(a) Observe that G acts on K in a natural way, as a subgroup of $\mathrm{Sym}(K)$. Prove that the orbits of G on K are the sets $R_f := \{\alpha \in K : f(\alpha) = 0\}$, where f ranges over the irreducible polynomials $\mathrm{Irr}(\beta, F, x)$ for $\beta \in K$.

(b) From Exercise 16.13, we know that K is a splitting field over F for some polynomial $f \in F[x] - F$ without repeated roots. Let R be the set of all roots of f in K. Show that R is a union of orbits, and that the action of G on R (see Exercise 19.8) induces an embedding $G \hookrightarrow \mathrm{Sym}(R)$.

Exercise 19.10. Let K be a finite Galois extension of F, and let $\alpha \in K$. Prove that $\mathrm{Irr}(\alpha, F, x) = \prod_{\gamma \in C}(x - \gamma)$, where C is the set of all conjugates of α over F in K. (See Definition 16.7 and Exercise 16.7.)

Exercise 19.11. Let $\alpha = 3\sqrt[4]{2} - \sqrt{2} + 4i\sqrt[4]{2} \in \mathbf{C}$. Find $\mathrm{Irr}(\alpha, \mathbf{Q}, x) =: f$. You may leave f in factored form. Hint: Use Exercises 19.10 and 16.25.

Exercise 19.12. Prove that if G is a group of order p^2, where p is prime, then G is abelian. Hint: Use Proposition 19.14.

Exercise 19.13. Prove the following slightly stronger version of Proposition 19.15: If G is a p-group of order p^r, then there is a tower of subgroups $\{c_G\} = G_0 < G_1 < \cdots < G_r = G$ such that $|G_i| = p^i$ for all $i \in \{0, 1, \ldots, r\}$.

Exercise 19.14. Let G be a group, and let $H \leq G$. Let $\mathcal{L}$ be the set of all left cosets of H in G.

(a) Prove that the operation $\cdot : G \times \mathcal{L} \to \mathcal{L}$ given by $a \cdot C = aC$ defines a group action of G on $\mathcal{L}$.

(b) Now suppose in addition that $[G : H] = n < \infty$. Show that the group action from part (a) above induces a group homomorphism $\sigma : G \to S_n$ whose kernel is a subgroup of H.

Exercise 19.15. In this exercise, you will prove the "full-strength" version of Sylow's Theorem:

Theorem 19.23 (Sylow's Theorem, Strong Form). *Let G be a finite group, and let p be a prime which divides the order of G. Let S be the set of all Sylow p-subgroups of G. Then:*

(i) $|S| \equiv 1 \pmod{p}$.

(ii) All elements of S are conjugate to each other in G.

(iii) Every p-subgroup of G is contained in some Sylow p-subgroup of G.

(a) Prove part (i) by picking any element $P \in S$ (which we can do because of the Weak Sylow Theorem, Theorem 19.17), and considering the action of P on S by conjugation. Hint: Lemmas 18.5 and 18.9 may be useful.

(b) Let $P, Q \in S$. Prove that there is an element $x \in G$ such that $Q = xPx^{-1}$ by considering the action of Q by conjugation on the set of all conjugates of P in G.

(c) Let H be a p-subgroup of G. Prove that H is contained in some element of S by considering the action of H on S by conjugation.

Exercise 19.16. Let G be a finite group.

(a) Let p be a prime which divides the order of G. Prove that if G has a unique Sylow p-subgroup H, then we must have $H \trianglelefteq G$.

(b) Suppose that G has a unique p-Sylow subgroup G_p for each prime p dividing the order of G, and that each G_p is cyclic. Prove that G is cyclic. (See Exercises 4.13 and 18.17.)

Exercise 19.17. Let p be an odd prime number. In this exercise, you will classify all groups of order $2p$, generalizing the result of Example 19.19 and incidentally proving that $S_3 \cong D_6$.

Prove that every group of order $2p$ is either cyclic or dihedral, by following the steps below:

(a) First suppose that G is an abelian group of order $2p$. Show that G must be cyclic.

(b) From now on, let G be a non-abelian group of order $2p$. Prove that G has exactly 1 element of order 1, $p - 1$ elements of order p, and p elements of order 2.

(c) Consider the action of G by conjugation on the set of elements of G of order p. Prove that every orbit has size 2.

(d) Use the previous parts of this exercise together with Exercise 10.6 to prove that $G \cong D_{2n}$.

20

Learning from Z

20.1 Introduction

Though we have ventured some way into abstract territory, we still have more to learn from the ring of ordinary integers, **Z**. The two notions we want to generalize in this chapter are fractions and unique factorization into primes.

20.2 Fractions

"G_d created the integers; Humans worked out all the rest."—Leopold Kronecker.

Following Kronecker's lead, let us imagine that we are starting with **Z**, ignorant of the field **Q**, but aware of the general concept of a field. How would we go about constructing a field around **Z**—that is, building **Q** from **Z**?

Our idea is to introduce fractions. But what *is* a "fraction"? To construct a field K containing **Z**, we at least need an element n^{-1} for each non-zero $n \in \mathbf{Z}$. Let us agree to use customary fraction notation, writing $1/n$ for n^{-1}.

Now we need to be able to multiply field elements, so for every pair m, $n \in \mathbf{Z}$ where $n \neq 0$, our field K must contain $m \cdot (1/n)$. Let us agree to write this element as m/n.

We still have not addressed the fact that our field K must also be closed under addition—and the new "summed" elements that we introduce must also have *their* inverses and products in K as well! But from our prior experience with **Q**, we anticipate that all of this will follow once we have the basic fraction form m/n with $m, n \in \mathbf{Z}$ and $n \neq 0$.

The most subtle point—again well-known from elementary-school mathematics—is that *two fractions which look different can represent the same number*. Indeed, we should have $m/n = a/b$ iff $(bn) \cdot (m/n) = (bn) \cdot (a/b)$ iff $bm = an$. This is just the usual cross-multiplication formula.

To summarize: it takes two ordinary integers to make a fraction, but some pairs of integers give rise to the same fraction. Now to make this precise, we can say that a fraction is an *equivalence class* of ordered pairs (m, n) with $m, n \in \mathbf{Z}$ and $n \neq 0$, where (m, n) is equivalent to (a, b) iff $an = bm$. Then we would agree to write m/n for the equivalence class of (m, n) under this equivalence relation.

We proceed to apply the ideas above by replacing **Z** with any domain R. Also, instead of always inverting *all* the non-zero elements of R, we will allow ourselves the freedom to choose a select subset U of elements for use in our denominators. Since multiplying two fractions should multiply their denominators, and since we don't want to allow division by zero, we will require the following property for U:

Definition 20.1. Let R be a commutative ring with 1, and let $U \subseteq R$. Then U is called *multiplicatively closed* if $1 \in U$, $0 \notin U$, and $\forall a, b \in U$, $a \cdot b \in U$.

Now we are ready to construct our fractions:

Lemma 20.2. *Let R be a domain, and let U be a multiplicatively closed subset of R. Define a relation $\sim$ on $R \times U$ by*

$$(a, u) \sim (b, v) \text{ iff } av = bu.$$

Then $\sim$ is an equivalence relation.

Proof. (i) [Show reflexivity] Let $f \in R \times U$. We can write $f = (a, u)$ with $a \in R$ and $u \in U$. Then $au = au$, so we have $(a, u) \sim (a, u)$, i.e., $f \sim f$.

(ii) [Show symmetry] Let $f, g \in R \times U$ and suppose that $f \sim g$. Write $f = (a, u)$ and $g = (b, v)$ with $a, b \in R$ and $u, v \in U$. Then $av = bu$. So $bu = av$, and therefore $g \sim f$.

(iii) [Show transitivity] Let $f, g, h \in R \times U$ and suppose that $f \sim g$ and $g \sim h$. Write $f = (a, u)$, $g = (b, v)$, and $h = (c, w)$ with $a, b, c \in R$ and $u, v, w \in U$. Then we have $av = bu$ and $bw = cv$. Multiplying the first equation by w and the second by v, we find $avw = buw$ and $buw = cuv$. Therefore, $avw = cuv$. We can re-write this as $v \cdot (aw - cu) = 0$. Since R is a domain and since $v \in U \subseteq R - \{0\}$ (by definition of a multiplicatively closed set), we have $aw - cu = 0$, and so $f \sim h$, as required. $\square$

Notation 20.3. We write a/u for the equivalence class of (a, u) in Lemma 20.2, and we call a/u a *fraction* with *numerator* a and *denominator* u.

Proposition 20.4. *Let R be a domain, and let U be a multiplicatively closed subset of R. Then the set of all fractions a/u with $a \in R$ and $u \in U$ forms a domain under the operations*

$$(a/u) + (b/v) = (av + bu)/(uv) \text{ and} \tag{20.1}$$
$$(a/u) \cdot (b/v) = (ab)/(uv) \tag{20.2}$$

for $a, b \in R$ and $u, v \in U$.

Proof. Since a fraction is an equivalence class of ordered pairs of elements of R, we must first show that the operations $+$ and $\cdot$ on fractions are well-defined. So suppose that $a/u = \tilde{a}/\tilde{u}$ and $b/v = \tilde{b}/\tilde{v}$ with $a, \tilde{a}, b, \tilde{b} \in R$ and $u, \tilde{u}, v, \tilde{v} \in U$. Then we have $a\tilde{u} = \tilde{a}u$ and $b\tilde{v} = \tilde{b}v$. So $\tilde{u}\tilde{v}(av + bu) = (a\tilde{u})v\tilde{v} + (b\tilde{v})u\tilde{u} = (\tilde{a}u)v\tilde{v} + (\tilde{b}v)u\tilde{u} = uv(\tilde{a}\tilde{v} + \tilde{b}\tilde{u})$. Thus, $(av + bu)/(uv) = (\tilde{a}\tilde{v} + \tilde{b}\tilde{u})/(\tilde{u}\tilde{v})$. So $+$ is well-defined. Also, $(ab)(\tilde{u}\tilde{v}) = (a\tilde{u})(b\tilde{v}) = (\tilde{a}u)(\tilde{b}v) = (\tilde{a}\tilde{b})(uv)$, so $(ab)/(uv) = (\tilde{a}\tilde{b})/(\tilde{u}\tilde{v})$. Thus $\cdot$ is also well-defined.

Exercise 20.3 asks for a proof that these fractions form a commutative ring S with $0_S = 0_R/1_R$ and $1_S = 1_R/1_R$. To complete the proof of the proposition, first suppose that $f, g \in S$ and $fg = 0_S$. [Show $f = 0_S$ or $g = 0_S$.] Then we can write $f = q/d$ and $g = r/e$ with $q, r \in R$ and $d, e \in U$. So $fg = (qr)/(de) = 0_R/1_R$. Therefore $(qr) \cdot 1_R = 0_R \cdot (de)$, so $qr = 0_R$. Since R is a domain by hypothesis, this forces $q = 0_R$ or $r = 0_R$; without loss of generality, suppose the former. Then we have $q \cdot 1_R = 0_R = 0_R \cdot d$, so $f = q/d = 0_R/1_R = 0_S$.

Finally, assume for a contradiction that $0_S = 1_S$. Then we have $0_R/1_R = 1_R/1_R$, so $0_R \cdot 1_R = 1_R \cdot 1_R$, and $0_R = 1_R$, contradicting that R is a domain. This completes the proof. □

Definition 20.5. Let R be a domain, and let U be a multiplicatively closed subset of R. The ring of fractions a/u with $a \in R$ and $u \in U$ is called the *ring of fractions of R with denominators from U* or the *localization of R at U*, and is denoted $R[U^{-1}]$. That is, we set

$$R[U^{-1}] := \{u/u \;:\; a \in R, \; u \in U\},$$

with ring operations as given in Proposition 20.4.

It is natural to try to identify the fraction $r/1_R$ with the element $r \in R$. The following result allows us to do that.

Lemma 20.6. *Let R be a domain, and let U be a multiplicatively closed subset of R. Then $R \hookrightarrow R[U^{-1}]$ via the natural map $r \mapsto r/1_R$.*

Proof. It is easy to see that the function $\nu : R \to R[U^{-1}]$ defined by $\nu(r) = r/1_R$ is a ring homomorphism. We verify that ν is injective by showing that its kernel is trivial (see Exercise 11.23). So suppose that $r \in R$ and $\nu(r) = 0$. This means that $r/1_R = 0_R/1_R$, and so $r \cdot 1_R = 0_R \cdot 1_R$. Thus $r = 0_R$, as desired. □

Remark 20.7. Because of Lemma 20.6, we may choose to identify R with its image in $R[U^{-1}]$ under the natural map. That is, we can think of R as a subring of $R[U^{-1}]$. We illustrate this point of view in the following result, which captures the fundamental idea that localizations create more units.

Lemma 20.8. *Let R be a domain, and let U be a multiplicatively closed subset of R. Then, viewing U as a subset of $R[U^{-1}]$, we have $U \subseteq (R[U^{-1}])^{\times}$.*

Proof. For $u \in U$, we have $1_R/u \in R[U^{-1}]$, and certainly $u \cdot (1_R/u) = u/u = 1_R/1_R = 1$ in $R[U^{-1}]$. □

We want to say that the localization of R at U is the "most efficient" way to invert everything in U—that is, turn everything in U into a unit. Reminiscent of the universal property of direct products, we can express this by saying that the localization $R[U^{-1}]$ is a mandatory first stop on any route which inverts everything in U:

Lemma 20.9 (Universal Property of Localization). *Let R be a domain, and let U be a multiplicatively closed subset of R. If $\sigma : R \to S$ is a ring homomorphism such that $\sigma(U) \subseteq S^\times$, then σ factors through the localization of R at U; more specifically, there is a unique ring homomorphism $\omega : R[U^{-1}] \to S$ such that the following diagram commutes:*

$$R \xrightarrow{\;\;\nu\;\;} R[U^{-1}] \dashrightarrow^{\;\exists!\omega\;} S$$
$$\sigma$$

$$(20.3)$$

where $\nu : a \mapsto a/1_R$ is the natural map.

Proof. Suppose that $\sigma : R \to S$ is a ring homomorphism with $\sigma(U) \subseteq S^\times$. Set $T = R[U^{-1}]$. We first suppose there is a ring homomorphism $\omega : T \to S$ such that $\omega \circ \nu = \sigma$. Then for any $r \in R$, we have $\omega(r/1_R) = \omega(\nu(r)) = \sigma(r)$. In particular, when $u \in U$, we have $\omega(u/1_R) = \sigma(u)$. Let $N = \nu(U)$. Then we have $N \subseteq T^\times$, $\omega(N) \subseteq S^\times$, and $1_T \in N$. Let $M = \langle N \rangle \leq T^\times$ be the group generated by N under multiplication. By Exercise 20.4, the restriction $\omega|_M : M \to S^\times$ is a group homomorphism with respect to multiplication. So for any fraction $r/u \in T$ with $r \in R$ and $u \in U$, we have $\omega(r/u) = \omega((r/1_R) \cdot (1_R/u)) = \omega(r/1_R) \cdot \omega(1_R/u) = \sigma(r) \cdot \omega((u/1_R)^{-1}) = \sigma(r) \cdot (\omega(u/1_R))^{-1} = \sigma(r) \cdot (\sigma(u))^{-1}$. This shows that ω is uniquely determined by σ.

To finish the proof, we attempt to define a function $\omega : T \to S$ by the formula $\omega(r/u) = \sigma(r) \cdot (\sigma(u))^{-1}$, and show that ω is a well-defined ring homomorphism satisfying $\omega \circ \nu = \sigma$. So suppose that $r/u = \tilde{r}/\tilde{u}$ with $r, \tilde{r} \in R$ and $u, \tilde{u} \in U$. Then we have $r\tilde{u} = \tilde{r}u$. So $\sigma(r\tilde{u}) = \sigma(\tilde{r}u)$. Since σ is a ring homomorphism and since $\sigma(U) \subseteq S^\times$, we have $\sigma(r)\sigma(\tilde{u}) = \sigma(\tilde{r})\sigma(u)$ and $\sigma(r)(\sigma(u))^{-1} = \sigma(\tilde{r})(\sigma(\tilde{u}))^{-1}$. This shows that ω is well-defined.

Next, let $f, g \in T$, and write $f = a/v$ and $g = b/w$ with $a, b \in R$ and $v, w \in U$. Then

$$\begin{aligned}
\omega(f + g) &= \omega(\, (aw + bv)/(vw)\,) \\
&= \sigma(aw + bv)(\sigma(vw))^{-1} \\
&= \sigma(aw)(\sigma(vw))^{-1} + \sigma(bv)(\sigma(vw))^{-1} \\
&= \sigma(a)\sigma(w)(\sigma(v)\sigma(w))^{-1} + \sigma(b)\sigma(v)(\sigma(v)\sigma(w))^{-1} \\
&= \sigma(a)\sigma(w)(\sigma(v))^{-1}(\sigma(w))^{-1} + \sigma(b)\sigma(v)(\sigma(v))^{-1}(\sigma(w))^{-1} \\
&= \sigma(a)(\sigma(v))^{-1} + \sigma(b)(\sigma(w))^{-1} \\
&= \omega(f) + \omega(g).
\end{aligned}$$

And $\omega(fg) = \omega((ab)/(vw)) = \sigma(ab)(\sigma(vw))^{-1} = \sigma(a)\sigma(b)(\sigma(v))^{-1}(\sigma(w))^{-1}$ $= \sigma(a)(\sigma(v))^{-1} \cdot \sigma(b)(\sigma(w))^{-1} = \omega(f)\omega(g)$. Therefore, ω is a ring homomorphism.

Finally, we note that since $1_R \in U$, we have $\sigma(1_R) \in S^\times$, so σ restricts to a multiplicative group homomorphism from $\{1_R\}$ to $S^\times$ (by Exercise 20.4). Thus we have $\sigma(1_R) = 1_S$. So if $r \in R$, then we have $\omega(\nu(r)) = \omega(r/1_R)$ $= \sigma(r) \cdot (\sigma(1_R))^{-1} = \sigma(r) \cdot (1_S)^{-1} = \sigma(r)$, so $\omega \circ \nu = \sigma$, as desired. $\qquad\square$

Remark 20.10. The reader is encouraged to use the universal property expressed by Lemma 20.9 to the greatest extent possible when proving results about localizations, and correspondingly to avoid explicit formulas or manifestations of localizations (such as the fraction representation) where possible. The result of Exercise 20.5 shows that this is a reasonable approach.

The following result captures the notion that the more denominators we allow, the bigger the resulting ring. The extreme cases on either side occur when $U = \{1\}$ and when $U = R - \{0\}$.

Proposition 20.11. *Let R be a domain, and let U, V be multiplicatively closed subsets of R with $U \subseteq V$. Then we have*

(1) $R[U^{-1}] \hookrightarrow R[V^{-1}]$ via the map $a/u \mapsto a/u$;

(2) If $U = \{1_R\}$, then the natural map $\nu : R \to R[U^{-1}]$, $a \mapsto a/1_R$ is a ring isomorphism;

(3) If $V = R - \{0_R\}$, then $R[V^{-1}]$ is a field.

Proof. (1) First, identify R as a subring of both $R[U^{-1}]$ and $R[V^{-1}]$ via the natural maps $\nu_U : R \to R[U^{-1}]$ and $\nu_V : R \to R[V^{-1}]$ (using Lemma 20.6). Then we have $U \subseteq V \subseteq R[V^{-1}]$, so ν_V factors as $\nu_V = \omega \circ \nu_U$ for a unique ring homomorphism $\omega : R[U^{-1}] \to R[V^{-1}]$ (by Lemma 20.9). We have $\omega(a/1_R) = \omega(\nu_U(a)) = \nu_V(a) = a/1_R$ for $a \in R$; so $\omega(a/u) = \omega(a \cdot u^{-1})$ $= \omega(a) \cdot \omega(u^{-1}) = \omega(a) \cdot (\omega(u))^{-1}$ (using Exercise 20.4) $= (a/1_R) \cdot (u/1_R)^{-1}$ $= a/u$ for $a \in R$ and $u \in U$. Finally, we note that ω is injective, since if $a/u = 0 = 0_R/1_R$ in $R[V^{-1}]$, then $a = 0_R$, and so $a/u = 0$ in $R[U^{-1}]$ as well.

(2) Let $U = \{1_R\}$. We know that ν is an injective ring homomorphism already (from Lemma 20.6). But a typical element of $R[U^{-1}]$ is of the form $a/1_R = \nu(a)$ with $a \in R$, so ν is also surjective.

(3) Let $V = R - \{0_R\}$. We know that $R[V^{-1}]$ is a domain already (from Proposition 20.4). Let $f \in R[V^{-1}] - (0)$. Then we can write $f = a/b$ with $a \in R$ and $b \in R - \{0_R\}$. Since $f \neq 0$, we must have $a \neq 0_R$, and thus $b/a \in R[V^{-1}]$. We have $f \cdot (b/a) = (ab)/(ab) = 1_R/1_R = 1$ in $R[V^{-1}]$. Therefore, $f \in (R[V^{-1}])^\times$, as required. $\qquad\square$

Remark 20.12. Rather than use the universal property of Lemma 20.9 to prove part (1) of Proposition 20.11, it may be quicker to give a more elementary proof. But note that when dealing with fractions, even over domains, some care is needed. For example, consider the fractions a/u on either side of the expression "$a/u \mapsto a/u$" in part (1) of the proposition. These two fractions are, in general, very different objects! (See Exercise 20.2.)

Corollary 20.13. *Every domain can be embedded in a field.*

Proof. Combine Lemma 20.6 with part (3) of Proposition 20.11. □

Definition 20.14. Let R be a domain. The field $R[U^{-1}]$, where $U = R - \{0\}$, is called the *field of fractions* of R.

Notation 20.15. The field of fractions of the domain R is denoted $k(R)$.

Example 20.16. The field of fractions of $\mathbf{Z}$ is $\mathbf{Q}$. If we just want to invert 3, then the smallest set we can take for U is $U = \{3^j : j \in \mathbf{N}\}$. With this choice for U, the localization of $\mathbf{Z}$ at U is $Z[U^{-1}] = \{a/3^j : a \in \mathbf{Z}, j \in \mathbf{N}\}$. On the other hand, suppose we would like to invert as much as possible *except* 3 by localizing $\mathbf{Z}$ at some appropriate multiplicative set V. In this case, we have to realize that if we include any multiple of 3, say 12, in V, then we are already allowing 3 as a denominator too, since $4/12 = 1/3$. So we must exclude all multiples of 3 from V. With this in mind, we try $V = \mathbf{Z} - (3)$. This set V is indeed multiplicatively closed, since if $a, b \in \mathbf{Z} - (3)$ then $ab \in \mathbf{Z} - (3)$: this just says, taking the contrapositive, that $3 \mid (ab)$ implies $3 \mid a$ or $3 \mid b$—which says precisely that 3 is *prime* (see Exercise 20.1). When we localize $\mathbf{Z}$ at this set V, we get the set of rational numbers whose denominators (in lowest terms) do not have 3 as a factor: $\mathbf{Z}[V^{-1}] = \{a/b : a, b \in \mathbf{Z}, 3 \nmid b\}$.

Example 20.17. Let D be a domain, and consider the polynomial ring $R = D[x]$. By Lemma 14.10, R is a domain. So we may form the field of fractions K of R, namely,

$$K = \{f/g : f, g \in D[x] \text{ and } g \neq 0\}.$$

If D is not just a domain but a *field*, then we have a special designation for this field of fractions:

Definition 20.18. Let F be a field. Then the field of fractions of $F[x]$ is called the *field of rational functions over F* (in the variable x), and is denoted $F(x)$.

Remark 20.19. There is a very satisfactory theory of localization over any commutative ring with 1, though it is somewhat more complicated than the theory over a domain. After completing the present text, the reader may wish to examine [3, Chapter 2].

20.3 Unique Factorization

One of the nicest and most familiar features of the ring **Z** is the unique factorization of integers into primes, expressed in the Fundamental Theorem of Arithmetic:

Theorem 20.20 (Fundamental Theorem of Arithmetic). *Let n be an integer, $n > 1$. Then there exist unique prime integers $1 < p_1 \leq p_2 \leq \cdots \leq p_r$ such that we can write $n = \prod_{i=1}^{r} p_i$.*

As we have mentioned, the use of the order relation $<$ (and $\leq$) does not carry over to general commutative rings with 1. Even in the ring $\mathbf{Z}$, we have allowed that for every positive prime p, the number $-p$ is just as good a prime. In general, the presence of units will get in the way of a truly unique factorization; what salvaged the situation in $\mathbf{Z}$ was the fact that the only units are 1 and -1, and we were able to recover from this by requiring all our factors to be greater than 1. But given the unavoidable presence of units, we compromise with the following definitions.

Definition 20.21. Let R be a commutative ring with 1, and let $a, b \in R$. We say that a and b are *associates* in R if $\exists u \in R^\times$ such that $a = ub$.

Definition 20.22. Let R be a domain. Then R is called a *Unique Factorization Domain*, or *UFD*, if every non-zero non-unit element of R factors into irreducible elements of R which are unique up to association. That is, R is called a UFD iff both

(UFD1) $\forall a \in R - R^\times - (0)$, $\exists a_1, \ldots, a_r \in R$ such that $a_1, \ldots, a_r$ are irreducible in R and $a = \prod_{i=1}^{r} a_i$; and

(UFD2) $\forall a \in R - R^\times - (0)$, if $a_1, \ldots, a_r, b_1, \ldots, b_s$ are irreducible elements of R such that $a = \prod_{i=1}^{r} a_i = \prod_{i=1}^{s} b_i$, then $s = r$, and, after a suitable reordering of the indices, a_i is an associate of b_i for every $i : 1 \leq i \leq r$.

In order to see what is involved in being a UFD, we shall address properties (UFD1) and (UFD2) separately, and find what could go wrong.

To violate (UFD1), there would need to be an element $a \in R - R^\times - (0)$ which does not factor into irreducibles. But then a itself must not be irreducible, which implies that $a = a_1 a_1'$ where $a_1, a_1' \in R - R^\times - (0)$; and at least one of a_1 or a_1', say a_1, does not factor into irreducibles, or else a itself would too. But now a_1 must factor into $a_2 a_2'$, where a_2 does not factor into irreducibles; and so on. So we get a sequence of factors $a_1, a_2, a_3, \ldots$ of a, where $a_i = a_{i+1} a_{i+1}'$ for all $i \geq 0$; for convenience, we set $a_0 = a$. Remembering our old adage "To divide is to contain" (see Exercise 12.10), we can rephrase this in terms of ideals, as follows: we have a chain of principal ideals

$$(a_0) \subseteq (a_1) \subseteq (a_2) \subseteq \cdots$$

in R. Moreover, because each a_{i+1}' is a non-unit and each a_{i+1} is non-zero, we can say $(a_i) = (a_{i+1} a_{i+1}') \neq (a_{i+1})$, by Lemma 12.16. Thus, to violate (1), there must be an infinitely long strictly increasing chain of principal ideals in R. This motivates the following definition.

Definition 20.23. Let R be a commutative ring with 1. We say that R satisfies the *ascending chain condition on principal ideals* if there is no infinite strictly increasing chain of principal ideals

$$(a_0) \subset (a_1) \subset (a_2) \subset \cdots$$

in R; equivalently, if every ascending chain of principal ideals in R

$$(a_0) \subseteq (a_1) \subseteq (a_2) \subseteq \cdots$$

must *stabilize*: that is, there must exist a number $N \in \mathbf{N}$ such that $(a_i) = (a_{i+1})$ for all $i \geq N$.

Remark 20.24. We often express this last condition by saying that $(a_i) = (a_{i+1})$ for all sufficiently large i, or for $i >> 0$.

Remark 20.25. There is a closely related condition which is immensely important in commutative algebra. Namely, a commutative ring with 1 is called *noetherian* if every ascending chain of ideals of R (not necessarily principal) stabilizes. Notice that the noetherian condition is stronger than the ascending chain condition on principal ideals; nonetheless, many important families of rings satisfy the noetherian condition.

Next we investigate how part (2) of the UFD definition can fail. To violate (2), we would need to have irreducibles $a_1, \ldots, a_r$, $b_1, \ldots, b_s$ in R such that $\prod_{i=1}^{r} a_i = \prod_{i=1}^{s} b_i$, but at least one of the a_i's, say a_1, is not an associate of any of the b_i's (see Exercise 20.13 for the reason we don't need to consider separately the possibility that $r \neq s$). Now a_1 divides $\prod_{i=1}^{s} b_i$ in R, but we must not have $a_1 \mid b_i$ in R for any i, since that would force a_1 and b_i to be associates (by Exercise 20.11). This would be impossible if a_1 were *prime* (by Exercise 20.12). Therefore, to violate (2), there must be an irreducible element of R which is not prime. It turns out that the conditions we were led to in this discussion are both necessary and sufficient:

Theorem 20.26. *Let R be a domain. Then R is a UFD iff R satisfies the ascending chain condition on principal ideals and every irreducible element of R is prime.*

Proof. ($\Rightarrow$): Suppose R is a UFD. Assume for a contradiction that $(a_0) \subset (a_1) \subset (a_2) \subset \cdots$ is a strictly increasing chain of principal ideals of R. Then for $i \geq 1$, we must have $a_i \neq 0$, since (a_i) properly contains (a_0); and (using Exercise 12.6) $a_i \notin R^\times$, since (a_i) is a proper subset of R. By property (UFD1), we can factor each a_i (for $i \geq 1$) into some number r_i of irreducibles in R, and by (UFD2), r_i does not depend on which factorization of a_i we choose. Now for any $i \geq 1$, since $(a_i) \subset (a_{i+1})$, we can write

$$a_i = f_i a_{i+1} \tag{20.4}$$

for some $f_i \in R - R^\times - (0)$. Factoring each side of Equation 20.4 into irreducibles shows that $r_i > r_{i+1}$ for $i \geq 1$. Now each r_i is a positive integer for

$i \geq 1$, yet we have $r_1 > r_2 > r_3 > \cdots$, which is impossible. This contradiction shows that R satisfies the ascending chain condition for principal ideals.

Next, let a be an irreducible element of R, and suppose that $b, c \in R$ and $a \mid bc$ in R, so that $(a) \supseteq (bc)$. Then $a \neq 0$ and $a \notin R^{\times}$ (by definition of irreducible). [Show: $a \mid b$ or $a \mid c$.] If $b \in R^{\times}$, then we have $(bc) = (c)$, so that $(a) \supseteq (c)$, and $a \mid c$, so we are done. Similarly, we are done if $c \in R^{\times}$. So we suppose that b and c are non-units. Likewise, we may suppose that $b, c \neq 0$, since certainly $a \mid 0$. Now we can write $bc = ad$ with $d \in R$, since $a \mid bc$ in R; and we can factor b and c into irreducibles in R, by (UFD1). Write $b = \prod_{i=1}^{r} b_i$ and $c = \prod_{i=1}^{s} c_i$ with b_i, c_i irreducible in R and $r, s \geq 1$. We must have $d \neq 0$, since $b, c \neq 0$ and R is a domain. Now we have

$$b_1 \cdots b_r \cdot c_1 \cdots c_s = ad. \tag{20.5}$$

If $d \in R^{\times}$, then ad is irreducible in R, by Exercise 12.3; so the number of irreducible factors on the left and right sides of Equation 20.5 is not the same, contradicting (UFD2). Therefore, $d \notin R^{\times}$. So we can factor d into irreducibles; write $d = \prod_{i=1}^{q} d_i$, where each d_i is irreducible in R. Then we may apply (UFD2) to the equation

$$b_1 \cdots b_r \cdot c_1 \cdots c_s = a \cdot d_1 \cdots d_q \tag{20.6}$$

to conclude that a is an associate of one of the b_i's or one of the c_i's. Therefore, either $a \mid b$ or $a \mid c$, as desired.

($\Leftarrow$): The argument preceding the statement of the theorem proves this direction. $\qquad\square$

We note that every field is vacuously a UFD, since there are no elements other than units and 0 (and thus no irreducible elements either). After fields, we have said that PIDs are the nicest type of ring we have studied so far; and **Z** is both a PID and a UFD; but it is still a little amazing that *all* PIDs enjoy unique factorization:

Proposition 20.27. *Every PID is a UFD.*

Proof. Let R be a PID. Suppose that $(a_0) \subseteq (a_1) \subseteq (a_2) \subseteq \cdots$ is an ascending chain of principal ideals of R. Set $I = \bigcup_{i=0}^{\infty}(a_i)$. Then I is an ideal of R (by Exercise 20.14), so, as R is a PID, we can write $I = (a)$ for some $a \in R$. Now $a \in I$, so (by definition of union) we must have $a \in (a_j)$ for some j. This forces $(a) \subseteq (a_j)$. But $(a) = I$, so the chain is stable after index j: that is, we have $(a_i) = I$ for all $i \geq j$.

Next, suppose that a is an irreducible element of R. Let $b, c \in R$, and suppose that a divides bc in R. [Show: $a \mid b$ or $a \mid c$.] [Idea: We want to show that either a and b have a as a common factor, or else a and c have a as a common factor; generating an ideal using two elements should bring out a common factor: $(gf, hf) \subseteq (f)$.] Let $I = (a, b)$ be the ideal of R generated by a and b. Since R is a PID, we can write $I = (d)$ for some $d \in R$. Now $a \in I$

by construction, so $d \mid a$ in R, and we have $a = dd'$ for some $d' \in R$. Since a is irreducible in R, this forces $d \in R^\times$ or $d' \in R^\times$. In the former case, we have $I = (d) = R$; in the latter case, we have $(a) = (dd') = (d)$ (by Lemma 12.16) $= I$. Similarly, working with the ideal $J = (a, c)$, we find that either $J = R$ or $J = (a)$. Now if $I = (a)$, then we have $b \in (a)$, so $a \mid b$ in R, as desired; similarly, we are done if $J = (a)$. So assume for a contradiction that $I = J = R$. Then $1 \in I$ and $1 \in J$, so we can write (using Lemma 12.15) $1 = r_1 a + r_2 b = s_1 a + s_2 c$ for some $r_1, r_2, s_1, s_2 \in R$. Thus we have

$$
\begin{aligned}
1^2 &= (r_1 a + r_2 b) \cdot (s_1 a + s_2 c) \\
&= a(r_1 s_1 a + r_1 s_2 c + r_2 s_1 b) + r_2 s_2 bc \\
&\in (a),
\end{aligned}
$$

the last inclusion following from the fact that a divides bc in R. So $1 \in (a)$, which forces $(a) = R$. But this implies (by Exercise 12.6) that $a \in R^\times$, contradicting the definition of irreducible. $\qquad\square$

Corollary 20.28. *If F is a field, then the polynomial ring $F[x]$ is a UFD.*

Proof. By Theorem 14.22, $F[x]$ is a PID, hence $F[x]$ is a UFD by Proposition 20.27. $\qquad\square$

We can get a lot more out of Proposition 20.27 by starting with Corollary 20.28 and combining it with our newfound knowledge that every domain is really part of a field (Corollary 20.13). The key results relating fractions to unique factorization go back to Gauss, who worked with **Z** and **Q**. But to do this, we first bring another familiar concept from **Z** to a general UFD: the notion of *greatest common divisor*, or gcd.

Recall that for two positive integers a and b, we may define $\gcd(a, b) = \max\{c \in \mathbf{Z} : c \mid a \text{ and } c \mid b\}$. Again, this definition is not directly portable to a general UFD, because we cannot express the notion of *max*—it depends on the ordering in **Z**. But there is another (better!) way to express gcd, which is familiar to those who have dabbled in elementary number theory. Namely, we can write a and b using a common (finite) set of primes, $a = \prod_{i=1}^{r} p_i^{d_i}$ and $b = \prod_{i=1}^{r} p_i^{e_i}$, with $d_i, e_i \in \mathbf{N}$; note that some of the exponents may be zero. Then we have $\gcd(a, b) = \prod_{i=1}^{r} p_i^{\min\{d_i, e_i\}}$. For the sake of uniformity, we can even write the factorizations of a and b as products over *all* the positive primes, as in

$$
a = \prod_{p \,:\, \text{prime}} p^{c_p},
$$

if we are willing to accept that an infinite product of 1's is 1.

Now in a general UFD R, given a non-zero element $a \in R$, we want to write something like $a = \prod_p p^{e_p}$, where p runs over all of the prime elements of R, and e_p is the highest power of p which divides a. Two problems emerge. First, we don't want to include two primes in our factorization if they are associates of each other in R, any more than we want to include both 3 and

−3 in the same factorization of a positive integer. We can remove this problem by taking a product over all prime *ideals* of the form (p) instead of over all prime elements p of R. The second problem is that we will end up with an infinite product, in general. This is fine, since only finitely many exponents will be positive; we ignore the repeated factors of the form $(p)^0$, since multiplying by 1 should have no effect.

To make these ideas work, we first need to know how to multiply two ideals.

Definition 20.29. Let R be a commutative ring with 1, and let I, J be two ideals of R. Then the *ideal product* of I and J is defined to be

$$I \cdot J := (\{a \cdot b \ : \ a \in I, \ b \in J\}),$$

the ideal of R generated by the products of elements of I with elements of J. We define $I^0 := (1) = R$, and we define I^n recursively as $I^n := I^{n-1} \cdot I$ for $n \geq 1$. In case R is a UFD and I is principal, we also define $I^\infty = (1)$, if $I = R$; and $I^\infty = (0)$, if $I \neq R$. (For some motivation for this last definition, see Exercise 20.16.)

Warning 20.30. In general, the setwise product will be a proper subset of the ideal product of two ideals, even though the notation alone does not distinguish between these two types of product. However, in the cases of interest to us—where all ideals in question are *principal*—then Lemma 20.32 assures us that the two types of product give the same result.

Lemma 20.31. *Let R be a commutative ring with 1, and let I, J be ideals of R. Then we have*

$$I \cdot J \subseteq I \cap J, \ and$$
$$R = I^0 \supseteq I^1 = I \supseteq I^2 \supseteq I^3 \supseteq \cdots \supseteq I^\infty.$$

Proof. If $a \in I$ and $b \in J$, then $ab \in I$ by (Id2) for I, and $ab \in J$ by (Id2) for J, so $ab \in I \cap J$. Since $I \cdot J$ is the smallest ideal which contains all such products, and $I \cap J$ is an ideal (by Exercise 11.8), we have $I \cdot J \subseteq I \cap J$. The second statement follows immediately from the first, after checking the special cases $I^1 = I$ and $I^\infty \subseteq I^n$ for any $n \in \mathbf{N}$. □

Lemma 20.32. *Let R be a commutative ring with 1. Let $I = (a)$, $J = (b)$ be two principal ideals of R, with $a, b \in R$. Then we have $I \cdot J = (ab)$, and $I^n = (a^n)$ for all $n \in \mathbf{N}$. Furthermore, the ideal product is associative and commutative.*

Proof. Exercise 20.15. □

Next, we want to get our hands on the exponents in prime factorizations. As promised, we will work with principal prime ideals instead of prime elements of a UFD. Thus, it makes sense to replace the maximum exponent e such that $p^e \mid a$ with the maximum exponent e such that $(p)^e \supseteq (a)$.

Definition 20.33. Let R be a UFD. Let $a \in R - (0)$, and let $\mathfrak{p}$ be a principal prime ideal of R. We set $\exp_{\mathfrak{p}}(a) = \max\{e \in \mathbf{N} \ : \ \mathfrak{p}^e \supseteq (a)\}$, and we call e the *exponent of* $\mathfrak{p}$ *in* a. We take $\exp_{\mathfrak{p}}(0) = \infty$.

Remark 20.34. In moving from prime elements to principal prime ideals in a UFD, we are picking up exactly one new ideal, namely the zero ideal, (0). This follows from Exercises 12.2 and 12.12. The element 0 straddles the line between being prime and non-prime, as it were, but we defined 0 as a non-prime, while (0) is prime.

We now formally state how we are to handle the kinds of infinite products of ideals which we shall encounter.

Convention 20.35. In an infinite product of ideals of a commutative ring with 1:

(i) If any factor is (0), then the product is (0).

(ii) We remove factors of the form (1); if this results in an empty product of ideals, then we interpret the empty product as (1).

An indication that we are on the right track is given by the following result.

Proposition 20.36. *Let R be a UFD, and let $a \in R$. Then we have*

$$(a) = \prod_{\mathfrak{p} \in \mathfrak{P}} \mathfrak{p}^{e_{\mathfrak{p}}}, \tag{20.7}$$

where $\mathfrak{P}$ is the set of all principal prime ideals of R, and $e_{\mathfrak{p}} = \exp_{\mathfrak{p}}(a)$. If $a \neq 0$, then all but finitely many of the exponents are 0, and furthermore, this representation of (a) is unique: that is, if $(a) = \prod_{\mathfrak{p} \in \mathfrak{P}} \mathfrak{p}^{f_{\mathfrak{p}}}$ where $f_{\mathfrak{p}} \in \mathbf{N}$ and $f_{\mathfrak{p}} = 0$ for all but finitely many $\mathfrak{p}$, then $f_{\mathfrak{p}} = \exp_{\mathfrak{p}}(a)$ for all $\mathfrak{p} \in \mathfrak{P}$.

Proof. If $a = 0$, then we have $\exp_{\mathfrak{p}}(a) = \infty$ for all $\mathfrak{p} \in \mathfrak{P}$, from Definition 20.33, so the product on the right-hand side of Equation 20.7 is the zero ideal according to Definition 20.29 and Convention 20.35 (i). So from now on, we suppose that $a \neq 0$.

We next treat the case when $a \in R^{\times}$. Let $\mathfrak{p} \in \mathfrak{P}$. We have $\mathfrak{p} \neq R$ by definition of prime ideal, and $(a) = R$ by Exercise 12.6. So $\mathfrak{p} \not\supseteq (a)$, and $\exp_{\mathfrak{p}}(a) = 0$. The desired formula now follows from Convention 20.35 (ii). As for uniqueness, this follows from Lemma 20.31 and the fact that a prime ideal must be proper.

Finally, suppse that $a \in R - R^{\times} - (0)$. By (UFD1), we can write $a = \prod_{j=1}^{r} a_j$ where each a_j is irreducible in R. By Exercise 20.10, the relation of association, $\sim$, is an equivalence relation on R, so we can find a finite number of elements $p_1, \ldots, p_n \in R$ such that each a_j is an associate of exactly one p_i. Each p_i is irreducible in R by Exercise 12.3, hence p_i is prime in R by Theorem 20.26. For $1 \leq i \leq n$, set $e_i = |\{j \in \mathbf{N} \ : \ 1 \leq j \leq r \text{ and } a_j \sim p_i\}|$. Then by definition of association and Lemma 12.16, we have $(a_j) = (p_i)$ for the unique i such that $a_j \sim p_i$. Let $\mathfrak{p}_i = (p_i)$ for $i \in \{1, \ldots, n\}$. By Exercise

12.12, we have $\mathfrak{p}_i \in \mathfrak{P}$. From Lemma 20.32 we have

$$(a) = \left(\prod_{j=1}^{r} a_j \right) = \prod_{j=1}^{r} (a_j) = \prod_{i=1}^{n} (p_i)^{e_i} = \prod_{i=1}^{n} \mathfrak{p}_i^{e_i} = \prod_{\mathfrak{p} \in \mathfrak{P}} \mathfrak{p}^{e_\mathfrak{p}},$$

where $e_\mathfrak{p} = e_i$ if $\mathfrak{p} = \mathfrak{p}_i$ for some i, and 0 otherwise. Now that we know that such a formula exists, suppose that $(a) = \prod_{\mathfrak{p} \in \mathfrak{P}} \mathfrak{p}^{f_\mathfrak{p}}$ where $f_\mathfrak{p} \in \mathbf{N}$ and $f_\mathfrak{p} = 0$ for all but finitely many $\mathfrak{p}$. Using Lemma 20.31, we see that for each $\mathfrak{p}$, we have $\mathfrak{p}^{f_\mathfrak{p}} \supseteq (a)$, so $\exp_\mathfrak{p}(a) \geq f_\mathfrak{p}$. On the other hand, if for some $\mathfrak{p}$ we had $\mathfrak{p}^{f_\mathfrak{p}+1} \supseteq (a)$, then we would have $p_\mathfrak{p}^{f_\mathfrak{p}+1} \mid a$ in R, where $p_\mathfrak{p}$ is a generator of $\mathfrak{p}$ in R. We demonstrate that this is impossble. Assume for a contradiction that $a = g \cdot p_\mathfrak{q}^{f_\mathfrak{q}+1}$ for some $g \in R$ and $\mathfrak{q} \in \mathfrak{P}$. Note that we may write $a = u \cdot \prod_{\mathfrak{p} \in \mathfrak{P}} p_\mathfrak{p}^{f_\mathfrak{p}}$ for some $u \in R^\times$. So we have $u^{-1} g p_\mathfrak{q} = \prod_{\mathfrak{p} \neq \mathfrak{q}} p_\mathfrak{p}^{f_\mathfrak{p}}$, so $p_\mathfrak{q}$ divides $\prod_{\mathfrak{p} \neq \mathfrak{q}} p_\mathfrak{p}^{f_\mathfrak{p}}$ in R. Since $p_\mathfrak{q}$ is prime in R, then $p_\mathfrak{q} \mid p_\mathfrak{p}$ in R for some $\mathfrak{p} \neq \mathfrak{q}$, by Exercise 20.12. So $p_\mathfrak{p}$ and $q_\mathfrak{q}$ are associates by Exercise 20.11, thus $\mathfrak{p} = \mathfrak{q}$, a contradiction. Therefore, $f_\mathfrak{p} = \exp_\mathfrak{p}(a)$ for all $\mathfrak{p} \in \mathfrak{p}$. This concludes the proof. $\qquad \square$

Remark 20.37. A birds'-eye view of the significance of unique factorization is that it allows us to turn a non-zero ring element a into a collection of integers— namely, the exponents in the prime factorization of (a). In this process, we lose the information about the units of the ring. Remembering that "a logarithm is an exponent," these exponents may be viewed as "discrete logarithms." Now logarithms transform multiplication into addition (recall Example 7.11). For a frighteningly beautiful realization of these ideas, see Exercise 20.23.

Now we are ready to define gcd in a general UFD.

Definition 20.38. Let $a, b \in R$, where R is a UFD. Let $\mathfrak{P}$ be the set of all principal prime ideals of R. For $\mathfrak{p} \in \mathfrak{P}$, let $d_\mathfrak{p}$ and $e_\mathfrak{p}$ denote the exponent of $\mathfrak{p}$ in a and b, respectively, with $d_\mathfrak{p}, e_\mathfrak{p} \in \mathbf{N} \cup \{\infty\}$. The *greatest common divisor* of a and b in R is the ideal

$$\gcd(a, b) := \prod_{\mathfrak{p} \in \mathfrak{P}} \mathfrak{p}^{\min\{d_\mathfrak{p}, e_\mathfrak{p}\}} \subseteq R.$$

More generally, for any non-empty set $A \subseteq R$, define

$$\gcd(A) := \prod_{\mathfrak{p} \in \mathfrak{P}} \mathfrak{p}^{\min\{e_\mathfrak{p}(a) \ : \ a \in A\}},$$

where $e_\mathfrak{p}(a)$ is the exponent of $\mathfrak{p}$ in a.

Remark 20.39. By Lemma 20.32 and Convention 20.35, every gcd is a principal ideal.

Notation 20.40. We write $\gcd(a_1, a_2, \ldots, a_n)$ to mean $\gcd(\{a_1, a_2, \ldots, a_n\})$.

Definition 20.41. Let R be a UFD, and let $f = \sum_{i=0}^{d} a_i x^i \in R[x]$. The *content* of f is the principal ideal

$$\text{cont}(f) := \gcd(a_0, a_1, \ldots, a_d) \subseteq R.$$

Lemma 20.42 (Gauss' Lemma). *Let R be a UFD, and let $f, g \in R[x]$. Then we have $\text{cont}(fg) = \text{cont}(f) \cdot \text{cont}(g)$.*

Proof. Write $f = \sum_{i=0}^{m} a_i x^i$ and $g = \sum_{i=0}^{n} b_i x^i$ with $a_i, b_i \in R$. Write $\text{cont}(f) = (c)$, $\text{cont}(g) = (d)$, and $\text{cont}(fg) = (e)$ with $c, d, e \in R$. The result follows directly from Exercise 20.18 (with $a = 0$) if f or g is 0, so we suppose that $f, g \neq 0$, and consequently (by Exercise 20.22) $c, d, e \neq 0$. Let p be a prime element of R, and let $a = \exp_{(p)}(c)$ and $b = \exp_{(p)}(d)$. Then we have $p^a \mid a_i$ and $p^b \mid b_i$ in R for all i. It follows that $p^a \mid f$ and $p^b \mid g$ in $R[x]$. Therefore, p^{a+b} divides fg in $R[x]$. So $\exp_{(p)}(e) \geq a + b$. On the other hand, let $j = \max\{i : p^{a+1} \nmid a_i\}$ and $k = \max\{i : p^{b+1} \nmid b_i\}$. Then we have $p^{a+1} \mid a_i$ for $i > j$, and $p^{b+1} \mid b_i$ for $i > k$. Now the coefficient of x^{j+k} in fg is $c_{j+k} = \sum_{i=0}^{j+k} a_i b_{j+k-i}$, and every term in this sum except for $a_j b_k$ is divisible by p^{a+b+1}. So we can write $c_{j+k} = p^{a+b+1} \cdot r + p^{a+b} s$ for some $r, s \in R$ with $p \nmid s$. It follows that $p^{a+b+1} \nmid c_{j+k}$. Therefore, $\exp_{(p)}(e) \leq a + b$. We have shown that $\exp_{(p)}(e) = \exp_{(p)}(c) + \exp_{(p)}(d)$ for every prime element p of R. It follows from Proposition 20.36 that (e) and (cd) have the same prime factorization in R, and $(e) = (cd)$, as desired. $\square$

Corollary 20.43. *Let R be a UFD, let $K = k(R)$ be the field of fractions of R, and let $f \in R[x]$. If $f = f_1 f_2$ with $f_1, f_2 \in K[x]$, then there exist $\tilde{f}_1, \tilde{f}_2 \in R[x]$ such that $\tilde{f}_i$ is an associate of f_i in $K[x]$ and $f = \tilde{f}_1 \tilde{f}_2$.*

Proof. The case $f = 0$ being easy, we suppose that $f \neq 0$. Suppose that $f = f_1 f_2$ where $f_1, f_2 \in K[x]$. There exist $d_1, d_2 \in R - (0)$ such that $d_i f_i \in R[x]$; for example, we can take d_i to be the product of the denominators of the coefficients of f_i. So $d_1 d_2 f = (d_1 f_1) \cdot (d_2 f_2)$ is a factorization of $d_1 d_2 f$ in $R[x]$. By Gauss' Lemma, together with Exercise 20.18, we have

$$\text{cont}(d_1 d_2 f) = (d_1 d_2)\text{cont}(f) = \text{cont}(d_1 f_1) \cdot \text{cont}(d_2 f_2). \tag{20.8}$$

Write $(c_i) = \text{cont}(d_i f_i)$, with $c_i \in R$, and set $\tilde{g}_i = d_i f_i / c_i$. Then $\tilde{g}_i \in R[x]$ (by Exercise 20.24), and $(d_1 d_2) \mid (c_1 c_2)$ in R. So we can write $c_1 c_2 = c d_1 d_2$ for some $c \in R$. Now $d_1 d_2 f = (d_1 f_1)(d_2 f_2) = (c_1 \tilde{g}_1)(c_2 \tilde{g}_2) = c d_1 d_2 \tilde{g}_1 \tilde{g}_2$, so $f = c \tilde{g}_1 \tilde{g}_2$. We take $\tilde{f}_1 = c \tilde{g}_1$ and $\tilde{f}_2 = \tilde{g}_2$ to complete the proof. $\square$

Corollary 20.44. *Let R be a UFD, let $K = k(R)$ be the field of fractions of R, and let $f \in R[x] - R$. If f is irreducible in $R[x]$, then f is irreducible in $K[x]$.*

Proof. Suppose that f is irreducible in $R[x]$. Assume for a contradiction that f is not irreducible in $K[x]$. Then we can write $f = f_1 f_2$ for some $f_i \in K[x] - K$.

By Corollary 20.43, there are elements $\tilde{f}_1, \tilde{f}_2 \in R[x]$ such that $\tilde{f}_i$ is an associate of f_i in $K[x]$ and $f = \tilde{f}_1 \tilde{f}_2$. But since units in $K[x]$ have degree zero, we have $\deg(\tilde{f}_i) = \deg(f_i) \geq 1$, which contradicts the irreducibility of f in $R[x]$. $\square$

Proposition 20.45. *If R is a UFD, then so is the polynomial ring $R[x]$.*

Proof. Suppose that R is a UFD, and let $S = R[x]$. Assume for a contradiction that $(f_0) \subset (f_1) \subset (f_2) \subset \cdots$ is a strictly increasing chain of principal ideals in S. Let $c_i = \operatorname{cont}(f_i)$. Since f_{i+1} divides f_i in S, Gauss' Lemma yields $c_0 \subseteq c_1 \subseteq c_2 \subseteq \cdots$. By Theorem 20.26, this sequence of principal ideals of R must stabilize; say $c_i = (c)$ for $i >> 0$, with $c \in R$. Also, we have $\deg(f_1) \geq \deg(f_2) \geq \deg(f_3) \geq \cdots \geq 0$ since f_{i+1} divides f_i in S. Since all of these degrees are natural numbers, this sequence of degrees must eventually stabilize; say $\deg(f_i) = d$ for all $i >> 0$. So for large enough i, we have $f_{i+1} = a_i \cdot f_i$ with $\deg(a_i) = 0$, hence $a_i \in R$. Taking contents yields $(c) = (a_i) \cdot (c) = (a_i \cdot c)$ for $i >> 0$. It follows that $a_i \in R^\times$, so $(f_{i+1}) = (f_i)$, a contradiction. Therefore, S satisfies the ascending chain condition on principal ideals.

Next, let f be an irreducible element of $R[x]$. [Show: f is prime in $R[x]$]

Case 1: $\deg(f) = 0$. Then $f \in R$. If $f = ab$ with $a, b \in R$, then viewing $a, b \in R[x]$, by irreducibility of f in $R[x]$ we have (without loss of generality) $a \subset (R[x])^\times = R^\times$. So f is also irreducible in R; hence f is prime in R by Theorem 20.26. Now suppose that f divides gh in $R[x]$. Then we can write $gh = fw$ for some $w \in R[x]$. Write $\operatorname{cont}(g) = (\gamma)$ and $\operatorname{cont}(h) = (\delta)$ with $\gamma, \delta \in R$. Then we have $(\gamma) \cdot (\delta) = (\gamma\delta) = (f) \cdot \operatorname{cont}(w)$, so f divides $\gamma\delta$ in R. Without loss of generality, $f \mid \gamma$ in R. By Exercise 20.24 (a), we have $f \mid g$ in $R[x]$; hence, f is prime in $R[x]$ in this case.

Case 2: $\deg(f) \geq 1$. Let $K = k(R)$ be the field of fractions of R. Note that we have $R[x] \leq K[x]$. By Corollary 20.44, f is irreducible in $K[x]$. Since $K[x]$ is a UFD (by Corollary 20.28), then f is also prime in $K[x]$. Now suppose that $g, h \in R[x]$ and f divides gh in $R[x]$. Then certainly f divides gh in $K[x]$, and as f is prime in $K[x]$, then without loss of generality we may suppose that f divides g in $K[x]$. So we can write $g = ft$ with $t \in K[x]$. There exists $d \in R - (0)$ such that $dt \in R[x]$; for example, take d to be the product of the denominators of the coefficients of t. Then $dg = f \cdot (dt)$ is a factorization of dg in $R[x]$. By Gauss' Lemma, we have $(d) \cdot \operatorname{cont}(g) = \operatorname{cont}(dg) = \operatorname{cont}(f) \cdot \operatorname{cont}(dt)$. Now $\operatorname{cont}(f) = (1)$ by Exercise 20.24(c). Therefore, $(d) \supseteq \operatorname{cont}(dt)$. It now follows from Exercise 20.24(a) that d divides dt in $R[x]$, so $t \in R[x]$. Since $g = ft$, we have shown that $f \mid g$ in $R[x]$. Thus, f is prime in $R[x]$. So by Proposition 20.26, $R[x]$ is a UFD. $\square$

We can apply Proposition 20.45 repeatedly by starting with a UFD and adding one variable after another; what we get is a polynomial ring in many variables, or "multivariate" polynomial ring. The notation for such rings is what we would expect (see also Exercise 14.11):

Definition 20.46. Let R be a commutative ring with 1. The *polynomial ring over R in the variables $x_1, \ldots, x_n$* is defined recursively as $R[x_1, \ldots, x_n] := (R[x_1, \ldots, x_{n-1}])[x_n]$.

Corollary 20.47. *If R is a UFD, then so is the polynomial ring $R[x_1, \ldots, x_n]$.*
□

Remark 20.48. The order in which we adjoin variables does not affect the structure of the resulting polynomial ring: we understand $R[x, y]$ and $R[y, x]$ to be the same ring. It is often useful, however, to consider a formal representation of a multivariate polynomial in terms of the order in which we adjoin the variables. For example, we can consider $f := x^2y - x^3 + xy \in \mathbf{Z}[x, y]$ as the polynomial $(-1)x^3 + (y)x^2 + (y)x \in (\mathbf{Z}[y])[x]$ or as $(x^2 + x)y + (-x^3) \in (\mathbf{Z}[x])[y]$. Accordingly, we speak of the *degree in x* and the *degree in y*. Here, we have $\deg_x(f) = 3$ and $\deg_y(f) = 1$.

20.4   Exercises

Exercise 20.1. Let R be a domain and let I be an ideal of R. Prove that $R - I$ is a multiplicative set iff I is prime.

Exercise 20.2. Let $U = \mathbf{Z} - (2)$ and $V = \mathbf{Z} - (0)$. Describe as explicitly as you can the fraction $2/3$ as an element of $\mathbf{Z}[U^{-1}]$ and also as an element of $\mathbf{Z}[V^{-1}]$, using the definition of a fraction as a certain set of ordered pairs. In particular, show that $2/3$ formally means two different things in these two rings.

Exercise 20.3. Under the hypotheses of Proposition 20.4, show that the set of fractions a/u with $a \in R$ and $u \in U$ forms a commutative ring S under the given operations, with $0_S = 0_R/1_R$ and $1_S = 1_R/1_R$. You may use the fact that the operations are well-defined.

Exercise 20.4. Let T and S be commutative rings with 1, and let $\tau : T \to S$ be a ring homomorphism. Let $N \subseteq T^\times$ be such that $1_T \in N$ and $\tau(N) \subseteq S^\times$. Let $M = \langle N \rangle$ be the subgroup of $T^\times$ generated by N.
  (a) Prove that $\tau(M) = \langle \tau(N) \rangle \leq S^\times$.
  (b) Prove that $\tau|_M : M \to S^\times$ is a group homomorphism.

Exercise 20.5. Prove that the property of localization described in Lemma 20.9 characterizes the localization in the following sense. Let R be a domain, U a multiplicatively closed subset of R, and T a domain with a ring homomorphism $\tau : R \to T$ such that $\tau(U) \subseteq T^\times$. Suppose that for every ring homomorphism $\sigma : R \to S$ such that $\sigma(U) \subseteq S^\times$, there is a unique ring homomorphism $\omega : T \to S$ such that $\omega \circ \tau = \sigma$. Prove that $T \cong R[U^{-1}]$.

Exercise 20.6. Let R be a domain, and let U be a multiplicatively closed subset of R such that $U \subseteq R^\times$. Prove that $R[U^{-1}] \cong R$.

Exercise 20.7. Let R be a domain, and let U, V be multiplicatively closed subsets of R with $U \subseteq V$. Prove that $(R[U^{-1}])[V^{-1}] \cong R[V^{-1}]$.

Exercise 20.8. Let F be a field of characteristic 0. Prove that $\mathbf{Q} \hookrightarrow F$.

Exercise 20.9. Let D be a domain, and let $K = k(D)$ be the field of fractions of D. Prove that $k(D[x]) \cong K(x)$.

Exercise 20.10. Let R be a commutative ring with 1. Define a relation $\sim$ on R by $a \sim b$ iff a and b are associates in R. Prove that $\sim$ is an equivalence relation.

Exercise 20.11. Let R be a domain, and let a and b be irreducible elements of R. Prove that $a \mid b$ in R iff a is an associate of b in R.

Exercise 20.12. Let R be a commutative ring with 1 and let a be a non-zero non-unit of R. Prove that a is a prime element of R iff for all $a_1, a_2, \ldots, a_n \in R$ with $n \geq 1$, we have $a \mid (a_1 a_2 \cdots a_n) \implies \exists i$ such that $a \mid a_i$.

Exercise 20.13. Let R be a domain, and suppose that whenever $a_1, \ldots, a_r, b_1, \ldots, b_s$ are irreducible elements of R with $r, s \geq 1$ such that $\prod_{i=1}^{r} a_i = \prod_{i=1}^{s} b_i$, then each a_i is an associate of some b_j. Prove by induction on $r + s$ that we must have $r = s$ under these circumstances.

Exercise 20.14. Let R be a commutative ring with 1, and let $I_0 \subseteq I_1 \subseteq I_2 \subseteq \cdots$ be an ascending chain of ideals of R. Set $I = \bigcup_{i=0}^{\infty} I_i$. Prove that I is an ideal of R.

Exercise 20.15. Prove Lemma 20.32.

Exercise 20.16. Let R be a UFD, and let I be a proper principal ideal of R. Prove that $\bigcap_{n=1}^{\infty} I^n = (0)$. (This motivates the definition of I^∞ as (0).) Hint: Assume for a contradiction that $0 \neq b \in I^n$ for all $n \in \mathbf{Z}^+$, and consider the prime factorizations of b and of a generator a of I.

Exercise 20.17. Let F be a field. Let R be the set of all expressions of the form $a_0 + a_1 x^{1/2^n} + a_2 x^{2/2^n} + \cdots + a_r x^{r/2^n}$ with $a_i \in F$ and $r, n \in \mathbf{N}$. Convince yourself that R forms a domain under the natural operations $+$ and $\cdot$, and prove that R does *not* satisfy the ascending chain condition on principal ideals.

Exercise 20.18. Let R be a UFD.
  (a) For any $a \in R$, prove that $\gcd(a) = (a)$; viewing a as a constant polynomial in $R[x]$, conclude that $\text{cont}(a) = (a)$.
  (b) For any non-empty set $A \subseteq R$, prove that $\gcd(A \cup \{0\}) = \gcd(A)$.

Exercise 20.19. Let R be a UFD and let A be a non-empty subset of R. Let $b \in R$ be a generator for $\gcd(A)$.
  (a) Prove that if $c \in R$ and $c \mid a$ for all $a \in A$, then $(c) \supseteq \gcd(A) = (b)$.
  (b) Prove that for all $a \in A$, we have b divides a in R.
  (c) Let $\tilde{A} = \{\tilde{a} \in R : a = \tilde{a}b$ for some $a \in A\}$. Prove that $\gcd(\tilde{A}) = (1)$. (Note that if $b \neq 0$ then we can write $\tilde{A} = \{a/b : a \in A\}$.)
  (d) Prove that if $f \in R[x]$, $c \in R$, and $c \mid f$ in $R[x]$, then $(c) \supseteq \text{cont}(f)$.

Exercise 20.20. Let R be a UFD, and let $\alpha \in k(R)$. Prove that we can write $\alpha = n/d$ for some $n, d \in R$ with $d \neq 0$ and $\gcd(n, d) = (1)$. Hint: Exercise 20.19 may help.

Exercise 20.21 (A UFD which is not a PID). Let F be a field, and let $R = F[x, y]$. Let $I = (x, y)$ be the ideal of R generated by the variables x and y. Prove that I is not principal. Hint: Assume for a contradiction that $\exists f \in R$ such that $(f) = I$, and consider the degree of f in the variables x and y separately.

Exercise 20.22. Let R be a UFD, and let $A \subseteq R$ be a non-empty subset of R.
  (a) Prove that we have $\gcd(A) \supseteq (A)$.
  (b) Prove that $\gcd(A) = (A)$ iff (A) is a principal ideal.
  (c) Find an example to show that we may have $\gcd(A) \neq (A)$.

Exercise 20.23. Let R be a UFD, and let $K = k(R)$ be the field of fractions of R.
  (a) For a principal prime ideal $\mathfrak{p}$ of R and a fraction $f = a/b \in K$ with $a, b \in R - (0)$, define $\exp_\mathfrak{p}(f) = \exp_\mathfrak{p}(a) - \exp_\mathfrak{p}(b)$. Show that this extension of exp is well-defined.
  (b) Prove that for every $f \in K$, we have $f \in R$ iff $\exp_\mathfrak{p}(f) \geq 0$ for all principal prime ideals $\mathfrak{p}$ of R.
  (c) Let $\mathfrak{P}$ be the set of all non-zero principal prime ideals of R. Define a function $\omega : K^\times \to \oplus_{\mathfrak{p} \in \mathfrak{P}}(\mathbf{Z}, +)$ by $f \mapsto (\exp_\mathfrak{p}(f))_{\mathfrak{p} \in \mathfrak{P}}$. Prove that ω is a surjective group homomorphism with $\ker(\omega) = R^\times$. (See Exercise 17.12 for this notation.)

Exercise 20.24. Let R be a UFD, and let $f \in R[x] - (0)$. Write $\mathrm{cont}(f) = (c)$ with $c \in R$. Let $K = k(R)$ be the field of fractions of R.
  (a) Note that $c \neq 0_R$ and $1/c \in K$, so that $(1/c) \cdot f \in K[x]$, and prove that in fact $(1/c) \cdot f \in R[x]$. Conclude that c divides f in $R[x]$.
  (b) Observe that as $c \in R \leq K \leq K[x]$, we can form the fraction $f/c \in K(x)$. Convince yourself that $f/c = (1/c) \cdot f$ under the identification of $K[x]$ as a subring of $K(x)$.
  (c) Prove that if f is irreducible in $R[x]$ and $\deg(f) \geq 1$, then $\mathrm{cont}(f) = (1)$.

Exercise 20.25 (A domain which is not a UFD). Let $R = \{a + bi\sqrt{6} : a, b \in \mathbf{Z}\}$.
  (a) Show that $R \leq \mathbf{C}$. Conclude (e.g., from Exercise 11.11) that R is a domain.
  (b) Show that $\mathbf{Q}[i\sqrt{6}] =: K$ is (isomorphic to) the field of fractions of R.
  (c) Show that K is a finite Galois extension of $\mathbf{Q}$, and find the elements of $G := \mathrm{Gal}(K/\mathbf{Q})$ explicitly.
  (d) Define a function $\mathcal{N} : K \to K$ by $\mathcal{N}(\alpha) = \prod_{\sigma \in G} \sigma(\alpha)$. Prove that $\forall \alpha, \beta \in K$, we have $\mathcal{N}(\alpha\beta) = \mathcal{N}(\alpha)\mathcal{N}(\beta)$.
  (e) Prove that in fact we have $\mathcal{N}(K) \subseteq \mathbf{Q}$.
  (f) Find an explicit formula for $\mathcal{N}$, and prove that $\mathcal{N}(R) \subseteq \mathbf{N}$.
  (g) Prove that for all $\alpha \in R$, we have $\alpha \in R^\times$ iff $\mathcal{N}(\alpha) \in \mathbf{Z}^\times$. Use this to show that $R^\times = \{-1, 1\}$.

(h) Prove that 5 is irreducible in R but not prime in R. (The facts proved above are useful here, or you can prove this directly.) Conclude that R is not a UFD.

Exercise 20.26. Let R be a UFD, and let $K = k(R)$ be the field of fractions of R. Let F be an extension field of K. We say that an element $\alpha \in F$ is *integral over* R if $f(\alpha) = 0$ for some non-zero monic polynomial $f \in R[x]$.

(a) Prove that if α is integral over R and $\alpha \in K$, then $\alpha \in R$.

(b) (For those who know about eigenvalues) Suppose that M is a square matrix with entries in $\mathbf{Z}$, and someone shows you a list of purported eigenvalues for M which includes the value 1.371. Can this value be exact?

21

The Problems of the Ancients

21.1 Introduction

Until this point, our goal has been to generalize the concepts found in ordinary number systems, and study the resulting structures. We have only occasionally used these structures to shed light on our original number systems. In the present chapter, we shift our focus, and aim our theory squarely at certain questions about numbers and geometry which were first raised many centuries ago.

21.2 Constructible Numbers

Before the invention of algebra, and even before the modern (arabic) representation of numbers existed, there was geometry. In this section, we consider the problem of so-called straightedge and compass constructions: we are given a straight, rigid bar of length 1, together with a compass. We may think of our "compass" as a long string together with a pencil which can be tied to one end of the string. By holding the other end of the string in place, we can draw a circle. The question is, which real numbers occur as the lengths of the line segments which can be constructed in a finite number of steps? Informally, we define the set of *constructible numbers* to be

$$\mathfrak{C} = \{\alpha \in \mathbf{R} \mid \pm\alpha \text{ can be constructed with straightedge and compass}\}.$$

Our first goal is to wrangle the physical operations which can be performed by the straightedge and compass into mathematical operations. We will have to come to an agreement about exactly which operations are allowed. Let us agree on the following:

(1) Given two points, we can draw a line segment between them—imagine stretching the string taut between the points and using the pencil and straightedge, one unit at a time.

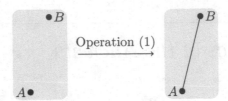

(2) Given that a line segment of length α can be constructed, and given a point P on some line segment, we can use the straightedge and compass to extend the line segment from the point P by a length of α units in either direction.

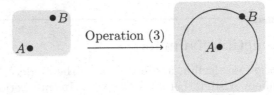

(3) Given two points A and B, we can place the two ends of the compass at these points and draw the circle with center A which passes through B.

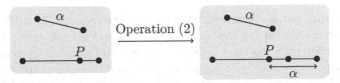

(4) We can identify the point where two given line segments intersect, the point(s) where two given circles intersect, and the point(s) where a given circle intersects a given line segment, when these intersection points exist.

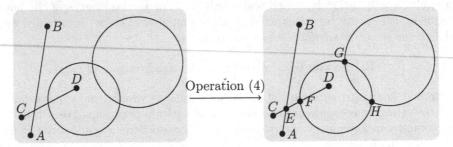

Notice that we do not allow "random" line segments or circles to be drawn: all of the operations listed above assume that we are given points, circles, or line segments which already exist, and proceed in a deterministic fashion. But clearly we need to have some kind of starting place; so we allow the one-time use of the following operation:

(0) To begin with, we can run the pencil along the straightedge to produce a line segment of length 1.

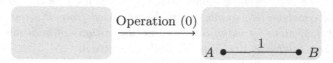

As a first illustration of what can be done with these operations, we prove the following useful result.

Lemma 21.1. *Given a line segment ℓ and a point $P \in \ell$, we can construct a line segment $\tilde{\ell}$ perpendicular to ℓ at P.*

Proof. Note that we must have begun our operations by constructing a line segment of length 1 using Operation (0). We can therefore use Operation (2) to construct two distinct line segments $\overline{PQ_1}$ and $\overline{PQ_2}$ which lie along ℓ and have length 1. Using Operation (4), we identify the points Q_1 and Q_2.

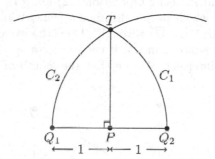

FIGURE 21.1: Constructing a line segment perpendicular to a given segment through a given point

Next, we use Operation (3) to draw the circle C_1 with center Q_1 which passes through Q_2. Then we draw the circle C_2 centered at Q_2 and passing through Q_1. The intersection points of these two circles are identifiable by Operation (4); choose one of them and call it T. Finally, the line segment $\overline{PT}$ can be drawn using Operation (1), and this segment has the desired properties; see Figure 21.1. $\square$

Now we return to the question of which real numbers can be constructed by straightedge and compass. The main insight here is that our operations are (basically) just linear and quadratic in nature. Algebraically, linear operations will allow us to do elementary arithmetic—what we now recognize as *field operations*. Quadratic operations allow us to take square roots—form *quadratic extension fields*.

Claim 21.2. (I) 1 *is constructible.*

(II) *If a and b are constructible numbers, then so are $a + b$, $-a$, and $a \cdot b$; further, if $a \neq 0$, then $1/a$ is constructible too.*

(III) *If a is a constructible number and $a \geq 0$, then $\sqrt{a}$ is also constructible.*

(IV) *All constructible numbers can be produced from the number 1 in a finite number of steps by using the algebraic operations of addition, negation, multiplication, multiplicative inverse, and square root.*

Proof. (I). Operation (0) gives us a line segment of length 1.

(II). We treat the case when $a, b > 0$. We have agreed that saying $-a$ is constructible means the same thing as saying that a is constructible; thus $-a$ is indeed constructible.

Starting with a line segment $\overline{A_1 A_2}$ of length a and a line segment $\overline{B_1 B_2}$ of length b, we can extend $\overline{A_1 A_2}$ from A_2 in the direction away from A_1 by length b, using Operation (2); this gives a line segment of length $a + b$.

To construct $a \cdot b$, take a line segment $\overline{AB}$ of length 1 and construct a perpendicular segment ℓ passing through B (see Figure 21.2), which is possible by Lemma 21.1. Since a is constructible, we may identify a point C on ℓ which lies at distance a from B, using Operations (2) and (4). Let $\lambda = \overline{AC}$. Next, identify the point D which lies at distance b from A in the direction of B. Draw a perpendicular Λ to $\overline{AD}$ through D (on the same side of $\overline{AD}$ as C). By extending the segments λ and Λ, if necessary, we are guaranteed to get a point E where they intersect. We see that the length of the segment $\overline{DE}$ is $a \cdot b$.

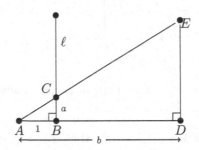

FIGURE 21.2: Constructing $a \cdot b$ given a and b

The construction of $1/a$ is left to the reader as Exercise 21.2.

(III). Finally, we show how to construct $\sqrt{a}$ (see Figure 21.3, where the case $a > 3$ is illustrated). First suppose that $a > 1$. Since a is constructible, so is $(a - 1)/2$, by parts (I) and (II) above. So we can construct a line segment $\overline{AC}$ containing an intermediate point B such that both $\overline{AB}$ and $\overline{BC}$ have length $(a - 1)/2$ (we use here that $a > 1$). Similarly, $(a + 1)/2$ is constructible, so we can identify a point D at distance $(a + 1)/2$ from A in the direction of C, and a point E at distance $(a + 1)/2$ from C in the direction of A. We draw the circle with center A passing through the point D, and the circle with center C passing through the point E. These two circles intersect at a

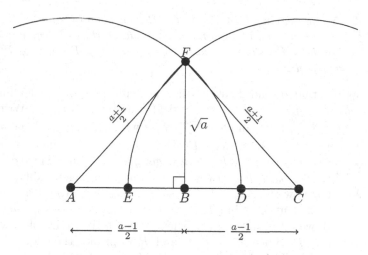

FIGURE 21.3: Constructing $\sqrt{a}$ given a

point F on one side of $\overline{AC}$. The Pythagorean Theorem applied to the right triangle $\triangle ABF$ shows that the length of $\overline{BF}$ is $\sqrt{a}$. If $a < 1$, then we use the construction above to find $\sqrt{1/a}$, and then construct the multiplicative inverse of this number.

The verification of (IV) is left as Exercise 21.3. □

Observe that if a and b are elements of some field F, then the results of any of the operations in parts (I) or (II) of Claim 21.2 also lie in F; while performing the square root operation in part (III) gives a result which is at worst in a quadratic extension of F. Thus, we would like to assert that any constructible number lies inside a tower of successive quadratic extensions of $\mathbf{Q}$, and conversely. We choose to make this assertion as a formal definition of constructibility:

Definition 21.3. A real number α is *constructible* if there is a tower of fields $\mathbf{Q} = F_0 \leq F_1 \leq F_2 \leq \cdots \leq F_n \leq \mathbf{R}$ such that $[F_{i+1} : F_i] = 2$ for all $i \in \{0, 1, \dots, n-1\}$ and $\alpha \in F_n$.

Lemma 21.4. *If α is constructible, then $[\mathbf{Q}[\alpha] : \mathbf{Q}] = 2^r$ for some $r \in \mathbf{N}$.*

Proof. Suppose α is constructible, and let $K = \mathbf{Q}[\alpha]$. Let F_n be as in the definition of constructible. Then $[F_n : \mathbf{Q}] = 2^n$ (using Exercise 15.9 and Lemma 15.26). Since $\alpha \in F_n$, we have $K \leq F_n$. Now we can say $[F_n : K] \cdot [K : \mathbf{Q}] = 2^n$, so $[K : \mathbf{Q}]$ is also a power of 2. □

To determine whether a number α is constructible, it will be useful to be able to find its irreducible polynomial over $\mathbf{Q}$, since this tells us the degree of $\mathbf{Q}[\alpha]$ over $\mathbf{Q}$. It is therefore a good time to discuss a criterion for irreducibility.

Lemma 21.5 (Eisenstein's Criterion). *Let R be a domain, and let $f = x^n + a_{n-1}x^{n-1} + \cdots + a_1 x + a_0$ be a monic polynomial in $R[x]$ of degree $n \geq 1$. If there is a prime ideal P of R such that $a_0, a_1, \ldots, a_{n-1} \in P$ but $a_0 \notin P^2$, then f is irreducible in $R[x]$.*

Proof. Suppose that f and P are as in the hypothesis of the lemma, and that $f = gh$ with $g, h \in R[x]$. Let $\ell = \deg(g)$ and $m = \deg(h)$. Write $f = \sum_{j=0}^n a_j x^j$, $g = \sum_{j=0}^\ell b_j x^j$, and $h = \sum_{j=0}^m c_j x^j$ with $a_j, b_j, c_j \in R$ and $a_n = 1$, $a_j \in P$ for $j : 0 \leq j \leq n-1$, and $a_0 \notin P^2$. We have $b_\ell \cdot c_m = 1$ (since f is monic), so we may replace g by $c_m g$ and h by $b_\ell h$ to get g and h monic. Since $a_0 = b_0 c_0$ and $a_0 \notin P^2$, either $b_0 \notin P$ or $c_0 \notin P$; without loss of generality, suppose $b_0 \notin P$. We know now that $c_m = 1 \notin P$, so it makes sense to set $k = \min\{j \mid c_j \notin P\}$. The coefficient of x^k in gh is $a_k = b_0 c_k + b_1 c_{k-1} + \cdots + b_{k-1} c_1 + b_k c_0$, where, as usual, we fill in missing coefficients with 0. By choice of k, we have $c_0, c_1, \ldots, c_{k-1} \in P$. Since $b_0, c_k \notin P$ and P is prime, then $b_0 c_k \notin P$. It follows that $a_k \notin P$. Therefore, $k = n = \deg(f)$. This forces $\deg(h) = n$ and $\deg(g) = 0$. Since g is monic, then $g = 1 \in (R[x])^\times$. Thus f is irreducible in $R[x]$. $\qquad\square$

Now let us start to confront some of the problems of the ancients. One of the classic questions in geometry asks whether it is possible to "double the cube." By this we mean: given a constructible number $\ell > 0$, representing the side length of a cube, is it possible to construct a number d such that a cube with side length d has exactly twice the volume of the original cube?

Theorem 21.6. *It is impossible to double the cube using straightedge and compass.*

Proof. Let ℓ be a positive constructible number. We solve $d^3 = 2\ell^3$ to get $d = \ell\sqrt[3]{2}$. Assume for a contradiction that d is constructible. Then the ratio $r = d/\ell = \sqrt[3]{2}$ is also constructible. Now r is a root of the polynomial $f(x) := x^3 - 2 \in \mathbf{Z}[x]$. Using Eisenstein's Criterion with $R = \mathbf{Z}$ and $P = (2)$, we see that f is irreducible in $\mathbf{Z}[x]$. Since $\mathbf{Z}$ is a UFD, then f is also irreducible in $\mathbf{Q}[x]$, by Corollary 20.44. So $f = \mathrm{Irr}(r, \mathbf{Q}, x)$ by Exercise 15.2, and $[\mathbf{Q}[r] : \mathbf{Q}] = 3$ by Lemma 15.25. But 3 is not an integer power of 2. Thus, by Lemma 21.4, r is not constructible. This contradiction completes the proof. $\qquad\square$

To adapt some lines of Vergil (Aeneid II:195–198):

Thus abstract algebra prevailed,
Where two thousand years of earlier math had failed.

One of the most famous geometric problems of all time is the problem of trisecting an angle. We interpret this problem as follows. Given two line segments which meet at an angle θ, construct a pair of line segments which meet at angle $\theta/3$. This question is not strictly within the realm of algebra for us yet; we need to rephrase it using the definition of constructible numbers.

So suppose that we have been able to construct two line segments $s_1 = \overline{PQ}$ and $s_2 = \overline{PR}$ which intersect at a point P, making an angle θ. Then we can drop a perpendicular from Q to s_2, using straightedge and compass, as follows. First, construct a number r such that r is greater than the distance from Q to s_2; this can always be done. Then draw the circle C with center Q and radius r. Let A and B be the intersection points of C with the line containing s_2. Draw the two circles with centers A and B and passing through B and A, respectively. These circles intersect each other at two points D and E. The line segment $\overline{DE}$ passes through Q and is perpendicular to s_2. Let F be the intersection point of $\overline{DE}$ and s_2. Then the number $|\overline{PF}| / |s_1| = \cos(\theta)$ is constructible. Conversely, if $\cos(\theta)$ is constructible, then it is not hard to see that we can reverse this process to draw two line segments which meet at angle θ. This motivates the following.

Definition 21.7. We will say that a number $\theta \in \mathbf{R}$ is *constructible as an angle* (or, more briefly, that the angle θ is constructible) if the number $\cos(\theta)$ is constructible.

Theorem 21.8. *It is impossible to trisect the angle with straightedge and compass. More precisely, there exists a constructible angle θ such that the angle $\theta/3$ is not constructible.*

Proof. Let $\theta = 60° = \pi/3$ radians. Then $\cos(\theta) = 1/2$. We recall the trigonometric identity

$$\cos(3\phi) = 4\cos^3(\phi) - 3\cos(\phi).$$

(See Exercise 21.7 if you are not familiar with this identity.) Substituting $\pi/9$ for ϕ gives $\cos(\pi/3) = 4\cos^3(\pi/9) - 3\cos(\pi/9)$. Therefore, $\cos(\pi/9)$ is a root of the polynomial $g := 4x^3 - 3x - 1/2$. Let $h = 2g = 8x^3 - 6x - 1 \in \mathbf{Z}[x]$. Unfortunately, when we divide h by 8 to get a monic polynomial, the result does not have integer coefficients, so we cannot apply Eisenstein's Criterion directly. In order to use Eisenstein's Criterion, we employ a trick to get a nice monic polynomial in $\mathbf{Z}[x]$. Namely, set $\alpha = \cos(\pi/9)$ and $\beta = 2\alpha - 1$. From the equation $8\alpha^3 - 6\alpha - 1 = 0$, we see that $(\beta + 1)^3 - 3(\beta + 1) - 1 = 0$, so β is a root of $f := x^3 + 3x^2 - 3$. Now f is irreducible in $\mathbf{Z}[x]$ by Eisenstein's Criterion, so as in the proof of Theorem 21.6, we see that $[\mathbf{Q}[\beta] : \mathbf{Q}] = 3$. Notice that $\mathbf{Q}[\beta] = \mathbf{Q}[\alpha]$. Therefore, α is not a constructible number, hence $\pi/9$ is not a constructible angle. $\qquad\square$

21.3 Constructible Regular Polygons

Next we turn to the subject of polygons. Recall that a polygon is called *regular* if all its sides have the same length. We will ask which regular polygons can be constructed by straightedge and compass.

Suppose that P is a regular polygon with n sides ($n \geq 3$); we call P a *regular n-gon*. Then P has n vertices $v_1, \ldots, v_n$; we index the vertices in counterclockwise order. These vertices lie on a circle C. Let s_i be the line segment from the center of C to the point v_i. Then the angle between s_i and s_{i+1} (in radians) is $2\pi/n$. If we can construct this angle, then we can draw a regular n-gon, and conversely. This motivates the following.

Definition 21.9. Let $n \in \mathbf{N}$, $n \geq 3$. To say that we can construct a regular n-gon with straightedge and compass means that $2\pi/n$ is constructible as an angle.

Thus, we want to know for which n the number $\cos(2\pi/n)$ is constructible. This amounts to asking about the structure of the field extension $\mathbf{Q}[\cos(\theta)]/\mathbf{Q}$, where $\theta = 2\pi/n$. In turn, this involves understanding the polynomials satisfied by cosines—in other words, trigonometric identities.

We saw in our study of the angle trisection problem that we can express $\cos(3\alpha)$ as a polynomial of degree 3 in $\cos(\alpha)$. It would seem that some kind of generalized "multiple-angle formula" could help us here, since $\cos(n\theta) = \cos(2\pi) = 1$ is a nice number, and we expect $\cos(n\theta)$ to be a polynomial in $\cos(\theta)$. While this is true, it is also true that the multiple-angle formulas grow more and more complicated as the multiple, n, increases. In order to simplify the tangle of equations and achieve a deeper understanding of trigonometric identities, we recall the formula

$$e^{i\alpha} = \cos(\alpha) + i\sin(\alpha). \tag{21.1}$$

We may view this formula as a grand unification of the natural exponential function and the trigonometric functions by way of the field of complex numbers. The basic identities satisfied by the natural exponential function e^z are much simpler than our trig identities, which is why we prefer to work with exponentials. We can recover cosines from exponentials using the following consequence of Formula 21.1:

$$\cos(\alpha) = \frac{e^{i\alpha} + e^{-i\alpha}}{2}. \tag{21.2}$$

So it makes sense to study the complex number $e^{i\theta}$.

Notation 21.10. For an integer $n \geq 1$, we let $\zeta_n = e^{2\pi i/n}$, and we call ζ_n a *primitive n^{th} root of unity*.

The name is justified by the following result.

Lemma 21.11. *For any integer $n \geq 1$, the number ζ_n is a root of the polynomial $x^n - 1 \in \mathbf{Q}[x]$, but is not a root of $x^m - 1$ for any integer $m \in \{1, 2, \ldots, n-1\}$. In particular, ζ_n is algebraic over $\mathbf{Q}$.*

Proof. We have $\zeta_n^n = (e^{2\pi i/n})^n = e^{2\pi i} = 1$. Let $m \in \mathbf{N} - \{0\}$, and suppose that $\zeta_n^m = 1$. Then $1 = \zeta_n^m = e^{2m\pi i/n} = \cos(2m\pi/n) + i\sin(2m\pi/n)$. Therefore $\cos(2m\pi/n) = 1$, so $2m/n$ is an even integer. This forces n to divide m in $\mathbf{Z}$, and establishes the result. $\square$

We can restate the previous result in the language of group theory:

Corollary 21.12. *For an integer $n \geq 1$, the complex number ζ_n has order n as an element of $\mathbf{C}^\times$.* □

The complex numbers $\zeta_n, \zeta_n^2, \ldots, \zeta_n^n$ lie on the unit circle with center 0 in the complex plane, and are themselves the vertices of a regular n-gon. Thus, they cut the unit circle into n equal pieces. The prefix *cyclo* comes from the Greek word for circle (as in *cycle* or *bicycle*), and the suffix *tomic* comes from the Greek word for cut (as in *atomic*—cannot be cut, or *appendectomy*—a procedure to cut out an appendix).

Definition 21.13. For an integer $n \geq 1$, we call $\mathbf{Q}[\zeta_n]$ the nth *cyclotomic field*. We define $\Phi_n(x) := \mathrm{Irr}(\zeta_n, \mathbf{Q}, x)$ and call $\Phi_n(x)$ the nth *cycolotomic polynomial*.

Now it is easy to find all the roots of $x^n - 1$:

Lemma 21.14. *The complex numbers $1 = \zeta_n^0, \zeta_n, \zeta_n^2, \ldots, \zeta_n^{n-1}$ are all the distinct roots of the polynomial $x^n - 1$ in $\mathbf{C}$.*

Proof. For any $j \in \mathbf{Z}$, we have $(\zeta_n^j)^n = \zeta_n^{nj} = (\zeta_n^n)^j = 1^j = 1$, so ζ_n^j is a root of $x^n - 1$. The group $\langle \zeta_n \rangle$ generated by ζ_n in $\mathbf{C}^\times$ has order n, by Corollary 21.12, and so $\zeta_n^0, \ldots, \zeta_n^{n-1}$ are distinct, from Lemma 5.6. A non-zero polynomial over a field cannot have more roots than its degree, by Theorem 14.25. □

Corollary 21.15. *For any integer $n \geq 1$, the polynomial $x^n - 1 \in \mathbf{Q}[x]$ factors as*

$$x^n - 1 = \prod_{j=0}^{n-1} (x - \zeta_n^j). \tag{21.3}$$

Proof. The linear polynomials $x - \zeta_n^j$ are irreducible in $\mathbf{C}[x]$ by Exercise 14.6, hence prime by Corollary 20.28 and Theorem 20.26. Each one divides $x^n - 1$ by the Root-Factor Theorem. No two of these polynomials are associates of each other for $0 \leq j \leq n - 1$, so by unique factorization in $\mathbf{C}[x]$, the entire product on the right-hand side of Equation 21.3 divides $x^n - 1$. The quotient has degree 0 and is monic, so must be 1.

Alternatively, both sides of Equation 21.3 are zero at ζ_n^j when $0 \leq j \leq n-1$, and are monic of degree n; so their difference is a polynomial of degree at most $n - 1$ with at least n distinct roots in $\mathbf{C}$. Hence this difference must be 0 by Theorem 14.25. □

What is the relationship of the cyclotomic field $\mathbf{Q}[\zeta_n]$ to the field $\mathbf{Q}[\cos(2\pi/n)]$? The reader is encouraged to play with Equation 21.2 and investigate this question before reading on.

Proposition 21.16. *For any integer $n \geq 3$, we have $\mathbf{Q}[\cos(\theta)] \leq \mathbf{Q}[\zeta_n]$ and $[\mathbf{Q}[\zeta_n] : \mathbf{Q}[\cos(\theta)]] = 2$, where $\theta := 2\pi/n$.*

Proof. Set $F = \mathbf{Q}[\cos(\theta)]$ and $K = \mathbf{Q}[\zeta_n]$. By Equation 21.2, we have

$$\cos(\theta) = (\zeta_n + \zeta_n^{-1})/2 \in K, \qquad (21.4)$$

so $F \leq K$. Since $F \leq \mathbf{R}$ but $\zeta_n \notin \mathbf{R}$ (since $n \geq 3$), we have $\zeta_n \notin F$, so $K \neq F$. Therefore $[K : F] > 1$, by Exercise 15.5. On the other hand, we can multiply Equation 21.4 by $2\zeta_n$ to see that ζ_n is a root of the polynomial $f := x^2 - 2\cos(\theta)x + 1 \in F[x]$, so $\deg(\mathrm{Irr}(\cos(\theta), F, x)) \leq 2$. Therefore $[K : F] \leq 2$ by Lemma 15.25. $\qquad \square$

The extraordinary fact that the roots of $x^n - 1$ are all powers of ζ_n leads to a particularly nice set of properties of the cyclotomic field $\mathbf{Q}[\zeta_n]$.

Proposition 21.17. *Let $n \in \mathbf{N}$, $n \geq 1$. The cyclotomic field $\mathbf{Q}[\zeta_n]$ is a finite Galois extension of $\mathbf{Q}$, and the Galois group $G_n := Gal(\mathbf{Q}[\zeta_n]/\mathbf{Q})$ is abelian. Every element $\sigma \in G_n$ sends ζ_n to ζ_n^m for a unique $m \in \{0, 1, 2, \ldots, n-1\}$; and for this m, we have $\gcd(m, n) = 1$. This gives an embedding $G_n \hookrightarrow (\mathbf{Z}/n\mathbf{Z})^\times$, $\sigma \mapsto m + n\mathbf{Z}$.*

Proof. Set $K = \mathbf{Q}[\zeta_n]$. By Proposition 16.23, $K/\mathbf{Q}$ is a separable extension. By Theorem 15.21 together with Corollary 21.15, $L := \mathbf{Q}[\zeta_n^0, \zeta_n^1, \ldots, \zeta_n^{n-1}]$ is the splitting field of $x^n - 1$ in $\mathbf{C}$. So L is normal over $\mathbf{Q}$ by Proposition 16.25. Now $\zeta_n \in L$, so $K \leq L$; but also $\zeta_n^j \in K$ for all $j \in \mathbf{N}$, so $L \leq K$. Therefore, $L = K$. So K is Galois over $\mathbf{Q}$. Next, let $\sigma \in G_n$, and set $H = \langle \zeta_n \rangle \leq \mathbf{C}^\times$. Let $\tau = \sigma|_H$. Then $\tau : H \to \mathbf{C}^\times$ is a group homomorphism, e.g. by Exercise 20.4. By Proposition 7.4, $\tau(H)$ consists of nth roots of 1, hence $\tau(H) \subseteq H$. Since σ is injective, so is τ; as H is finite, this forces $\tau(H) = H$. So $\tau \in \mathrm{Aut}(H)$. But H is cyclic of order n, hence $H \cong \mathbf{Z}_n := (\mathbf{Z}/n\mathbf{Z}, +)$. Thus $\mathrm{Aut}(H) \cong \mathrm{Aut}(\mathbf{Z}_n) \cong (\mathbf{Z}/n\mathbf{Z})^\times$ by Exercise 12.18. So we get a group homomorphism $\omega : G_n \to (\mathbf{Z}/n\mathbf{Z})^\times$. Tracing through the details of the various maps in question shows that $\omega(\sigma) = m + n\mathbf{Z}$ where $\sigma(\zeta_n) = \zeta_n^m$, as required. Finally, if $\omega(\sigma) = 1 + n\mathbf{Z}$, then $\sigma(\zeta_n) = \zeta_n$, and so $\sigma = \mathrm{id}_K$. Thus, ω is an embedding by Lemma 9.18. $\qquad \square$

We can now reduce the problem of the constructibility of regular polygons to a fairly basic question about cyclotomic fields.

Proposition 21.18. *For any integer $n \geq 3$, the regular n-gon is constructible if and only if $[\mathbf{Q}[\zeta_n] : \mathbf{Q}] = 2^r$ for some $r \in \mathbf{N}$.*

Proof. ($\Rightarrow$) : Suppose that the regular n-gon is constructible. This means by definition that the number $\cos(2\pi/n)$ is constructible. So $[\mathbf{Q}[\cos(2\pi/n)] : \mathbf{Q}] = 2^k$ for some $k \in \mathbf{N}$, by Lemma 21.4. Therefore $[\mathbf{Q}[\zeta_n] : \mathbf{Q}] = 2^{k+1}$, by Proposition 21.16.

($\Leftarrow$) : Let $K := \mathbf{Q}[\zeta_n]$, and suppose that $[K : \mathbf{Q}] = 2^r$ with $r \in \mathbf{N}$. Let $F = \mathbf{Q}[\cos(2\pi/n)]$, and let $G = \mathrm{Gal}(K/\mathbf{Q})$. From Proposition 21.16, we see that $[F : \mathbf{Q}] = 2^{r-1}$. By the Fundamental Theorem of Galois Theory (FTGT), we have $F = K^H$ for some $H \leq G$. By Proposition 21.17, G is abelian. So $H \trianglelefteq G$ (by Exercise 11.4), and F is Galois over $\mathbf{Q}$ by FTGT. Let $A = \text{Gal}(F/\mathbf{Q})$. Then $|A| = [F : \mathbf{Q}]$ by FTGT. So A is a finite 2-group. By Exercise 19.13, there is a tower of subgroups $\{e\} = A_0 < A_1 < \cdots < A_{r-1} = A$ with $[A_i : A_{i-1}] = 2$ for all $i \in \{1, 2, \ldots, r-1\}$. By FTGT, this tower of subgroups corresponds to a tower of fields $\mathbf{Q} = F_{r-1} < \cdots < F_1 < F_0 = F$ with $[F_i : F_{i+1}] = 2$ for all $i \in \{0, 1, \ldots, r-2\}$. Therefore, $\cos(2\pi/n)$ is constructible. $\square$

By Proposition 21.18, the problem of the constructibility of regular n-gons has been reduced to finding the degree of the irreducible polynomial of ζ_n over $\mathbf{Q}$, i.e, the degree of $\Phi_n(x)$. Since the polynomial $x^n - 1$ is in the kernel of evaluation at ζ_n, we must have $\Phi_n \mid (x^n - 1)$ in $\mathbf{Q}[x]$. So let us try our hand at factoring $x^n - 1$ for some small values of n. We have

$$x^1 - 1 = x - 1; \tag{21.5}$$

$$x^2 - 1 = (x - 1)(x + 1); \tag{21.6}$$

$$x^3 - 1 = (x - 1)(x^2 + x + 1). \tag{21.7}$$

But how can we be sure that this last factor, $f := x^2 + x + 1$, is irreducible in $\mathbf{Q}[x]$? Since f is just quadratic, we can "brute-force" the question by using the quadratic formula, or we could use Exercise 14.7 together with the Rational Root Theorem (see Exercise 21.4). But for a result that generalizes, we prefer to use Eisenstein's Criterion (EC). Since EC does not apply to f directly, we must be clever. Let $y = x + 1$, and consider $g := f(y) = y^2 + y + 1 - (x + 1)^2 + (x + 1) + 1 = x^2 + 3x + 3$. Now g is irreducible in $\mathbf{Z}[x]$ by Eisenstein's Criterion, but how does that help with f? The key here is the realization that the function

$$\Omega : \mathbf{Z}[x] \to \mathbf{Z}[x], \quad h(x) \mapsto h(x + 1)$$

is an automorphism of $\mathbf{Z}[x]$. This allows us to say that f is irreducible in $\mathbf{Z}[x]$ iff $g = \Omega(f)$ is. (Compare the argument made in the proof of Theorem 21.8.) Thus, Equation 21.7 gives a factorization of $x^3 - 1$ into irreducibles in $\mathbf{Z}[x]$.

The same ideas can be used to factor the related polynomial $x^9 - 1$. Every cube root of 1 is also a ninth root of 1, so it makes sense that $x^3 - 1$ divides $x^9 - 1$ in $\mathbf{Q}[x]$. In fact, we have $(x^9 - 1)/(x^3 - 1) = x^6 + x^3 + 1 = (x^3)^2 + x^3 + 1 =: h(x)$. Again it is convenient to set $y = x + 1$. We find $h(y) = (y^3)^2 + y^3 + 1 = ((x + 1)^3)^2 + (x + 1)^3 + 1 \equiv (x^3 + 1)^2 + (x^3 + 1) + 1 \pmod{3} \equiv x^6 \pmod{3}$. The constant term of $h(y)$ is just 3, so h is irreducible by EC. We make these ideas precise using the following definition and results.

Definition 21.19. Let R be a commutative ring with 1, and let $\mathfrak{A}$ be an ideal of R. Let $b, c \in R$. Then we write $b \equiv c \pmod{\mathfrak{A}}$ to mean $c - b \in \mathfrak{A}$. We read the former expression as "b is congruent to c modulo the ideal $\mathfrak{A}$."

Lemma 21.20. *Let R be a commutative ring with 1, and let p be a prime in* $\mathbf{Z}$. *Let* $\mathfrak{p} = pR$, *the ideal of R generated by the image of p under the natural map* $\chi : \mathbf{Z} \to R$ *(see Lemma 16.20). Then for all $a, b \in R$, we have*

$$(a + b)^p \equiv a^p + b^p \pmod{\mathfrak{p}}. \tag{21.8}$$

More generally, if $a_1, \ldots, a_r \in R$, then we have

$$(a_1 + \cdots + a_r)^p \equiv a_1^p + \cdots + a_r^p \pmod{\mathfrak{p}}. \tag{21.9}$$

Proof. By the Binomial Theorem in R (Exercise 16.19), we have $(a + b)^p = \sum_{k=0}^{p} \binom{p}{k} \cdot a^k b^{p-k}$. If $1 \leq k \leq p - 1$, then certainly $p \nmid k!$ and $p \nmid (p - k)!$, but $p \mid p!$, so $p \mid \binom{p}{k}$. This establishes Equation (21.8). Equation (21.9) follows immediately by induction on r.   □

The definition of the characteristic of a ring makes the following result immediate:

Corollary 21.21. *Let p be a prime in $\mathbf{Z}$, and let R be a commutative ring with 1 such that $char(R) = p$. Then the map $\tau : R \to R$, $a \mapsto a^p$ is a ring homomorphism. In particular, for all $a_1, \ldots, a_r \in R$, we have $(a_1 + \cdots + a_r)^p = a_1^p + \cdots + a_r^p$.*   □

Proposition 21.22. *Let $q = p^r$, where p is a positive prime integer and r is a positive integer. Then*

$$\Phi_q(x) = \frac{x^{p^r} - 1}{x^{p^{r-1}} - 1}. \tag{21.10}$$

Proof. Let $f(x) = (x^{p^r} - 1)/(x^{p^{r-1}} - 1) \in \mathbf{Q}(x)$. Set $z = x^{p^{r-1}}$, so that

$$f = (z^p - 1)/(z - 1) = z^{p-1} + z^{p-2} + \cdots + z + 1 \in \mathbf{Z}[x]. \tag{21.11}$$

Let $\Omega : \mathbf{Q}[x] \to \mathbf{Q}[x]$ be the function given by $g(x) \mapsto g(x + 1)$ for $g \in \mathbf{Q}[x]$, and let $\mathfrak{p} = p \cdot \mathbf{Z}[x]$. Note that p is irreducible in $\mathbf{Z}[x]$, which is a UFD by Proposition 20.45, so p is prime in $\mathbf{Z}[x]$ by Theorem 20.26, hence $\mathfrak{p}$ is a prime ideal by Exercise 12.12. We have $\Omega(f) = ((x+1)^{p^{r-1}})^{p-1} + ((x+1)^{p^{r-1}})^{p-2} + \cdots + (x + 1)^{p^{r-1}} + 1 \equiv (z + 1)^{p-1} + (z + 1)^{p-2} + \cdots + (z + 1) + 1 \pmod{\mathfrak{p}}$ by Lemma 21.20. So $\Omega(f) \equiv ((z + 1)^p - 1)/((z + 1) - 1) \pmod{\mathfrak{p}}$. Now set $\tilde{f} := ((z+1)^p - 1)/z \equiv z^p/z \equiv z^{p-1} \equiv x^{p^{r-1}(p-1)} \pmod{\mathfrak{p}}$. Then $\Omega(f) \equiv \tilde{f} \equiv x^{p^{r-1}(p-1)} \pmod{\mathfrak{p}}$, and the constant term of $\Omega(f)$ is $(\Omega(f))(0) = f(1) = p$ (by Equation 21.11). So $\tilde{f}$ is irreducible in $\mathbf{Z}[x]$ by Eisenstein's Criterion, hence irreducible in $\mathbf{Q}[x]$ by Corollary 20.44. Since $\Omega \in \text{Aut}(\mathbf{Q}[x])$ (e.g. by Exercise 21.5), then f is also irreducible in $\mathbf{Q}[x]$ by Exercise 21.6. Let $g = x^{p^{r-1}} - 1$. We have

$$x^{p^r} - 1 = fg. \tag{21.12}$$

Let $\zeta_q = e^{2\pi i/q}$, as usual. Evaluating both sides of Equation (21.12) at $x = \zeta_q$ gives $0 = f(\zeta_q) \cdot g(\zeta_q)$. Now $g(\zeta_q) \neq 0$ by Lemma 21.11, so $f(\zeta_q) = 0$. Since f is monic and irreducible in $\mathbf{Q}[x]$, it follows from Exercise 15.2 that $f = \text{Irr}(\zeta_q, \mathbf{Q}, x) =: \Phi_q(x)$.   □

Corollary 21.23. *Let q be a prime power. Then* $\deg(\Phi_q(x)) = \phi(q)$, *and we have* $\mathrm{Gal}(\mathbf{Q}[\zeta_q]/\mathbf{Q}) \cong (\mathbf{Z}/q\mathbf{Z})^\times$ *via the embedding of Proposition 21.17.*

Proof. Write $q = p^r$ where p is prime in $\mathbf{Z}$ and $p, r \in \mathbf{N}$. We have

$$
\begin{aligned}
|\mathrm{Gal}(\mathbf{Q}[\zeta_q]/\mathbf{Q})| &= [\mathbf{Q}[\zeta_q] : \mathbf{Q}] && \text{(by FTGT, Theorem 16.32)} \\
&= \deg(\mathrm{Irr}(\zeta_q, \mathbf{Q}, x)) && \text{(by Lemma 15.25)} \\
&= \deg(\Phi_q(x)) && \text{(by definition of } \Phi_q) \\
&= p^r - p^{r-1} && \text{(by Proposition 21.22)} \\
&= \phi(p^r) && \text{(by Exercise 17.7)} \\
&= |(\mathbf{Z}/q\mathbf{Z})^\times| && \text{(by definition of } \phi(q)).
\end{aligned}
$$

Since this quantity is finite, the embedding must be a bijection. $\qquad\square$

It turns out that Corollary 21.23 is true for any positive integer n, not just for prime powers q. (See Exercise 21.15 for a proof of this statement which relies on a well-known theorem from number theory.) However, even without this more general result we already know enough about the structure of cyclotomic fields to settle the question of which regular polygons are constructible.

Corollary 21.24. *Let n be a positive integer. Suppose that* $n = \prod_{i=1}^r p_i^{e_i}$ *is the prime factorization of n. Then* $\mathrm{lcm}(\{\phi(p_i^{e_i}) \ : \ 1 < i < r\})$ *divides* $\deg(\Phi_n(x))$, *which in turn divides* $\phi(n)$.

Proof. Pick a prime $p = p_i$ dividing n, and let $q = p_i^{e_i}$, $m = n/q$. Then we have $\zeta_q = \zeta_n^m$, from which we see that $\mathbf{Q}[\zeta_q] \le \mathbf{Q}[\zeta_n]$. So $[\mathbf{Q}[\zeta_q] : \mathbf{Q}]$ divides $[\mathbf{Q}[\zeta_n] : \mathbf{Q}]$ (by Lemma 15.26 and Exercise 15.9). But $[\mathbf{Q}[\zeta_q] : \mathbf{Q}] = \phi(q)$ (see the proof of Corollary 21.23), while $[\mathbf{Q}[\zeta_n] : \mathbf{Q}] = \deg(\Phi_n)$. This establishes the first divisibility statement. For the second, note that the order of $\mathrm{Gal}(\mathbf{Q}[\zeta_n]/\mathbf{Q})$ divides the order of $(\mathbf{Z}/n\mathbf{Z})^\times$ by Proposition 21.17 and Lagrange's Theorem. $\qquad\square$

Theorem 21.25. *Let n be a positive integer, $n \ge 3$. Then a regular n-gon is constructible by straightedge and compass iff $\phi(n)$ is a power of 2. This occurs iff the prime factorization of n has the form* $n = 2^a \cdot \prod_{i=1}^r p_i$, *where $a \in \mathbf{N}$ and the p_i are distinct odd primes such that each $p_i - 1$ is a power of 2.*

Proof. From Corollary 21.24 and Exercise 17.7, we see that a prime divides $\deg(\Phi_n)$ iff it divides $\phi(n)$. Since $\deg(\Phi_n) = [\mathbf{Q}[\zeta_n] : \mathbf{Q}]$, the first part of the result now follows from Proposition 21.18. For the second part, we can always write $n = 2^a \cdot \prod_{i=1}^r p_i^{e_i}$, where $a, e_i \in \mathbf{N}$, $e_i \ge 1$, and the p_i are distinct odd primes. Then $\phi(n) = 2^{a-1} \cdot \prod_{i=1}^r p_i^{e_i - 1}(p_i - 1)$ (by Exercise 17.7). The desired conclusion now follows by inspection. $\qquad\square$

Remark 21.26. An odd prime $p \in \mathbf{N}$ such that $p-1$ is a power of 2 is known as a *Fermat prime*. The only known Fermat primes are 3, 5, 17, 257, and 65537.

Example 21.27. The smallest Fermat prime is 3, so Theorem 21.25 guarantees that a regular 3-gon is constructible. Indeed, we have $\cos(2\pi/3) = -1/2 \in \mathbf{Q}$.

Example 21.28. The second-smallest Fermat prime is 5. To construct $\alpha :=$ $\cos(2\pi/5)$, we start with the tower of fields $\mathbf{Q} \leq \mathbf{Q}[\alpha] =: F \leq \mathbf{Q}[\zeta_5] =: K$. The index between the last two fields in this tower is 2 (by Proposition 21.16), while the overall index of $\mathbf{Q}$ in K is $\phi(5) = 4$ (by Corollary 21.23). It follows that $[F : \mathbf{Q}] = 2$. We have $\mathrm{Irr}(\zeta_5, \mathbf{Q}, x) = (x^5 - 1)/(x - 1) = x^4 + x^3 + x^2 + x + 1$ (by Proposition 21.22), so the fundamental relation satsfied by ζ_5 over $\mathbf{Q}$ is

$$\zeta_5^4 + \zeta_5^3 + \zeta_5^2 + \zeta_5 + 1 = 0. \tag{21.13}$$

Of course, we also have $\zeta_5^5 = 1$ (e.g. by Lemma 21.11); so $\zeta_5^a = \zeta_5^b$ whenever $a \equiv b \pmod 5$ (by Exercise 5.7).

Again by Corollary 21.23, we have $\mathrm{Gal}(K/\mathbf{Q}) := G \cong (\mathbf{Z}/5\mathbf{Z})^\times$. Now $(\mathbf{Z}/5\mathbf{Z})^\times$ is cyclic, generated by 2 (and by 3, but let's use 2). The isomorphism of Corollary 21.23 therefore gives $G = \langle \sigma \rangle$, where $\sigma(\zeta_5) = \zeta_5^2$ and, in general, $\sigma\left(\sum_{j=0}^k c_j \zeta_5^j\right) = \sum_{j=0}^k c_j \zeta_5^{2j}$ for $c_j \in \mathbf{Q}$ and $k \in \mathbf{N}$. By the Fundamental Theorem of Galois Theory (FTGT), the field F is the fixed field of a subgroup $H \leq G$ of order 2. There is only one such subgroup, namely $H := \langle \sigma^2 \rangle$. Again by FTGT, we have $L := \mathrm{Gal}(F/\mathbf{Q}) \cong G/H$, via the map $\omega \mapsto \omega|_F$. The set of conjugates of α over $\mathbf{Q}$ in $\mathbf{C}$ is therefore $C := \{\tau(\alpha) : \tau \in G\}$ $= \{\alpha, \sigma(\alpha)\}$. We have $\alpha = (\zeta_5 + \zeta_5^{-1})/2$, while $\sigma(\alpha) = (\zeta_5^2 + \zeta_5^{-2})/2$. So $C = \{(\zeta_5 + \zeta_5^{-1})/2, (\zeta_5^2 + \zeta_5^{-2})/2\}$. We calculate

$$
\begin{aligned}
\mathrm{Irr}(\alpha, \mathbf{Q}, x) &= \prod_{\gamma \in C}(x - \gamma) && \text{(by Exercise 19.10)} \\
&= (x - \alpha)(x - \sigma(\alpha)) \\
&= (x - (\zeta_5 + \zeta_5^{-1})/2)(x - (\zeta_5^2 + \zeta_5^{-2})/2) \\
&= x^2 - (\zeta_5 + \zeta_5^{-1} + \zeta_5^2 + \zeta_5^{-2})x/2 + (\zeta_5^3 + \zeta_5^{-1} + \zeta_5 + \zeta_5^{-3})/4 \\
&= x^2 - (\zeta_5 + \zeta_5^4 + \zeta_5^2 + \zeta_5^3)x/2 + (\zeta_5^3 + \zeta_5^4 + \zeta_5 + \zeta_5^2)/4 \\
&= x^2 - (-1)x/2 + (-1)/4 && \text{(by Equation 21.13)} \\
&= x^2 + x/2 - 1/4.
\end{aligned}
$$

Therefore, $\alpha^2 + \alpha/2 - 1/4 = 0$, so $\alpha = (-1/2 \pm \sqrt{1/4 - 4 \cdot 1 \cdot (-1/4)})/2$ $= (-1/2 \pm \sqrt{5/4})/2 = -1/4 \pm \sqrt{5}/4$. Since $\alpha > 0$, we have $\alpha = (\sqrt{5} - 1)/4$. (See Figure 21.4.)

Example 21.29. Let us step through the details of constructing a regular 15-gon by straightedge and compass. More accurately speaking, we will show how to write $\cos(2\pi/15)$ using only rational numbers, arithmetic operations, and square roots. Since $15 = 3 \cdot 5$, Theorem 21.25 assures us that this is possible. However, in this situation we are somewhat at a loss as to the Galois group $G := \mathrm{Gal}(K/\mathbf{Q})$, with $K := \mathbf{Q}[\zeta_{15}]$. We know that $G \hookrightarrow (\mathbf{Z}/15\mathbf{Z})^\times$, but we are unwilling to use the result of Exercise 21.15 giving $G \cong (\mathbf{Z}/15\mathbf{Z})^\times$, since this

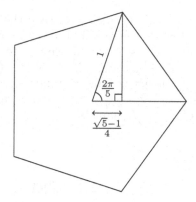

FIGURE 21.4: A regular 5-gon, aka pentagon

relies on a theorem of Dirichlet which is beyond the scope of this text. Our strategy (as so often in mathematics) is to find a procedure which "should" work *if* in fact our hunch is correct that this embedding is an isomorphism.

So set $\Gamma := (\mathbf{Z}/15\mathbf{Z})^\times$, and write ζ for ζ_{15}. We suspect that $|G| = |\Gamma| = 8$ and thus that every element of K has a unique expression of the form $\sum_{j=0}^{7} c_j \zeta^j$ with $c_j \in \mathbf{Q}$ (by Lemma 15.25). From Exercise 17.7, we have $\Gamma \cong (\mathbf{Z}/3\mathbf{Z})^\times \oplus (\mathbf{Z}/5\mathbf{Z})^\times$ via the map $\omega : a + 15\mathbf{Z} \mapsto (a + 3\mathbf{Z}, a + 5\mathbf{Z})$. Wouldn't it be nice if we could represent an element of K using ζ_3 and ζ_5 instead of ζ, and make this representation mesh with the action of the direct summands of Γ? Well, we can! On the one hand, we have $\zeta_3 = \zeta^5$ and $\zeta_5 = \zeta^3$, so $\mathbf{Q}[\zeta_5, \zeta_3] \leq \mathbf{Q}[\zeta]$; while on the other hand, we have $\zeta = \zeta_5^2 \cdot \zeta_3^{-1} \in \mathbf{Q}[\zeta_5, \zeta_3]$. So $\mathbf{Q}[\zeta] = \mathbf{Q}[\zeta_5, \zeta_3]$. Set $L := \mathbf{Q}[\zeta_5]$. We know from Corollary 21.23 that $[L : \mathbf{Q}] = 4$, and thus $L = \{\sum_{j=0}^{3} c_j \zeta_5^j : c_j \in \mathbf{Q}\}$. Now $K = L[\zeta_3]$, and ζ_3 is a root of $\mathrm{Irr}(\zeta_3, \mathbf{Q}, x) = (x^3 - 1)/(x - 1) = x^2 + x + 1$. So $\deg(g) \leq 2$, where $g := \mathrm{Irr}(\zeta_3, L, x)$. It follows that either $K = L$ or else every element of K has a unique expression of the form $\ell_1 + \ell_2 \zeta_3$ with $\ell_1, \ell_2 \in L$ (again by Lemma 15.25). In either case, we have $K = \{\ell_1 + \ell_2 \zeta_3 : \ell_1, \ell_2 \in L\}$ (although in the former case, this expression is not unique). To summarize, we have

$$K = \{c_0 + c_1\zeta_5 + c_2\zeta_5^2 + c_3\zeta_5^3 + (d_0 + d_1\zeta_5 + d_2\zeta_5^2 + d_3\zeta_5^3)\zeta_3 : c_j, d_k \in \mathbf{Q}\}. \quad (21.14)$$

Let us verify that the action of G on K is compatible with our factored point of view. If $\sigma \in G$ is an automorphism such that $\sigma(\zeta) = \zeta^a$, then we have $\sigma(\zeta_3) = \sigma(\zeta^5) = \zeta^{5a} = \zeta_3^a$ and $\sigma(\zeta_5) = \sigma(\zeta^3) = \zeta^{3a} = \zeta_5^a$. Conversely, if $\sigma \in G$ is such that $\sigma(\zeta_3) = \zeta_3^t$ and $\sigma(\zeta_5) = \zeta_5^u$, then $\sigma(\zeta) = \zeta^r$, where r is the unique integer modulo 15 such that $r \equiv t \pmod 3$ and $r \equiv u \pmod 5$. Thus, an element $(t + 3\mathbf{Z}, u + 5\mathbf{Z}) \in \Gamma$, when viewed as an element of G, has the action $\zeta_3 \mapsto \zeta_3^t$, $\zeta_5 \mapsto \zeta_5^u$. Beautiful!

Now we are ready to calculate. Set $\alpha = \cos(2\pi/15) = (\zeta + \zeta^{-1})/2$. We first write α in the form indicated in Equation 21.14. We have $\alpha = (\zeta + \zeta^{-1})/2 = (\zeta_5^2\zeta_3^{-1} + \zeta_5^{-2}\zeta_3)/2 = (\zeta_5^2\zeta_3^2 + \zeta_5^3\zeta_3)/2 = (\zeta_5^2(-1 - \zeta_3) + \zeta_5^3\zeta_3)/2 = -\frac{1}{2}\zeta_5^2 +$

$(-\frac{1}{2}\zeta_5^2 + \frac{1}{2}\zeta_5^3)\zeta_3$. Now if in fact $|G| = |\Gamma| = 8$, then the conjugates of α over L should be α itself and $\sigma(\alpha)$, where σ is the automorphism of K over $\mathbf{Q}$ which fixes ζ_5 but not ζ_3: that is, σ would send ζ_3 to ζ_3^2 and ζ_5 to ζ_5. We would have $\sigma(\alpha) = -\frac{1}{2}\zeta_5^2 + (-\frac{1}{2}\zeta_5^2 + \frac{1}{2}\zeta_5^3)\zeta_3^2 = -\frac{1}{2}\zeta_5^3 + (\frac{1}{2}\zeta_5^2 - \frac{1}{2}\zeta_5^3)\zeta_3$. But if $|\Gamma| < 8$, then such a map σ is not well-defined; nevertheless, no one can stop us from setting $\beta = -\frac{1}{2}\zeta_5^3 + (\frac{1}{2}\zeta_5^2 - \frac{1}{2}\zeta_5^3)\zeta_3$ and computing what should be the irreducible polynomial of α over L, namely

$$\begin{aligned}
f &:= (x - \alpha)(x - \beta)\\
&= x^2 - (\alpha + \beta)x + (\alpha\beta)\\
&= x^2 + (\zeta_5^2 + \zeta_5^3)x/2 + (\zeta_5^5 + (-\zeta_5^4 + 2\zeta_5^5 - \zeta_5^6)\zeta_3 + (-\zeta_5^4 + 2\zeta_5^5 - \zeta_5^6)\zeta_3^2)/4\\
&= x^2 + (\zeta_5^2 + \zeta_5^3)x/2 + (-\zeta_5^5 + \zeta_5^4 + \zeta_5^6 + 0\cdot\zeta_3)/4\\
&= x^2 + (\zeta_5^2 + \zeta_5^3)x/2 + (-2 - \zeta_5^2 - \zeta_5^3)/4.
\end{aligned}$$

Set $\gamma := \zeta_5^2 + \zeta_5^3$. Then $\gamma \in L$, and γ is fixed by complex conjugation $\zeta_5 \mapsto \zeta_5^{-1}$, so the conjugates of γ over $\mathbf{Q}$ are γ and $\tau(\gamma)$, where $\tau(\zeta_5) = \zeta_5^2$. Now $\tau(\gamma) = \zeta_5^4 + \zeta_5^6 = -1 - \zeta_5^2 - \zeta_5^3$. We compute $(x-\gamma)(x-\tau(\gamma)) = x^2 + x + (-\zeta_5^2 - \zeta_5^3 - \zeta_5^4 - \zeta_5^5 - \zeta_5^5 - \zeta_5^6) = x^2 + x - 1$. Therefore $\gamma = (-1 \pm \sqrt{5})/2$. Glancing at the positions of ζ_5^2 and ζ_5^3 in the complex plane shows (see Figure 21.5) that $\zeta_5^2 + \zeta_5^3 < 0$. Therefore, $\gamma = (-1 - \sqrt{5})/2$. So we have $f = x^2 - (1+\sqrt{5})x/4 + (-3+\sqrt{5})/8$.

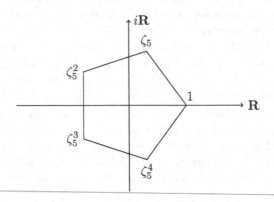

FIGURE 21.5: the fifth roots of 1 in $\mathbf{C}$

Since α is a root of f, we use the quadratic formula to find α. The discriminant of f is $((1 + \sqrt{5})/4)^2 - 4\cdot 1\cdot(-3+\sqrt{5})/8 = (6 + 2\sqrt{5})/16 - (-3+\sqrt{5})/2 = (15 - 3\sqrt{5})/8$. So $\alpha = \left((1+\sqrt{5})/4 \pm \sqrt{(15 - 3\sqrt{5})/8}\right)/2 = (1 + \sqrt{5} \pm \sqrt{30 - 6\sqrt{5}})/8$. Now $1+\sqrt{5} < \sqrt{30 - 6\sqrt{5}}$ iff $6+2\sqrt{5} < 30-6\sqrt{5}$ iff $8\sqrt{5} < 24$ iff $\sqrt{5} < 3$, which is true. Therefore, as $\alpha = \cos(2\pi/15) > 0$, we have

$$\cos(2\pi/15) = \frac{1 + \sqrt{5} + \sqrt{30 - 6\sqrt{5}}}{8}. \tag{21.15}$$

Remark 21.30. In fact, it is not hard to show that $[K : \mathbf{Q}] = 8$ in Example 21.29. For suppose not; then $[K : \mathbf{Q}] = 4$, $K = L = \mathbf{Q}[\zeta_5]$, and $\mathrm{Gal}(K/\mathbf{Q}) \cong (\mathbf{Z}/5\mathbf{Z})^\times \cong \mathbf{Z}_4$. We would have $(-1 + i\sqrt{3})/2 = \zeta_3 \in K$. But $\mathbf{Z}_4$ has a unique subgroup of order 2, corresponding by Galois theory to a unique quadratic extension of $\mathbf{Q}$ in K; yet both $\mathbf{Q}[\sqrt{5}]$ and $\mathbf{Q}[i\sqrt{3}]$ are subfields of K which are quadratic over $\mathbf{Q}$, and these fields are clearly not equal, since one is in $\mathbf{R}$ but the other is not. We purposely avoided this argument in that example in order to show how one can make do with only partial information.

Remark 21.31. The reader is encouraged to work on Example 21.29 using the unfactored group $(\mathbf{Z}/15\mathbf{Z})^\times$ and the cyclotomic polynomial Φ_{15} instead of the factored form $(\mathbf{Z}/3\mathbf{Z})^\times \oplus (\mathbf{Z}/5\mathbf{Z})^\times$ and the polynomials Φ_3 and Φ_5, if only to convince yourself that the factored approach is easier.

21.4 Exercises

Exercise 21.1. Notice that in Operations (1), (2), and (3), we implicitly assumed that our compass can be extended to an arbitrary distance—that our string is infinitely long. Would the set $\mathfrak{C}$ of constructible numbers change if we assumed instead that the string had a finite length ℓ? If so, would $\mathfrak{C}$ depend on ℓ?

Exercise 21.2. Prove that if a is constructible and $a \neq 0$, then $1/a$ is also constructible. This finishes the proof of part (II) of Claim 21.2.

Exercise 21.3. Prove part (IV) of Claim 21.2. Suggestion: Impose Cartesian coordinates so that the initial line segment has endpoints at $(0,0)$ and $(1,0)$. Induct on the total number of operations performed, including multiplicity, and at each step, establish that the coordinates of every identified point are themselves constructible numbers.

Exercise 21.4. Let R be a UFD, and let $f(x) = \sum_{i=0}^{n} a_i x^i \in R[x]$ with $a_i \in R$ and $a_n \neq 0$. Let K be the field of fractions of R. Prove that if $\alpha \in K$ and $f(\alpha) = 0$, then we can write $\alpha = a/b$ where $a, b \in R$, $b \neq 0$, and $a \mid a_0$, $b \mid a_n$ in R. (The special case of this result when $R = \mathbf{Z}$ is known as the *Rational Root Theorem*.) Hint: Exercise 20.20 may be useful.

Exercise 21.5. Let R be a commutative ring with 1, and let $S = R[x]$. Let $h \in S$.

(a) Prove that there is a unique ring homomorphism $\omega : S \to S$ over R such that $\omega(x) = h$. (It is natural to denote $\omega(f)$ by $f \circ h$ or $f(h)$ for $f \in S$; see Exercise 14.15.)

(b) Suppose now that R is a field. Prove that $\omega \in \mathrm{Aut}(S)$ iff $\deg(h) = 1$.

Exercise 21.6. Let R be a domain and let $\sigma \in \mathrm{Aut}(R)$. Let $f \in R$. Prove that f is irreducible in R iff $\sigma(f)$ is irreducible in R.

Exercise 21.7. Use the identities $\cos(a+b) = \cos(a)\cos(b) - \sin(a)\sin(b)$ and $\sin(a+b) = \sin(a)\cos(b) + \cos(a)\sin(b)$ to prove the "triple-angle formula" $\cos(3\phi) = 4\cos^3(\phi) - 3\cos(\phi)$.

Exercise 21.8. Let n be an integer with $n \geq 3$. Set $F := \mathbf{Q}[\cos(2\pi/n)]$ and $K := \mathbf{Q}[\zeta_n]$. Let $\sigma : \mathbf{C} \to \mathbf{C}$ be the complex conjugation map, $\sigma(a+bi) = a - bi$ for $a, b \in \mathbf{R}$. Let τ be the restriction of σ to K. Let $G = \mathrm{Gal}(K/F)$.
  (a) Prove that $\tau \in G$.
  (b) Let $T = \langle \tau \rangle \leq G$. Show that $|T| = 2$.
  (c) Prove that $F = K^T$ is the fixed field of T, and thus that we have $F = K \cap \mathbf{R}$.

Exercise 21.9. (a) Let $a, b \in \mathbf{Z}$, and let $g = \gcd(a, b)$. Prove that the ideal (a, b) is equal to the ideal (g) in $\mathbf{Z}$. Conclude that there exist $c, d \in \mathbf{Z}$ such that $ca + db = g$. (Compare Exercise 4.7.)
  (b) Let n be a positive integer, and suppose that n factors as $n = ab$, with $a, b \in \mathbf{N}$ and $\gcd(a, b) = 1$. Prove that $\mathbf{Q}[\zeta_n] = \mathbf{Q}[\zeta_a, \zeta_b]$.
  (c) Prove that if $n = \prod_{j=1}^r p_j^{e_j}$ where $p_1, \ldots, p_r$ are distinct positive prime integers and $e_j \in \mathbf{N}$, then $\mathbf{Q}[\zeta_n] = \mathbf{Q}[\zeta_{q_1}, \ldots, \zeta_{q_r}]$, where $q_j := p_j^{e_j}$.

Exercise 21.10. Find an explicit formula for $\cos(2\pi/n)$ for the following values of n. In each case, describe a tower of fields starting with $\mathbf{Q}$ and ending with $\mathbf{Q}[\cos(2\pi/n)]$ such that each successive extension has degree 2 over the previous field.
  (a) $n = 24$.
  (b) $n = 17$.

Exercise 21.11. How many nested levels of square roots would be needed in order to write $\cos(2\pi/65537)$ explicitly using only rational numbers, field operations, and square roots?

Exercise 21.12. Let n be a positive integer, and suppose that ρ is an nth root of 1 in $\mathbf{C}$ (not necessarily primitive); that is, $\rho^n = 1$. Let $K = \mathbf{Q}[\rho]$. Let m be the order of ρ in $K^\times$.
  (a) Prove that $K = \mathbf{Q}[\zeta_m]$, the mth cyclotomic field.
  (b) Prove that if $m > 2$, then $K \neq \mathbf{Q}$ and $[K : \mathbf{Q}]$ is even.

Exercise 21.13. Let K be a finite Galois extension field of F, and let $G = \mathrm{Gal}(K/F)$. Define a function $\mathcal{N} : K \to K$ by the formula $\mathcal{N}(\alpha) = \prod_{\sigma \in G} \sigma(\alpha)$ for $\alpha \in K$. (Compare Exercise 20.25.)
  (a) Prove that $\mathcal{N}(K) \subseteq F$. Note: we also write $\mathcal{N}_{K/F}$ for $\mathcal{N}$, and describe $\mathcal{N}$ as the "norm map from K down to F."
  (b) Prove that $\forall \alpha, \beta \in K$, $\mathcal{N}(\alpha\beta) = \mathcal{N}(\alpha) \cdot \mathcal{N}(\beta)$.
  (c) Suppose that $F \leq L \leq K$ and L is Galois over F. Prove that $\mathcal{N}_{K/F} = \mathcal{N}_{L/F} \circ \mathcal{N}_{K/L}$.

Exercise 21.14. (a) Let R be a UFD, and let $K = k(R)$ be the field of fractions of R. Let L be a finite extension field of K. Suppose that $\alpha \in L$ and α is integral over R (see Exercise 20.26). Prove that $\mathrm{Irr}(\alpha, F, x) \in R[x]$. Hint: Corollary 20.43 may be useful.

(b) Use part (a) above to conclude that $\Phi_n \in \mathbf{Z}[x]$ for all positive integers n.

(c) Let n be a positive integer and let $d = \deg(\Phi_n)$. Use part (b) above to prove that $\mathbf{Z}[\zeta_n] = \{a_0 + a_1\zeta_n + \cdots + a_{d-1}\zeta_n^{d-1} : a_i \in \mathbf{Z}\}$. Also note that this representation of an element of $\mathbf{Z}[\zeta_n]$ is unique.

Exercise 21.15. In this exercise, you will make use of the following result to find the cyclotomic polynomial Φ_n for any positive integer n.

Theorem 21.32 (Dirichlet's Theorem on Primes in Arithmetic Progressions). *Let m, n be two positive integers such that $\gcd(m, n) = 1$. Then there are infinitely many primes p such that $p \equiv m \pmod{n}$.*

Let n be a positive integer, and suppose that $\sum_{j=0}^{r} c_j \zeta_n^j = 0$ with $r \in \mathbf{N}$ and $c_j \in \mathbf{Z}$. Let $m \in \{1, 2, \ldots, n\}$ be such that $\gcd(m, n) = 1$.

(a) Suppose that q is a prime such that $q \equiv m \pmod{n}$. Show that we have $\sum_{j=0}^{r} c_j \zeta_n^{mj} \in q \cdot \mathbf{Z}[\zeta_n]$. Hint: take the qth power of the original relation.

(b) Use Dirichlet's Theorem (Theorem 21.32) to prove that $\sum_{j=0}^{r} c_j \zeta_n^{mj} = 0$. Hint: Exercise 21.14 may be helpful.

(c) Prove that ζ_n^m is a root of Φ_n. Deduce that $\deg(\Phi_n) \geq \phi(n)$.

(d) Use Corollary 21.24 to conclude that $\deg(\Phi_n) = \phi(n)$ and that

$$\Phi_n(x) = \prod_{\substack{1 \leq a \leq n \\ \gcd(a,n)=1}} (x - \zeta_n^a).$$

Observe that this allows us to generalize Corollary 21.23 from q to n. For a proof of Dirichlet's Theorem (using techniques from analysis), see [1]. For a self-contained algebraic proof of the result of part (d) of this exercise, see e.g. [5, Chapter 33].

Exercise 21.16. This exercise gives a practical formula to compute $\Phi(n)$ that does not involve complex roots of 1.

(a) Let n be a positive integer. Use Exercise 21.15 to prove that

$$\Phi_n(x) = (x^n - 1) / \prod_{\substack{k|n \\ 1 \leq k < n}} \Phi_k(x). \qquad (21.16)$$

(b) Use Equation 21.16 together with Proposition 21.22 to find explicit formulas for Φ_n when $n \leq 6$.

Exercise 21.17. Let n be a positive integer. Prove that $n = \sum_{\substack{1 \leq f \leq n \\ f|n}} \phi(f)$.

Exercise 21.18 (A Finite Multiplicative Subgroup of a Field is Cyclic). Let F be a field, and let G be a finite subgroup of $F^\times$. Set $n = |G|$.

(a) Prove that if $H \leq G$ and $|H| = f$, then every element of H is a root of the polynomial $x^f - 1 \in F[x]$.

(b) Deduce that for every factor f of n, there is at most one subgroup of G of order f.

(c) Use Exercise 18.3 and part (b) above to prove that for every factor f of n, there are at most $\phi(f)$ elements of G of order f.

(d) Prove that for every factor f of n, there are exactly $\phi(f)$ elements of G of order f. Taking $f = n$, proceed to deduce that G is cyclic. Hint: every element of G has some order f dividing n; use the result of Exercise 21.17.

Exercise 21.19. Let F be a field with char$(F) = p > 0$. Let n be a positive integer, and let p^e be the highest power of p which divides n. Set $m := n/p^e$ and $f := x^n - 1 \in F[x]$. Let K be a splitting field for f over F.

(a) Let $g := x^m - 1 \in F[x]$. Prove that K is also a splitting field for g over F. Hint: use Corollary 21.21.

(b) Prove that K is a Galois extension of F. (You may find Exercise 16.13 useful.)

Exercise 21.20. Let F be a field. Let n be a positive integer, and let K be a splitting field over F for the polynomial $x^n - 1 \in F[x]$. (We may think of K/F as a generalization of a cyclotomic field extension.)

(a) Let R be the set of all roots of $x^n - 1$ in K. Prove that R is a finite subgroup of $K^{\times}$.

(b) Use Exercise 21.18 to deduce that $R = \langle \zeta \rangle$ for some $\zeta \in K^{\times}$.

(c) Show that we have $K = F[\zeta]$.

(d) Set $r = |R|$. Prove that K is a finite Galois extension field of F, and that $G := \mathrm{Gal}(K/F) \hookrightarrow (\mathbf{Z}/r\mathbf{Z})^{\times}$ (you can mostly imitate the proof of Proposition 21.17, but use Exercise 21.19 to help in case char$(F) > 0$). Conclude that G is abelian.

(e) Suppose that ω is an nth root of 1 in some extension field of F. Deduce that $L := F[\omega]$ is a Galois extension of F, and $\mathrm{Gal}(L/F)$ is abelian. Hint: Embed L in a splitting field of $x^n - 1$ over F.

Exercise 21.21. This exercise continues Exercise 16.24. Let K be a finite field, and set $p = \mathrm{char}(K)$. Let F be the prime subfield of K.

(a) Define a map $\sigma : K \to K$ by the formula $a \mapsto a^p$. Use Corollary 21.21 to help prove that $\sigma \in \mathrm{Gal}(K/F)$.

(b) Prove that $\mathrm{Gal}(K/F) = \langle \sigma \rangle$.

Exercise 21.22 (Fundamental Theory of Finite Fields). This exercise continues Exercise 21.21. Let $F = \mathbf{Z}/p\mathbf{Z}$, where p is prime. Let K be a splitting field for $f := x^{p^r} - x$ over F, and let $G := \mathrm{Gal}(K/F)$.

(a) Show that $|K| \geq p^r$, hence $|G| \geq r$. (Hint: does f have any repeated roots?)

(b) Let σ be the pth-power map of Exercise 21.21, which generates G. Set $\tau = \sigma^r \in G$. Let $\alpha \in K$ be a root of f. Prove that $\tau(\alpha) = \alpha$, and use this to show that $\tau = \mathrm{id}_K$. Conclude that $|G| \leq r$.

(c) Prove that a finite field of order n exists if and only if n is a prime power, and that any two finite fields of the same order are isomorphic.

22

Solvability of Polynomial Equations by Radicals

22.1 Radicals

In this chapter, we complete the original program of Evariste Galois, which aims to understand the solution of polynomial equations in a single variable x by means of formulas involving a combination of ordinary arithmetic and roots—square, cube, or higher. We will take our coefficients from a field F.

We start at the beginning. A linear equation

$$ax + b = 0, \qquad (22.1)$$

to be truly linear, must have $a \neq 0$; then the unique solution in the field F is

$$x = -\frac{b}{a}. \qquad (22.2)$$

A quadratic equation

$$ax^2 + bx + c = 0 \qquad (22.3)$$

has up to 2 distinct solutions, which are given by the quadratic formula

$$x = \frac{-b \pm \sqrt{b^2 - 4ac}}{2a}. \qquad (22.4)$$

Already, some issues appear here. First, we have no definition for a "square root" of an element in a general field F. In fact, the only situation when a square root has an unambiguous interpretation for us occurs when we take the square root of a non-negative real number, which by convention is another non-negative real number. Second, while we do know how to interpret an integer multiplied by a field element (for instance, $2a$ means $a + a$), we are in trouble in case $\mathrm{char}(F) = 2$, for then the denominator is 0. The latter problem is easy to side-step: we can take F to have characteristic different from 2.

The first problem can be smoothed over in a natural way: we can say that a square root (not "the" square root!) of an element $\alpha \in F$ is an element ρ of some extension field K of F with the property that $\rho^2 = \alpha$. An nth root can be defined similarly for any positive integer n. As the reader may be aware, the term "radical" is a synonym for "root." This leads us to the following definitions.

DOI: 10.1201/9781003252139-22

Definition 22.1. Let F be a field. An element ρ of an extension field of F is called a *(simple) radical over F* if $\rho^n \in F$ for some $n \in \mathbf{Z}^+$.

The next definition describes the kind of elements that can be found by nesting roots inside each other any finite number of times and also applying field operations.

Definition 22.2. Let F be a field. An element α of an extension field of F is called a *radical expression over F* if there is a tower of fields $F = F_0 \leq F_1 \leq \cdots \leq F_r$ such that $\alpha \in F_r$, and for all $i \in \{1, 2, \ldots, r\}$, we have $F_i = F_{i-1}[\rho_i]$ for some ρ_i which is a simple radical over F_{i-1}.

Example 22.3. Working in $\mathbf{R}$, it is easy to write down radical expressions over $\mathbf{Q}$. Two examples are $\alpha := 2 + \sqrt[5]{9}$ and $\beta := \sqrt[3]{\frac{7}{5} - \sqrt[4]{2 + \sqrt{2}}}$.

Example 22.4. Let $\alpha \in \mathbf{R}$ be a constructible number. By definition, this means that there exists a tower of fields $\mathbf{Q} = F_0 \leq F_1 \leq \cdots \leq F_n \leq \mathbf{R}$ such that $\alpha \in F_n$ and $[F_i : F_{i-1}] = 2$ for all $i \in \{1, \ldots, n\}$. It follows that $F_i = F_{i-1}[\rho_i]$ for some ρ_i which is a root of a quadratic polynomial $f_i = a_i x^2 + b_i x + c_i \in F_{i-1}[x]$. By choosing ρ_i instead to be a square root of the discriminant $D_i = b_i^2 - 4a_i c_i$ of f_i, we do not change the fields F_i; so we see that α is a radical expression over $\mathbf{Q}$. As the reader may have surmised by now, a constructible number is the same thing as a radical expression over $\mathbf{Q}$ in which all of the radical signs are just square roots (and everything is in $\mathbf{R}$).

Our parenthetical note at the end of Example 22.4 is of more than passing interest: it will be very convenient to consider radicals in some "big" field which contains everything in the picture, such as $\mathbf{R}$ in that example. To see what difficulties we wish to avoid, suppose that F is a field, α is a radical expression over F, and β is a simple radical over $F[\alpha]$. We certainly want to be able to prove that β is then a radical expression over F. Now we know there is a tower of fields $F = F_0 \leq F_1 \leq \cdots \leq F_r$ such that $\alpha \in F_r$ and $F_i = F_{i-1}[\rho_i]$, where each ρ_i is a simple radical over F_{i-1}. The natural argument is to consider the tower

$$F = F_0 \leq F_1 \leq \cdots \leq F_r \leq F_r[\beta]. \tag{22.5}$$

Unfortunately, the top step in this tower does not make sense. The trouble is that, since β and F_r do not belong to a common field, there is no way to adjoin β to F_r; we do not know how to perform addition, say, with β and an arbitrary element of F_r. Instead, as it is, we can only adjoin β to $F[\alpha]$. With a common superfield Ω, the situation is remedied. The two situations are illustrated in Figure 22.1. Fortunately, we understand fields well enough to fix this problem:

Lemma 22.5 (Existence of a Common Superfield). *Suppose that F is a field and $\alpha_1, \ldots, \alpha_k$ are radical expressions over F. Then there is an extension field Ω of F such that for each $j \in \{1, \ldots, k\}$, there is a tower of fields $F = F_{0,j} \leq F_{1,j} \cdots \leq F_{r(j),j} \leq \Omega$ with $\alpha_j \in F_{r(j),j}$, where $F_{i,j} = F_{i-1,j}[\rho_{i,j}]$ and $\rho_{i,j}$ is a simple radical over $F_{i-1,j}$.*

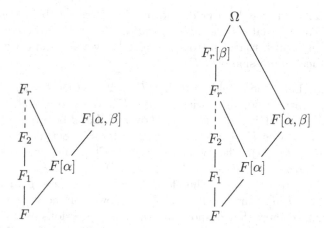

FIGURE 22.1: Field towers without (left) and with (right) a common superfield

Proof. We induct on k. By definition of radical expression, for each $j \in \{1, \ldots, k\}$ there is a tower of fields $L_{0,j} = F \le L_{1,j} \le \cdots \le L_{r(j),j}$ such that $\alpha_j \in L_{r(j),j}$ and $L_{i,j} = L_{i-1,j}[\rho_{i,j}]$, where $\rho_{i,j}$ is a simple radical over $L_{i-1,j}$. Thus, the conclusion is immediate when $k = 1$. So suppose inductively that there is an extension Ψ which contains all the requisite field towers for $\alpha_1, \ldots, \alpha_{k-1}$. Since each $\rho_{i,k}$ is algebraic over $L_{i-1,k}$, then $L_{r(k),k}$ is a finite extension of F. So by Lemma 16.10, there is an extension field Ω of Ψ which admits an embedding $L_{r(k),k} \hookrightarrow \Omega$ over F. We take $F_{i,k}$ to be the image of $L_{i,k}$ under this embedding. $\qquad\square$

Remark 22.6. A popular fix for the problem of the common superfield is to suppose in advance that all fields and elements in question lie within a single algebraically closed field. This is often convenient; the reader may wish to take this approach in what follows, and see how it simplifies some of the proofs. Instead of this approach, we shall make use of Lemma 16.10 when we need to, as in the proof of Lemma 22.5. But see Project 23.4 for a construction of an algebraically closed superfield in the case when all fields in question are finite, and Project 23.3 for a proof of the existence of an algebraically closed superfield in general, assuming the Axiom of Choice.

22.2 Solvable Polynomials

Next, we define what it means for a polynomial to be solvable by radicals: namely, that all its roots are radical expressions.

Definition 22.7. Let F be a field, and let $f \in F[x] - F$ be a non-constant polynomial. We say that f is *solvable by radicals* (over F) if there is an extension field K of F such that f factors completely over K, and every root of f in K is a radical expression over F.

Example 22.8. Let F be a field with $\operatorname{char}(F) \neq 2$, and let $f = ax^2 + bx + c \in F[x]$ be a quadratic polynomial, with $a, b, c \in F$ and $a \neq 0$. Set $D = b^2 - 4ac \in F$, and let L be a splitting field for $x^2 - D$ over F. Let d be a root of $x^2 - D$ in L. Note that $2a \neq 0$ in F, since $\operatorname{char}(F) \neq 2$ and $a \neq 0$. Set $\alpha = (-b+d)/(2a)$ and $\beta = (-b - d)/(2a)$. Then we have $\alpha, \beta \in F[d]$, and $(x - \alpha)(x - \beta) = x^2 - (\alpha + \beta)x + \alpha\beta = x^2 + (b/a)x + (b^2 - d^2)/(4a^2) = x^2 + (b/a)x + (4ac)/(4a^2) = x^2 + (b/a)x + (c/a) = f/a$. Therefore, $f = a(x - \alpha)(x - \beta)$, so f factors completely over $F[d]$. Further, $F \leq F[d]$ is a tower of the type required by Definition 22.2 to show that α and β are radical expressions over F. So every quadratic polynomial over a field of characteristic different from 2 is solvable by radicals.

Our main question is, which polynomials are solvable by radicals? Historically, people worked in characteristic 0, where (as a consequence of Example 22.8) all quadratic polynomials are solvable by radicals. But what about degree-3 ("cubic") polynomials? Is there a "cubic formula"? And what of higher degrees?

Since the definition of radical expression applies to *every* element in the top field F_r in Definition 22.2, it makes sense to give one more definition:

Definition 22.9. Let K be an extension field of F. We say that K is a *solvable* extension of F (or that K is *solvable over* F) if every element of K is a radical expression over F.

The next result says that a finite extension is solvable iff it is generated by radical expressions, iff there is a *single* tower of radical extensions containing it.

Lemma 22.10. *Let K be a finite extension field of F. Then the following are equivalent:*

(1) K is solvable over F;

(2) We can write $K = F[\alpha_1, \ldots, \alpha_m]$ where $\alpha_1, \ldots, \alpha_m \in K$ and each α_i is a radical expression over F;

(3) We can write $F = F_0 \leq F_1 \leq \cdots \leq F_r$ where $F_j = F_{j-1}[\rho_j]$, each ρ_j is a simple radical over F_{j-1}, and $K \leq F_r$.

Proof. (1) $\Rightarrow$ (2): Suppose that K is solvable over F. By Exercise 15.7, we can write $K = F[\alpha_1, \ldots, \alpha_m]$ for some $\alpha_1, \ldots, \alpha_m \in K$. Now each α_i is a radical expression over F, by Definition 22.9.

(2) $\Rightarrow$ (3): Suppose that such elements α_i $(1 \leq i \leq m)$ exist. Then by definition of radical expression, we can write $\alpha_i \in F[\rho_{i,1}, \ldots, \rho_{i,n(i)}]$ where $\rho_{i,j}$ is a simple radical over $F[\rho_{i,1}, \ldots, \rho_{i,j-1}]$. By Lemma 22.5, we may suppose that all the $\rho_{i,j}$'s lie in a common superfield Ω of F. Let

$L = F[\rho_{1,1}, \ldots, \rho_{1,n(1)}, \ldots, \rho_{m,1}, \ldots, \rho_{m,n(m)}]$. By adjoining the $\rho_{i,j}$'s to F in the order listed, we see that each $\rho_{i,j}$ is a simple radical over the previous field. Further, we have $\alpha_i \in L$ for each i, so $K \leq L$, and we have established (3).

(3) $\Rightarrow$ (1): This follows immediately from Definitions 22.2 and 22.9. $\square$

Corollary 22.11. *Let F be a field, and let $f \in F[x] - F$. Let K be a splitting field for f over F. Then f is solvable by radicals over F iff K is solvable over F.*

Proof. By Theorem 15.21, we have $K = F[\alpha_1, \ldots, \alpha_n]$, where α_i are the roots of f in K. The result is now immediate from Lemma 22.10 and the fact that any two splitting fields for f over F are isomorphic over F (Exercise 16.9). $\square$

In order to clarify condition (3) in Lemme 22.10, we make the following definition.

Definition 22.12. Let K be a finite extension field of F. We say that K is a *pure radical extension* of F if we can write $F = F_0 \leq F_1 \leq \cdots \leq F_r$ where $F_j = F_{j-1}[\rho_j]$, each ρ_j is a simple radical over F_{j-1}, and $K = F_r$.

Thus, a finite extension is solvable iff it is a subfield of a pure radical extension. Next we ask, what do we get if we form a radical expression out of radical expressions? The answer is, another radical expression:

Lemma 22.13 (Transitivity of Solvability). *Let F, K, and L be fields with $F \leq K \leq L$. Then L is solvable over F iff both L is solvable over K and K is solvable over F.*

Proof. ($\Rightarrow$): Suppose that L is solvable over F. Let $\alpha \in L$. By definition of radical expression, there is a tower of fields $F = F_0 \leq F_1 \leq \cdots \leq F_r$ such that $\alpha \in F_r$, and for all $i \in \{1, 2, \ldots, r\}$, we have $F_i = F_{i-1}[\rho_i]$ for some ρ_i which is a simple radical over F_{i-1}. Since F_r is a finite extension of F, then we also have $[F_r : F[\alpha]] < \infty$. Using Lemma 16.10, we can find an embedding $F_r \hookrightarrow \Omega$ over $F[\alpha]$ for some extension field Ω of L; so without loss of generality, we suppose that F_r and L are both contained in a common superfield. Set $K_0 = K$ and $K_i = K_{i-1}[\rho_i]$ for $i \in \{1, 2, \ldots, r\}$. Then inductively, we see that $F_i \leq K_i$ for each i. Now some power of ρ_i is in F_i, hence also in K_i; so ρ_i is a simple radical over K_i. Since $\alpha \in F_r \leq K_r$, we have shown that α is a radical expression over K. Therefore, L is solvable over K. To see that K is solvable over F, let $\beta \in K$, and just observe that $\beta \in L$, so β is a radical expression over F: Definition 22.2 does not actually use the field from which β was chosen.

($\Leftarrow$): Suppose that L is solvable over K and K is solvable over F. Let $\alpha \in L$. Then there is a tower of fields $K = K_0 \leq K_1 \leq \cdots \leq K_r$ such that $\alpha \in K_r$ and for all $i \in \{1, 2, \ldots, r\}$, we have $K_i = K_{i-1}[\rho_i]$ for some ρ_i which is a simple radical over K_{i-1}. So we have $K_i = K[\rho_1, \ldots, \rho_i]$. We proceed by induction on i to prove that K_i is solvable over F, which is equivalent (via Lemma 22.10) to proving that $\rho_1, \ldots, \rho_i$ are radical expressions over F. When

$i = 0$, this statement is vacuously true. So inductively suppose that $0 \leq i < r$ and K_i is solvable over F. Now $a := \rho_{i+1}^n \in K_i$ for some $n \in \mathbf{Z}^+$, so a is a radical expression over F by inductive hypothesis. So there is a tower of fields $F = F_0 \leq F_1 \leq \cdots \leq F_t$ such that $a \in F_t$, and for all $j \in \{1, 2, \ldots, t\}$, we have $F_j = F_{j-1}[r_j]$ for some r_j which is a simple radical over F_{j-1}. Since F_t is a finite extension of F, then we also have $[F_t : F[a]] < \infty$; so by Lemma 16.10, there is an extension field Ω of K_i such that $F_t \hookrightarrow \Omega$ over $F[a]$. We suppose without loss of generality that in fact $F_t \leq \Omega$. The tower of fields $F_0 \leq F_1 \leq \cdots \leq F_t \leq F_t[\rho_{i+1}]$ shows that ρ_{i+1} is a radical expression over F. This completes the induction. Now since $\alpha \in K_r$ and K_r is solvable over F, then α is a radical expression over F. This completes the proof. $\square$

22.3   Solvable Groups

Let us start with the simplest type of (non-trivial) solvable extension and investigate its structure. So suppose that ρ is a simple radical over a field F; thus, $\rho^n =: \alpha \in F$ for some $n \in \mathbf{Z}^+$. We see that ρ is a root of the polynomial $f(x) := x^n - \alpha \in F[x]$. But rather than looking merely at $F[\rho]$, we strongly suspect that we will be much better off by considering a normal closure L of $F[\rho]$ over F. For then (at least, if f has no repeated roots), L is Galois over F, and we may be able to use group theory to study our fields.

What are the roots of f? Well, ρ is one root. An examination of Example 15.27 might lead us to suspect that every root of f has the form ρ multiplied by an nth root of 1. In fact, this is correct. To get our normal closure, our idea is to first adjoin all necessary roots of 1, and then adjoin our original radical.

Proposition 22.14. *Let F be a field, and let ρ be a simple radical over F, with $\rho^n = \alpha \in F$ and $n \in \mathbf{Z}^+$. Let $f = x^n - \alpha \in F[x]$. Let L be a normal closure of $F[\rho]$ over F. Let R be the set of all n^{th} roots of 1 in L, and set $K = F[R]$. Then:*

(1) $K = F[\zeta]$ for some $\zeta \in R$;
(2) $L = K[\rho]$;
(3) If $\mathrm{char}(F)$ does not divide n, then L is Galois over F.

Proof. The case $\rho = 0$ being easy, we suppose that $\rho \neq 0$. Note that R is a subgroup of $L^\times$. By the Root-Factor Theorem (Theorem 14.24), $|R| < \infty$, so by Exercise 21.18, R is cyclic. Let ζ be a generator of R; then we have (1).

Since $K \leq L$ and $\rho \in L$, we have $K[\rho] \leq L$. Let $g = \mathrm{Irr}(\rho, F, x)$. We may have $g \neq f$, but since $f(\rho) = 0$, certainly g divides f in $F[x]$. Let r be any root of g in L. Then $f(r) = f(\rho) = 0$, so $r^n = \rho^n = \alpha$, and $(r/\rho)^n = 1$. Thus $r/\rho \in R$, so $r/\rho \in K \leq K[\rho]$. Since also $\rho \in K[\rho]$, we get $r \in K[\rho]$. Thus $K[\rho]$ contains all roots of g in L, and hence contains L (e.g. by Theorem 15.21 and

Proposition 16.25). This gives (2), and we note that L is a splitting field for g over F.

For (3), suppose that $\operatorname{char}(F) \nmid n$. Then $n \cdot 1_F \in F^\times$; and since $f'(x) = n \cdot x^{n-1} = (n \cdot 1_F)x^{n-1} \in F[x]$, the only possible root of f' is 0. But 0 is not a root of f. So by Proposition 16.15, f has no repeated roots, and therefore neither does g. Now L is Galois over F by Exercise 16.13. $\qquad \square$

To understand the structure of the field extension L/F under the hypotheses of Proposition 22.14, Galois theory tells us to look at the structure of the corresponding group $G := \operatorname{Gal}(L/F)$, and vice versa. The intermediate field K makes a nice way station.

Lemma 22.15. *Let F be a field, and let $f = x^n - \alpha \in F[x]$ be a polynomial with $n \geq 1$ and $\alpha \in F$. Let ρ be a root of f in some extension of F, and L be a normal closure of $F[\rho]$ over F. Suppose that $\operatorname{char}(F) \nmid n$. Then L is Galois over F, and there is a subgroup H of $G := \operatorname{Gal}(L/F)$ such that H is abelian, $H \trianglelefteq G$, and G/H is abelian.*

Proof. By Proposition 22.14 we can say that $L = K[\rho]$, where $K = F[\zeta]$ for some $\zeta \in L$ with $\zeta^n = 1$. Proposition 22.14 also gives that L is Galois over F. Let $G = \operatorname{Gal}(L/F)$, and let $H = \operatorname{Gal}(L/K)$. Then $H \leq G$ (by Lemma 16.29). Now K is Galois over F, and $\operatorname{Gal}(K/F)$ is abelian, by Exercise 21.20; so $H \trianglelefteq G$ and G/H is abelian, by the Fundamental Theorem of Galois Theory.

Let $\sigma \in H$. As in the proof of Proposition 22.14, all roots of f in L are of the form $\zeta^k \rho$, where $k \in \mathbf{N}$. Since $f \in F[x] \leq K[x]$, and $\sigma \in \operatorname{Gal}(L/K)$, then σ must send a root of f in L to another root of f (e.g. by Exercise 16.4). So $\sigma(\rho) = \zeta^k \rho$ for some $k \in \mathbf{N}$. Since $\zeta \in K$, we have $\sigma(\zeta) = \zeta$. Now let $\tau \in H$; then similarly, we have $\tau(\rho) = \zeta^\ell \rho$ for some $\ell \in \mathbf{N}$, and $\tau(\zeta) = \zeta$. So $\sigma\tau(\rho) = \sigma(\zeta^\ell \rho) = \sigma(\zeta^\ell)\sigma(\rho) = \zeta^\ell \sigma(\rho) = \zeta^\ell \zeta^k \rho = \zeta^{\ell+k}\rho = \zeta^{k+\ell}\rho = \zeta^k \zeta^\ell \rho = \zeta^k \tau(\rho) = \tau(\zeta^k)\tau(\rho) = \tau(\zeta^k \rho) = \tau\sigma(\rho)$. Since $L = K[\rho]$, every element of H is determined by its action on ρ. So $\sigma\tau = \tau\sigma$. But σ and τ were arbitrary elements of H, so H is abelian. $\qquad \square$

One observation we can now make about a normal closure L of $F[\rho]$ over F (where ρ is a simple radical over F) is that L itself is a solvable extension of F. This is because L is composed of two steps, where in each step we adjoin a simple radical: first adjoining a root of 1, and then adjoining ρ itself. In particular, a normal closure of the simple radical extension $F[\rho]$ is a pure radical extension of F. Can we generalize this result? Is a normal closure of any pure radical extension also pure radical? Let's look at a small example. What are the conjugates of $\beta := \sqrt[5]{3 + \sqrt[7]{2}}$ over $\mathbf{Q}$? If our hope is realizable, then the conjugates should also be radical expressions over $\mathbf{Q}$. Since β is a root of $f := x^5 - (3 + \sqrt[7]{2})$, we may be tempted to conjecture that the conjugates of β are the roots of $x^5 - (3 + \zeta_7^j \sqrt[7]{2})$ for $j \in \{0, 1, \ldots, 6\}$: that is, we just take roots of the polynomials formed by conjugating the coefficients of f. In fact, this argument works.

Lemma 22.16. *Let K be a pure radical extension of F, and let L be a normal closure of K over F. Then L is also a pure radical extension of F.*

Proof. We can write $F = F_0 \le F_1 \le \cdots \le F_r$ where $F_i = F_{i-1}[\rho_i]$, ρ_i is a simple radical over F_{i-1}, and $K = F_r$. We induct on r. When $r = 0$, then $L = K = F$, which is certainly a pure radical extension of F. Inductively, we have that the normal closure L' of F_{r-1} over F inside L is a pure radical extension of F. Since ρ_r is a simple radical over F_{r-1}, we have $\alpha := \rho_r^n \in F_{r-1}$ for some $n \in \mathbf{Z}^+$. Now F_r is algebraic over F (since F_r is a finite extension of F), so we can set $g := \mathrm{Irr}(\rho_r, F, x)$. Let R be the set of all roots of g in L. By Exercise 22.3, we have that $L'[R]$ is normal over F. Certainly $L'[R] \le L$. On the other hand, L is the smallest extension of K inside L which is normal over F, by definition of normal closure; but $K = F_r = F_{r-1}[\rho_r] \le L'[\rho_r] \le L'[R]$. Therefore $L = L'[R]$. Let $\beta \in R$. [Show that β is a root of $x^n - \tilde{\alpha}$ for some conjugate $\tilde{\alpha}$ of α.] Set $h(x) = \mathrm{Irr}(\alpha, F, x)$ and $\tilde{\alpha} = \beta^n$. [Idea: $(x - \alpha)$ divides h, i.e. $(x - \rho_r^n)$ divides h; so ρ_r will be a root of $x^n - \alpha$. Replace x by x^n in h.] Set $\omega(x) = h(x^n) \in F[x]$. Then $\omega(\rho_r) = h(\rho_r^n) = h(\alpha) = 0$. It follows that g divides ω in $F[x]$. Since $g(\beta) = 0$, we must have $\omega(\beta) = 0$ as well; therefore, $h(\beta^n) = 0$, so $\tilde{\alpha}$ is a root of h. But $\alpha \in F_{r-1} \le L'$, and L' is normal over F, so $\tilde{\alpha}$ is also in L'. Since β is a root of $x^n - \tilde{\alpha}$, then β is a simple radical over L'. So every element of R is a simple radical over L'. Since $L = L'[R]$, and L' is a pure radical extension of F, then we can adjoin the finitely many elements of R one at a time in any order to L' to see that L is also a pure radical extension of F. $\qquad\square$

To understand the group theory corresponding to solvable extension fields, we will start with a solvable extension which is also Galois. We embed this extension in a pure radical extension field realized as a tower of simple radical extensions, and replace each step in this tower with its normal closure. Our original solvable extension will correspond via Galois Theory to a quotient of the overall Galois group. The catch is that the characteristic of our fields cannot divide *any* of the degrees of the radicals in order for our machinery to work properly. So we will make the simplifying assumption that we are in characteristic 0.

Proposition 22.17. *Let F be a field of characteristic 0, and let K be a finite solvable Galois extension of F. Set $G = \mathrm{Gal}(K/F)$. Then there is a tower of groups $\{e_G\} = G_0 \le G_1 \le \cdots \le G_n$ such that G is isomorphic to a quotient of G_n and for all $i \in \{1, \ldots, n\}$, we have $G_{i-1} \trianglelefteq G_i$ and G_i/G_{i-1} is abelian.*

Proof. By Lemma 22.10, we can realize K as a subfield of a pure radical extension L of F; and by Lemma 22.16, without loss of generality we suppose that L is normal over F. We can write $F = F_0 \le F_1 \le \cdots \le F_r = L$ where $F_j = F_{j-1}[\rho_j]$, each ρ_j is a simple radical over F_{j-1}, and $K \le L$.

Construct a new tower of fields $L_0 \le L_2 \le \cdots \le L_{2r} \le L$ as follows. Set $L_0 = F$, and for $j \ge 1$ let L_{2j} be the normal closure of $L_{2(j-1)}[\rho_j]$ over $L_{2(j-1)}$ in L. We see inductively that $F_j \le L_{2j}$ for each j. Now $L_{2r} \le L$, but also $L = F_r \leq L_{2r}$, so $L_{2r} = L$. Because $\operatorname{char}(F) = 0$ and L is normal over F, then L is Galois over F. By Lemma 22.15, we can say that $L_{2j}/L_{2(j-1)}$ is Galois and $\Gamma_{2j} := \operatorname{Gal}(L_{2j}/L_{2(j-1)})$ decomposes as $\{e\} \leq \Gamma_{2j-1} \leq \Gamma_{2j}$, where Γ_{2j-1} is abelian, $\Gamma_{2j-1} \trianglelefteq \Gamma_{2j}$, and $\Gamma_{2j}/\Gamma_{2j-1}$ is abelian. For $j \in \{1, \ldots, r\}$, set $L_{2j-1} = L_{2j}^{\Gamma_{2j-1}}$, the fixed field of Γ_{2j-1}, and set $G_j = \operatorname{Gal}(L/L_{2r-j})$ for $j \in \{0, 1, \ldots, 2r\}$. Then we have $\{e\} = G_0 \leq G_1 \leq \cdots \leq G_{2r}$. Basic Galois theory (see Exercise 22.8) shows that $G_{j-1} \trianglelefteq G_j$ and $G_j/G_{j-1} \cong \Gamma_j/\Gamma_{j-1}$ (if j is even) or Γ_j (if j is odd). Set $H = \operatorname{Gal}(L/K)$. By the Fundamental Theorem of Galois Theory, we have $G \cong \operatorname{Gal}(L/F)/H = G_{2r}/H$, which completes the proof. $\qquad\square$

Proposition 22.17 gives us a framework for translating the notions of a pure radical extension and a solvable extension from field theory into group theory. Namely, we could define a finite group G to be "pure radical" if there is a tower of subgroups as in that proposition with G at the top, and a "solvable" group to be a quotient of a pure radical group. But it turns out that there is no need for two separate definitions: with groups, the two notions are equivalent. Thus, we make the following definition.

Definition 22.18. Let G be a finite group. We say that G is *solvable* if there exists a tower of subgroups $\{e_G\} = G_0 \leq G_1 \leq \cdots \leq G_n = G$ such that for all $i \in \{1, \ldots, n\}$, we have $G_{i-1} \trianglelefteq G_i$ and G_i/G_{i-1} is abelian.

Remark 22.19. Although the property of being a subgroup is transitive—that is, if $A \leq B$ and $B \leq C$, then $A \leq C$—the property of being a *normal* subgroup is not transitive: see Exercise 16.25.

Next we demonstrate the equivalence referred to above: we don't get any new groups by taking quotients of solvable groups.

Lemma 22.20. *A quotient of a solvable group is solvable. That is, if G is a solvable group and Q is isomorphic to a quotient of G, then Q is also solvable.*

Proof. Let G be a solvable group, and write $\{e_G\} = G_0 \leq G_1 \leq \cdots \leq G_n = G$ where for all $j \in \{1, \ldots, n\}$, we have $G_{j-1} \trianglelefteq G_j$ and G_j/G_{j-1} is abelian. Suppose that there is a surjective group homomorphism $\nu : G \to Q$, which by the Fundamental Theorem of Group Homomorphisms is the same thing as saying that Q is isomorphic to a quotient of G. Set $Q_j = \nu(G_j)$. Then we have $\{e_Q\} = Q_0 \leq Q_1 \leq \cdots \leq Q_n = Q$ (using Theorem 7.13 (i)). Furthermore, we have $Q_j \trianglelefteq Q_{j+1}$ by Exercise 7.8, and Q_{j+1}/Q_j is isomorphic to a quotient of G_{j+1}/G_j by Exercise 22.10. A quotient of an abelian group is abelian (by Exercise 11.5), so we are done. $\qquad\square$

Corollary 22.21. *Let F be a field of characteristic 0, and let K be a finite solvable Galois extension of F. Then $\operatorname{Gal}(K/F)$ is a solvable group.*

Proof. Combine Proposition 22.17 with Lemma 22.20. $\qquad\square$

Corollary 22.21 implies that if a polynomial in characteristic 0 is solvable by radicals, then the Galois group of its splitting field must be solvable. The natural question now is whether the converse holds. Again it makes sense to start with the simplest possible non-trivial field extension K/F in terms of the Galois group G—namely, when G is cyclic. Even in this restricted situation, we can make the situation still simpler by requiring G to have prime order: for then, G has no intermediate subgroups, and K/F has no intermediate fields.

So suppose that we have $K = F[\alpha]$, where α is a simple radical over F, the extension K/F is Galois, and $G = \text{Gal}(K/F)$ has prime order p. How can we recover α from K and F? We know that the conjugates of α over F are of the form $\zeta^j \alpha$ where ζ is a root of 1 in K. We also know that G is cyclic, say $G = \langle \sigma \rangle$. Thus, we could look for an element $\alpha \in K$ such that $\sigma(\alpha) = \zeta^j \alpha$ for some $j \in \mathbf{Z}^+$.

Because G has order p, we know that $\sigma^p = e_G$, the identity function on K. Let us now indulge in a wild flight of fancy for the rest of the paragraph. You might agree that the identity function behaves as a sort of 1. With our knowledge of cyclotomic polynomials, it is tempting to rewrite the preceding equation as $\sigma^p - 1 = 0$; then factoring this equation via Corollary 21.15, we would find $\prod_{j=0}^{p-1}(\sigma - \zeta^j) = 0$, where $\zeta^p = 1$. If we were in a domain, we could conclude that $\sigma - \zeta^j = 0$ for some j, and thus that for any $\alpha \in K$, we have $\sigma(\alpha) = \zeta^j \alpha$, as desired. But this argument cannot possibly work as stated, since it seems to conclude that *all* elements of K are simple radicals over F.

Amazingly, we shall be able to take the ideas from the preceding paragraph and make them work sufficiently well to establish the converse we seek. The key is to find a suitable ring which contains both σ and things that behave like multiplication by the field elements 1, ζ, ..., ζ^{p-1}. By viewing σ in its role as a *linear transformation*, all this is possible. The reader is encouraged to review Exercises 13.15 through 13.18 at this point. For these exercises tell us that the set of all linear transformations from a vector space to itself naturally forms a ring, and the center of this ring is the set of all transformations which come from scalar multiplication by elements of the base field.

Proposition 22.22. *Let K/F be a finite Galois extension of prime degree $p = [K : F]$. Suppose that $char(F) \neq p$. Then $K \leq F[\zeta, \alpha]$ where ζ is a root of $x^p - 1$ and $\alpha^p \in F[\zeta]$. In particular, K is a solvable extension of F.*

Proof. [We want to factor $x^p - 1$, so we must introduce p^{th} roots of 1, which may not be in F or even in K.] Let L be a splitting field for $x^p - 1$ over K. By Exercise 21.20, we have $L = K[\zeta]$ for some $\zeta \in L$ such that all the roots of $x^p - 1$ in L are powers of ζ. Let $\tilde{F} = F[\zeta]$.

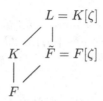

Since $x^p - 1 = (x - 1)(x^{p-1} + x^{p-2} + \cdots + x + 1)$, we see that either $\zeta = 1$ or ζ is a root of $x^{p-1} + x^{p-2} + \cdots + x + 1$. In either case, we have $[\tilde{F} : F] = \deg(\mathrm{Irr}(\zeta, F, x)) \leq p - 1$. Since $[K : F] = p$ by hypothesis, then p divides $[L : F]$, and since $p \nmid [\tilde{F} : F]$, we must have $p \mid [L : \tilde{F}]$. By Proposition 19.13, we know that L is Galois over $\tilde{F}$ and $\tilde{G} := \mathrm{Gal}(L/\tilde{F}) \hookrightarrow \mathrm{Gal}(K/F)$. It follows that $[L : \tilde{F}] = p = |\tilde{G}|$.

Let R be the ring of all linear transformations from L to L as an $\tilde{F}$-vector space. Then we have $\tilde{G} \subseteq R$ (e.g. by Exercise 16.22). Let $C = C(R)$ be the center of R; then we have $C = \{\mu_a : a \in \tilde{F}\}$, where μ_a is the multiplication-by-a map from L to L; that is, $\mu_a(b) = ab$ for $b \in L$. Since $|\tilde{G}| = p$, which is prime, then $\tilde{G}$ is cyclic by Theorem 8.20. Write $\tilde{G} = \langle \sigma \rangle$. By Exercise 14.12, the ring $S := \{\sum_{j=0}^n c_j \sigma^j : c_i \in C, n \in \mathbf{Z}^+\}$ is a commutative subring of R. Thus, we can consider the evaluation-at-σ map $\varepsilon_\sigma : C[x] \to S$, and we have $S = C[\sigma]$. Now $C \cong \tilde{F}$ (as rings) via the map $\mu_a \mapsto a$, by Exercise 13.18. This map extends to an isomorphism $C[x] \cong \tilde{F}[x]$ (by Lemma 16.1), so we can compose with ε_σ to get a ring homomorphism $\tau : \tilde{F}[x] \to S$, $\sum_{j=0}^n c_j x^j \mapsto \sum_{j=0}^n \mu_{c_j} \sigma^j$. Writing 1 for 1_F, we have $\mu_1 = \mathrm{id}_L = e_{\tilde{G}}$, so $\tau(x^p - 1) = \sigma^p - \mu_1 = 0_S$ (by Corollary 8.25). Since $\mathrm{char}(F) = \mathrm{char}(\tilde{F}) \neq p$, the polynomial $x^p - 1$ has no repeated roots in $\tilde{F}$ (by Proposition 16.15), so the order of ζ in $\tilde{F}^\times$ is p and not 1. Thus $x^p - 1$ factors in $\tilde{F}[x]$ as $x^p - 1 = \prod_{j=0}^{p-1}(x - \zeta^j)$. So we have $\tau(x^p - 1) = \prod_{j=0}^{p-1}(\sigma - \mu_{\zeta^j}) = 0_S$. Set $\omega_k = \prod_{j=0}^k (\sigma - \mu_{\zeta^j}) \in S$ for $k \in \{0, 1, \ldots, p-1\}$. Let $\beta \in L - \tilde{F}$; we know such an element exists because $[L : \tilde{F}] = p > 1$ (using Exercise 15.5). If $\sigma(\beta) = \beta$, then $\beta \in L^{\{\sigma\}}$ (see Exercise 16.23) $= L^{\langle \sigma \rangle} = L^{\tilde{G}} = \tilde{F}$, which is false. Therefore, $\sigma(\beta) \neq \beta$, and $\omega_0(\beta) \neq 0$. Set $n = \max\{k \in \mathbf{N} : \omega_k(\beta) \neq 0\}$; we have $n < p - 1$, since $\omega_{p-1} = 0$. Set $\alpha = \omega_n(\beta)$. Now by construction we have $0 = \omega_{n+1}(\beta) = (\sigma - \mu_{\zeta^{n+1}})(\omega_n(\beta)) = (\sigma - \mu_{\zeta^{n+1}})(\alpha) = \sigma(\alpha) - \zeta^{n+1}\alpha$. So $\sigma(\alpha) = \zeta^{n+1}\alpha$. It follows that $\sigma(\alpha^p) = (\sigma(\alpha))^p = \zeta^{p(n+1)}\alpha^p = \alpha^p$ (using that $\zeta^p = 1$). Since α^p is fixed by σ, we see as before that $\alpha^p \in \tilde{F}$. Now since $1 \leq n + 1 \leq p - 1$, we have $\zeta^{n+1} \neq 1$, and thus $\sigma(\alpha) \neq \alpha$. It follows by Galois Theory that $\alpha \notin \tilde{F}$. Since $[L : \tilde{F}]$ is prime, and $\tilde{F} < \tilde{F}[\alpha] \leq L$, we must have $\tilde{F}[\alpha] = L$. This completes the proof. $\square$

When a field extension is Galois of prime degree, at least in characteristic 0, then just knowing the Galois group structure is enough to guarantee that the field extension is solvable. How does this help with a general solvable Galois group? We show next how any solvable group can be relentlessly broken down into extensions of prime order.

Proposition 22.23. *Let G be a non-trivial finite group. Then G is solvable iff there exists a tower of subgroups $\{e_G\} = G_0 \leq G_1 \leq \cdots \leq G_n = G$ such that for all $i \in \{1, \ldots, n\}$, we have $G_{i-1} \unlhd G_i$ and G_i/G_{i-1} is cyclic of prime order.*

Proof. ($\Rightarrow$): From the definition of solvable group, there is a tower $\{e_G\} = G_0 \leq G_1 \leq \cdots \leq G_n = G$ such that for all $i \in \{1, \ldots, n\}$, we have $G_{i-1} \unlhd G_i$ and G_i/G_{i-1} is abelian. Since G is non-trivial, we may assume that all the inclusions here are strict. If the result were true for all non-trivial finite abelian groups, then we could lift each tower for G_i/G_{i-1} up to G, and they would fit together to form a tower with the desired properties (see Exercises 7.17 and 19.3). So we only need to consider the case when G is abelian. We proceed by induction on $|G|$. In case $|G| = 2$, then G is already cyclic of prime order, so we take $n = 1$ and $G_1 = G$. Inductively, suppose that the result holds for abelian groups of order less than $|G|$. Let p be a prime which divides $|G|$. By Cauchy's Theorem, there is an element $g \in G$ of order p. Let $H = \langle g \rangle$. Then we have $H \unlhd G$ (since G is abelian), and G/H is also abelian, of order less than $|G|$. If G/H is trivial, then G is already cyclic of prime order, so we are done. Otherwise, by inductive hypothesis, G/H has a tower of the desired type, which we lift to G and insert after the tower $\{e_G\} \leq H$ to finish the proof.

($\Leftarrow$): Since every cyclic group is abelian, this direction is immediate from the definition of solvable group. $\qquad\square$

Now we are ready to characterize those polynomials over a field of characteristic 0 which are solvable by radicals, solely in terms of the group-theoretic properties of their Galois groups.

Theorem 22.24. *Let F be a field of characteristic 0, and let $f \in F[x] - F$. Let K be a splitting field for f over F, and set $G = \mathrm{Gal}(K/F)$. Then f is solvable by radicals iff G is a solvable group.*

Proof. ($\Rightarrow$): Suppose that f is solvable by radicals. Then by Corollary 22.11, K is a solvable extension of F. So by Corollary 22.21, G is solvable.

($\Leftarrow$): Suppose that G is a solvable group. By Proposition 22.23, there is a tower of groups $\{e_G\} = G_0 \leq G_1 \leq \cdots \leq G_n = G$ such that for all $j \in \{1, \ldots, n\}$, we have $G_{j-1} \unlhd G_i$ and G_j/G_{j-1} is cyclic of prime order. Set $F_j = K^{G_{n-j}}$, the fixed field of G_{n-j}. Then we have $F = F_0 \leq F_1 \leq \cdots \leq F_n = K$, and (by Exercise 22.8) each F_j is Galois over F_{j-1} with $\mathrm{Gal}(F_j/F_{j-1}) \cong G_{n-j+1}/G_{n-j}$. By Proposition 22.22, we know that each F_j is a solvable extension of F_{j-1}. Now applying Lemma 22.13 repeatedly, we see that K is a solvable extension of F. So by Corollary 22.11, f is solvable by radicals. $\qquad\square$

22.4   Galois Groups in the Generic Case

In light of Theorem 22.24, the question of which polynomials are solvable by radicals (in characteristic 0) is intimately related to the question of the solvability of Galois groups. If we want to go beyond the quadratic formula to a cubic formula which gives the general solution to a cubic polynomial, for instance, then we need to understand which groups occur as Galois groups of the splitting field of a cubic polynomial.

So let us summarize some relevant results which we have already seen in previous chapters. Let F be a field of characteristic 0, let $f \in F[x] - F$, and let K be a splitting field for f over F. Let R be the set of all roots of f in K, and set $G = \text{Gal}(K/F)$. We know from Exercise 19.9 that $G \hookrightarrow \text{Sym}(R)$ via the action of G on the roots of f; an element of G is determined by what it does to these roots. We also know that $|R| \leq \deg(f) =: n$, so that $\text{Sym}(R) \cong S_{|R|} \hookrightarrow S_n$ (by Exercises 9.5 and 9.6). Thus we have $G \hookrightarrow S_n$. It is natural to ask whether this embedding can ever be an isomorphism: that is, whether every possible permutation of the roots can yield an automorphism of K over F.

In order for this to happen, we first must have $|R| = n$. Write $R = \{r_1, \ldots, r_n\}$. Now if $r_1, \ldots, r_n$ were independent variables in a polynomial ring, then we would be able to permute the r_j's at will—all the r_j's would look alike in a very strong sense—and we could realize every element of S_n as an automorphism. Let us try this approach.

Proposition 22.25. *Let C be a field and let n be a positive integer. Let $S = C[r_1, \ldots, r_n]$ be the polynomial ring in the n independent variables r_1, $\ldots$, r_n with coefficients in C. Set $K = k(S) = C(r_1, \ldots, r_n)$, the field of fractions of S. Let $f = \prod_{j=1}^{n}(x - r_i) \in K[x]$. Write $f = x^n + \sum_{j=0}^{n-1} a_j x^j$ with $a_j \in S$, and set $F = C(a_0, \ldots, a_{n-1})$, so that $f \in F[x]$. Then K is a splitting field for f over F, and we have $\text{Gal}(K/F) \cong S_n$.*

Proof. By inspection, f factors completely over K. Suppose that L is a field such that f factors completely over L and $F \leq L \leq K$. Then $C \leq L$ and L must contain each r_i, so we have $S = C[r_1, \ldots, r_n] \leq L$. Therefore, $K := k(S) \leq k(L) = L$. So K is a splitting field for f over F. Since f has no repeated roots (we assumed that r_1 through r_n were algebraically independent over C), then K/F is Galois, by Exercise 16.13. Let $R = \{r_1, \ldots, r_n\}$, and let $\sigma \in \text{Sym}(R)$. By Exercise 14.11, there is a unique ring homomorphism $\tilde{\sigma} : S \to S$ over C such that $\tilde{\sigma}(r_j) = \sigma(r_j)$ for all j. By applying the Universal Property of Localization (Lemma 20.9) to the composite map $S \xrightarrow{\tilde{\sigma}} S \hookrightarrow k(S) = K$, we see that $\tilde{\sigma}$ extends to a ring homomorphism $\overline{\sigma} : K \to K$. It is easy to see that $\overline{\sigma}$ is bijective (its inverse is induced by σ^{-1}), so $\overline{\sigma} \in \text{Aut}(K)$. Let $s : K[x] \to K[x]$ be the extension of $\overline{\sigma}$ to $K[x]$ of Lemma 16.1. Since s fixes x and permutes the r_j's, we have $s(f) = f$. From this it follows that s fixes each a_j, and thus s

fixes every element of F; hence so does $\overline{\sigma}$. Thus, $\overline{\sigma} \in \mathrm{Gal}(K/F) =: G$. Also, we have $\overline{\sigma}|_R = \sigma$. The map $G \hookrightarrow \mathrm{Sym}(R)$, $\tau \mapsto \tau|_R$ of Exercise 19.9 is therefore surjective. Hence $G \cong \mathrm{Sym}(R) \cong S_n$. $\qquad\square$

In a sense, the polynomial f of Proposition 22.25 is the most "generic" monic polynomial of degree n; if f is solvable by radicals, then we may suspect that any degree-n polynomial is too. This hunch is borne out in the following results.

Lemma 22.26. *Let G be a finite group. Then:*
(1) If $H \le G$ and G is solvable, then H is also solvable.
(2) Suppose $N \trianglelefteq G$. Then G is solvable iff both N and G/N are solvable.

Proof. (1): By Cayley's Theorem together with Exercise 9.6, we have $G \hookrightarrow S_n$ for some n. By Proposition 22.25 with $C = \mathbf{Q}$, there is a finite Galois extension K/F, with $\mathrm{char}(F) = 0$, such that $S := \mathrm{Gal}(K/F) \cong S_n$. Without loss of generality, we suppose $H \le G \le S$. Set $A = K^G$ and $B = K^H$. Then $F \le A \le B \le K$. Now K is Galois over A, with $\mathrm{Gal}(K/A) = G$, and similarly, $\mathrm{Gal}(K/B) = H$. By Exercise 16.13, K is a splitting field over A for some polynomial $f \in A[x]$. Since G is solvable, then by Theorem 22.24, f is solvable by radicals over A; so by Corollary 22.11, K is a solvable extension of A. By Lemma 22.13, K is also a solvable extension of B. By Exercise 16.13 again, we can find a polynomial $g \in B[x]$ such that K is a splitting field for g over B, and then by Corollary 22.11 g is solvable by radicals over B. Therefore H is a solvable group by Theorem 22.24.

The proof of (2) is left to the reader as Exercise 22.4. $\qquad\square$

Remark 22.27. There is a more direct, purely group-theoretic proof of Lemma 22.26 (see Exercise 22.5). But the proof given above shows how a universal family of groups (the symmetric groups, in this case) can translate "obvious" facts about radicals to not-quite-so-obvious facts about groups, via Galois theory.

Corollary 22.28. *Let n be a positive integer. The following are equivalent:*
(1) Every polynomial of degree n over a field of characteristic 0 is solvable by radicals;
(2) S_n is a solvable group.

Proof. ($\Rightarrow$): Suppose (1) holds. By Proposition 22.25, there is a field F of characteristic 0 and a polynomial $f \in F[x] - F$ of degree n such that $G := \mathrm{Gal}(K/F) \cong S_n$, where K is a splitting field for f over F. Now f is solvable by radicals (by (1)), so G is a solvable group by Theorem 22.24.

($\Leftarrow$): Suppose that S_n is solvable. Let F be a field of characteristic 0, and let $f \in F[x]$ be a polynomial of degree n. Let K be a splitting field for f over F, and let R be the set of all roots of f in K. Then K is Galois over F, and we have $G := \mathrm{Gal}(K/F) \hookrightarrow \mathrm{Sym}(R) \hookrightarrow S_n$. So G is isomorphic to a subgroup of S_n. Since S_n is solvable by hypothesis, then G is also solvable, by Lemma 22.26. So f is solvable by radicals, by Theorem 22.24. $\qquad\square$

Remark 22.29. For a counterexample to the biconditional of Corollary 22.28 in a field of positive characteristic, see Exercise 22.18.

22.5   Which Groups Are Solvable?

It is high time that we look into methods for determining whether a given group is solvable, and in particular which of the symmetric groups S_n are solvable. In the definition of solvable group, we need a normal subgroup whose corresponding quotient group is abelian. Let G be a finite group. Recall from Example 10.7 and Definition 10.8 that the commutator subgroup of G has exactly these properties, and moreover is in a sense optimal in this regard. By forming commutator subgroups repeatedly, we can decide whether G is solvable; that is the content of the following result.

Lemma 22.30. *Let G be a finite group. Form a sequence $G_0, G_1, G_2, \ldots$ by setting $G_0 = G$ and taking G_j to be the commutator subgroup of G_{j-1} for each $j \geq 1$. This sequence must stabilize, and G is solvable iff we have $G_n = \{e_G\}$ for all sufficiently large n.*

Proof. From the definition of the commutator subgroup, for each $j \geq 1$ we have $G_{j+1} \trianglelefteq G_j$ and G_j/G_{j+1} is abelian. So $G_0 \supseteq G_1 \supseteq \cdots$. Since G is finite, the sequence must stabilize at some G_m: that is, we must have $G_n = G_m$ for all $n \geq m$. We proceed to prove the second statement of the result.

($\Rightarrow$): We prove the contrapositive. So suppose that $G_m \neq \{e_G\}$. Then $G_{m+1} = G_m$ by construction. Assume for a contradiction that G is solvable. Since $G_m \leq G$, then G_m is also solvable, by Lemma 22.26. So there is a sequence of subgroups of G_m,

$$\{e_G\} = H_0 \leq H_1 \leq H_2 \leq \cdots \leq H_r = G_m,$$

such that $H_j \trianglelefteq H_{j+1}$ and H_{j+1}/H_j is abelian for each $j \in \{0, \ldots, r-1\}$. We may remove any instances where $H_j = H_{j+1}$ and suppose that $H_j < H_{j+1}$ for each j (this uses the fact that G_m is non-trivial). Thus, in particular, we have $H_{r-1} \triangleleft G_m$ and G_m/H_{r-1} is abelian. Now since $G_{m+1} = G_m$, the commutator subgroup of G_m is itself, and the largest abelian quotient of G_m is the trivial group; but G_m/H_{r-1} is a non-trivial abelian quotient of G_m, a contradiction. This shows that G is not solvable.   ●

($\Leftarrow$): Suppose that $G_m = \{e_G\}$ for some m. Then we have $\{e_G\} = G_m \leq G_{m-1} \leq \cdots \leq G_0 = G$, with $G_j \trianglelefteq G_{j-1}$ for each $j \in \{1, \ldots, m\}$, and each G_{j-1}/G_j is abelian. This is what it means to say that G is solvable.   □

Example 22.31. S_1 is a trivial group, so the sequence $S_1 \leq S_1$ satisfies the conditions of Definition 22.18, and S_1 is solvable. This corresponds to the fact that a linear equation is solvable over a field.

Example 22.32. S_2 has order 2, which is prime, so S_2 is abelian. Hence the sequence $\{e\} \leq S_2$ shows that S_2 is solvable. Similarly, we see that every abelian group is solvable. We can see this directly using Definition 22.18, or from Lemma 22.30, since the commutator subgroup of an abelian group is trivial.

Let's go farther: Proposition 22.22 is a constructive result, which gives a formula for finding a generator of a solvable field extension starting with any element from outside of the ground field. If we apply this result to a root of an irreducible quadratic polynomial $f = ax^2 + bx + c \in F[x]$, where F is a field with $\mathrm{char}(F) \neq 2$, then the quadratic formula should materialize out of the mist. So let K be a splitting field for f over F, and let $\rho \in K$ be a root of f. Then $G := \mathrm{Gal}(K/F)$ has order 2, and is generated by an element σ. We have $f = a(x - \rho)(x - \sigma(\rho))$. Therefore, $\rho + \sigma(\rho) = -b/a$ and $\rho\sigma(\rho) = c/a$. Let ζ be a primitive root of $x^2 - 1$ in F, i.e., $\zeta = -1$, and set $\alpha = \omega_0(\rho) = (\sigma - \mu_{\zeta^0})(\rho) = \sigma(\rho) - \rho$. We should have $\alpha^2 \in F$; so we compute $\alpha^2 = (\sigma(\rho) - \rho)^2 = \sigma(\rho^2) - 2\rho\sigma(\rho) + \rho^2 = (-b/a)^2 - 4\rho\sigma(\rho)$ $= (b/a)^2 - 4c/a = (b^2 - 4ac)/a^2$. Indeed, $\alpha^2 \in F$. Since $\rho + \sigma(\rho) = -b/a$, we have $\rho = \frac{1}{2}((\rho+\sigma(\rho))+(\rho-\sigma(\rho))) = \frac{1}{2}(-b/a-\alpha) = \frac{1}{2}(-b/a\pm\sqrt{(b^2 - 4ac)/a^2})$
$$= \frac{-b \pm \sqrt{b^2 - 4ac}}{2a}.$$

Example 22.33. We have $|S_3| = 6$. By Sylow's Theorem, S_3 has a subgroup H of order 3. Now H must be abelian since 3 is prime; and also we must have $H \trianglelefteq S_3$ since $[S_3 : H] = 2$ (using Exercise 8.7). Furthermore, S_3/H has order 2, so must also be abelian. Thus the sequence $\{e\} \leq H \leq S_3$ shows that S_3 is solvable according to Definition 22.18. So a "cubic formula" should exist.

Example 22.34. Let $G = S_4$. Then $|G| = 24$, so G has at least one subgroup of order 8, by Sylow's Theorem.

First suppose that there is a unique Sylow 2-subgroup H of G. In this case, since any conjugate of H is also a subgroup of G of order 8, we conclude that $H \trianglelefteq G$. Now H is a 2-group, so H is solvable by Exercise 22.6. The quotient G/H has order 3, so must be cyclic, and therefore also solvable. By Lemma 22.26, we can say that G itself is solvable.

On the other hand, suppose that G has at least two distinct subgroups H and K of order 8. Set $L = H \cap K$. Then $24 = |G| \geq |HK| = |H| \cdot |K| / |H \cap K|$ (by Lemma 18.9) $= 64/|L|$, so $|L| \geq 64/24 = 8/3$. Also, we have $|L|$ divides 8 by Lagrange's Theorem, and $|L| < 8$ since $H \neq K$. Therefore, $|L| = 4$. Now L is normal in both H and K by Exercise 8.7. Thus $H, K \leq N_G(L) \leq G$. By Lagrange's Theorem, $|N_G(L)|$ is a multiple of 8 and a factor of 24, but not equal to 8, since H and K are distinct subsets of $N_G(L)$ of order 8. Therefore, $|N_G(L)| = 24$, so $N_G(L) = G$ and $L \trianglelefteq G$. Now L is a 2-group, hence solvable; and G/L has order 6, so is solvable by Exercise 22.7. So by Lemma 22.26, G is solvable.

We conclude that S_4 is solvable.

Remark 22.35. In fact, S_4 has 3 distinct subgroups of order 8. But notice how we did not need to perform any calculations in S_4 in Example 22.34. Our argument shows that every group of order 24 is solvable!

Remark 22.36. We could have omitted Examples 22.31 through 22.33, using the knowledge that $S_1 \hookrightarrow S_2 \hookrightarrow S_3 \hookrightarrow S_4$; since S_4 is solvable, then Lemma 22.26 assures us that the smaller S_j's are solvable as well. But it is perhaps more instructive to start small and build up to S_4, as we chose to do.

Example 22.37. Every time we prove another S_n to be solvable, we are also proving (via Corollary 22.28) that every polynomial of degree n is solvable by radicals (in characteristic 0, at least). Historically, the quadratic formula was known since ancient times; the cubic formula was discovered around the start of the Italian Renaissance; and the general quartic (fourth-degree) polynomial was solved by radicals within another half-century after that. But the general quintic, i.e. fifth-degree, polynomial remained unsolved through the time of Abel and Galois, more than 250 years later. Knowing what we know, we shall take a different tack with S_5 than with the previous symmetric groups.

What's new in S_5? One new feature is that S_5 has 5-cycles. Let H be the subgroup of S_5 generated by 5-cycles:

$$H := \langle \{(a_1, a_2, a_3, a_4, a_5) : 1 \le a_i \le 5, a_i\text{'s are distinct}\} \rangle. \qquad (22.6)$$

Suppose that $\sigma = (a_1, \ldots, a_5)$ and $\tau = (b_1, \ldots, b_5)$ are 5-cycles. Our strategy is to find σ and τ such that the commutator $\sigma \tau \sigma^{-1} \tau^{-1}$ is yet another 5-cycle; if we can produce an arbitrary 5-cycle in this fashion, then the commutator subgroup of H will have to be H itself, which will stop the descending series of Lemma 22.30 in its tracks and force S_5 to be unsolvable. So we wish to solve $\sigma \tau \sigma^{-1} \tau^{-1} = \omega$, where ω is a 5-cycle. Somewhat arbitrarily, let us try $\omega = (b_1, b_4, b_2, b_3, b_5)$. Then we want

$$\sigma \tau \sigma^{-1} = \omega \tau = (b_1, b_3, b_2, b_5, b_4). \qquad (22.7)$$

Since $\omega \tau$ is another 5-cycle, and since any two 5-cycles are conjugate in S_5 (by Exercise 19.7(b)), it must be possible to solve this equation for σ; but the question is whether σ can be chosen to be a 5-cycle too. We recall from Exercise 19.7(a) that the conjugate of σ by τ is $\sigma \tau \sigma^{-1} = (\sigma(b_1), \ldots, \sigma(b_5))$. The reader is invited at this point to find a choice for σ which works. Because of the importance of the result, we shall state and prove it formally:

Proposition 22.38. *S_5 is not solvable.*

Proof. Let H be the subgroup of S_5 generated by 5-cycles:

$$H := \langle \{(a_1, a_2, a_3, a_4, a_5) : 1 \le a_i \le 5, a_i\text{'s are distinct}\} \rangle. \qquad (22.8)$$

Let C be the commutator subgroup of H. Let ω be an arbitrary 5-cycle in S_5, and write $\omega = (c_1, c_2, c_3, c_4, c_5)$. Set $\sigma = (c_1, c_3, c_5, c_4, c_2)$ and $\tau =$

$(c_1, c_3, c_4, c_2, c_5)$. We compute

$$\sigma\tau\sigma^{-1}\tau^{-1} = (\sigma(c_1), \sigma(c_3), \sigma(c_4), \sigma(c_2), \sigma(c_5))\tau^{-1} \qquad (22.9)$$

$$= (c_3, c_5, c_2, c_1, c_4)\tau^{-1} \qquad (22.10)$$

$$= (c_1, c_2, c_3, c_4, c_5) \qquad (22.11)$$

$$= \omega. \qquad (22.12)$$

Thus, $\omega \in C \leq S_5$. So C contains all 5-cycles in S_5, hence $H \leq C$, by construction of H. Now we also have $C \leq H$ (in fact, $C \trianglelefteq H$). Therefore, $C = H$. Thus, the series of commutator subgroups of H stabilizes at H. Since $H \neq \{e\}$, then H is not solvable, by Lemma 22.30. Since $H \leq S_5$, now Lemma 22.26 tells us that S_5 is not solvable either. $\qquad\square$

22.6   The Grand Finale

At last we can completely classify the degrees n for which every polynomial equation of degree n over a field of characteristic zero in a single variable is solvable by radicals.

Theorem 22.39. *Let n be a positive integer. If $n \leq 4$, then every polynomial of degree n over a field of characteristic 0 is solvable by radicals. But if $n \geq 5$, then there exists a field F of characteristic 0 and a polynomial f over F of degree n such that f is not solvable by radicals over F. In particular, the general fifth-degree polynomial is not solvable by radicals, even in characteristic 0.*

Proof. Corollary 22.28 together with Examples 22.31 through 22.34 show that polynomials of degree at most 4 are solvable by radicals over a field of characteristic 0. But if $n \geq 5$, then we have $S_5 \hookrightarrow S_n$, so Proposition 22.38 and Lemma 22.26 complete the argument. $\qquad\square$

22.7   Exercises

Exercise 22.1. Verify that the real number β in Example 22.3 is a radical expression over $\mathbf{Q}$ by finding an appropriate tower of field extensions starting with $\mathbf{Q}$ which satisfies the conditions of Definition 22.2.

Exercise 22.2. Suppose that F is a field, α is a radical expression over F, and β is a simple radical over $F[\alpha]$. Prove that β is a radical expression over F.

Exercise 22.3. Let F be a field, let K be a finite normal extension field of

Exercises 281

F, and let $f \in F[x] - (0)$ be a non-zero polynomial. Suppose that L is an extension field of K such that f factors completely in L, and let R be the set of all roots of f in L. Prove that $K[R]$ is normal over F.

Exercise 22.4. Prove part (2) of Lemma 22.26 using the Galois correspondence (in the same spirit as the proof of part (1) given in the text).

Exercise 22.5. In this exercise, you will give an alternative proof of Lemma 22.26, using "pure" group theory only. Suppose that G is a finite group.

(a) Suppose that G is solvable, and let $H \leq G$. Let $\{e_G\} = G_0 \leq G_1 \leq \cdots \leq G_r = G$ be a tower of subgroups of G such that $G_{i-1} \trianglelefteq G_i$ and G_i/G_{i-1} is abelian. Show that $H_i := H \cap G_i$ has the same properties, and thus that H is solvable.

(b) Suppose that $N \trianglelefteq G$ and both N and G/N are solvable. Use the definition of a solvable group together with Exercises 7.17 and 19.3 to prove that G is solvable.

Exercise 22.6. Prove that every p-group is solvable, where p is a prime number. Hint: Proposition 19.14 may be helpful.

Exercise 22.7. Let p be a prime number. Prove that every group of order $2p$ is solvable.

Exercise 22.8. Let K/F be a finite Galois extension. Set $G = \text{Gal}(K/F)$. Suppose that $F \leq F_1 \leq F_2 \leq K$. Let $G_j = \text{Gal}(K/F_j)$ for $j \subset \{1, 2\}$. Prove that F_2 is Galois over F_1 iff $G_2 \trianglelefteq G_1$, and in that case we have $\text{Gal}(F_2/F_1) \cong G_1/G_2$. (You will only need basic Galois theory from Chapter 16.)

Exercise 22.9. Let F be a field of characteristic 0, and let K be a finite normal extension field of F. Prove or disprove that if K is solvable over F, then K is a pure radical extension of F.

Exercise 22.10. Let $\sigma : G \to H$ be a surjective group homomorphism. Suppose that $N \trianglelefteq G$, and set $K = \sigma(N)$. Note that $K \trianglelefteq H$ by Exercise 7.8. Let $\nu : H \to H/K$ be the natural map, and set $\omega = \nu \circ \sigma$. Prove that ω induces a surjective group homomorphism from G/N to H/K. Hint: See Exercises 7.9 and 9.15.

Exercise 22.11. Compute the coefficients $a_0, \ldots, a_{n-1}$ of the polynomial f from Proposition 22.25 in terms of the variables $r_1, \ldots, r_n$, in the cases $n = 2$ and $n = 3$. (These polynomials are known as the *elementary symmetric functions* in the r_j's.)

Exercise 22.12 (S_n is generated by transpositions). Let $n \in \mathbf{Z}^+$. A *transposition* in S_n is defined to be any 2-cycle; that is, any element $\sigma \in S_n$ such that in cycle notation we have $\sigma = (a, b)$ for some $a, b \in \{1, 2, \ldots, n\}$ with $a \neq b$.

(a) Prove that for any r-cycle $(x_1, x_2, \ldots, x_r)$ with $r \geq 2$, we have

$$(x_1, x_2, \ldots, x_r) = (x_{r-1}, x_r)(x_{r-2}, x_r) \cdots (x_1, x_r).$$

Thus, an r-cycle is a product of $r - 1$ transpositions.

(b) Let $T = \{\sigma \in S_n : \sigma \text{ is a transposition}\}$. Prove that $\langle T \rangle = S_n$. (See Exercise 19.6.)

Exercise 22.13 (The Alternating Group A_n). Let $n \in \mathbf{Z}^+$, and let $\tau = (a, b)$ be a transposition in S_n (see Exercise 22.12).

(a) Suppose that a and b both belong to an r-cycle c, say $c = (x_1, \ldots, x_r)$ with $x_i = a$, $x_j = b$, and $1 \le i < j \le r$. Prove that we have

$$c\tau = (x_1, x_2, \ldots, x_i, x_{j+1}, x_{j+2}, \ldots, x_r)(x_{i+1}, x_{i+2}, \ldots, x_j),$$

the product of two disjoint cycles of lengths $r - (j - i)$ and $j - i$.

(b) Suppose that a and b belong to two disjoint cycles c and d, say $c = (x_1, \ldots, x_r)$ and $d = (y_1, \ldots, y_s)$ with $x_i = a$ and $y_j = b$. Note that without loss of generality, we can suppose that $j = s$. Prove that we then have

$$cd\tau = (x_1, x_2, \ldots, x_i, y_1, y_2, \ldots, y_s, x_{i+1}, x_{i+2}, \ldots, x_r),$$

a single $(r + s)$-cycle.

(c) Define a function $f : S_n \to \mathbf{Z}$ by the formula $f(\sigma) = \sum_{j=1}^k (\ell_j - 1)$, where the cycle structure of σ is $\mathrm{cs}(\sigma) = (\ell_1, \ldots, \ell_k)$ (see Exercise 19.7). Use the previous parts of this exercise to prove that for any $\sigma \in S_n$ and any transposition $\tau \in S_n$, we have $f(\sigma\tau) = f(\sigma) \pm 1$, and thus $f(\sigma\tau) \equiv f(\sigma) + f(\tau)$ (mod 2). Use Exercise 22.12 to conclude that f induces a homomorphism $\pi : S_n \to (\mathbf{Z}/2\mathbf{Z}, +)$, $\pi(\sigma) = f(\sigma) + 2\mathbf{Z}$.

(d) Define $A_n := \{\sigma \in S_n : \sigma$ can be written as a product of an even number of transpositions$\}$. Prove that $A_n = \ker(\pi)$ and that if $n \ge 2$, we have $[S_n : A_n] = 2$. Note that we have $A_n \trianglelefteq S_n$. We call A_n the *alternating group on the set* $\{1, \ldots, n\}$.

(e) Suppose that r is an odd integer and $3 \le r \le n$. Let H_r be the subgroup of S_n generated by all r-cycles. Prove that $H_r = A_n$. Thus, the group H considered in the proof of Proposition 22.38 is none other than A_5, and we can conclude that A_5 is not solvable.

Exercise 22.14 (An inseparable extension). Let $F = \mathbf{Z}/p\mathbf{Z}$ be the field of integers modulo p, where p is prime in $\mathbf{Z}$. Let $K = F(t)$ be the field of rational functions over F in the variable t. Let $f = x^p - t \in K[x]$, and let L be a splitting field for f over K.

(a) Prove that f is irreducible in $K[x]$.

(b) Let α be a root of f in L. Prove that $f = (x - \alpha)^p$. Conclude that $f = \mathrm{Irr}(\alpha, K, x)$ and thus that L is not separable over K; we say that L is an *inseparable* extension of K. Note that L is a solvable, normal extension of K, but L is not Galois over K.

Exercise 22.15. Suppose that K/F is any finite extension of finite fields (that is, $F \le K$ are finite fields and $[K : F] < \infty$). Prove that K is a solvable extension of F.

Exercise 22.16. Suppose that F is a field, and α is a radical over F; say $\alpha^n \in F$ with $n \in \mathbf{Z}^+$. Also suppose that the polynomial $x^n - 1$ factors completely over F.

(a) Let $f = x^n - \alpha^n \in F[x]$, and suppose that $f = f_1 f_2$ with $f_j \in F[x]$. Set $d_j = \deg(f_j)$. Prove that we have $\alpha^{d_j} \in F$.

(b) Let $m = \min\{j \in \mathbf{Z}^+ : \alpha^j \in F\}$. Prove that m divides n and that $\operatorname{Irr}(\alpha, F, x) = x^m - \alpha^m$.

(c) Suppose that $m = ab$ with $a, b \in \mathbf{Z}$ and $1 < a, b < m$. Let $\beta = \alpha^a$. Prove that $\operatorname{Irr}(\beta, F, x) = x^b - \beta^b$ and $\operatorname{Irr}(\alpha, F[\beta], x) = x^a - \alpha^a$.

(d) Use the preceding parts of this exercise to prove that if $m > 1$ and $m = \prod_{j=1}^r p_j$, where each p_j is prime, then we can write $F = F_0 < F_1 < \cdots < F_r = F[\alpha]$ with $F_j = F_{j-1}[\alpha^{m_j}]$, $m_j := m/(p_1 p_2 \cdots p_j)$, $[F_j : F_{j-1}] = p_j$, and $\operatorname{Irr}(\alpha_j, F_{j-1}, x) = x^{p_j} - \alpha^{m_j p_j}$.

Exercise 22.17. Suppose that K is a pure radical extension field of F, with $\operatorname{char}(F) = p > 0$. Also suppose that for each positive integer d, the polynomial $x^d - 1$ factors completely over F.

(a) Prove that we can write $F = L_0 < L_1 < \cdots < L_t = K$, where $L_j = L_{j-1}[\alpha_j]$ and $\operatorname{Irr}(\alpha_j, L_{j-1}, x) = x^{p_j} - \alpha_j^{p_j}$ for some elements $\alpha_j \in K$ and primes p_j. (Use Exercise 22.16.)

(b) Suppose that $p_j = p$ and $p_{j+1} = q \neq p$. Set $\alpha = \alpha_j$ and $\beta = \alpha_{j+1}$, so that $L_j = L_{j-1}[\alpha]$, $L_{j+1} = L_j[\beta]$, $\operatorname{Irr}(\alpha, L_{j-1}, x) = x^p - \alpha^p$, and $\operatorname{Irr}(\beta, L_j, x) = x^q - \beta^q$. Note that we have $\beta^q = f(\alpha)$ for some polynomial $f \in L_{j-1}[x]$. Set $b = f(\alpha^p)$. Prove that we have $\operatorname{Irr}(\beta^p, L_{j-1}, x) = x^q - b$, $\operatorname{Irr}(\beta, L_{j-1}[\beta^p], x) \in \{x^p - \beta^p, x - \beta\}$, and $\operatorname{Irr}(\alpha, L_{j-1}[\beta], x) \in \{x^p - \alpha^p, x - \alpha\}$. Use this to show that without loss of generality, we may suppose that all of the degree-p extensions in the tower of L_i's occur in a consecutive sequence at the top of the tower.

(c) Use Lemma 15.25 and Exercise 14.11 to show that every element $\gamma \in K$ can be written as $\gamma = h(\alpha_1, \ldots, \alpha_t)$ for a unique polynomial $h \in F[x_1, \ldots, x_t]$ such that $\deg_{x_j}(h) < p_j$ for each j.

(d) Let $S = \{j : 1 \leq j \leq t \text{ and } p_j \neq p\}$. Let $\gamma \in K$, and let $S_r = S \cup \{1, 2, \ldots, t - r\}$ for $r \in \mathbf{N}$. Prove by induction on r that we have $\gamma^{p^r} \in F[\{\alpha_j : j \in S_r\}]$. Hint: Use Corollary 21.21.

(e) Prove that for all $\gamma \in K$, the index $[F[\gamma^{p^t}] : F]$ is not a multiple of p.

Exercise 22.18. Let F be an algebraically closed field of characteristic 2. Let $K = F(t)$ be the field of rational functions over F in the variable t, and let $f = x^2 + x + t \in K[x]$. Let L be a splitting field for f over K, and let α be a root of f in L.

(a) Prove that L is a Galois extension of K with $[L : K] = 2$, so that $\operatorname{Gal}(L/K) \cong S_2$ is a solvable group.

(b) Prove by induction on n that we have $\alpha^{2^n} = \alpha + t + t^2 + t^4 + \cdots + t^{2^{n-1}}$ for all $n \in \mathbf{N}$. Use this formula to show that $K[\alpha^{2^n}] = L$ for all $n \in \mathbf{N}$.

(c) Prove that L is not a solvable extension of K. (Exercise 22.17 may be useful.)

Exercise 22.19. If the roots of a monic polynomial are algebraically independent over a given field, then it makes sense that the coefficients of that polynomial should also be algebraically independent over that field. But prove this. Specifically, under the hypotheses of Proposition 22.25, prove that $C[a_0, \ldots, a_{n-1}]$ is a polynomial ring in n variables over C; that is,

$\{a_0, \ldots, a_{n-1}\}$ is algebraically independent over C. Suggestion: Form a legitimate polynomial ring $C[A_0, \ldots, A_{n-1}] =: T$, let $h = x^n + \sum_{j=0}^{n-1} A_j x^h \in T[x]$, let Λ be a splitting field for h over $k(T)$, let $R_1, \ldots, R_n$ be the roots of h in Λ, and consider the map sending r_j to R_j over C (see Exercise 14.11).

Exercise 22.20. Under the hypotheses of Proposition 22.25, use the following steps to prove that $S_0 := S \cap F = C[a_0, \ldots, a_{n-1}] =: \Gamma$.

(a) Prove that $\Gamma \subseteq S_0$.

(b) Prove that for each $d \in \mathbf{N}$, the sets

$$H_d := \{w \in S_0 \ : \ \deg_{\text{tot}}(t) = d \text{ for every term } t \text{ of } w\}$$

and

$$\tilde{H}_d := \{y \in \Gamma \ : \ \deg_{\text{tot}}(t) = d \text{ for every term } t \text{ of } y\}$$

are C-vector spaces, where $\deg_{\text{tot}}$ denotes the total degree, that is, the sum of the degrees in $r_1, \ldots, r_n$. Note: a polynomial in which every term has the same total degree is called *homogeneous*.

(c) Prove that the set $\{a_0^{e_0} \cdots a_{n-1}^{e_{n-1}} \ : \ e_j \in \mathbf{N}, \sum(n-j)e_j = d\}$ is a basis of $\tilde{H}_d$ over C. Note: Exercise 22.19 may be useful here.

(d) Prove that the set $\{\psi(r_1^{e_1} \cdots r_n^{e_n}) \ : \ e_j \in \mathbf{N}, e_1 \leq \cdots \leq e_n, \sum e_j = d\}$ spans H_d over C, where $\psi \ : \ S \to \Gamma$ is given by the formula $\psi(w) = \sum_{\sigma \in G} \sigma(w)$ and $G := \text{Gal}(K/F)$.

(e) Show that the sets $\{(e_1, \ldots, e_n) \ : \ e_j \in \mathbf{N}, \sum(n-j)e_j = d\}$ and $\{(e_1, \ldots, e_n) \ : \ e_j \in \mathbf{N}, e_1 \leq \cdots \leq e_n, \sum e_j = d\}$ have the same size. Note: this exercise would be much easier if we knew that every element of S were integral over Γ, but this is a bit beyond our scope.

Exercise 22.21. A curious student once observed that when we derived the quadratic formula in Example 22.32, we assumed that our quadratic polynomial was *irreducible*, and yet the quadratic formula, as derived in Example 22.8, works for *all* quadratic polynomials over a field of characteristic different from 2; indeed, we say *the* quadratic formula, implying there is just one! Why should this be true, and does it generalize to cubics and quartics? We explore these questions in this exercise.

Adopt the hypotheses of Proposition 22.25, with the additional assumptions that C is algebraically closed and of characteristic 0. Further, suppose that f is solvable by radicals over C, which we now know is equivalent to saying $n \leq 4$. Let $g \in C[x]$ be monic of degree n, and write $g = x^n + \sum_{j=0}^{n-1} \alpha_j x^j$ with $\alpha_j \in C$. Write $F = F_0 \leq F_1 \leq \cdots \leq F_m = K$ where F_j/F_{j-1} is Galois of order p_j, a prime number, which is possible since K/F is a solvable Galois extension. Let $S_j = S \cap F_j$.

(a) Prove that for all $\sigma \in \text{Gal}(F_j/F_{j-1})$, we have $\sigma(S_j) \subseteq S_j$.

(b) Prove that every element of S_j is a C-linear combination of p_j^{th} roots of elements of S_{j-1}, with each radical in S_j. Hint: Exercise 14.9 will be useful; look at the technique used in the proof of Proposition 22.22.

(c) Write $g = \prod_{j=1}^{n}(x - \beta_j)$ with $\beta_j \in C$. Let $\tau \ : \ S \to C$ be the ring homomorphism over C sending r_j to β_j. Prove that $\tau(a_j) = \alpha_j$.

(d) Deduce that we can write a root of g using the "same" radical expression that works for f, only substituting the coefficients of g for those of f. (Exercise 22.20 may help here.)

Exercise 22.22. Derive a "cubic formula," i.e., a formula for the roots of a general cubic polynomial involving radical expressions. You may suppose that the coefficients of the polynomial lie in a field whose characteristic does not divide 6.

23

Projects

23.1 Gyrogroups

Prerequisites. Chapter 3 at a minimum; it would be better to have gone through Chapter 9 to understand the notion of automorphism in the context of group theory.

When we chose the axioms that define groups (or other algebraic structures such as rings and vector spaces), what led us to these particular choices? What if we had chosen a different set of axioms? Experience has shown that the group axioms make a good choice because of the wide variety of applications (so many things turn out to be groups) combined with the deep consequences (groups have a lot of structure). But no one can stop us from selecting an other set of axioms and naming a new kind of algebraic structure. Many such structures have been investigated, and in general each new kind of structure gets its own terminology and its own theory.

In this project, we explore a type of structure called a *gyrogroup* which was proposed around 1992. Before we can give the definition of a gyrogroup, we need to discuss automorphisms of objects more general than groups.

Definition 23.1. Let $+$ be a binary operation on a set G. An *automorphism* of $(G, +)$ is a bijective function $\alpha : G \to G$ such that $\forall x, y \in G, \alpha(x + y) = \alpha(x) + \alpha(y)$. The set of all automorphisms of $(G, +)$ is written $\mathrm{Aut}(G, +)$.

Task 23.1.1. Suppose that $\alpha, \beta \in \mathrm{Aut}(G, +)$. Prove that α^{-1} and $\beta \circ \alpha$ are also in $\mathrm{Aut}(G, +)$.

We next present the original definition of a gyrogroup as stated by Ungar in [15]. We follow this work in using the notation $+$ for the binary operation, even though this operation may not be commutative. We also agree to write the image of an element z under an automorphism α as αz instead of $\alpha(z)$, to reduce the number of parentheses needed.

Definition 23.2. A *gyrogroup* is a triple $(G, +, g)$ where $+$ is a binary operation on G and $g : G^2 \to \mathrm{Aut}(G, +)$, satisfying the following axioms for all $x, y, z \in G$:

    G1. $x + y \in G$
    G2a. $x + (y + z) = (x + y) + g(x, y)z$

DOI: 10.1201/9781003252139-23

G2b. $(x + y) + z = x + (y + g(y, x)z)$

G3. $x + y = g(y, x)(y + x)$

G4. $(G, +)$ has an identity element, 0

G5. x has an inverse element, written $-x$

G6. $g(0, y) = \mathrm{id}_G$, the identity function on G

G7. $g(x + y, y) = g(x, y)$

G8. $g(x, y)z = -(x + y) + (x + (y + z))$

Remark 23.3. The $+$ operation of a gyrogroup is not required to be associative. Indeed, $+$ is associative if $g(x, y)$ is the identity function on G for all x, y, and we shall see that the converse also holds.

When we define a new kind of structure, we usually want to give the fewest axioms possible to get the job done. In addition to saving space, this allows us to establish more easily that a given object satisfies these axioms. It is also pleasing to understand when some axioms follow logically from others and thus are redundant. In the case of gyrogroups, it turns out that some of the original axioms are redundant. To prove this, we will define a "weak gyrogroup" to be a gyrogroup with some of the axioms removed, and then you will prove that every weak gyrogroup is actually a gyrogroup.

Definition 23.4. A *weak gyrogroup* is a triple $(G, +, g)$ where $+$ is a binary operation on G and $g : G^2 \to \mathrm{Aut}(G, +)$, satisfying Axioms G2a, G3, G4, G5, and G7.

Notice that we removed Axioms G1, G2b, G6, and G8 from the list of gyrogroup axioms when we defined a weak gyrogroup. Thus every gyrogroup is also a weak gyrogroup, and our goal is to establish the converse.

To start with, observe that Axiom G1 says that G is closed under $+$, which is guaranteed from the hypothesis that $+$ is a binary operation on G. Thus, we automatically have:

Lemma 23.5. *Every weak gyrogroup satisfies Axiom G1.* $\quad\square$

We can also guarantee the uniqueness of the 0 element based on Lemma 3.33:

Lemma 23.6. *In a weak gyrogroup, the identity element is unique.* $\quad\square$

A natural next question is whether inverse elements are unique in a weak gyrogroup. Looking at the corresponding result for groups, Lemma 3.34, we may be disappointed to see that the proof relies on the associative property, which we don't expect to have in a weak gyrogroup. But we will not let this stop us yet. The key idea is to first establish a cancellation law:

Task 23.1.2. Prove that the *Left Cancellation Law* is true in a weak gyrogroup: that is, prove

Lemma 23.7 (Left Cancellation Law). *Let G be a gyrogroup. Suppose that $a, b, c \in G$ and $c + a = c + b$. Then $a = b$.*

Hint: Use Axiom G2a for appropriate choices of x, y, and z, together with the fact that automorphisms are bijective.

We get the following corollary immediately:

Corollary 23.8. *In a weak gyrogroup, inverses are unique. More precisely, let G be a weak gyrogroup and let $a \in G$. If $b, c \in G$ and $a + b = 0 = a + c$, then $b = c$.* □

Now we are justified in using the notation $-a$ for *the* inverse of a in a weak gyrogroup. The following task should also be straightforward now.

Task 23.1.3. Let $a \in G$, where G is a weak gyrogroup. Prove that $-(-a) = a$.

We can further exploit Left Cancellation to get another Gyrogroup Axiom:

Task 23.1.4. Let G be a weak gyrogroup. Prove Axiom G6 for G. Hint: Use Axiom G2a and Left Cancellation.

From this, we get the following:

Corollary 23.9. *Let G be a weak gyrogroup, and let $a \in G$. Then we have $g(-a, a) = id_G$.*

Proof. We have $g(-a, a) = g(-a + a, a)$ by Axiom G7. Now use Axiom G6 (which is justified by Task 23.1.4) to get the desired conclusion. □

We are now in a position to begin to solve equations in a weak gyrogroup. To solve for x in the equation

$$a + x = b,$$

it is natural to add $-a$ on the left to both sides. The next task shows that this always works.

Task 23.1.5. Let G be a weak gyrogroup, and let $a, b \in G$. Prove that for all $x \in G$, we have $-a + (a + x) = x$. Conclude that the equation $a + x = b$ has the unique solution $x = -a + b$. Note that you must show that this choice of x does satisfy the given equation, and is the only possible solution.

The fact that the $+$ operation need not be associative can present challenges to our notation as well as to our habits of thinking (since we have perhaps become used to all of our operations being associative). Recall the discussion of parenthesization from Chapter 3 beginning with Expression 3.2, and also in Exercise 3.10. To make our notation clearer, we will use the left-addition functions defined by elements of G, as in Cayley's Theorem (Theorem 10.9):

Definition 23.10. Let G be a set with a binary operation $+$. For an element $x \in G$, we define the *left-addition function* associated with x to be the function $\pi_x : G \to G$ given by the formula $\pi_x(y) = x + y$ for $y \in G$.

With this notation, for example, we can write $x + (y + z)$ as

$$\pi_x \pi_y(z), \tag{23.1}$$

where the (unwritten) operation between π_x and π_y is composition of functions. Notice that $(x + y) + z$ can be written as $\pi_{x+y}(z)$, but not in general as $\pi_x \pi_y(z)$ unless $+$ is associative. To practice using this notation, we start with an easy task.

Task 23.1.6. Let G be a weak gyrogroup. Prove that $\pi_0 = \text{id}_G$, the identity function on G.

In the context of group theory, the functions π_x were permutations of the underlying set of the group. Next you will prove the same property for weak gyrogroups.

Task 23.1.7. Let G be a weak gyrogroup. Prove that for all $a \in G$, we have $\pi_a \pi_{-a} = \pi_{-a} \pi_a = \text{id}_G$. (You should find the result of Task 23.1.5 helpful.) Conclude that each π_a is a permutation of the set G, and $\pi_{-a} = \pi_a^{-1}$.

Now that we know the π functions are invertible, we can establish another gyrogroup axiom.

Task 23.1.8. Let G be a weak gyrogroup. Show that for all $x, y \in G$, we have $\pi_x \pi_y = \pi_{x+y} g(x, y)$ by using Axiom G2a. Solve for $g(x, y)$ in this equation, and use the result to prove that Axiom G8 holds in G.

We are also able to establish our earlier statement from Remark 23.3 about when the $+$ operation is associative, even in a weak gyrogroup:

Task 23.1.9. Let G be a weak gyrogroup. Prove that $+$ is associative iff for all $x, y \in G$, we have $g(x, y) = \text{id}_G$.

The following task helps to explain why we use additive notation in gyrogroups.

Task 23.1.10. Let G be a set with an associative binary operation $+$. Prove that the following are equivalent:

(1) $(G, +)$ is an abelian group;
(2) $(G, +, \text{id}_G)$ is a gyrogroup;
(3) $(G, +, g)$ is a gyrogroup for some function $g : G^2 \to \text{Aut}(G)$.

We will next establish a few identities to help prepare for the proof of Axiom G2b, which is the only axiom left.

Task 23.1.11. Let G be a weak gyrogroup.
(a) Prove that for all $x, y \in G$, we have

$$\pi_{(x+y)+y}^{-1} \pi_{x+y} = \pi_{x+y}^{-1} \pi_x.$$

(Axiom G7 and the formula for $g(x, y)$ from Task 23.1.8 may be helpful.)
(b) Use the result of part (a) above to prove that for all $x, y \in G$, we have

$$\pi_{b+(a+b)} = \pi_b \pi_a \pi_b, \tag{23.2}$$

where $a = -x$ and $b = x + y$.

(c) Prove that for all $a, b \in G$, there exist $x, y \in G$ such that $a = -x$ and $b = x + y$. Conclude that Equation 23.2 is true for all $a, b \in G$.

(d) Evaluate Equation 23.2 at an element $c \in G$ to show that for all $a, b, c \in G$, we have

$$(b + (a + b)) + c = b + (a + (b + c)). \tag{23.3}$$

Task 23.1.12. Let G be a weak gyrogroup.

(a) Prove that Axiom G2b holds in G if and only if for all $x, y \in G$, we have $g(y, x)g(x, y) = \mathrm{id}_G$. Hint: In Axiom G2a, we can replace z with $g(y, x)z$.

(b) Use the previous results to prove that Axiom G2b does hold in G.

Putting together what we have proved, we can now state:

Proposition 23.11. *Every weak gyrogroup is a gyrogroup.* $\qquad\square$

Remark 23.12. In [14], Sabinin noted that the original axioms proposed by Ungar in [15] contained redundancies. The techniques used in [14], however, will likely not be familiar to the reader without a background in an area of algebra known as loop theory; see [9] for this. Thus the present project is an attempt to accomplish the main goals of [14] without the use of loop theory.

23.2 Kaleidoscopes

Prerequisites. Chapter 5.

Since the "Kaleidoscope Principle" is a theme of this book, we owe it to ourselves to understand how actual kaleidoscopes work. It's all done with mirrors! To start with, we will therefore study the physics and optics of mirrors, but with a very limited point of view, and from a mathematical framework.

The setting is a 3-dimensional space $\mathbf{R}^3$. We assume that the following objects exist in our space:

(1) **The ground**, which we assume is an infinite plane; for convenience, we take the ground to be the plane $z = 0$, i.e., the x, y-plane, which we write as $\mathbf{R}^2$. We may think of each point of the ground as having a color.

(2) **Rays of light**. We assume that these are points (physicists call them *photons*) which travel in a straight line until they hit other objects.

(3) **A viewer**. We assume the viewer is located at a fixed position above the ground. We really only care about one of the viewer's eyes which we assume is observing the ground; this corresponds to the fact that a typical kaleidoscope only lets the viewer peek into the instrument with one eye at a time. We take the viewer's eye position to be $(0, 0, h)$ for a fixed number $h > 0$.

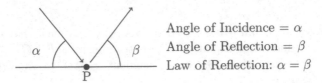

FIGURE 23.1: The Law of Reflection: side view

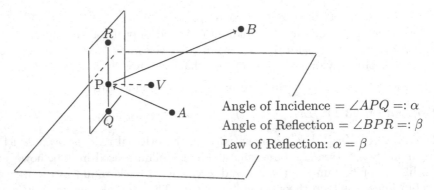

FIGURE 23.2: The Law of Reflection: perspective view

(4) **Mirrors.** We assume that each mirror is of the form $T \times (0, h]$ where $T \subset \mathbf{R}^2$. That is, each mirror is the vertical extension of a 2-dimensional shape just above the ground up until eye-level. For now, we will restrict T to be a line segment, so the mirror is a vertical rectangle.

The way that light behaves when it hits a mirror is described by the *Law of Reflection*, which says that the angle of incidence equals the angle of reflection (see Figure 23.1). More specifically, suppose that a photon hits a flat mirror at point P; let α be the angle made by the incoming light ray with the mirror; let A be a point on the incoming light ray (not equal to P), and let V be a point directly above P with respect to the mirror (that is, the line segment $\overline{VP}$ is perpendicular to the mirror). Then the outgoing light ray also makes angle α with the mirror, and lies in the plane containing the points A, V, and P (see Figure 23.2).

Sometimes the effect of a mirror is described by saying that the viewer's eye (or brain) is "fooled" into thinking that the view they see in the mirror represents an actual scene behind the mirror, a scene from a "mirror world," because the brain expects light to travel in a straight line and is not programmed to understand mirrors. But this idea of an actual scene existing behind the mirror is very useful (although we stop short of believing that, like Alice, one can step through the looking-glass). It is not just people's brains that see this effect: if for example we put a camera in place of the viewer's eye, then the camera would record the scene as if the mirror world existed; mathematically, it produces the same result. To show this is your first task:

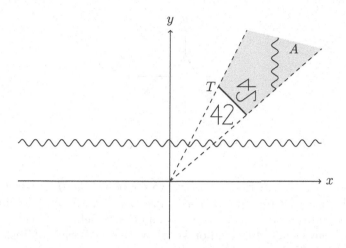

FIGURE 23.3: The Mirror-World Effect: single mirror; the region A is shaded

Task 23.2.1. Suppose that our scene S contains just one mirror $M = T \times (0, h]$, where T is a line segment. Let A be the subset of $\mathbf{R}^2$ which is behind the mirror with respect to the viewer: that is, the union of the rays from the origin to T minus the line segments from the origin to T; see Figure 23.3. Let ℓ be the line containing T, and let ρ . $\mathbf{R}^2 \to \mathbf{R}^2$ be reflection about the line ℓ. Define f · $\mathbf{R}^2 \to \mathbf{R}^2$ by

$$f(P) - \begin{cases} P, & \text{if } P \notin A; \\ \rho(P), & \text{if } P \subset A. \end{cases}$$

Show that a light ray traveling on the line from the viewer's eye toward a point $P \in \mathbf{R}^2$ will end up hitting the point $f(P)$ on the ground. Conclude that the viewer will see exactly the same view if we remove the mirror M and replace the color of each point $P \in \mathbf{R}^2$ with the color of $f(P)$. Note that light seen by the viewer must actually travel in the opposite direction, from the ground to the eye, but light paths are reversible! Also note that in Figure 23.3, the "snaky" horizontal line and the "42" are in the "real" world, directly viewed by the observer, while the vertical snake and the backwards vertical 42 are reflected images. In particular, note that the reflected snake ends abruptly at the boundaries of A; this corresponds to the fact that the function f is not continuous.

To get more interesting effects, we can use more than one mirror at a time. With multiple mirrors, a light ray may be reflected (or "bounce") multiple times on its journey from the ground to the viewer. This lets us see the real power of the mirror world concept. Note that our assumptions imply that for any finite number of mirrors, any given light ray will bounce only finitely many times between the ground and the viewer's eye.

Task 23.2.2. Suppose that the scene S contains any finite number of mirrors. Let $\mathcal{M}$ be the set of all mirrors M which can be seen directly by the viewer:

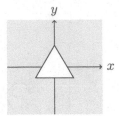

FIGURE 23.4: Standard kaleidoscope configuration

that is, the line segments from the origin to the points on $\mathcal{M}$ intersect no other mirrors (in case of a partially blocked mirror, we can divide the mirror into two smaller mirrors and then use this definition). Define a function f as in Task 23.2.1 but using $\mathcal{M}$ instead of M. Show that the result of Task 23.2.1 is still true if in addition to replacing colors, we also replace mirrors according to the function f: so we place a mirror above point Q iff a mirror occurs above point $f(Q)$. In other words, the mirror world can contain mirrors too!

We are now ready to discuss kaleidoscopes. A standard kaleidoscope can be described within our framework as a group of 3 mirrors that form an equilateral triangle centered at the origin when viewed from above (see Figure 23.4). In the figure, only the white area inside the triangle is directly viewed by the observer; the shaded area will consist entirely of reflections from this central area. To be definite, we take the equilateral triangle to have sides of length 1, so the height of the triangle is $\sqrt{3}/2$. By using the result of Task 23.2.2 repeatedly, we can understand the visual effects that this configuration of mirrors will produce.

Task 23.2.3. Show that the kaleidoscope shown on the left side of Figure 23.5 is equivalent to the configuration shown on the right side of that figure. Note that the 3 triangular regions outside of the original triangle in the right-hand illustration represent the part of the final image produced by light rays which took exactly 1 bounce. The line segments at the boundary of the shaded region represent mirrors; reflected images have not been shown in the shaded region. The image "42" is for illustrative purposes only; your solution should be general, giving a function k that plays the role of the function f in Task 23.2.1; here, you need only define k in the unshaded region.

Task 23.2.4. Continue Task 23.2.3 to define a function k on the region corresponding to 2 bounces; 3 bounces; and 4 bounces (by now you should get the big picture). Show that the final image corresponds to a partition of the plane into equilateral triangles, each of which is a (multiple) reflection of the original triangle; and that the number of bounces taken to reach a point $P \in \mathbf{R}^2$ is equal to the number of times that the line from the origin to P intersects the edges of these triangles (you may ignore points on these edges; technically we have a "tiling" of the plane instead of a partition, since the edges cause a slight problem).

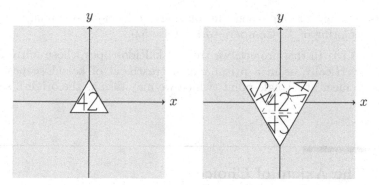

FIGURE 23.5: Standard kaleidoscope image: with no bounces (left) and after one bounce (right)

Task 23.2.5. If you have studied (pre)calculus, then show that the final function k from Task 23.2.4 is continuous. This is one reason that the standard kaleidoscope uses the configuration that it does. However, show that a non-standard kaleidoscope made using 6 mirrors arranged in a regular hexagon centered at the origin results in a discontinuous k function; this has an unpleasant visual effect.

Task 23.2.6. Use the result of Task 23.2.4 to prove that the function σ : $\mathbf{R}^2 \to \mathbf{R}^2$ given by the formula

$$\sigma(x, y) = (x + 1.5, y + \sqrt{3}/2)$$

is a symmetry of the final kaleidoscope image with respect to $\mathbf{R}^2$. Then show the same of the function

$$\tau(x, y) = (x + 1.5, y - \sqrt{3}/2).$$

Task 23.2.7. Let $C = (-1/2, -\sqrt{3}/6)$, the bottom-left corner of the original triangle. Define a new u, v-coordinate system in the plane by translating C to the origin: that is, by setting $u = x + 1/2$ and $v = y + \sqrt{3}/6$. Prove that the dihedral group D_6 is in the symmetry group of the kaleidoscope image in the u, v coordinate system. (This author has noticed that when actually looking through a kaleidoscope, the viewer's eye seems drawn to a corner point such as C as a natural center instead of the origin, perhaps because of this symmetry!)

Task 23.2.8. Adopt the u, v-coordinates of Task 23.2.7. Let α and β be the functions σ and τ of Task 23.2.6 expressed in terms of u and v. Let H be the subgroup of $\mathrm{Sym}(\mathbf{R}^2)$ generated by α, β, and D_6. Let X be the region of $\mathbf{R}^2$ inside the kaleidoscope's original triangle (including the triangle itself). Let $Y = \{f(a) \mid f \in H \text{ and } a \in X\}$. Prove that $Y = \mathbf{R}^2$. Conclude that if X is colored in any manner whatsoever, then the image in the kaleidoscope will include H in its symmetry group, and conversely, if a coloring of $\mathbf{R}^2$ includes

H in its symmetry group, then the coloring is a kaleidoscope image for some choice of coloring of X. Compare to Exercise 5.9.

Remark 23.13. In this project, we treated kaleidoscopes whose mirrors were arranged vertically, as in a prism. For a treatment of kaleidoscopes whose mirrors all meet at a single point (which we may take as the origin), see [6].

23.3   The Axiom of Choice

Prerequisites. This project may be undertaken at any point, but to understand all of the applications presented, the reader should have completed through Chapter 17.

This project explains an axiom at the foundations of mathematics, and applies it to algebra. Roughly speaking, we resort to the Axiom of Choice (or its equivalents) when we are dealing with things that are "too infinite" to realize explicitly. The subject matter in this project is "deep" in the sense that a complete understanding requires a knowledge of the standard axioms of set theory. Since we are primarily studying algebra and not set theory or logic, we will concentrate on the applications instead of on the basic questions of logical consistency and independence among axioms; in fact, we do not even attempt to give a list of standard set axioms.

It turns out that there are several different statements which are logically equivalent to the Axiom of Choice (AoC for short), each of which has its own peculiar flavor and patterns of use. We shall begin with the statement known as Zorn's Lemma, as this leads to the others fairly easily. We need some preliminary definitions.

Definition 23.14. Let $\sim$ be a relation on a set S (in the sense of Definition 8.3). We say that $\sim$ is *anti-symmetric* if for all $a, b \in S$, if $a \sim b$ and $b \sim a$, then $b = a$.

We remind the reader that the notions of *reflexive* and *transitive* have been defined earlier (Definition 8.4).

Definition 23.15. A *partial order* on a set S is a relation $\leq$ which is reflexive, transitive, and anti-symmetric. A set with a partial order is called a *partially ordered set*, or *poset* for short. We will use the symbol $\leq$ by default for a partial order. We also use the symbol $<$ in the usual way, so $a < b$ means $a \leq b$ and $a \neq b$.

Definition 23.16. A *total order* on a set S is a partial order $\leq$ such that for all $a, b \in S$, we have either $a < b$, or $a = b$, or $b < a$. (Note that by definition of partial order and $<$, at most one of these statements can be true!)

Task 23.3.1. Prove that the following relations are partial orders:

1. The usual less-than-or-equal relation $\leq$ on the set $\mathbf{R}$;

2. The subset relation $\subseteq$ (also known as the *inclusion* relation) on the power set $\mathcal{P}(S)$ of a set S (by definition, $\mathcal{P}(S)$ is the set of all subsets of S);

3. The restriction of any partial order on a set S to any subset T of S (note that technically, we must restrict to T^2, not T itself).

Definition 23.17. An element m of a poset S is called *maximal* if there exists no element a of S such that $m < a$. An element $M \in S$ is called a *maximum* element of S if for all $a \in S$ we have $a \leq M$.

Task 23.3.2. Prove that in any poset, there is at most one maximum element, and if M is a maximum then M is also maximal. Show by example that a poset may have any number of maximal elements.

Definition 23.18. A *chain* is a totally ordered subset of a poset. More precisely, if $\leq$ is a partial order on S, then a chain in S is a subset $C \subseteq S$ such that the restriction of $\leq$ to C is a total order.

Definition 23.19. Let S be a poset, and let $T \subseteq S$. An element u of S is called an *upper bound* for T if for every $a \in T$ we have $a \leq u$. (Note that this is different from saying that u is a maximum element of T, since we may have $u \in S - T$.)

Task 23.3.3. Consider the set of all real numbers $\mathbf{R}$ with the usual partial order $\leq$. Find a subset T of $\mathbf{R}$ such that T has an upper bound in $\mathbf{R}$, but there is no upper bound for T in T itself.

Now we are ready to state Zorn's Lemma. It is a matter of tradition (but somewhat unfortunate) that this statement has been labeled a "lemma" instead of an axiom.

Axiom 23.20 (Zorn's Lemma). Suppose that S is a non-empty partially ordered set, and that every chain in S has an upper bound in S. Then S contains a maximal element.

Remark 23.21. We often apply Zorn's Lemma to a set S whose elements are sets partially ordered by inclusion. In these situations, a typical strategy is to show that the union of a chain in S is also an element of S. Then this union provides the required upper bound for the chain.

Task 23.3.4. Let R be a non-trivial commutative ring with 1. Use Zorn's Lemma to prove that R has at least one maximal ideal, by following the steps below.

1. Let S be the set of all proper ideals of R, partially ordered by inclusion. Why is S non-empty?

2. Let C be a chain in S. Prove that $\cup_{I \in C} I \in S$. (Compare Exercise 20.14, and note that here, our chain is not indexed by the natural numbers; in fact, our chain may be uncountably infinite!)

3. Use Zorn's Lemma to conclude that S contains a maximal element M with respect to inclusion, which is the same thing as a maximal ideal of R.

Often, a property of interest requires the existence of finitely many objects which (together) satisfy specified conditions. The following task illustrates an important example of this phenomenon.

Task 23.3.5. Let V be a vector space over a field F, and let $T \subseteq V$. Prove that T is linearly independent over F if and only if every finite subset of T is linearly independent over F.

The following task provides results which are useful in the situation above.

Task 23.3.6. Let S a set with a total ordering $\leq$, and let $T \subseteq S$.
   (a) Prove that T is totally ordered by the restriction of $\leq$.
   (b) Prove that if T is finite, then T has a maximum element; use induction on $|T|$ (do not use Zorn's Lemma here).

From part (b) of Task 23.3.6, we can see the relevance of our earlier comment that AoC is most useful in infinite situations.

Task 23.3.7. Let V be a vector space over a field F. Use Zorn's Lemma to prove the existence of a basis of V, as follows.
   1. Let S be the set of all linearly independent subsets of V, partially ordered by inclusion. Why is S non-empty?
   2. Let C be a chain in S. Prove that $\cup_{T \in C} T \in S$.
   3. Use Zorn's Lemma and Exercise 13.7 to complete the proof.

We state the result of Task 23.3.7 as a proposition:

Proposition 23.22. *Assuming the Axiom of Choice, every vector space has a basis.*

The reader may be wondering by now exactly what the Axiom of Choice itself says. The following is one common formulation of AoC; the reader may encounter slight variations in different sources, but these should all be logically equivalent to each other.

Axiom 23.23 (Axiom of Choice (AoC)). Let S be a set whose elements are non-empty sets. Then there exists a function f with domain S such that for all $x \in S$ we have $f(x) \in x$.

The AoC thus lets us "choose" one element from every set in a given collection of sets, hence its name; we call the function f a *choice function* for S. One immediate consequence is the following:

Proposition 23.24. *Assuming the Axiom of Choice, for every subgroup H of a group G, there exists a complete set of left coset representatives for H in G.*

Proof. Let $H \leq G$ be groups. Apply the Axiom of Choice to the set $\mathcal{L}$ of all left cosets of H in G to get a function $f : \mathcal{L} \to G$ with the property that for all $C \in \mathcal{L}$ we have $f(C) \in C$. Then the image $\{f(C) : C \in \mathcal{L}\}$ is a complete set of left coset representatives for H in G. $\qquad\square$

Next we sketch a proof relating Zorn's Lemma to AoC. To accomplish such a proof on a firm footing, we should really work within the branch of mathematics known as axiomatic set theory, which is outside of our scope; we do note, for example, that one of the standard axioms of set theory guarantees the existence of the union of any two given sets, which is used in the proof below. For a good introduction to set theory based on the axiomatic approach, see [4].

Proposition 23.25. *Zorn's Lemma implies the Axiom of Choice.*

Proof. Let S be a set whose elements are non-empty sets. Let $\mathcal{F}$ be the set of all functions f whose domain $D(f)$ is a subset of S, and such that $f(x) \in x$ for every $x \in D$. Note that $\mathcal{F}$ is non-empty since we have $\emptyset \in \mathcal{F}$. Partially order $\mathcal{F}$ by restriction: that is, $f \le g$ iff $D(f) \subseteq D(g)$ and $f = g|_{D(f)}$. We leave it to the reader to verify that the union of any chain in $\mathcal{F}$ is again in $\mathcal{F}$ (treating a function formally as a set of ordered pairs; see Definition 1.16). Therefore, Zorn's Lemma gives a maximal element $f \in \mathcal{F}$. Let $D = D(f)$ be the domain of f. Assume for a contradiction that $D \ne S$. Then since $D \subseteq S$, there exists $A \in S - D$; and further, there exists $a \in A$ since $A \ne \emptyset$. Now let $g = f \cup \{(A, a)\}$, and notice that $g \in \mathcal{F}$ and $f < g$, contradicting that f is maximal. This completes the proof. $\qquad\square$

Remark 23.26. As indicated earlier, the converse of Proposition 23.25 is also true.

We now return to an assertion made back on page 10.

Proposition 23.27. *Let $\mathcal{C} = (S_i)_{i \in \mathcal{I}}$ be an indexed collection of non-empty sets, where the index set $\mathcal{I}$ is non-empty. Assuming the Axiom of Choice, there is a direct product P of $\mathcal{C}$ in the category of sets, and we have $P \ne \emptyset$.*

Task 23.3.8. Prove Proposition 23.27. Suggestion: Let $T_i = \{(i, a) \; : \; a \in S_i\}$, and let P be the set of all choice functions for $\{T_i \; : \; i \in \mathcal{I}\}$.

Next, we guide the reader through the proof of yet another important result which depends on AoC.

Definition 23.28. Let F be a field. An *algebraic closure* of F is an extension field K of F such that K is algebraic over F and K is algebraically closed.

Task 23.3.9. Let F be a field. Use Zorn's Lemma to prove that an algebraic closure of F exists. Suggestion: Let S be the set of all algebraic extension fields of F (technically, we could restrict the underlying sets of these extension fields to guarantee that S is not a "proper class," but we shall ignore this). Partially order S by the subfield relation, $\le$. Show that the union of a chain in S is again in S. Prove by contradiction that any maximal element of S is algebraically closed.

Task 23.3.10. Let F be a field, and suppose that K is an algebraic closure of F. Prove that there is no algebraically closed field L such that $F \le L < K$. (You should not need AoC or its equivalents to do this.)

Remark 23.29. Although the Axiom of Choice is now widely accepted and usually passes without much remark, it does have consequences which are non-intuitive to some people. Perhaps the most notorious of these consequences is the so-called *Banach-Tarski Paradox*, which tells us that an ordinary, solid sphere in $\mathbf{R}^3$ can be partitioned into finitely many subsets such that a finite sequence of spatial translations and rotations of these subsets can produce a solid sphere of twice the radius of the original. Although it has been termed a "paradox," in fact this is a theorem which has been proved rigorously; the "trick" is that the proof uses AoC, and the subsets which are produced cannot have any well-defined volume individually! For a very accessible account of this result with plenty of background material included, see [16].

Aside from the Banach-Tarski Paradox, some people are not satisfied to use AoC because it produces a result "non-constructively"—that is, without providing a formula or algorithm for the choice function. To get a sense of this complaint, the reader may wish to try to find a complete set of coset representatives for $(\mathbf{Q}, +)$ in $(\mathbf{R}, +)$; such a thing is guaranteed to exist by AoC, but writing down an explicit description seems difficult. For a fairly thorough and mathematically detailed account of the history of AoC, see [12].

23.4   Some Category Theory

Prerequisites. To complete this project, it is recommended to have finished through Chapter 21 of the text. The reader who has finished Chapter 13 should be able to complete through Task 23.4.2.

We have previously used the words *category*, *object*, and *morphism* without attempting to give precise definitions. In this project, we give a more formal treatment to these notions, and then give a corrected definition of *isomorphism* which applies to any category.

Definition 23.30. A *category* is an ordered pair (Ob, Mor) satisfying the following axioms:

(1) Ob is a collection of sets, called *objects*;

(2) Mor is a collection of sets, called *morphisms*; further, for every A and B in Ob, there is a set $\mathrm{Mor}(A, B)$, called the set of *morphisms from A to B*, and Mor is the union of all these sets;

(3) For every object A there is a distinguished element of $\mathrm{Mor}(A, A)$ called the *identity morphism on A* and written Id_A.

(4) For every A, B, and C in Ob, and for every $f \in \mathrm{Mor}(A, B)$ and every $g \in \mathrm{Mor}(B, C)$, there is a morphism $h \in \mathrm{Mor}(A, C)$ called the *composition of g with f* and written $h = g \circ f$;

(5) For every A and B in Ob, and for every $f \in \mathrm{Mor}(A, B)$, we have $f \circ \mathrm{Id}_A = f$ and $\mathrm{Id}_B \circ f = f$;

(6) For every $f \in \text{Mor}(A, B)$, $g \in \text{Mor}(B, C)$, and $h \in \text{Mor}(C, D)$, we have $(h \circ g) \circ f = h \circ (g \circ f)$ (morphisms are associative).

Notation 23.31. Often we name our category. If we have a category $\mathcal{C}$, then we may write $\text{Ob}(\mathcal{C})$ for the collection of all objects in this category. We may also write $\text{Mor}_{\mathcal{C}}(A, B)$ instead of $\text{Mor}(A, B)$ when we want to note which category a morphism comes from, and we write $\text{Mor}(\mathcal{C})$ for the collection of *all* morphisms of the category $\mathcal{C}$.

Remark 23.32. Some care is needed to talk about categories correctly. For example, in the category of groups, $\mathcal{G}$, the collection $\text{Ob}(\mathcal{G})$ is the collection of all groups. It turns out that this collection is *too big to be a set!* Instead, it is what set theorists and logicians call a "proper class." Similarly, $\text{Mor}(\mathcal{G})$ is also a proper class. We can get into logical paradoxes if we are sloppy about these notions.

Task 23.4.1. Describe each of the following using the language of categories, and verify that the Category Axioms are satisfied. You will be able to do most of the work by simply finding the appropriate results from earlier chapters of this text. (i) Groups; (ii) Rings; (iii) Vector Spaces.

Remark 23.33. In all of our examples so far, a morphism from A to B has been a *function* from A to B, with certain special properties. But nothing in the Category Axioms requires this to be true in general; in principle, a morphism from A to B can be any set at all. However, in most categories we encounter, a morphism from A to B will at least include a function $f : A \to B$ as a component of the morphism, if not the entire morphism itself. In any case, we use the notation $f : A \to B$ to mean that $f \in \text{Mor}(A, B)$, even when f is not simply a function from A to B.

Notice that the Category Axioms do not mention "isomorphism." Instead, the concept of an isomorphism is defined from the basic ingredients of categories given in the axioms. Until now, we have defined an isomorphism to be a bijective homomorphism (or a bijective linear transformation, in the case of vector spaces); but this is not a good definition of isomorphism in more general categories, especially when you remember that a morphism does not even need to be a function! The correct general definition of isomorphism is given below.

Definition 23.34. Let $\mathcal{C}$ be a category. Let A, B be objects of $\mathcal{C}$. An *isomorphism* from A to B is a morphism $f : A \to B$ such that there exists a morphism $g : B \to A$ satisfying $g \circ f = \text{Id}_A$ and $f \circ g = \text{Id}_B$.

Task 23.4.2. Prove that our previous definitions of "isomorphisms" (as bijective morphisms) are equivalent to Definition 23.34 in the cateogories of groups, rings, and vector spaces. Again, previous results should be helpful.

To introduce the reader to a new category-theoretic concept (called a *direct limit*) and simultaneously prepare the way for an example of a category in which a bijective morphism is *not* the same thing as an isomorphism, we will work to understand fields of prime characteristic somewhat better.

Let p be an ordinary prime integer. We have seen that $\mathbf{F}_p := \mathbf{Z}/p\mathbf{Z}$ is a field, and that we can extend $\mathbf{F}_p$ to get a finite field of any order which is a power of p, by taking a splitting field of the polynomial $x^a - x$ for an appropriate value of a. Furthermore, we saw that for each positive integer n, there is a *unique* field K_n with $|K_n| = p^n$, up to isomorphism in the category of rings.

Task 23.4.3. Find justifications in the text for all statements made in the previous paragraph. Then use these statements to help show that for all positive integers m and n there is an embedding $e_{m,mn} : K_m \hookrightarrow K_{mn}$.

Looking at the previous task, it is tempting to ask whether all of these embeddings are "going somewhere": is there a big field K in which all of the K_n are embedded? If the maps $e_{m,mn}$ were simply *inclusions*, then we would want to take K to be the union of the K_n. Now an embedding is just as good as an inclusion, except that we need to identify elements of the embedded object with (differently-labeled) elements of the bigger object; so there *ought* to be a way of making something like a union out of these ingredients. Before we can accomplish this, however, we would like to ensure that the maps $e_{m,mn}$ are consistent with each other. That is the purpose of the following task.

Task 23.4.4. (a) Prove that the number of distinct embeddings of K_n into K_{mn} is equal to m, using Galois theory.

(b) For each $n \in \mathbf{Z}^+$, fix a generator ω_n for $K_n^\times$. Prove that there is a unique embedding $\sigma_{m,mn} : K_m \hookrightarrow K_{mn}$ such that $\sigma_{m,mn}(\omega_m) = \omega_{mn}^q$, where $q = (p^{mn} - 1)/(p^m - 1)$.

(c) Prove that the maps $\sigma_{m,mn}$ are compatible in the sense that we have $\sigma_{\ell m, \ell mn} \circ \sigma_{\ell, \ell m} = \sigma_{\ell, \ell mn}$ for all $\ell, m, n \in \mathbf{Z}^+$.

Definition 23.35. Use the notation of Task 23.4.4. Let

$$S = \bigcup_{n=1}^{\infty} K_n,$$

and define a relation $\sim$ on S by $a \sim b$ iff $a \in K_m$, $b \in K_{mn}$, and $\sigma_{m,mn}(a) = b$ for some $m, n \in \mathbf{Z}^+$, or vice-versa (with the roles of a and b switched, so that $\sim$ is symmetric). The *direct limit* of the fields K_n with respect to the maps $\sigma_{m,mn}$ is the set K of equivalence classes of S under $\sim$.

Task 23.4.5. Prove that $\sim$ really is an equivalence relation on S. Then prove that K is a field under appropriately defined addition and multiplication. Remember to establish that these operations are well-defined and that K is closed under these operations!

Task 23.4.6. Prove that for every positive integer n, we have $K_n \hookrightarrow K$ as a field embedding; let L_n denote the image of K_n in K. Then prove that $K = \cup_{n=1}^{\infty} L_n$. Use these results to show that K is infinite, and that every element of K is algebraic over the prime subfield of K.

Task 23.4.7. Prove that K is algebraically closed.

Task 23.4.8. Define a function $f : K \to K$ by the formula $f(a) = a^p$. Prove that $f \in \text{Aut}(K)$.

As the reader has probably guessed, the concept of a direct limit can be defined in much greater generality. A definition in the category of rings is given next.

Definition 23.36. Let $\mathcal{I}$ be a partially ordered set (see Definition 23.15) such that every pair $\{i, j\} \subseteq \mathcal{I}$ has an upper bound in $\mathcal{I}$, and let $\{R_i\}_{i \in \mathcal{I}}$ be a collection of rings indexed by $\mathcal{I}$. Suppose that whenever $i, j \in \mathcal{I}$ with $i \leq j$, we have a ring homomorphism $\sigma_{i,j} : R_i \to R_j$, and that these homomorphisms are compatible in the sense that whenever $i \leq j \leq k$ we have $\sigma_{j,k} \circ \sigma_{i,j} = \sigma_{i,k}$. Then a *direct limit* of the rings R_i with respect to the maps $\sigma_{i,j}$ is a ring R together with ring homomorphisms $e_i : R_i \to R$ such that for every ring T with ring homomorphisms $\tau_i : R_i \to T$ compatible with the maps $\sigma_{i,j}$, there exists a ring homomorphism $\omega : R \to T$ making the following diagram commute:

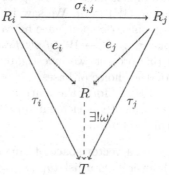

$$(23.4)$$

Note: when we say that the τ_i are compatible with the $\sigma_{i,j}$, we mean that the outer triangle in Diagram 23.4 commutes.

Task 23.4.9. Verify that the field K constructed in Definition 23.35 is a direct limit of the rings K_n according to Definition 23.36, taking $\mathcal{I} = \mathbf{Z}^+$, with the partial order on $\mathcal{I}$ defined by divisibility in $\mathbf{Z}$ (instead of the usual $\leq$ relation), and the maps $\sigma_{m,mn}$ from Task 23.4.4.

Task 23.4.10. Consider the category $\mathcal{C}$ whose objects are fields, and whose morphisms are ring homomorphisms given by *polynomial* functions from a field to itself. To be more precise: if A and B are two *distinct* fields, then we take $\text{Mor}(A, B) = \emptyset$. We take $\text{Mor}(A, A)$ to be the set of all ring homomorphisms $f : A \to A$ such that there exists a single polynomial $g \in A[x]$ satisfying $f(a) = g(a)$ for all $a \in A$. (This may seem artificial, but this kind of map is important in the area of math called algebraic geometry.)

Let f and L_n be as in Tasks 23.4.8 and 23.4.6, respectively. Prove f is a morphism in the category $\mathcal{C}$, and that f is a bijective function. Also prove that for every positive integer n, the restriction of f to L_n is an isomorphism in the category $\mathcal{C}$. But prove that f itself is *not* an isomorphism in $\mathcal{C}$. Thus, in the category $\mathcal{C}$, a bijective morphism is not the same thing as an isomorphism.

For a somewhat more in-depth discussion of category theory that is still not *too* abstract, see [3, Appendix 5]. For a treatment of direct limits and more in this vein, see [10, Chapter III]. For one of the great classics in general category theory, see [11]. Also, note that the subject of homological algebra is intertwined with category theory: specifically, with a nice type called *abelian categories*. For good sources on homological algebra, see [17] or [13].

23.5   Linear Algebra: Change of Basis

Prerequisites. Chapter 13 (including the exercises on linear transformations).

Let V and W be finite-dimensional vector spaces over the same field F, with $\dim_F(V) = n$ and $\dim_F(W) = m$. We saw in Exercises 13.15 through 13.18 that there is a one-to-one correspondence between matrices in $M_{m,n}(F)$ and linear transformations $\sigma : V \to W$, *if we first choose a basis B of V and a basis C of W*. In this project, we investigate exactly how the matrix representing σ depends on the choice of bases.

As noted in Exercise 13.15, to form the matrix of a linear transformation requires a knowledge of how our basis elements are to be ordered. Therefore, we give the following formal definition:

Definition 23.37. Let V be a vector space of dimension $n < \infty$ over a field F. An *ordered basis* of V over F is a list $(b_1, \ldots, b_n)$ of elements of V such that the set $\{b_1, \ldots, b_n\}$ is a basis of V over F.

Task 23.5.1. Prove that the elements listed in an ordered basis must be distinct.

We also formalize the idea that an ordered basis provides a coordinate system for a vector space:

Definition 23.38. Let V be a vector space over F of dimension $n < \infty$, and let $B = (b_1, \ldots, b_n)$ be an ordered basis of V over F. Let $v \in V$. Then the *coordinate vector of v with respect to B* is the element $c = (c_1, \ldots, c_n) \in F^n$, where c_i are the unique elements of F such that we have $v = \sum_{i=1}^n c_i b_i$.

Next, we rephrase the results of the aforementioned exercises to emphasize a useful point of view.

Task 23.5.2. Let V be a vector space over a field F of finite dimension n.

(a) Let $B = (b_1, \ldots, b_n)$ be an ordered basis of V over F. Prove that B has the following property: For every choice of vectors $v_1, \ldots, v_n \in V$ (with repetition allowed), there is a unique linear transformation $\sigma : V \to V$ such that $\sigma(b_i) = v_i$ for all i.

(b) Prove that a list $B = (b_1, \ldots, b_n)$ with $b_i \in V$ is an ordered basis of V over F if and only if B has the property stated in part (a) above. Compare this to the results in Example 7.12 and Exercise 14.10. The common theme in these three situations is the "freeness" of vector spaces, free groups, and polynomial rings.

Our approach to studying the effect of basis choice on the matrix of a linear transformation is to take two choices of ordered basis for the same vector space and compare the resulting matrices. From this point on, we fix the following notation:

Let $\sigma : V \to W$ be a linear transformation, where $\dim_F(V) = n$ and $\dim_F(W) = m$. Let $B = (b_1, \ldots, b_n)$ and $B' = (b'_1, \ldots, b'_n)$ be ordered bases of V over F. Let $C = (c_1, \ldots, c_m)$ be an ordered basis of W over F. Let M be the matrix of σ with respect to B and C, and let M' be the matrix of σ with respect to B' and C.

Task 23.5.3. Let M_i be the ith column of M. Confirm that M_i is the coordinate vector of $\sigma(b_i)$ with respect to C. Make a similar statement for M'_i, the ith column of M'.

Task 23.5.4. Let $\alpha : V \to V$ be the unique linear transformation sending b_i to b'_i. Prove that α is a vector space isomorphism.

Next we further fix the following notation:

Let α be the isomorphism of Task 23.5.4. Let A be the matrix of α with respect to B (recall that this means we choose B as the basis for V in both of its roles as domain and codomain of α).

Task 23.5.5. Prove that $A \in \text{GL}_n(F)$. (Use the result of Task 23.5.4.) More precisely, show that A^{-1} is the matrix of the linear transformation sending b'_i to b_i, with respect to the basis B.

Task 23.5.6. Let $v \in V$, and let c be the coordinate vector of v with respect to B. View c as a column vector, that is, as an element of $M_{m,1}(F)$. Prove that Ac is the coordinate vector of v with respect to B', viewed as a column vector.

Task 23.5.7. Prove that $M' \cdot A = M$, hence $M' = M \cdot A^{-1}$.

Now that we have explored how a change of basis in the domain affects the matrix of σ, let us see what happens when we change the basis of the codomain. To this end, let $C' = (c'_1, \ldots, c'_m)$ be another ordered basis of W over F. Let $\gamma : W \to W$ be the unique linear transformation sending c_i to c'_i, and let G be the matrix of γ with respect to C. Let M'' be the matrix of σ with respect to B and C'.

Task 23.5.8. Prove that $M'' = G \cdot M$.

Next, we allow *both* bases to change:

Task 23.5.9. Let N be the matrix of σ with respect to B' and C'. Prove that $N = G \cdot M \cdot A^{-1}$.

Now that we have understood the general change-of-basis formula, we turn to the important special case when $W = V$.

Task 23.5.10. Suppose that $W = V$, $C = B$, and $C' = B'$. Thus $\sigma : V \to V$, and the matrices representing σ are square. Show that in this situation we have $\alpha = \gamma$ and $G = A$, so that

$$N = A \cdot M \cdot A^{-1}. \tag{23.5}$$

Definition 23.39. Let V be a vector space over a field F, with $\dim_F(V) = n < \infty$. Let B and B' be ordered bases of V over F, and let $A \in M_n(F)$ be the matrix whose i^{th} column is the coordinate vector of the ith element of B' with respect to B. Then we call A the *change-of-basis matrix* from B to B'.

Equation 23.5 is perhaps the most important change-of-basis formula. We recognize this formula as saying that N is the conjugate of M by A in the group $\mathrm{GL}_n(F)$ (see Definition 7.15). From Exercise 9.9, we know that N is the image of M under an automorphism, and hence we expect that N and M should "look the same" with respect to every aspect of the group $\mathrm{GL}_n(F)$. This makes sense, since in fact N and M represent *exactly the same* linear transformation, only with respect to different choices of basis! What is more, our new knowledge from linear algebra sheds light on the meaning of conjugation: to apply the conjugate of M by A is to first change coordinates from B' to B (by applying A^{-1}); then to apply our function σ using the coordinate system of B (by multiplying by M); and finally to change coordinates back from B to B' (by applying A). The result, N, provides the way to apply the same function σ using the coordinate system of B' instead of B.

Task 23.5.11. Let $Q \in \mathrm{GL}_n(F)$ be any invertible n by n matrix over F. Prove that Q is the change-of-basis matrix from B to B'' for some basis B'' of V over F.

Definition 23.40. Two matrices which are related by Equation 23.5 are called *similar*. We write $M \sim N$ to mean that there exists an invertible matrix A satisfying Equation 23.5.

23.6 Linear Algebra: Determinants

Prerequisites. Chapter 13 and Project 23.5.

Recall that a *matrix* is a (finite) grid of elements which are called the *entries* of the matrix. For now, the entries will belong to a field, and we will only be

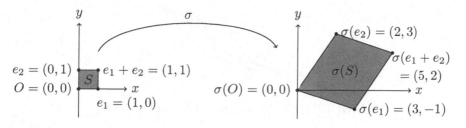

FIGURE 23.6: Applying the linear transformation σ to the unit square S (left) results in the parallelogram $\sigma(S)$ (right)

concerned with *square* matrices. Let us fix a field F and a finite-dimensional vector space V over F with $\dim_F(V) = n$. We saw in Exercises 13.15 through 13.18 that every linear transformation from V to V can be represented as an $n \times n$ matrix that depends on a choice of basis of V over F. We saw in Project 23.5 that changing the basis we use for V amounts to conjugating this matrix by an invertible matrix.

Recall that F^n is a vector space over F under componentwise operations (see Example 13.4 and Exercise 13.12). By re-writing elements of F^n as columns instead of rows, we will view each column of a matrix $M \in M_n(F)$ as a "column vector" in F^n.

Our motivation in this project comes from the special case when $F = \mathbf{R}$, the field of real numbers. Specifically, we will develop a theory motivated by the following question: by what factor does a linear transformation $\sigma : \mathbf{R}^n \to \mathbf{R}^n$ change the area (or volume, or generalized volume) of a region in $\mathbf{R}^n$? Consider as an example the matrix

$$M = \begin{pmatrix} 3 & 2 \\ -1 & 3 \end{pmatrix} \in M_2(\mathbf{R}). \tag{23.6}$$

Suppose that we use the basis $B = (e_1, e_2)$ to interpret M as a linear transformation σ, where we write elements of $\mathbf{R}^2$ as column vectors, with

$$e_1 = \begin{pmatrix} 1 \\ 0 \end{pmatrix} \text{ and } e_2 = \begin{pmatrix} 0 \\ 1 \end{pmatrix}.$$

Then we can read the value of $\sigma(e_i)$ as the ith column of M. Finally, by identifying the point (a, b) with the column vector

$$\begin{pmatrix} a \\ b \end{pmatrix},$$

we can display the effect of σ on the square $[0, 1] \times [0, 1] \subset \mathbf{R}^2$. The result is shown in Figure 23.6.

To help with the next task, we recall that the area of a parallelogram is equal to its length multiplied by its height (see Figure 23.7). We can use any side of the parallelogram as the "length" side.

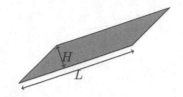

FIGURE 23.7: Area of parallelogram $= L \cdot H$

Task 23.6.1. Show that the area of the parallelogram $\sigma(S)$ in Figure 23.6 is equal to 11, which is also equal to $|M|$ according to the formula given in Exercise 4.10.

Task 23.6.2. Show that for any linear transformation $\sigma \; : \; \mathbf{R}^2 \to \mathbf{R}^2$, the image of the unit square $[0, 1] \times [0, 1]$ will be a parallelogram, a line segment, or the origin. What can you say about when each case occurs?

Our strategy will be to use three properties of area, suitably generalized to n-dimensional space, as axioms for something we shall call a "determinant." Thus, a determinant is a generalized version of area. The importance of the concept of area suggests that determinants will be quite useful; indeed, determinants appear in critical places in both algebra and analysis.

Definition 23.41. Let F be a field and let $n \in \mathbf{Z}^+$. A *determinant* is a function det $: \; M_n(\mathbf{F}) \to \mathbf{F}$ which satisfies the following three axioms:

    1. $\det(I) = 1$, where I is the $n \times n$ identity matrix.

    2. If M is an $n \times n$ matrix, $\alpha \in F$, and N is the matrix obtained by replacing a column c of M by the column vector αc, then we have $\det(N) = \alpha \cdot \det(M)$.

    3. If M is an $n \times n$ matrix, c_i and c_j are the ith and j^{th} columns of M, respectively, with $i \neq j$, and N is the matrix obtained from M by replacing column c_i by $c_i + c_j$, then we have $\det(N) = \det(M)$.

Task 23.6.3. Show geometrically that the Determinant Axioms are true for $F = \mathbf{R}$ and $n = 2$ if we replace det by the area of the parallelogram formed from the column vectors of a 2×2 real matrix, in the case $\alpha > 0$.

Remark 23.42. In Axiom 2, when $F = \mathbf{R}$, we may have $\alpha < 0$. In light of Axiom 1, we see that a determinant can be negative. Sometimes the determinant is referred to as a "signed area" or "signed volume" because of this fact. It turns out to be better to allow negative real values for the determinant than to force the determinant to be positive all the time; for example, general fields do not have a notion of positive or negative!

We shall see that there is a unique determinant function which satisfies our three axioms. To start with, you will establish a few consequences of the axioms, after we introduce the concept of a so-called "diagonal" matrix.

Definition 23.43. An $n \times n$ square matrix M is called *diagonal* if $M_{i,j} = 0$ whenever $i \neq j$; here, as usual, $M_{i,j}$ denotes the entry in row i, column j of M. That is, a matrix is diagonal if its only non-zero entries occur on the main diagonal of M. We denote a diagonal matrix by $\text{diag}(d_1, d_2, \ldots, d_n)$, where the d_i are the diagonal entries, in order from upper-left to lower-right.

Remark 23.44. Visually, a diagonal matrix looks as follows:

$$
\begin{pmatrix}
d_1 & 0 & 0 & \cdots & 0 \\
0 & d_2 & 0 & \cdots & 0 \\
0 & 0 & d_3 & \cdots & 0 \\
\vdots & \vdots & \vdots & \ddots & \vdots \\
0 & 0 & 0 & \cdots & d_n
\end{pmatrix}.
$$

Diagonal matrices are the simplest types of matrices after scalar matrices.

Task 23.6.4. Let det be a determinant function. Prove that det has the following properties:

4. If any column of the matrix M is zero (that is, its entries are all 0), then $\det(M) = 0$.

5. If $M = \text{diag}(d_1, d_2, \ldots, d_n)$, then $\det(M) = d_1 \cdot d_2 \cdots \cdots d_n$.

6. (a) If we replace the ith column c_i of M by the column vector $c_i + \alpha \cdot c_j$, for any $\alpha \in F$ and any column c_j with $j \neq i$, then the determinant does not change.

(b) More generally, if we replace c_i by $c_i + \sum_{j \neq i} \alpha_j \cdot c_j$ for any elements $\alpha_j \in F$, then the determinant does not change. Suggestion: Use part (a) inductively.

7. If we switch any two columns of a matrix, then the determinant is multiplied by -1.

Next we will clarify a small but important issue about the definition of linear dependence. Originally, we defined linear dependence as a property of a *set* of vectors (Definition 13.22). But now, we wish to talk about whether the column vectors of a matrix are linearly dependent. If there are duplicate columns, then our present definition is not quite what we need. For example, if $M \in M_2(F)$ has two non-zero but identical columns, then the *set* of columns of M is linearly independent, but we would like to say that the *list* of columns of M is linearly *dependent*.

Definition 23.45. Let V be a vector space over a field F, and let $L = (v_1, \ldots, v_n)$ be an ordered list of elements of V. Then L is *linearly dependent* over F if there exist elements $a_1, \ldots, a_n \in F$, not all 0, such that $\sum_{i=1}^{n} a_i v_i = 0$. Otherwise, L is called *linearly independent* over F.

Task 23.6.5. (a) Prove that a list $L = (v_1, \ldots, v_n)$ of vectors is linearly dependent iff either the set $\{v_1, \ldots, v_n\}$ is linearly dependent, or the list contains duplicates, that is, there exist $i \neq j$ with $v_i = v_j$.

(b) Prove that if any sublist of a list L is linearly dependent, then so is L (define "sublist" appropriately to make this work!).

Task 23.6.6. Prove that every determinant function has the following property:

> 8. If the list of column vectors of M is linearly dependent over F, then we have $\det(M) = 0$.

Now that the concept of linear dependence in F^n has entered the picture, we should establish a basic fact for use later.

Task 23.6.7. Let F be a field and let $n \in \mathbf{Z}^+$. Let j be an integer such that $1 \le j \le n$. Set $W = \{(a_1, \ldots, a_n) \in F^n \mid a_i = 0 \text{ for } 1 \le i \le j\}$. Prove that $W \le F^n$ and $\dim_F(W) = n - j$.

Task 23.6.8. Let $M \in M_n(F)$ be a matrix with columns $c_1, \ldots, c_n$. Suppose that $(c_1, \ldots, c_n)$ is linearly independent over F. Show that, starting with M, we can use Properties 6(a) and 7 repeatedly to produce a diagonal matrix $D = \operatorname{diag}(d_1, \ldots, d_n)$ where each diagonal entry d_i is non-zero. Conclude that $\det(M) \ne 0$.

In view of Determinant Property 8 and Task 23.6.8, we have the following result:

Proposition 23.46. *Let $M \in M_n(F)$. Then $\det(M) = 0$ if and only if the list of columns of M is linearly dependent over F.*

Remark 23.47. The result of Proposition 23.46 is consistent with the interpretation of the determinant as a generalized area; for the determinant to be zero, the matrix columns must not span all of F^n, but instead span a smaller-dimensional space. For example, in $\mathbf{R}^2$ the "area" of a line segment or a point is 0, while the area of a true 2-dimensional parallelogram is non-zero.

Definition 23.48. Let $M \in M_n(F)$. Then M is called *singular* if $\det(M) = 0$; otherwise, M is called *non-singular*.

Task 23.6.9. Let $M \in M_n(F)$, and suppose that $\omega : V \to V$ is a linear transformation such that M is the matrix of ω with respect to some basis of V over F. Prove that M is singular iff $\ker(\sigma) \ne 0$.

Next, we build an explicit formula for a determinant function.

Task 23.6.10. Let $M \in M_n(F)$ be a square matrix with column list $(c_1, \ldots, c_n)$. Let N be the matrix obtained from M by replacing column number i by the column vector $\sum_{j=1}^{n} a_j c_j$, where $a_j \in F$. Prove that we have $\det(N) = a_i \cdot \det(M)$.

Task 23.6.11. Prove that every determinant function has the following property:

> 9. Let M be a square matrix whose i^{th} column can be written as a sum, $u + v$. Let M' and M'' be the matrices obtained from M by

replacing the ith column by u and by v, respectively. Then we have $\det(M) = \det(M') + \det(M'')$.

Task 23.6.12. Let $M \in M_n(F)$ be a square matrix. Let $\mathcal{F}$ be the set of all functions $f : [n] \to [n]$, where $[n] = \{1, 2, \ldots, n\}$. For $f \in \mathcal{F}$, let M_f denote the matrix whose entry in row i, column $f(i)$ is the same as the corresponding entry in M, with all other entries zero. Use Determinant Property 9 repeatedly to prove the following formula:

$$\det(M) = \sum_{f \in \mathcal{F}} \det(M_f).$$

Task 23.6.13. Prove that, in the notation of Task 23.6.12, we have $\det(M_f) = 0$ if f is not bijective. Conclude that we have

$$\det(M) = \sum_{f \in S_n} \det(M_f).$$

We are not far from our goal of obtaining an explicit determinant formula. The next step is to convert the matrices which appear in Task 23.6.13 into diagonal matrices by repeatedly switching pairs of columns. To that end, we make the following definition.

Definition 23.49. Let $f \in S_n$. Let $E \in M_n(F)$ be the matrix all of whose entries are 1. Let $s(f)$ be the minimum number of switches of column pairs needed to transform E_f into the identity matrix I_n. Then the *sign* of f is $\text{sign}(f) = (-1)^{s(f)}$.

Task 23.6.14. Prove that for all $f \in S_n$, we have $0 \le s(f) \le n - 1$. In particular, conclude that $s(f)$ exists.

Task 23.6.15. Prove that every determinant function satisfies the following formula:

10. $\det(M) = \displaystyle\sum_{f \in S_n} \text{sign}(f) \cdot M_{1,f(1)} M_{2,f(2)} \cdots M_{n,f(n)}$, where as usual $M_{i,j}$ denotes the entry of M in row i, column j.

Conclude that there is at most one determinant function for a given positive integer n and field F.

To prove the *existence* of a determinant function, we must show that the formula in Determinant Property 10 satisfies the three Determinant Axioms. The main difficulty here is to show that Definition 23.49 is strong enough to handle *any* sequence of column switches that brings E_f to the identity matrix, not just a minimal such sequence. Until we know this, we will denote the function on the right-hand side of Determinant Property 10 by Δ.

Task 23.6.16. Prove that the function $\Psi : S_n \to GL_n(F)$ given by the formula $\Psi(f) = E_f$ is an embedding of groups.

Task 23.6.17. Let $f \in S_n$, and let N be the matrix obtained from E_f by switching column number a with column number b for some $a \neq b$. Let $h \in S_n$ be the permutation which switches a and b while leaving all other elements of $[n]$ fixed. Prove that $N = E_g$ where $g = fh$. Note: such a permutation h is called a *transposition*; see also Exercises 22.12 and 22.13.

Task 23.6.18. For a permutation $f \in S_n$, define

$$t(f) = |\{(i,j) \in [n]^2 \;:\; i < j \text{ and } f(i) > f(j)\}|,$$

the number of pairs which f puts "out of order." Prove that if h is a transposition in S_n, then we have $t(fh) \equiv 1 + t(f) \pmod{2}$. Suggestion: Let $a < b$ be the elements of $[n]$ switched by h. Compare f to fh and show that, except for the pair (a, b) itself, the pairs which change from in-order to out-of-order or vice versa naturally come in pairs.

Task 23.6.19. Let $f \in S_n$, and suppose that f can be written as a product of k transpositions. Use Task 23.6.18 to prove that we must have $(-1)^k = (-1)^{t(f)}$. Conclude that, in particular, $\text{sign}(f) = (-1)^{t(f)}$.

Task 23.6.20. Prove that Δ satisfies Determinant Property 7.

Task 23.6.21. Prove that the function Δ defined by Determinant Property 10 satsfies the three Determinant Axioms. Conclude that there is a unique determinant function. Suggestion: First show that Δ satisfies Determinant Axioms 1 and 2 as well as Determinant Property 9. Then use the previous results to establish the final Determinant Axiom 3.

Finally, we will show that the formula from Exercise 4.10(a) generalizes to matrices of any size. Instead of trying to prove this result using explicit formulas, which would be messy, we will take advantage of the Determinant Axioms.

Task 23.6.22. Let $A, B \in M_n(F)$. Use the steps below to prove the following property:

  11. $\det(A \cdot B) = \det(A) \cdot \det(B)$.

(a) First suppose that $\det(A) = 0$. Prove that $\det(A \cdot B) = 0$ too by using the correspondence between matrices and linear transformations.

(b) Now suppose that $\det(A) = \alpha \neq 0$, so $\alpha \in F^\times$. For any matrix $M \in M_n(F)$, let $\delta(M) = \det(A \cdot M) \cdot \alpha^{-1}$. Show that δ satisfies the three determinant axioms. Conclude that $\delta = \det$, hence $\det(A \cdot B) = \det(A) \cdot \det(B)$, as desired.

At last, we can realize our original goal of measuring the factor by which (generalized) area increases under a linear transformation. The next task confirms our intuition that this factor, the determinant, should be a property of a linear transformation itself, independent of any choice of basis.

Task 23.6.23. Let V be a vector space over a field F, and let $\sigma : V \to V$ be a linear transformation. Use the results of Project 23.5 and Task 23.6.22 to prove that the determinant of a matrix of σ does not depend on which basis of V we choose. Thus, we can define $\det(\sigma)$ without reference to a basis.

We next present another remarkable property of determinants. First we introduce a new matrix operation that switches the rows and columns of a matrix.

Definition 23.50. Let $M \in M_{m,n}(F)$. The *transpose* of M is the matrix $N \in M_{n,m}(F)$ such that for all i, j we have $N_{i,j} = M_{j,i}$. We write $N = M^T$. We say that M is *symmetric* if we have $M^T = M$.

Task 23.6.24. Use Determinant Property 10 to prove that for any square matrix M we have $\det(M^T) = \det(M)$.

We saw in Determinant Property 5 that the determinant of a diagonal matrix is easy to compute: it is just the product of the diagonal entries. Now that we understand determinants better, we can generalize that result to matrices of a somewhat less special form which is often attainable in practice.

Definition 23.51. Let $M \in M_n(F)$. Then M is called an *upper triangular* matrix if we have $M_{i,j} = 0$ whenever $i > j$, and M is called *lower triangular* if $M_{i,j} = 0$ whenever $i < j$. We say that M is *triangular* if M is either upper triangular or lower triangular.

Task 23.6.25. Prove that the determinant of a triangular matrix is equal to the product of its diagonal entries.

23.7   Linear Algebra: Eigenvalues

Prerequisites. Chapter 13 and Project 23.6.

"Since we lack the time to carry out this experiment in practice, Eigenfactor uses mathematics to simulate this process." [2]

Let V be a finite-dimensional vector space over a field F, with $\dim_F(V) = n \geq 1$. In this project we again focus on linear transformations $\sigma : V \to V$ from the vector space to itself. We know that there can be many different matrix representations of the same linear transformation σ, depending on our choice of a basis of V over F. Our goal is to find the "best" basis to get the simplest possible matrix of σ.

We have seen that the simplest types of square matrices are, in order, the identity matrix; scalar matrices; and diagonal matrices. Now we know that if the matrix of a linear transformation with respect to some basis is a scalar matrix, then the matrix with respect to *every* basis will be that same scalar matrix (see Exercise 13.18). Thus we cannot hope to simplify the matrix form of a linear transformation from a non-scalar matrix to a scalar matrix.

Therefore, we will ask for the next simplest thing: given σ, is there a basis of V over F for which the matrix of σ is diagonal?

What would it mean to say that the matrix M of σ with respect to the basis $B = \{b_1, \ldots, b_n\}$ is the diagonal matrix $\mathrm{diag}(d_1, \ldots, d_n)$? Well, this means that

$$\sigma(b_i) = 0 \cdot b_1 + 0 \cdot b_2 + \cdots + 0 \cdot b_{i-1} + d_i \cdot b_i + 0 \cdot b_{i+1} + \cdots + 0 \cdot b_n = d_i \cdot b_i \quad (23.7)$$

for all $i \in \{1, 2, \ldots, n\}$. In other words, the effect of σ on each basis vector is simply to multiply that vector by a certain scalar. With this motivation, we make the following definitions.

Definition 23.52. An *eigenvalue* of a linear transformation $\sigma : V \to V$ is an element $\lambda \in F$ such that we have $\sigma(v) = \lambda v$ for some vector $v \in V - \{0\}$. We call such a vector v an *eigenvector* of σ.

Definition 23.53. A linear transformation $\sigma : V \to V$ is called *diagonalizable* if there is a basis B of V over F such that the matrix of σ with respect to B is diagonal.

Remark 23.54. Note the requirement in Definition 23.52 that $v \neq 0$. Indeed, our motivating situation required v to be an element of a basis of V, so v can only be the zero vector if $V = \{0\}$, which is forbidden by our assumption that $\dim_F(V) \geq 1$. If we allowed $v = 0$ in this definition, then *every* scalar would be an eigenvalue, since we must have $\sigma(0) = 0 = \lambda \cdot 0$ for all $\lambda \in F$. On the other hand, we do allow λ to be 0, and this is an important special case.

Remark 23.55. The words "eigenvalue" and "eigenvector" come from German, where *eigen* is a root used to denote ownership or belonging. The sense of the word "eigenvalue" is a value which is closely associated with, or characteristic of, the given linear transformation.

Task 23.7.1. Let $\sigma : V \to V$ be a linear transformation. Prove that σ is diagonalizable if and only if there is a basis of V over F which consists of eigenvectors of σ.

Definition 23.56. A basis consisting of eigenvectors of a given linear transformation is called an *eigenbasis*.

As we have come to expect, important concepts in a given algebraic category are often compatible with the natural structures of that category. The next task gives one example of this phenomenon, and shows that we have not yet exhausted the potential applications of the "eigen" prefix!

Task 23.7.2. Let V be a vector space over a field F. Let $\sigma : V \to V$ be a linear transformation, and let $\lambda \in F$. Let $V_\lambda = \{0_V\} \cup \{a \in V \mid \sigma(a) = \lambda a\}$. Prove that $V_\lambda \leq V$. We call V_λ the *eigenspace* of V corresponding to the eigenvalue λ (with respect to σ). Note that V_λ is defined even if λ is *not* actually an eigenvalue of σ, and in this case we have $V_\lambda = \{0_V\}$.

Until now, we have viewed a matrix as a particular, concrete way to represent a given linear transformation. But if we instead start with a matrix $M \in M_n(F)$ without any other information, then there is a natural way to construct a linear transformation from M. We need a definition before explaining how to do so.

Definition 23.57. Let V be the vector space formed from the set F^n using componentwise operations. The *standard basis* of V is the set $\{e_1, \ldots, e_n\}$ where the ith component of e_i is 1 and the other components are 0.

Task 23.7.3. Verify that a "standard basis" really is a basis.

Task 23.7.4. Let $V = F^n$ under componentwise operations; let B be the standard basis of V; and take $\sigma(a) = Ma$ for $a \in V$, where we view elements of V as column vectors. Prove that in this situation, $\sigma : V \to V$ is a linear transformation, and M is the matrix of σ with respect to the basis B. Thus we may speak of the eigenvalues, eigenvectors, diagonalizability, etc. of a square matrix.

Definition 23.58. If $\sigma : F^n \to F^n$ is a linear transformation, then the matrix of σ with respect to the standard basis of F^n is called the *standard matrix* of σ.

Task 23.7.5. Let

$$M = \begin{pmatrix} 3 & -2 \\ -7 & 5 \end{pmatrix}.$$

Show that M is diagonalizable when viewed as an element of $M_2(\mathbf{R})$, but not when viewed as an element of $M_2(\mathbf{Q})$.

Task 23.7.5 illustrates that a given matrix may become diagonalizable if we extend the base field. But there are some matrices which are not diagonalizable no matter how much we extend the base field, as you will see in the following task.

Task 23.7.6. Let

$$M = \begin{pmatrix} 0 & -9 \\ 1 & 6 \end{pmatrix}.$$

Show that M is not diagonalizable over *any* extension field of $\mathbf{Q}$.

Computing the eigenvalues of a given square matrix without a guiding theory may quickly become a nightmare of simultaneous equation-solving. To discover a good point of view, we first use a result from the Determinants project.

Task 23.7.7. Let $M \in M_n(F)$. Use the result of Task 23.6.9 to prove that 0 is an eigenvalue of M iff $\det(M) = 0$.

After this modest beginning, we need two more ideas before we will arrive at the desired insight into eigenvalues. The first idea allows us to generalize the result of Task 23.7.7 from 0 to *any* eigenvalue λ, by manipulating the equation $Mv = \lambda v$.

Task 23.7.8. Let $M \in M_n(F)$ and let $\lambda \in F$. Use basic matrix algebra together with the result of Task 23.7.7 to prove that λ is an eigenvalue of M iff $\det(\lambda I - M) = 0$.

Now we only need one more idea to reach the great insight we are seeking. To fully realize the potential of the result of Task 23.7.8, our idea is to replace the field element λ with a variable x. From the explicit determinant formula in Determinant Property 10, we see that $\det(xI - M)$ is a polynomial in $F[x]$; thus, *the roots of this polynomial should be precisely the eigenvalues of the matrix M*. The only trouble with this reasoning is that our theory of matrices has only allowed the entries of a matrix to come from a *field*, not from a more general ring such as a polynomial ring. One way to fix this problem is to use the field of rational functions $F(x)$ as our starting point instead of F itself. (In fact, there is a natural theory of matrices over any commutative ring with 1.)

Task 23.7.9. Let F be a field and $n \in \mathbf{Z}^+$. Let K be any extension field of F. Show that $M_n(F) \leq M_n(K)$ (as rings).

Definition 23.59. Let $M \in M_n(F)$, where F is a field. View F as a subring of $F(x)$, the field of rational functions over F. The *characteristic polynomial of M* is the element $\chi_M(x) = \det(xI - M) \in F(x)$.

Task 23.7.10. Prove that we have $\chi_M(x) \in F[x]$, justifying the use of the word "polynomial" in the definition of $\chi_M(x)$. More specifically, prove that $\chi_M(x)$ is monic and of degree n.

Task 23.7.11. Let $M \in M_n(F)$. Prove that the eigenvalues of M in F are precisely the roots of $\chi_M(x)$ in F. Note that some authors are willing to allow eigenvalues to come from any extension field L of F without mentioning L.

Bundled up in its characteristic polynomial are many of the key secrets of a matrix. In many cases, knowing $\chi_M(x)$ is enough to know that M is diagonalizable; understanding this is our next goal. We start with the result that eigenvectors corresponding to different eigenvalues are linearly independent.

Task 23.7.12. Let $M \in M_n(F)$. Let $\lambda_1, \ldots, \lambda_k$ be k distinct elements of F, and suppose that for all $i \in \{1, \ldots, k\}$, v_i is an eigenvector of M with eigenvalue λ_i. Prove that $(v_1, \ldots, v_k)$ is linearly independent over F. Suggestion: Induct on k.

Task 23.7.13. Let $M \in M_n(F)$, and suppose that $\chi_M(x)$ has n distinct roots in F. Prove that M is diagonalizable over F.

Task 23.7.14. Prove that similar matrices (see Definition 23.40) have the same characteristic polynomials.

Task 23.7.15. Suppose that F is an algebraically closed field, and let $M \in M_2(F)$. Prove that M is similar to a lower triangular matrix,

$$M \sim L = \begin{pmatrix} L_{1,1} & 0 \\ L_{2,1} & L_{2,2} \end{pmatrix}.$$

Show that we have $\chi_M(x) = (x - L_{1,1})(x - L_{2,2})$, and thus that M is diagonalizable if $L_{1,1} \neq L_{2,2}$. Show that the set T of all 2 by 2 lower triangular matrices over F is a vector subspace of $M_2(F)$ of dimension 3 over F. Conclude that the set of all non-diagonalizable matrices in T is a subset of a 2-dimensional subspace of T, and is thus "small" compared to T itself.

We conclude this project with some applications of eigenvalues.

Task 23.7.16. Define a sequence $a_1, a_2, \ldots$ by $a_1 = a_2 = 1$ and $a_{j+2} = a_{j+1} + a_j$ for all $j \in \mathbf{Z}^+$. Find a matrix $M \in M_2(\mathbf{R})$ such that

$$M \cdot (a_j, a_{j+1})^T = (a_{j+1}, a_{j+2})^T$$

for all $j \in \mathbf{Z}^+$. Find the eigenvalues of M; you should find that they are two distinct real numbers λ_1 and λ_2 where we can write $0 < \lambda_1 < 1 < \lambda_2$. Deduce that we can write $M \sim \operatorname{diag}(\lambda_1, \lambda_2)$, hence that there exist real constants c_1, c_2 such that for all $j \in \mathbf{Z}^+$ we have $a_j = c_1 \lambda_1^j + c_2 \lambda_2^j$. Use the known values of a_1, a_2, λ_1, and λ_2 to solve for c_1 and c_2. Since $0 < \lambda_1 < 1$, note that we have $a_j \approx c_2 \lambda_2^j$ for large values of j; evaluate both sides for a few values of $j \geq 3$ and compare. This illustrates how eigenvalues, especially the largest ones, control the long-term behavior of linear systems. Note that the sequence $a_1, a_2, \ldots$ is the famous *Fibonacci sequence*.

Task 23.7.17. As another example of how eigenvalues and eigenvectors occur in practice, consider a space divided into n regions R_1 through R_n. Suppose that at any given time, each region R_i contains a certain quantity q_i of something: perhaps q_i measures the number of people in region R_i, in millions. We make the following assumptions:

(1) time is divided into intervals of a fixed length;

(2) in any single time interval, the population in region R_i will move to the other regions (including region R_i itself) in fixed proportions; and

(3) the "universe" $U = \cup_{j=1}^n R_i$ forms a closed system: that is, no one can move into U from outside of U, or vice versa.

For example, it could be that 23.1% of the people in region R_1 move from region R_1 to region R_4 in each time interval. In general, let $M_{i,j}$ be the fraction of people moving from region R_j to region R_i in each time interval. Let M be the $n \times n$ matrix whose entries are the numbers $M_{i,j}$. Note that we should have $\sum_{i=1}^n M_{i,j} = 1$ for each $j \in \{1, \ldots, n\}$. Use this fact to prove that 1 is an eigenvalue of M. Then prove that 1 is the only possible real eigenvalue of M whose eigenvectors have non-negative entries, as our assumptions require. Note that an eigenvector with eigenvalue 1 corresponds to a "steady-state" distribution of people throughout the n regions, that is, a distribution which will not change over time. Find such an eigenvector for the particular matrix

$$M = \begin{pmatrix} 0.3 & 0.25 & 0.4 \\ 0.5 & 0.3 & 0.1 \\ 0.2 & 0.45 & 0.5 \end{pmatrix}.$$

We note that the type of model described in this task was the starting point for Larry Page and Sergey Brin's PageRank® algorithm, which was the foundation of the Google search engine when it was launched commercially in 1998; in this case, the regions represent web pages.

In case there is still any doubt whether the material in this project is important, we note that eigenvectors appear in the list of axioms of quantum mechanics, one of the fundamental theories of physics that describes the nature of matter. Namely, the state of an object is described by a vector; each measurable property p of the object has an associated linear transformation t_p; and when an observation of the value of p is made, the state of the object instantly changes to become an eigenvector of t_p whose eigenvalue is precisely the measured value of p. Just before the measurement is made, the state of the object can be a more general vector, a linear combination of eigenvectors, which indicates that the object does not have a definite single value for property p until we measure it—one of the weird facts of quantum physics!

Many other applications of eigenvalues exist, including one referenced in the quotation at the start of this project, namely, a ranking algorithm for academic journals. In this situation, published articles form the "regions" and citations form the "movement" across regions. The goal is to measure the impact that any given article makes within its own subject area.

23.8   Linear Algebra: Rotations

Prerequisites. Chapter 13 and Project 23.6.

For which collections of vectors in $\mathbf{R}^n$ is it possible to rotate the vectors about the origin simultaneously so that they all have non-negative coordinates? To be specific, let us use the standard basis for $\mathbf{R}^n$ over $\mathbf{R}$, and represent a vector as a column of n real numbers with respect to this basis. In order to work on this question, we first need to agree what we mean by a "rotation." At least in dimensions $n \leq 3$, the reader is familiar with the geometric idea of rotation about a point or about an axis; we will not discuss these notions for now, except to state that we would like our rotations to send the origin of $\mathbf{R}^n$ to itself. Since a rotation would seem to involve "curving" as a basic necessity, some readers may find it slightly disconcerting that every rotation which fixes the origin should be a *linear transformation*; but the mystery clears up when we realize that, after all, a rotation should send lines to lines.

We would like to define a "rotation" to be a linear transformation from $\mathbf{R}^n$ to itself which preserves distances and angles; this leaves us to find an algebraic way to measure both of these quantities. To build our intuition, let's start with rotations in $\mathbf{R}^2$, where we already have some experience from the discussion of rotational symmetries in Chapter 5. We will consider the word "rotation" to be somewhat informal, and later replace it with a suitable general definition and a new term.

Task 23.8.1. Let α be a real number. Verify that Equation 5.1 represents the counterclockwise rotation $R =: R_\alpha$ about the origin by α radians, even if α does not have the special form $2\pi/n$ where $n \in \mathbf{Z}^+$. Then prove that R_α is a linear transformation from $\mathbf{R}^2$ to $\mathbf{R}^2$, and find the standard matrix of R_α.

Next, we will investigate linear transformations which preserve distances. We recall that the distance from $(0,0)$ to (v_1, v_2) in $\mathbf{R}^2$ is $\sqrt{v_1^2 + v_2^2}$. If the reader has not studied distance formulas in higher dimensions, then it may be surprising to see how the two-dimensional distance formula generalizes: instead of taking a sum of cubes followed by a cube root in three dimensions, we just keep adding squares of the components and taking the square root.

Task 23.8.2. Show geometrically that the distance from $(0,0,0)$ to (v_1, v_2, v_3) in $\mathbf{R}^3$ is equal to $\sqrt{v_1^2 + v_2^2 + v_3^2}$. Suggestion: Construct a right triangle in the x, y-plane with vertices $(0,0,0)$, $(v_1, 0, 0)$, and $(v_1, v_2, 0)$ and another right triangle perpendicular to the x, y-plane with vertices $(0,0,0)$, $(v_1, v_2, 0)$, and (v_1, v_2, v_3).

The following definition captures the general distance formula using the notion of length; the idea is that the length of a vector is equal to the distance from the tail to the head. Also note that since we are viewing our vectors as columns, the transpose of a vector is a row; this is more convenient to fit on the page.

Definition 23.60. Let $v = (v_1, \ldots, v_n)^T \in \mathbf{R}^n$. The *length* of v is the real number

$$|v| := \sqrt{v_1^2 + v_2^2 + \cdots + v_n^2}. \tag{23.8}$$

The *distance* between v and another vector $w \in \mathbf{R}^n$ is the real number $|v - w|$.

Now we can start to be more precise about the desired properties of a general rotation.

Definition 23.61. Let $t : \mathbf{R}^n \to \mathbf{R}^n$ be a linear transformation. We say that t *preserves lengths* if for all $v \in \mathbf{R}^n$, we have $|t(v)| = |v|$. We say that t *preserves distances* if for all $v, w \in \mathbf{R}^n$, we have $|t(v - w)| = |v - w|$.

Task 23.8.3. Show that a linear transformation from $\mathbf{R}^n$ to itself preserves distances iff it preserves lengths.

Even though Task 23.8.3 suggests that we should focus on length-preserving transformations, since their definition only involves one arbitrary choice of vector instead of two, it turns out that there is useful information to be learned by expanding the formula for distances:

Task 23.8.4. Let $v = (v_1, \ldots, v_n), w = (w_1, \ldots, w_n) \in \mathbf{R}^n$. Show that we have

$$|v - w|^2 = |v|^2 + |w|^2 - 2p(v, w) \tag{23.9}$$

where $p(v, w) = \sum_{i=1}^n v_i w_i$. Deduce that if $R : \mathbf{R}^n \to \mathbf{R}^n$ is a linear transformation that preserves lengths, then for all $v, w \in \mathbf{R}^n$ we must have $p(Rv, Rw) = p(v, w)$. Also note that we have $p(v, w) = v^T w = w^T v$, interpreting a 1×1 matrix as a real number.

The function which we labeled p in Task 23.8.4 above is evidently important: it must be preserved by any linear transformation which preserves distances or lengths. Therefore we will study the properties of p more closely. We begin by assigning p its traditional (though unimaginative) name; also, we generalize from $\mathbf{R}$ to an arbitrary field since the formula for p still makes sense there.

Definition 23.62. Let F be a field. The *dot product* on F^n is the function $\cdot : F^n \times F^n \to F$ given by the formula $v \cdot w = w^T v$.

Remark 23.63. The dot product is also known as the *scalar product*, since its range consists of scalars, i.e., field elements.

Task 23.8.5. Let F be a field, and let $\cdot$ be the dot product on F^n. Prove that for all $u, v, w \in F^n$ and all $c \in F$, the following properties hold:

IP1. $u \cdot (v + w) = u \cdot v + u \cdot w$ and $(u + v) \cdot w = u \cdot w + v \cdot w$.

IP2. $c(u \cdot v) = (cu) \cdot v = u \cdot (cv)$.

IP3. $u \cdot v = v \cdot u$.

The reader may be wondering why we used the prefix "IP" in naming the properties of the dot product. This is because we will use the same properties as axioms to define a more general type of function called an *inner product*.

Definition 23.64. Let V be a vector space over a field F. An *inner product* on V is a function $\cdot : V^2 \to F$ satisfying properties IP1 and IP2. An inner product is called *symmetric* if it also satisfies IP3.

Task 23.8.6. Let V be an n-dimensional vector space over a field F, with ordered basis $B = (b_1, \ldots, b_n)$.

(a) Suppose that $\cdot$ is an inner product on V. Let $A \in M_n(F)$ be the matrix whose entry in row i and column j is $a_{i,j} = b_i \cdot b_j$. Prove that for all $v, w \in V$, we have $v \cdot w = c_v^T A c_w$, where c_v and c_w denote the coordinate vectors representing v and w with respect to B (as column vectors). Further, prove that $\cdot$ is symmetric iff A is symmetric.

(b) Conversely, prove that if $A \in M_n(F)$, then the formula $v \cdot w = c_v^T A c_w$ defines an inner product on V, which is symmetric iff A is symmetric. Conclude that there is a bijective correspondence between inner products on V and $n \times n$ matrices over F, with symmetric inner products corresponding to symmetric matrices.

(c) Show that the matrix corresponding to the dot product on F^n with the standard basis is the identity matrix.

Definition 23.65. Let V be an n-dimensional vector space over a field F, with ordered basis $B = (b_1, \ldots, b_n)$. Let $\cdot$ be an inner product on V. Then the *matrix of* $\cdot$ *with respect to* B is the matrix $A \in M_n(F)$ given by the correspondence in Task 23.8.6.

Next we single out one additional property of the dot product, which allows us to recover lengths and distances:

Task 23.8.7. Show that for all $v \in \mathbf{R}^n$, we have $|v| = \sqrt{v \cdot v}$, where $\cdot$ is the dot product. Conclude that we can express the distance between any two vectors of $\mathbf{R}^n$ in terms of the dot product.

In exploring the consequences of preserving distances, we have not yet considered what it means to preserve angles. Luckily, the work we already did allows us to show that preserving angles is a consequence of preserving distances. Specifically, let us consider Equation 23.9. This equation reminds us of the Law of Cosines in trigonometry (see Figure 23.8). If we identify O

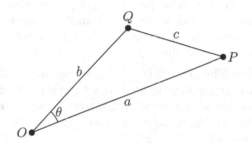

FIGURE 23.8: The Law of Cosines: $|c|^2 = |a|^2 + |b|^2 - 2|a||b|\cos(\theta)$

as the origin, and draw vector v from O to P and vector w from O to Q, then $v - w$ is the vector from Q to P, and we have $|v| = |a|$, $|w| = |b|$, and $|v - w| = |c|$. Thus everything fits perfectly, suggesting that we should have $v \cdot w = |v||w|\cos(\theta)$, where θ is the angle between v and w. Since geometry is outside of our scope, instead of accepting this equation at face value, we would like to use it to *define* the angle between two vectors: note that the remaining quantities can be computed using algebra! But to be careful, since the range of the cosine function is the interval $[-1, 1]$, we should first show that the ratio $v \cdot w / |v||w|$ always lies in this interval whenever $|v|$ and $|w|$ are not zero. That is, we would like to prove the following theorem:

Theorem 23.66. *Let* $v, w \in \mathbf{R}^n - \{0\}$, *where* $n \in \mathbf{Z}^+$. *Let* $\cdot$ *be the dot product on* $\mathbf{R}^n$. *Then we have*

$$-|v||w| \leq v \cdot w \leq |v||w|, \tag{23.10}$$

with equality on the left iff $w = -v$, *and equality on the right iff* $w = v$.

First we reduce to the case when both vectors have length 1.

Definition 23.67. A *unit vector* is a vector in $\mathbf{R}^n$ of length 1. The set of all unit vectors in $\mathbf{R}^n$ will be denoted UV(n).

Task 23.8.8. Prove that Theorem 23.66 is true if and only if it is satisfied for all unit vectors v and w.

Next we examine the form of unit vectors.

Task 23.8.9. (a) Show that $UV(1) = \{1, -1\}$, if we identify $\mathbf{R}^1$ with $\mathbf{R}$.

(b) Show that $UV(2) = \{(\cos(\theta), \sin(\theta)) : 0 \leq \theta < 2\pi\}$. Further, show that the function $f : [0, 2\pi) \to UV(2)$ given by the formula $\theta \mapsto (\cos(\theta), \sin(\theta))$ is bijective.

(c) Prove by induction that for all $n \geq 2$, UV_n is the set of all $u \in \mathbf{R}^n$ of the form

$$u = (\cos(\theta_1), \sin(\theta_1)\cos(\theta_2), \sin(\theta_1)\sin(\theta_2)\cos(\theta_3), \ldots, \sin(\theta_1)\cdots\sin(\theta_{n-1}))$$

where $\theta_i \in [0, 2\pi)$; that is, $u = (u_1, u_2, \ldots, u_n)$ where $u_1 = \cos(\theta_1)$, $u_2 = \sin(\theta_1)\cos(\theta_2)$, and in general, $u_j = \sin(\theta_1)\sin(\theta_2)\cdots\sin(\theta_{j-1})\cos(\theta_j)$ if $1 \leq j < n$, with final component $u_n = \sin(\theta_1)\sin(\theta_2)\cdots\sin(\theta_{n-1})$. Note that we no longer have uniqueness of the angles θ_i when $n > 2$.

Task 23.8.10. Prove Theorem 23.66 by induction on the dimension n. You will only need basic trigonometric identities together with the results of the previous tasks; you may also find useful the fact the a real-valued linear function whose domain is a closed interval must have its extreme values at the endpoints of the domain.

We are now in a position to measure the angle between two non-zero vectors in $\mathbf{R}^n$.

Definition 23.68. Let $v, w \in \mathbf{R}^n - \{0\}$. The *angle between v and w* is the unique real number θ in the interval $[0, \pi]$ such that $\cos(\theta) = (v \cdot w)/(|v|\,|w|)$, where $\cdot$ is the dot product.

A special case of interest is when two vectors meet at right angles:

Definition 23.69. Let $v, w \in \mathbf{R}^n$. We say that v and w are *orthogonal* (to each other) if $v \cdot w = 0$.

Remark 23.70. When v and w are non-zero, then to say that v and w are orthogonal means that the angle between them is $\pi/2$. But by our definition, the zero vector is orthogonal to every vector, even though angles are not defined in this case.

The following definition is now amply motivated:

Definition 23.71. Let $t : \mathbf{R}^n \to \mathbf{R}^n$ be a linear transformation. We say that t *preserves angles* if t is injective and for all $v, w \in \mathbf{R}^n - \{0\}$, we have

$$\frac{t(v) \cdot t(w)}{|t(v)|\,|t(w)|} = \frac{v \cdot w}{|v|\,|w|}, \tag{23.11}$$

where $\cdot$ is the dot product.

Task 23.8.11. Let $t : \mathbf{R}^n \to \mathbf{R}^n$ be a linear transformation. Prove that if t preserves distances, then t preserves angles.

Now we have established the following result:

Proposition 23.72. *Let* t : $\mathbf{R}^n \to \mathbf{R}^n$ *be a linear transformation. If* t *preserves lengths, then* t *preserves distances and angles.*

As promised, now that we are in a position to formally define the idea of a "rotation," we will introduce a new word:

Definition 23.73. A linear transformation t : $\mathbf{R}^n \to \mathbf{R}^n$ is called an *orthogonal transformation* (of $\mathbf{R}^n$) if t preserves lengths.

Proposition 23.72 is stated in plain language, at what we might call a "conceptual" level, but we have also built enough theory to reformulate this result in a single, elegant equation.

Task 23.8.12. Let t be an orthogonal transformation of $\mathbf{R}^n$. Let $B = (e_1, \ldots, e_n)$ be the ordered standard basis of $\mathbf{R}^n$, and let $\cdot$ be the dot product.
  (a) Verify that for all $i, j \in \{1, \ldots, n\}$, we have

$$e_i \cdot e_j = \begin{cases} 1, & \text{if } i = j; \\ 0, & \text{if } i \neq j. \end{cases}$$

  (b) Let A be the matrix of t with respect to B. Recall that the columns of A are the coordinate vectors of $t(e_i)$ with respect to B. Verify that the entry in row i and column j of the product $A^T A$ is the value $t(e_i) \cdot t(e_j)$.
  (c) Conclude that $A^T A = I$, the $n \times n$ identity matrix.
  (d) Now let s : $\mathbf{R}^n \to \mathbf{R}^n$ be a linear transformation whose standard matrix C satisfies the equation $C^T C = I$. Prove that for all $i, j \in \{1, \ldots, n\}$, we have $s(e_i) \cdot s(e_j) = e_i \cdot e_j$. Use the properties of the dot product to extend this formula to an arbitrary pair of vectors in $\mathbf{R}^n$. Use these results to complete the proof of Proposition 23.74 below.

Proposition 23.74. *Let* t : $\mathbf{R}^n \to \mathbf{R}^n$ *be a linear transformation, and let* A *be the standard matrix of* t. *Then* t *is an orthogonal transformation if and only if*
$$A^T A = I. \tag{23.12}$$

Because Equation 23.12 makes sense when A is a square matrix over *any* field, we use that equation to define a type of matrix which we may view as a generalization of a rotation matrix:

Definition 23.75. Let F be a field and let $n \in \mathbf{Z}^+$. A matrix $A \in M_n(F)$ is called an *orthogonal matrix* if $A^T A = I$, where I is the $n \times n$ identity matrix in $M_n(F)$.

Now that we have a simple formula defining orthogonal matrices, we can start to explore their properties.

Task 23.8.13. Let A be an $n \times n$ orthogonal matrix with entries in a field F. Prove that we must have $\det(A) \in \{1_F, -1_F\}$. Conclude that $A \in \mathrm{GL}_n(F)$ and that $A^{-1} = A^T$.

Notation 23.76. The set of all $n \times n$ orthogonal matrices over a field F is denoted $O_n(F)$. The subset of $O_n(F)$ consisting of matrices with determinant 1 is denoted $SO_n(F)$.

As an important and naturally defined subset of a group, we expect that the set of all orthogonal matrices of a given size over a given field should be a subgroup of that group. This is true; to prove it, we first establish a property of transposes:

Lemma 23.77. *Let F be a field, and let A and B be matrices over F such that the product AB is defined (see Exercise 13.16). Then the product $B^T A^T$ is also defined, and we have $(AB)^T = B^T A^T$.*

Task 23.8.14. Prove Lemma 23.77.

Task 23.8.15. Let F be a field, and let $n \in \mathbf{Z}^+$. Prove that $O_n(F) \leq \mathrm{GL}_n(F)$. Then prove that $SO_n(F)$ is a normal subgroup of $O_n(F)$ of index 2 if $\mathrm{char}(F) \neq 2$. Note that if $\mathrm{char}(F) = 2$, then $SO_n(F) = O_n(F)$.

Now we are justified in naming these matrix groups: $O_n(F)$ is called the *orthogonal group* of $n \times n$ matrices over F, and $SO_n(F)$ is the *special orthogonal group*.

Task 23.8.16. Using the notation of Task 23.8.1, let M_α be the standard matrix of R_α. Show that we have

$$SO_2(\mathbf{R}) = \{M_\alpha \; : \; 0 \leq \alpha < 2\pi\}.$$

Thus, the special orthogonal group over $\mathbf{R}$ represents "true" rotations, at least in dimension two. Now let F be the flip about the x-axis defined by Equation 5.2, and let L be the standard matrix of F. Verify that $L \in O_2(\mathbf{R})$ but $L \notin SO_2(\mathbf{R})$. Conclude that $O_2(\mathbf{R}) = SO_2(\mathbf{R}) \cup L \cdot SO_2(\mathbf{R})$.

Remark 23.78. Task 23.8.16 gives an idea of how an orthogonal transformation is more general than a rotation. Notice that we can accomplish the flip F by performing a rotation in $\mathbf{R}^3$ of $180°$ about the x-axis, but the 2×2 matrix L fails to capture the information that the z-axis has also been flipped! This is where the "missing" -1 can be found that would allow F, and hence all of $O_2(\mathbf{R})$, to consist of actual rotations with determinant 1 only.

The next two tasks help to develop some insight about the orthogonal groups by examining their relationships to each other and to the symmetric groups S_n.

Task 23.8.17. Let F be a field, and let $n \in \mathbf{Z}^+$. Define a function $\sigma \; : \; O_n(F) \to M_{n+1}(F)$ by the formula

$$\sigma(M) = \left(\begin{array}{c|c} M & \begin{matrix} 0 \\ \vdots \\ 0 \end{matrix} \\ \hline 0 \cdots 0 & 1 \end{array} \right)$$

$$(23.13)$$

for $M \in O_n(F)$. Prove that we have $\sigma(M) \in O_{n+1}(F)$, and that σ is an embedding of $O_n(F)$ into $O_{n+1}(F)$.

Remark 23.79. The matrix $\sigma(M)$ in Task 23.8.17 is an example of what is called a "block-diagonal" matrix. The strategically placed 0s effectively separate the action of the top-left block, M, and the bottom-right block, 1, into the spaces F^n and F, where F is identified with its natural embedding into the last component of F^{n+1}.

Task 23.8.18. Let F be a field, and let $n \in \mathbf{Z}^+$. For $f \in S_n$, let E_f be the matrix whose entries in row i, column $f(i)$ are 1 for all $i \in \{1, 2, \ldots, n\}$, and whose other entries are 0, as in Task 23.6.16. Prove that for all $f \in S_n$, we have $E_f \in O_n(F)$. Conclude that the map Ψ from Task 23.6.16 is an embedding of S_n in $O_n(F)$. Note that this implies that we can permute the components of a vector in F^n however we like by applying an appropriate orthogonal matrix.

Next, we will explore some of the "power" of orthogonal transformations: given a vector $v \in \mathbf{R}^n$, for example, what are the possible output values Mv when $M \in O_n(\mathbf{R})$? We know that Mv must have the same length as v, but are there any other restrictions?

Task 23.8.19. Let $n \in \mathbf{Z}^+$ and let $v \in \mathbf{R}^n$. Prove that there exists $M \subset O_n(\mathbf{R})$ such that $Mv = |v| e_1$. Suggestion: Induct on n. The results of Tasks 23.8.17 and 23.8.18 may be useful here.

Task 23.8.19 shows that any vector can be moved so that it "points along the positive x-axis" by applying an appropriate orthogonal matrix (if we think of the first component of a vector as the x component). This agrees with our intuition from $O_2(\mathbf{R})$. Even more, our intuition suggests that any list of m vectors in $\mathbf{R}^n$ can be simultaneously rotated to "fit" into the first m components. To realize this idea, we will write the list of vectors as columns of a matrix C; then computing the product AC has the effect of simultaneously applying A to every column of C.

Task 23.8.20. Let $m, n \in \mathbf{Z}^+$ and let $C \in M_{m,n}(\mathbf{R})$. Prove by induction on m that there exists a matrix $A \in O_m(\mathbf{R})$ such that $(AC)_{j,k} = 0$ whenever $j > k$, where $(AC)_{j,k}$ denotes the entry of AC in row j, column k. Conclude as a special case the result of Proposition 23.80 below. Suggestion: Make use of the result of Task 23.8.19 and appropriate block-diagonal matrices.

Proposition 23.80. *For every $C \in M_n(\mathbf{R})$, there exists $A \in O_n(\mathbf{R})$ such that AC is an upper triangular matrix.*

Although not directly helpful to solving our main problem, the next result is important enough to state and prove, as it follows readily from our current knowledge.

Theorem 23.81. *Let $C \in GL_n(\mathbf{R})$, and let $(c_1, c_2, \ldots, c_n)$ be the list of columns of C. Then $|\det(M)| \leq |c_1| \cdot |c_2| \cdots \cdot |c_n|$, with equality iff c_i is orthogonal to c_j for all $i \neq j$.*

Task 23.8.21. Prove Theorem 23.81. Suggestion: Make use of Proposition 23.80 and Task 23.6.25.

Now that we know a little about rotations, we return to the original question raised at the opening of this project. Let's make things a bit easier by considering only a *finite* list of column vectors $\mathcal{L} = (c_1, \ldots, c_m)$ with $c_i \in \mathbf{R}^n$. The result of "rotating a vector c_i about the origin" we interpret as the vector Ac_i for some $A \in O_n(\mathbf{R})$. To consider the entire list $\mathcal{L}$ at once, we can, as suggested above, write each of the columns c_i from left to right to produce a matrix

$$C = \begin{pmatrix} c_1 & c_2 & \cdots & c_m \end{pmatrix} \in M_{n,m}(\mathbf{R}). \tag{23.14}$$

Then the results of simultaneously rotating all of the vectors in $\mathcal{L}$ using A are the columns of the matrix $AC \in M_{n,m}(\mathbf{R})$. So our question becomes: For which matrices $C \in M_{n,m}(\mathbf{R})$ does there exist a matrix $A \in O_n(\mathbf{R})$ such that the entries of AC are all non-negative?

To explore this question, let us look at the relationship between C and AC. Since A is an orthogonal matrix, we expect that each column of AC should have the same length (as a column vector) as the corresponding column of C, and furthermore that the dot product of any two columns of AC should be the same as the dot product of the corresponding columns of C. As we saw in Task 23.8.12, these dot products are just the entries of the matrices $(AC)^T(AC)$ and C^TC, respectively. Therefore, we expect these two matrices to be equal:

Task 23.8.22. Let $C \in M_{n,m}(\mathbf{R})$ and $A \in O_n(\mathbf{R})$. Prove that we must have $(AC)^T(AC) = C^TC$. (This is a straightforward calculation using the previous results, in particular Lemma 23.77.)

The result of Task 23.8.22 gives us a test which allows us to show easily, in some cases at least, that two $n \times m$ matrices are *not* related by an orthogonal transformation: namely, if $C^TC \neq D^TD$, then there can be no orthogonal matrix A such that $D = AC$. This is an example of an important phenomenon in mathematics, so we will discuss it further now. The next task helps to put the current situation in a more general framework.

Task 23.8.23. Define a relation $\sim$ on $S := M_{n,m}(\mathbf{R})$ by the condition $C \sim D$ iff there exists $A \in O_n(\mathbf{R})$ such that $D = AC$. Prove that $\sim$ is an equivalence relation on S. Define a function $f : S \to M_n(\mathbf{R})$ by the formula $f(C) = C^TC$. Prove that if $C, D \in S$ and C is in the same equivalence class as D, then $f(C) = f(D)$.

Again, the significance of the function f in Task 23.8.23 is that it provides an easy way to potentially show that two elements in a set are not related according to $\sim$: it is easy to compute C^TC and D^TD from C and D and compare these two matrices, but it may not be easy to compute every possible value of the matrix AC as A runs through all elements of $O_n(\mathbf{R})$ in order to compare each AC to D! In this situation, we say that f is an *invariant* of $\sim$. A general definition of this concept follows.

Definition 23.82. Let S be a set with an equivalence relation $\sim$. A function $f : S \to T$ (for some set T) is called an *invariant* of $\sim$ if for all $a, b \in S$ we have $a \sim b \implies f(a) = f(b)$. If $f : S \to T$ and $g : S \to U$ are two invariants of the same equivalence relation $\sim$, then we say that f is *finer* than g if for every $t \in T$, there exists $u \in U$ such that $f^{-1}(t) \subseteq g^{-1}(u)$. Instead of saying that f is finer than g, we may say that g is *coarser* than f.

Remark 23.83. A finer invariant gives more information, while a coarser invariant gives less. The finest possible invariants are those that have distinct values on different equivalence classes. If we define the "quotient set" $S/\sim$ to be the set of all equivalence classes of S under $\sim$, then an invariant of $\sim$ is the same thing as a function with domain S which naturally induces a function on $S/\sim$. In this view, the finest invariants are those that induce injective functions on $S/\sim$.

Task 23.8.24. Assume the notation of Task 23.8.23, with $m = n$. Define a function $g : M_n(\mathbf{R}) \to \mathbf{R}$ by the formula $g(C) = (\det(C))^2$. Show that g is an invariant of $\sim$, and that g is coarser than f.

Again we return to our main question. Our strategy will use the invariant f of Task 23.8.23, $f(C) = C^T C$. From now on, let $\sim$ denote the relation from that task.

Definition 23.84. Let $C \in M_{n,m}(\mathbf{R})$. We will say that C is *rotatable into a non-negative matrix* if $C \sim D$ for some matrix D with non-negative entries.

To give the reader an example of using the invariant f, we prove the following result:

Lemma 23.85. *Let $C \in M_{n,m}(\mathbf{R})$. If C is rotatable into a non-negative matrix, then all pairs of non-zero column vectors in C meet at angles of at most $90°$.*

Proof. Suppose that $C \sim D$ where D has non-negative entries. Then certainly $D^T D$ also has non-negative entries. But also, we have $f(C) = f(D)$, so $C^T C$ has non-negative entries. Notice that this condition says exactly that all pairs of column vectors in C meet at angles of at most $90°$ whenever the angles are defined, that is, for non-zero columns of C. $\square$

A natural question is whether the converse of Lemma 23.85 holds.

Task 23.8.25. Prove the converse of Lemma 23.85 in the case $n = 2$. Suggestion: Let $C \in M_{2,m}(\mathbf{R})$, and suppose that $C^T C$ has non-negative entries. Let $(c_1, \ldots, c_m)$ be the list of column vectors of C, and let $\theta_{j,k}$ be the angle between c_j and c_k, ignoring any columns which are zero. Look at the maximum value of $\theta_{j,k}$ and the two columns where it occurs; find an orthogonal matrix which moves one of these two columns to the positive x-axis.

To make more progress on our question, we would like to better understand the invariant f. Given $C \in M_{n,m}(\mathbf{R})$, we can ask for solutions D to the equation $f(C) = f(D)$, and then seek non-negative examples for D within the solution set. A more "neutral" way of asking this question is: Given a matrix $M \in M_m(\mathbf{R})$, for which $A \in M_{n,m}(\mathbf{R})$ do we have $f(A) = M$? In other words, what is the pre-image $f^{-1}(M)$ of M under f? Perhaps the most basic version of this question is, when is $f^{-1}(M)$ non-empty? In order to simplify the question still further, we will restrict from now on to the case $m = n$. Thus we are asking: For which $M \in M_n(\mathbf{R})$ does there exist $C \in M_n(\mathbf{R})$ such that $C^T C = M$?

Our experience so far suggests that we might view $C^T C$ as the middle part of the formula for the length of the image of a vector under C: namely, $|Cv|^2 = (Cv)^T Cv = v^T C^T Cv$. This gives a necessary condition on any matrix M for which $f^{-1}(M)$ is non-empty:

Lemma 23.86. *Let $M \in M_n(\mathbf{R})$. If $f^{-1}(M)$ is non-empty, then we have $v^T M v \geq 0$ for every $v \in \mathbf{R}^n$.*

Task 23.8.26. Prove Lemma 23.86.

Notice that the condition in Lemma 23.86 involves the same formula that defines an inner product from a given square matrix (see Task 23.8.6). Thus we arrive at the following definition:

Definition 23.87. Let $\cdot$ be an inner product on $\mathbf{R}^n$. We say that $\cdot$ is *positive semi-definite* if for all $v \in \mathbf{R}^n$ we have $v \cdot v \geq 0$. We say that $\cdot$ is *positive definite* if $v \cdot v > 0$ for all $v \in \mathbf{R}^n - \{0\}$. We say that a matrix $M \in M_n(\mathbf{R})$ is positive semi-definite (respectively, positive definite) if the inner product corresponding to M with respect to the standard basis of $\mathbf{R}^n$ is positive semi-definite (respectively, positive definite).

What difference does the "semi-" make in a semi-definite matrix M? Of course, it allows some values of $v \cdot v$ to be 0 even when $v \neq 0$, where $\cdot$ is the inner product corresponding to M. As the reader may have suspected, this is related to the question of which vectors M sends to 0; that is, to whether M is singular.

Task 23.8.27. Let $M \in M_n(\mathbf{R})$. Prove that if M is singular, then M is not positive definite.

There is another property we can wrestle out of the special form $C^T C$. This property is suggested by the identity $v^T w = w^T v$.

Task 23.8.28. Let $C \in M_n(\mathbf{R})$. Prove that the matrix $C^T C$ is symmetric. Conclude that if $M \in M_n(\mathbf{R})$ and $f^{-1}(M)$ is non-empty, then M is symmetric.

Grouping together Lemma 23.86 and Task 23.8.28, we have the following:

Lemma 23.88. *Let $M \in M_n(\mathbf{R})$. If there is a matrix $C \in M_n(\mathbf{R})$ such that $C^T C = M$, then M is symmetric and positive semi-definite.*

Proposition 23.80 is quite strong: it "almost" gives a unique form for choosing representatives from each equivalence class of $M_n(\mathbf{R})$ under $\sim$. This proposition may be used in the following task. First we give a definition that isolates the upper-left corners of a square matrix.

Definition 23.89. Let $A \in M_n(F)$, where F is a field. For $k \in \{1, 2, \ldots, n\}$, the kth *principal minor* of A is the matrix $B \in M_k(F)$ such that $B_{i,j} = A_{i,j}$ for all $i, j \in \{1, 2, \ldots, k\}$.

Task 23.8.29. Let $C \in M_n(\mathbf{R})$, and let $A = C^T C$. Use Proposition 23.80 to prove that there is an upper triangular matrix $B \in M_n(\mathbf{R})$ such that for every $j \in \{1, 2, \ldots, n\}$, we have $A_j = B_j^T B_j$, where A_j and B_j denote the jth principal minors of A and of B, respectively. Use this result to help prove Lemma 23.90 below.

Lemma 23.90. *Let $A \in M_n(\mathbf{R})$. If $A = C^T C$ for some $C \in M_n(\mathbf{R})$, then we have $\det(A_j) \geq 0$ for all $j \in \{1, 2, \ldots, n\}$, where A_j is the jth principal minor of A. Furthermore, if $\det(A_j) = 0$ for some j, then we have $\det(A) = 0$.*

Now that attention has been called to the principal minors of a positive semi-definite matrix, we will bring minors to bear on the question of definite versus semi-definite.

Task 23.8.30. Let $A \in M_n(\mathbf{R})$ be positive semi-definite. Prove that if $\det(A_j) = 0$ for some $j \in \{1, 2, \ldots, n\}$, where A_j is the jth principal minor of A, then A is not positive definite.

Next we investigate a converse to the results of Tasks 23.8.30 and 23.8.29.

Task 23.8.31. Let $A \in M_n(\mathbf{R})$, and suppose that A is symmetric and $\det(A_j) > 0$ for all $j \in \{1, 2, \ldots, n\}$, where A_j is the jth principal minor of A. Prove that $A = C^T C$ for some $C \in GL_n(\mathbf{R})$. Suggestion: Show inductively that A_j is of the form $C(j)^T C(j)$ where $C(j) \in M_j(\mathbf{R})$ and $C(j)$ is upper triangular, with non-zero entries on the diagonal. Use the following steps to move the induction from j to $j + 1$. Show that we must have $\det(C(j)) \neq 0$. Look for a vector $v \in \mathbf{R}^j$ that has the appropriate dot products with the columns of $C(j)$. The final step of the induction is to show that $|v|^2 < A_{j+1,j+1}$ so that we may choose the $(j+1)^{th}$ component of v. For this purpose, let z be a square root of the real number $A_{j+1,j+1} - |v|^2$ in $\mathbf{C}$, and show that this assignment gives the desired matrix A_{j+1}; deduce that if $z \in \mathbf{C} - \mathbf{R}$ then $\det(A_{j+1}) < 0$, a contradiction.

Task 23.8.32. Prove Theorem 23.91 below. This should not involve much "new" work.

Theorem 23.91. *Let A be a symmetric matrix in $M_n(\mathbf{R})$. Then the following are equivalent:*
(i) $A = C^T C$ for some $C \in GL_n(\mathbf{R})$.
(ii) $\det(A_j) > 0$ for all $j \in \{1, 2, \ldots, n\}$, where A_j is the jth principal minor of A.

Finally, we can give a partial answer to our original question. Although not a complete answer, it is surprising enough to form a conclusion to this project.

Definition 23.92. Let $n \in \mathbf{R}^n$. The *non-negative orthant* of $\mathbf{R}^n$ is the subset of $\mathbf{R}^n$ whose elements have non-negative components; i.e., the set

$$\{(x_1, \ldots, x_n) \in \mathbf{R}^n \ : \ x_i \geq 0 \text{ for all } i\}.$$

Proposition 23.93. *There exists a list of n vectors in $\mathbf{R}^n$ such that every pair of vectors meets at an angle of at most $90°$, but such that these vectors cannot be simultaneously rotated into the non-negative orthant of $\mathbf{R}^n$. More formally, there exists a matrix $C \in M_n(\mathbf{R})$ such that $C^T C$ has non-negative entries, but such that there does not exist a matrix $A \in O_n(\mathbf{R})$ such that AC has non-negative entries.*

Proof. Let $n = 5$. Consider the matrix

$$M = M(\alpha) = \begin{bmatrix} 1 & 0 & 0 & \alpha & \alpha \\ 0 & 1 & 0 & \alpha & \alpha \\ 0 & 0 & 1 & \alpha & \alpha \\ \alpha & \alpha & \alpha & 1 & 0 \\ \alpha & \alpha & \alpha & 0 & 1 \end{bmatrix}$$

where α is a non-negative real number. Let $M_j(\alpha)$ denote the jth principal minor of M as a function of α. When $\alpha = 0$, then $M = I$, so we have $\det(M_j(0)) = 1$ for every $j \in \{1, \ldots, 5\}$. Now $\det(M_j(\alpha))$ is a polynomial function in α, so by continuity, there exists a small positive value of α such that $\det(M_j(\alpha)) > 0$ for each j. Fix such an α. By Theorem 23.91, we can write $M = C^T C$ for some matrix $C \in M_n(\mathbf{R})$. Assume for a contradiction that C has only non-negative entries. Then, since the first 3 column vectors of C are mutually orthogonal, at least one of these columns has exactly one non-zero entry; without loss of generality (in view of the symmetry of M with respect to the first 3 columns), the only non-zero entry in the first column of C is $C_{1,1} = 1$. But then $C_{1,4} = C_{1,5} = \alpha$, so columns 4 and 5 of C cannot be orthogonal, contradicting the fact that $M_{4,5} = 0$. $\qquad\square$

For further information on this topic, see the article [7].

23.9   Power Series

Prerequisites. Having completed the text through Chapter 20 is recommended to fully understand this material; the reader who has completed through Chapter 14 should be able to reach Task 23.9.8.

We have studied polynomials since our first (middle-school) algebra class, and devoted a whole chapter to them in this text (Chapter 14). Those readers of a more daring nature may have wondered what happens to polynomials if we allow the powers of the variable to go up forever: is there such a thing as an "infinite polynomial" which is an infinite sum of terms $a_0 + a_1 x + a_2 x^2 + \cdots$ and so on, without ending? Those who have studied calculus may have seen such things, and they are called *power series*. As with polynomials, however, we will be careful to distinguish between a *formal* power series, where we have a variable like x which is algebraically independent over the coefficient ring, versus a power series *function* whose "variable" actually represents, say, a real number. In a typical calculus class, these two different concepts are easily confused.

Definition 23.94. Let R be a commutative ring with 1. A *power series in the variable x with coefficients from R* is a sum

$$f = \sum_{j=0}^{\infty} a_j x^j$$

where $a_j \in R$. More precisely, we can represent f as a function $f : \mathbf{N} \to R$ from the set of natural numbers to the ring R, where $f(j) = a_j$. Unlike in the case of polynomials, we do not require the coefficients a_j to be zero for all $j >> 0$.

Notation 23.95. The set of all power series in x with coefficients from R is denoted $R[[x]]$.

Our immediate worry with power series is whether we can add and multiply them without having to compute an infinite sum or an infinite product. If we had to compute something like

$$a_0 + a_1 + a_2 + \cdots$$

with $a_j \in R$, then we would be in trouble! It turns out that everything is alright:

Task 23.9.1. Prove that addition and multiplication of power series can be defined as in Definition 14.1, but replacing the upper bounds of the sums by ∞, except that the upper bound in the inner sum in Equation 14.2 remains k. (This amounts to noting that in this definition, all of the sums *of elements of R* are finite sums.) Then prove the power series version of Lemma 14.8, namely:

Lemma 23.96. If R is a commutative ring with 1, then so is $R[[x]]$.

Finally, show that we have a natural embedding $R[x] \hookrightarrow R[[x]]$.

Task 23.9.2. Let R be a commutative ring with 1. Suppose that instead of using the natural numbers as powers of the variable x, we allow any integer power, to get things such as

$$\sum_{j=-\infty}^{\infty} a_j x^j$$

with $a_j \in R$. Let us call such things *doubly infinite series*. Show that we can still define addition of two doubly infinite series, but that the "natural" definition of multiplication of doubly infinite series fails to make sense in general.

Task 23.9.3. Let R be a commutative ring with 1, and let

$$f = \sum_{j=0}^{\infty} x^j \in R[[x]].$$

Calculate $(1 - x) \cdot f$ to prove that

$$f = (1 - x)^{-1} \text{ in } R[[x]]. \tag{23.15}$$

This result may not surprise the reader who is familiar with geometric series; what is new here for those who studied geometric series in algebra or calculus? First, we do not need to add the condition "if $|x| < 1$" to Equation 23.15; indeed, this condition does not even make sense here, since x is a variable, not necessarily a number. Instead, Equation 23.15 is an identity in the ring $R[[x]]$.

In calculus, we learn that we can add up an infinite series of real numbers only if the numbers get small (approach 0) as we add more and more of them. There is a similar way to think about the infinite sum in a power series; in this case, we can think of the higher and higher powers of x as being closer and closer to 0. In terms of ideals, this is more natural, since we have the following:

Task 23.9.4. Let R be a commutative ring with 1, and let $I = (x) = xR[[x]]$ be the ideal generated by x in the power series ring $R[[x]]$. Prove that for every positive integer n, we have $I^n = (x^n)$. Then prove that $\cap_{j=1}^{\infty} I^n = (0)$.

Definition 23.97. Let S be a commutative ring with 1, and suppose that I is an ideal of S such that $\cap_{j=1}^{\infty} I^n = (0)$. We say that S is *complete with respect to* I if for every sequence $a_1, a_2, \ldots$ of elements of S satisfying $a_{n+1} \equiv a_n$ mod I^n, there is a unique element $a \in S$ such that $a \equiv a_n$ mod I^n for all $n \in \mathbf{Z}^+$.

Notation 23.98. In the situation of Definition 23.97, we will write

$$a = \lim_{n \to \infty} a_n$$

and we say that a is the limit of the a_n.

Completeness is an important notion that occurs in both analysis and algebra, although the definition given above, in terms of ideals, is more common in algebra. The point of completeness is that a sequence of ring elements that "seems to be going somewhere" must in fact have a limiting value within the given ring.

Task 23.9.5. Let R be a commutative ring with $1 \neq 0$. Prove that $R[[x]]$ is complete with respect to (x), but that the corresponding statement for $R[x]$ is false.

Task 23.9.6 (Limit Laws). Prove that in a complete ring, limits commute with both addition and multiplication. More precisely, let S be a commutative ring with 1 which is complete with respect to an ideal I. Suppose that $\lim_{n \to \infty} a_n = a$ and $\lim_{n \to \infty} b_n = b$ in S. Prove that $\lim_{n \to \infty}(a_n + b_n) = a + b$ and $\lim_{n \to \infty}(a_n b_n) = ab$.

We have seen that polynomial rings do not have very many units, at least when the coefficient ring is a domain (Lemma 14.12). But in Task 23.9.3, we found that in any commutative ring with 1, the polynomial $x + 1$ is always a unit of the bigger ring of power series $R[[x]]$. Exactly what can we say about units in power series rings?

Task 23.9.7. Let R be a commutative ring with 1, and let $S = R[[x]]$. Prove that we have

$$S^\times = \{a_0 + a_1 x + a_2 x^2 + \cdots \in S \mid a_0 \in R^\times\}.$$

That is, a power series is a unit iff its constant coefficient is a unit, no matter what the higher terms look like. Suggestion: The harder direction is the $\Leftarrow$ implication. Suppose $f \in S$ has unit constant term. Prove by induction that for every $n \in \mathbf{N}$ there is a polynomial $g_n \in R[x]$ of degree at most n such that $f \cdot g_n \equiv 1 \mod (x^{n+1})$, with $g_{n+1} \equiv g_n \mod (x^n)$. Then use the Limit Laws.

From now on, we restrict the coefficient ring of our power series to be a field, in order to get some stronger results. When we move from polynomials to power series, we lose the concept of *degree*, since a single power series can involve arbitrarily large powers of x. But when we work over a field, then the *smallest* power of x in a power series plays a similar role:

Task 23.9.8. Let F be a field, and let $S = F[[x]]$. Let $a \in S - (0)$. Prove that there is a unique $n \in \mathbf{N}$ and a unique $u \in S^\times$ such that $a = x^n \cdot u$. Use this result to prove that S is a PID.

Notation 23.99. Let F be a field, and let $S = F[[x]]$. We denote the field of fractions of S by $F((x))$. We call $F((x))$ the field of *Laurent series* with coefficients in F.

Next, we introduce a standard notation for localizing when we "really" just need to invert a single non-zero ring element:

Notation 23.100. Let S be a domain, and let $a \in S - (0)$. Let $U = \{a^n \mid n \in \mathbf{N}\}$. We denote the localization $S[U^{-1}]$ by $S[a^{-1}]$.

Task 23.9.9. Let F be a field, and let $S = F[[x]]$. Prove that $F((x)) = S[x^{-1}] = \{\sum_{j=k}^{\infty} a_j x^j \mid k \in \mathbf{Z}, a_j \in F\}$. That is, a general Laurent series over F looks like a power series over F except that we can start summing at a *negative* power of x.

We conclude this project with two applications of power series: one to discrete mathematics, and another to calculus.

Task 23.9.10. Define a sequence $a_1, a_2, \ldots$ by $a_1 = a_2 = 1$ and $a_{n+2} = a_{n+1} + a_n$ for all $n \in \mathbf{Z}^+$ (compare to task 23.7.16). Let $f = \sum_{j=1}^{\infty} a_j x^j \in \mathbf{R}[[x]]$. Show that f satisfies the equation $f = x + x^2 f + xf$. Solve for f to see that f is actually a rational function, and write the result in the form

$$f = \frac{A}{x - \omega} + \frac{B}{x - \overline{\omega}},$$

where A and B are appropriate real numbers, and ω and $\overline{\omega}$ are the roots of the denominator of f. Then use the formula for the sum of a geometric series to find an explicit formula for the coefficients of f.

Finally, we introduce derivatives of power series.

Definition 23.101. Let R be a commutative ring with 1. For $f = \sum_{j=0}^{\infty} a_j x^j \in R[[x]]$, define the *derivative of f with respect to x* to be the power series

$$f' = \sum_{j=1}^{\infty} j \cdot a_j x^j.$$

Task 23.9.11. Prove that, just as for polynomials, the sum and product rules are true for derivatives of power series.

Task 23.9.12. Let F be a field of characteristic 0. Prove that there is a unique power series $f \in F[[x]]$ with constant term 1 such that $f' = f$, and find an explicit formula for f. Then compute f' from f to verify that $f' = f$. This power series will be familiar if you studied Taylor Series in calculus, as it is the power series representation about $x = 0$ of the natural exponential function e^x for $x \in \mathbf{R}$. Thus, power series can let us see directly the inner workings even of this mysterious function! In addition, you have just used power series to solve the *differential equation* $f' = f$. In general, a "differential equation" just means any equation involving derivatives.

To go any farther in a discussion of the real function e^x would be out of our scope. Unlike in the case of polynomial rings, there is no general way to *evaluate* a power series at a given ring element: in the case of power series, we would in general get an infinite sum of ring elements. This crosses the boundary from algebra to analysis. However, we cannot resist giving one last task that relates the function e^x to algebraic concepts:

Task 23.9.13. Let $f \in F[[x]]$ be the power series from Task 23.9.12. Prove that f is transcendental over the field of rational functions $F(x)$; thus we say that e^x is a *transcendental function*. Suggestion: assume for a contradiction that f is algebraic over $F(x)$; clear denominators from the irreducible polynomial of f over $F(x)$; and take derivatives until the constant term disappears. You need to show that the remaining terms do not disappear to arrive at a contradiction.

23.10   Quadratic Probing

Prerequisites. Having completed Chapter 17 is recommended.

In computer science, especially within data science, one of the most basic problems is how to store and retrieve data quickly. Given a sequence of m memory locations (also called *addresses*) consecutively numbered $0, 1, 2, \ldots,$ $m - 1$, and assuming that each data object fits into a single memory address, how can we decide where to store the next object that comes in? How should we proceed to search for a requested object?

A very simple solution is to store each new object in the next available memory address, keeping track of this last used address. But if we do this, then searching for a requested object would become slower and slower, on average, as the number of stored objects grew; in the worst case, where the requested object is never found (because it was never stored in the first place), the search time is proportional to the total number of stored objects. It turns out that there are much better solutions!

One trusted solution to the data storage-and-retrieval problem is known as *hashing*. In this approach, each incoming object is first converted into a number by applying a fixed function h, called the hash function, which has as ouput values the integers between 0 and $m - 1$. Then we try to store each incoming object x at memory address $h(x)$. Of course, if someone requests object x, we also know to look for x at the same location, $h(x)$. In this situation, the entire collection of memory from 0 to $m - 1$ is called a *hash table*. For later use, we set $R_m = \{0, 1, \ldots, m - 1\}$. (We note that our discussion is slightly simplified, and is relevant when we only care whether or not a given object exists in the hash table. In practice, we instead designate a fixed portion of the data objects as a *key*, and apply the hash function to the key. For example, the key of a "Person" object could be the name of the Person; then given only a name, we could search for the corresponding Person.)

The problem with hashing is that two different objects may have the same hash value; this is known as a *collision*, since both objects want to be stored at the same location. To resolve collisions, the usual solution is to find a memory location other than the preferred location $h(x)$ in which to store x. Specifically, we search for an unused location, starting with $h(x)$, and proceeding in a fixed order through the other memory locations until an unused address is found. In this procedure, we use the word *probe* instead of "search," and the order in which we probe for an unused address is given by a function

$$p \; : \; R_m^2 \to R_m,$$

called a *probing function*. To say $p(i, j) = k$ means that when an object x has hash value $h(x) = i$ and we are performing the jth probe for x, then we should look at address k. We note that the initial attempt at probing has the value $j = 0$, consistent with the choice of the set R_m. (Computer scientists, like logicians, begin counting at zero.) We require $p(i, 0) = i$ so that a probe for an object x will always start at location $i = h(x)$, as mentioned earlier; for this reason, we sometimes refer to $h(x)$ as the *home position* of object x.

Example 23.102. Suppose that $m = 5$ and we choose the probing function p given by the formula $p(i, j) = (i + j)\%5$, where the $\%$ operator means to take the remainder modulo 5. More precisely, $\%$ is a function from $\mathbf{Z} \times (\mathbf{Z} - \{0\})$ to $\mathbf{Z}$, with $a\%b = c$ where $c \equiv a \pmod{b}$ and $c \in R_b$. (So $\%$ is "almost" a binary operation on $\mathbf{Z}$, but we do not allow $b = 0$.) Further suppose that an object x has hash value $h(x) = 3$. Then when we attempt to add x to our hash table, we will probe for an empty memory location in the order $3, 4, 0, 1, 2$. We may not need to use this entire probing sequence, since we stop as soon as we find an empty location. If we are asked to retrieve x from the hash table, we will use the same probing sequence, but this time, instead of looking for an empty address, we look for an address which contains x. More generally, for any positive integer m, we can use the function $p(i, j) = (i + j)\%m$ for probing; this is called the *linear probing function*.

It turns out that linear probing creates some problems which can be largely avoided using alternative probing functions. Namely (for those readers who are interested), if we assume that the hash values of incoming objects are independent and uniformly distributed over the set R_m, then linear probing tends to create large "clusters" of consecutive occupied locations in the hash table; this will increase the variance of the time needed to store and retrieve objects. Therefore, we define a slightly more complicated type of probing function.

Definition 23.103. For a real number t, we write $\lfloor t \rfloor$ for the unique integer a such that $a \leq t < a+1$, and we call a the *floor* of t. The function $p : R_m^2 \to R_m$ given by the formula

$$p(i, j) = (i + (-1)^{j+1} \cdot \lfloor (j+1)/2 \rfloor^2)\%m$$

is called the *quadratic probing function* for a hash table of size m.

Although quadratic probing helps to reduce clustering, it introduces another problem: a quadratic probing function may not be surjective! A lack of surjectivity prevents quadratic probing from reaching some memory locations, which could cause a data storage attempt to fail even if the hash table is not full.

Task 23.10.1. Show by computation that the quadratic probing function is surjective when $m = 7$, but not surjective when $m = 5$.

Our main concern in this project is to prove the following result, which tells us which table sizes m give surjective quadratic probing functions.

Theorem 23.104 (Theorem Q). *Let m be a positive integer. Then the quadratic probing function is surjective for a hash table of size m iff one of the following conditions is true:*
    (1) $m = 1$ or $m = 2$;

(2) *m is prime and m % 4 = 3;*
(3) *m is even, m/2 is prime, and (m/2) % 4 = 3.*

The proof of Theorem Q requires some knowledge of how arithmetic works modulo m, or, even better, some knowledge of abstract algebra. For this reason, a complete proof is usually not given in computer science courses.

In what follows, we endow the set R_m with addition and multiplication modulo m, so that R_m is isomorphic to the quotient ring $\mathbf{Z}/m\mathbf{Z}$.

To understand when quadratic probing is surjective for a hash table of size m, we will break the question down into simpler questions ("reduce" the problem). Our first task just establishes the translation between probing a hash table on the one hand, and modular algebra on the other hand.

Task 23.10.2. Prove that quadratic probing is surjective for a hash table of size m iff for every possible home position $e \in R_m$, we have

$$\{e + j^2, e - j^2 \ : \ 0 \le j \le m - 1\} = R_m.$$

Since the set of perfect squares in R_m is evidently important here, we give it a name:

Notation 23.105. We set $S_m = \{j^2 \ : \ j \in R_m\}$. Naturally enough, we interpret $-S_m$ to mean $\{-a \ : \ a \in S_m\} = \{-j^2 \ : \ j \in R_m\}$.

The next reduction amounts to saying that quadratic probing modulo m works for every home position iff it works for home position 0. The idea of the proof is simply to shift everything by the value of the home position.

Task 23.10.3. Prove that quadratic probing is surjective for a hash table of size m iff $S_m \cup -S_m = R_m$.

Since S_m and $-S_m$ are subsets of R_m, which is a finite set of size m, our question amounts to asking whether the size of $S_m \cup -S_m$ is m; that is, we have proved the following lemma:

Lemma 23.106. *Quadratic probing is surjective for a hash table of size m iff $|S_m \cup -S_m| = m$.*

It may not come as a surprise that $-S_m$ has the same size as S_m, but it does require a proof:

Task 23.10.4. Prove that $|S_m| = |-S_m|$.

Combining the previous results enables us to complete the following task.

Task 23.10.5. Prove that quadratic probing reaches every cell of a hash table of size m iff $|S_m| = (m + |S_m \cap -S_m|)/2$; and if this happens, then $|S_m| \ge (m + 1)/2$. Suggestion: Use the formula $|A \cup B| = |A| + |B| - |A \cap B|$, which is true for any finite sets A and B.

Since $|R_m| = m$, Task 23.10.5 tells us that for quadratic probing to be surjective, over half of the elements of R_m must be squares. From your familiarity with ordinary perfect squares in $\mathbf{Z}$, this may seem like a bizarre condition: very few ordinary integers are perfect squares, and the density of the squares approaches 0 as numbers get bigger. But modulo m, it turns out that about half of all numbers are perfect squares when m is prime, and the density of squares diminishes as the number of prime factors of m increases. This will emerge from the following work.

Task 23.10.6. Let p be an odd prime, and let r be a positive integer. Let $n = p^r$, and let $d_n = |S_n|/n$, the "density" of the squares modulo n. Prove that we have $d_n = 1/2 + 1/(2n)$ if $r = 1$, and $d_n \leq 1/2$ if $r > 1$.

Suggestion: Consider the squaring function $q : R_n \to S_n$ given by the formula $q(a) = a^2$, and look at the number of pre-images in R_n of an element $a \in S_n$.

We also want to know what happens for powers of 2:

Task 23.10.7. Let $n = 2^r$ where r is a positive integer, and let $d_n = |S_n|/n$. Prove that $d_n = 1$ when $r = 1$, and $d_n \leq 1/2$ when $r > 1$.

We next use the Chinese Remainder Theorem (Theorem 17.11) to understand the density of the squares modulo m for arbitrary positive integers m.

Task 23.10.8 (Densities Multiply). Let m be a positive integer, and factor $m = \prod_{i=1}^k p_i^{r_i}$ with p_i distinct primes and r_i natural numbers. Let $f : R_m \to \prod_{i=1}^k \mathbf{Z}/p_i^{r_i}\mathbf{Z}$ be the map of Theorem 17.11. Prove that we have $f(S_m) = S_{p_1^{r_1}} \times \cdots \times S_{p_k^{r_k}}$ and $d_m = \prod_{i=1}^k d_{p_i^{r_i}}$.

We are now ready to prove a large piece of the forward direction of Theorem Q.

Task 23.10.9. Let m be a positive integer. Prove that if quadratic probing is surjective for a hash table of size m, then $m = 1$ or $m = 2$ or m is an odd prime or m is twice an odd prime.

We are getting close to our final result, but we still need to understand the relationship of S_p and $-S_p$ when p is an odd prime. Specifically, to apply Task 23.10.5 at "full strength," we want to know how much these two sets overlap.

Task 23.10.10. Let p be an odd prime. Prove that if $-1 \in S_p$, then $S_p = -S_p$; but if $-1 \notin S_p$, then $S_p \cap -S_p = \{0\}$.

It remains to determine for which odd primes p is -1 a square modulo p.

Task 23.10.11. Let p be an odd prime. Prove that $-1 \in S_p$ iff $p \equiv 1 \pmod 4$. Suggestion: Use the results of Exercises 12.15, 21.18, 18.4, and 8.11 on the group $R_p^\times$.

Finally we have enough tools to reach our goal.

Task 23.10.12. Prove Theorem Q. Suggestion: First verify the cases $m = 1$ and $m = 2$. By Task 23.10.9, we may suppose that $m = p$ or $m = 2p$ where p is an odd prime. Show that $-S_{2p} = -(S_2 \times S_p) = S_2 \times -S_p$. Thus when we move from $m = p$ to $m = 2p$ in the equation of Task 23.10.5, both sides of the equation are just multiplied by 2. So we see that quadratic probing works for

$m = p$ iff it works for $m = 2p$. Thus we only need to consider the case $m = p$. Consider the two cases $p \equiv 1 \pmod 4$ and $p \equiv 3 \pmod 4$ separately.

23.11 Euclidean Domains

Prerequisites. Having completed through Chapter 20 is recommended.

A Principal Ideal Domain (PID) is one of the nicest types of ring we have seen. Two examples of PIDs in our experience so far are the ring of integers $\mathbf{Z}$ and the ring of polynomials $F[x]$ where F is any field. In both of these cases, in order to prove that every ideal is principal, we used a "quotient-remainder" formula

$$g = qf + r$$

where g and f are given ring elements, and q and r are ring elements to be determined, where the "remainder" r is in some sense smaller than f. In the case of $\mathbf{Z}$, "smaller" has the usual sense of $<$ with integers, while in the case of $F[x]$, "smaller" meant having smaller degree. Thus in both cases, we were able to associate an ordinary number $s(a)$ with each (non-zero) ring element a, and show that the quotient-remainder formula always has a solution with $s(r) < s(f)$ (provided f and r are not 0). In this project, we will use this idea to define a new type of ring, called a *Euclidean Domain*; then we will study one particular example of a Euclidean Domain and use it to prove a theorem from classical number theory.

Definition 23.107. A *Euclidean Domain* is a domain R such that there exists a function $s : R - (0) \to \mathbf{N}$ with the following property: for all $f, g \in R - (0)$ there exist $q, r \in R$ such that both
  (i) $g = qf + r$, and
  (ii) either $r = 0$ or $s(r) < s(f)$.

Remark 23.108. The function s is called a *Euclidean function*; it is not formally part of a Euclidean Domain, and in fact, any given Euclidean Domain will have more than one Euclidean function.

Task 23.11.1. Verify that both $\mathbf{Z}$ and $F[x]$, where F is any field, are Euclidean Domains.

The next tasks establish the place of Euclidean Domains in the hierarchy of rings.

Task 23.11.2. Prove that every Euclidean Domain is a PID. You should find that the main idea from the proof of Theorem 14.22 can be applied here.

Task 23.11.3. Prove that if F is a field, then *every* function $s : F - (0) \to \mathbf{N}$ is a Euclidean function for F; hence F is a Euclidean domain.

Next, we will study a specific subring of **C**. This ring is named after the mathematician Carl F. Gauss.

Definition 23.109. The set of *Gaussian integers* is the set

$$\mathcal{G} = \{a + bi \ : \ a, b \in \mathbf{Z}\}$$

where i is as usual the imaginary unit in the field of complex numbers **C**.

Task 23.11.4. Show that we have $\mathcal{G} = \mathbf{Z}[i]$, the ring of integers adjoin i. Conclude that $\mathcal{G}$ is a subring of the field $\mathbf{Q}[i]$, and that every element of $\mathcal{G}$ has a *unique* representation in the form $a + bi$ with $a, b \in \mathbf{Z}$.

Our next goal is to prove that $\mathcal{G}$ is a Euclidean domain. First we look at the geometry of how $\mathcal{G}$ sits inside of **C**. Figure 23.9 shows a small portion of $\mathcal{G}$ near 0.

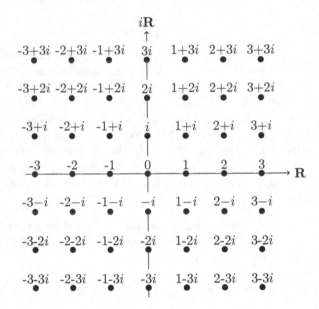

FIGURE 23.9: Geometric View of the set of Gaussian integers, $\mathcal{G}$

Looking at Figure 23.9, it seems natural to view $\mathcal{G}$ as the set of corner points (vertices) of the solid unit squares

$$\{x + yi \ : \ a \leq x \leq a + 1, b \leq y \leq b + 1\}$$

where $a, b \in \mathbf{Z}$. In fact, this point of view will be useful in proving that $\mathcal{G}$ is a Euclidean Domain; we now explore it further. These squares almost—but not quite—partition the complex plane **C**; the trouble is that their edges overlap. To fix this problem, we let

$$S_{a,b} = \{x + yi \ : \ a \leq x < a + 1, b \leq y < b + 1\}$$

for $a, b \in \mathbf{Z}$.

Task 23.11.5. Prove that the sets $S_{a,b}$ partition **C**.

What is the algebraic significance of Task 23.11.5? By adding an appropriate element of $\mathcal{G}$, we can move from any point of **C** to the square $S_{0,0}$, in a unique way. The following task expresses this idea precisely.

Task 23.11.6. Note that $(\mathcal{G}, +) \leq (\mathbf{C}, +)$, and prove that $S_{0,0}$ is a complete set of left coset representatives of $(\mathcal{G}, +)$ in $(\mathbf{C}, +)$.

Next we return to the definition of a Euclidean Domain. Observe that condition (i) of Definition 23.107 contains the expression qf, where f is given and q is an arbitrary ring element. We recognize this form as the general element of the principal ideal (f).

Task 23.11.7. Let R be a domain. Prove that R is a Euclidean Domain iff there is a function $s \; : \; R - (0) \to \mathbf{N}$ such that for all $f, g \in R - (0)$ there exists $r \in R$ with $g - r \in (f)$ and either $r = 0$ or $s(r) < s(f)$.

The result of Task 23.11.7 may be interpreted as saying that in a Euclidean Domain every ring element g is "close enough" to some element of (f), if we think of the function s as measuring some kind of distance. Therefore we will investigate the principal ideals $I = (f) = \mathcal{G}f$ of $\mathcal{G}$. To get an element of I, we compute $(x+iy) \cdot f$ with $x, y \in \mathbf{Z}$. Separating the real and complex components of the first factor, we recognize that $x \cdot f$ is a point on the line in **C** containing f and 0, namely, the line $\mathbf{R}f$. What about $iy \cdot f$? If the reader has not seen this before, you are in for a treat:

Task 23.11.8. Define a function $\mu_i \; : \; \mathbf{C} \to \mathbf{C}$ by the formula $z \mapsto iz$ for $z \in \mathbf{C}$. Prove that μ_i is a linear transformation of **C** considered as a vector space over **R**. Let $B = \{1, i\}$, the usual basis of **C** over **R**. Find the matrix of μ_i with respect to B, and use this to show that μ_i is the counterclockwise rotation by $90°$ about the origin in **C**.

Using Task 23.11.8, we see that the line $\mathbf{R}if$ is perpendicular to the line $\mathbf{R}f$. Furthermore, the points 0, f, if, and $f + if$ provide the corner points of a fundamental square with respect to (f) which plays the same role as $S_{0,0}$ does with respect to $\mathcal{G}$; see Figure 23.10.

Task 23.11.9. Let $f \in \mathcal{G} - (0)$, and let $I = (f) = \mathcal{G}f$. Let

$$S_f = \{(x + iy) \cdot f \; : \; 0 \leq x < 1, 0 \leq y < 1\} = S_{0,0} \cdot f.$$

Sketch the set S_f in Figure 23.10, and prove that S_f is a complete set of left coset representatives of $(I, +)$ in $(\mathbf{C}, +)$.

Our strategy to prove that $\mathcal{G}$ is a Euclidean Domain comes from understanding the geometry of (f). Recall that given an arbitrary $g \in \mathcal{G}$, we need to find an element of (f) which is "close" to g. In a general Euclidean Domain R, we may not have a geometric interpretation of R, and the meaning of "close" is only made clear when we can define the function s; but in the case of $\mathcal{G}$, we have a notion of distance in **C** which we will attempt to use in our definition of s.

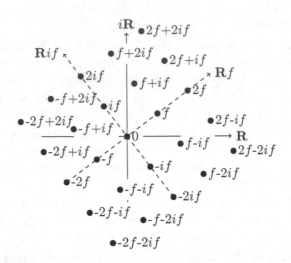

FIGURE 23.10: Geometric View of the principal ideal (f) of $\mathcal{G}$

Task 23.11.10. Define s : $\mathcal{G} \to \mathbf{N}$ by the formula $s(z) = |z|$, the usual absolute value of z as a complex number; that is, $s(a + bi) = \sqrt{a^2 + b^2}$. Note that we can interpret $|z|$ as the distance of z from 0. Let $f \in \mathcal{G} - (0)$, and let $g \in \mathcal{G}$. Use geometry (pictures!) to convince yourself that there exists an element $w \in (f)$ such that $s(g - w) < s(f)$. Then prove the existence of such a w rigorously, using algebra.

You have just proved:

Proposition 23.110. $\mathcal{G}$ *is a Euclidean Domain; the usual complex absolute value function* $s(z) = |z|$ *is a Euclidean function for* $\mathcal{G}$.

As a corollary of Proposition 23.110, we have that $\mathcal{G}$ is a PID, hence also a UFD—all because of a bit of geometry with squares! This unique factorization result has powerful consequences for the study of ordinary integers, as we shall see next.

Task 23.11.11. Define a function $\mathcal{N}$: $\mathcal{G} \to \mathcal{G}$ by the formula $\mathcal{N}$: $w \mapsto w \cdot \sigma(w)$, where σ is the complex conjugation function. Prove that the image of $\mathcal{N}$ is in $\mathbf{N}$. Use $\mathcal{N}$ to prove that if $a + bi$ divides c in $\mathcal{G}$ where $a, b, c \in \mathbf{Z}$, then $a^2 + b^2$ divides c^2 in $\mathbf{Z}$.

Task 23.11.12. Prove that we have $\mathcal{G}^{\times} = \{1, -1, i, -i\}$.

Task 23.11.13. Let $p \in \mathbf{Z}^+$ be an odd prime.
(a) Prove that the equation $a^2 + 1 \equiv 0 \pmod{p}$ has a solution $a \in \mathbf{Z}$ iff $p \equiv 1 \pmod 4$ (compare Task 23.10.11).
(b) Deduce that p is prime in $\mathcal{G}$ iff $p \equiv 3 \pmod 4$. Hint: $a^2 + 1$ factors in $\mathcal{G}$.

Task 23.11.14. Prove Theorem 23.111 (below).

Theorem 23.111. *All primes in* $\mathbf{Z}^+$ *which are congruent to 1 modulo 4 can be written as the sum of two squares. More precisely, let p be a positive prime integer such that $p \equiv 1 \pmod 4$. Then there exist unique positive integers a, b with $a < b$ such that $p = a^2 + b^2$.*

The quotient-remainder equation $g = qf + r$ lends itself to the study of the greatest common divisor $\gcd(f, g)$. The idea is that a ring element divides both f and g if and only if it divides both f and r. This happens in any UFD, even without a Euclidean function, as you will show next.

Task 23.11.15. Let R be a UFD, and let $f, g \in R-(0)$. Suppose that $g = qf+r$ where $q, r \in R$. Prove that we have $\gcd(f, g) = \gcd(r, f)$.

If we are in fact in a Euclidean Domain, then we can not only solve the quotient-remainder equation, but we can repeat the process until the remainder r becomes 0:

Algorithm 23.112 (Euclidean Algorithm). Let R be a Euclidean Domain with Euclidean function s. Let $f, g \in R - (0)$.

Step 1: Set $i = 0$, $a_i = g$, and $a_{i+1} = f$.

Step 2: Write $a_i = q_i a_{i+1} + a_{i+2}$ with $q_i, a_{i+2} \in R$ and either $a_{i+2} = 0$ or $s(a_{i+2}) < s(a_{i+1})$.

Step 3: Replace i by $i + 1$.

Step 4: If $a_{i+1} = 0$ then output a_i, else go to Step 2.

Task 23.11.16. Prove that the Euclidean Algorithm will always produce an output for any non-zero inputs f, g; that is, prove that the condition $a_{i+1} = 0$ will eventually become true. (This "halting property" is actually one of the requirements that an algorithm must satisfy by definition.) Furthermore, prove that the output value r has the property that $(r) = \gcd(f, g)$.

Remark 23.113. The Euclidean Algorithm is very fast—that is, requires very few steps to complete, including repetitions—in the case when $R = \mathbf{Z}$ and $s(n) = |n|$. Thus although it seems to be rather hard to factor a single integer, it is relatively easy to find the greatest common factor of two given integers. On the other hand, in the case of a polynomial ring over a field with s equal to the degree function, the Euclidean Algorithm can be much slower.

23.12 Resultants

Prerequisites. Having completed Project 23.11 is recommended before undertaking this project.

In this project we will apply a modified version of the Euclidean Algorithm to polynomials. The idea is that, unlike in the case of $\mathbf{Z}$, this algorithm has a fairly "predictable" effect on polynomials, so the final result in the most general case should be a well-defined object with some universal importance. We want the new algorithm to respect the coefficients of the original input polynomials so as never to require division by them. In order to carry out this plan, we will make two modifications to the Euclidean Algorithm: we will "slow it down" by changing Step 2 so that we are only allowing the "quotient" polynomials q_i to be single terms of the form cx^j; and we will cross-multiply appropriately at each iteration to avoid division.

Algorithm 23.114 (Modified Euclidean Algorithm). Given a field F and polynomials $f, g \in F[x] - (0)$.

Step 1: Set $i = 0$. Let $f_i = f$ and $g_i = g$ if $\deg(f) \leq \deg(g)$, else let $f_i = g$ and $g_i = f$.

Step 2(a): Write $g_i = bx^n +$ (terms of lower degree) and $f_i = ax^m +$ (terms of lower degree), where $a, b \in F^\times$.

Step 2(b): Let $h_i = ag_i - bx^{n-m}f_i$.

Step 2(c): If $\deg(h_i) < \deg(f_i)$, then let $g_{i+1} = f_i$ and $f_{i+1} = h_i$; else let $g_{i+1} = h_i$ and $f_{i+1} = f_i$.

Step 3: Replace i by $i + 1$.

Step 4: If $f_i = 0$ then output g_i, else go to Step 2(a).

Task 23.12.1. Apply the Modified Euclidean Algorithm (MEA) with $F = \mathbf{Q}$, $f = x^2 + 4x - 21$, and $g = 2x^3 - 4x^2 - 5x - 3$. You should find that the algorithm ends when $i = 4$ with the output value $g_4 = 85x - 255 = 85(x - 3)$. Then apply the original Euclidean Algorithm with $R = \mathbf{Q}[x]$ where s is the degree function, and the same f and g; compare the process and the results to the modified version.

Task 23.12.2. Let $f, g \in F[x] - (0)$, where F is a field and $\deg(g) \geq \deg(f)$.

(a) Note that because $F[x]$ is a PID, we always have $\gcd(f_i, g_i) = (f_i, g_i)$, the ideal of $F[x]$ generated by f_i and g_i, by Exercise 20.22. Prove that in every iteration of the Modified Euclidean Algorithm, throughout each part of Step 2 we will have $\deg(g_i) \geq \deg(f_i) \geq 0$; $\gcd(f_i, g_i) = \gcd(f_{i+1}, g_{i+1})$ in $F[x]$; and $\deg(f_{i+1}) + \deg(g_{i+1}) < \deg(f_i) + \deg(g_i)$ if $f_{i+1} \neq 0$. Conclude that the algorithm will eventually output a value g_k, and that we will have $\gcd(f, g) = (f, g) = (g_k)$ in $F[x]$.

(b) Suppose further that R is a domain with $R \leq F$ and that $f, g \in R[x]$. Prove that for each i such that f_i and g_i are defined, we have $f_i, g_i \in R[x]$. Also prove that f_i and g_i belong to the ideal I of $R[x]$ generated by f and g. In particular, conclude that the output value g_k is an element of I.

At this point, the reader may be wondering why we defined the Modified Euclidean Algorithm, since it seems to produce essentially the same result as the original Euclidean Algorithm but more slowly, and only in a special case! The key difference is to be found in part (b) of Task 23.12.2, which allows us to work within a chosen domain. By choosing this domain carefully, we will find next that the MEA can be used to solve systems of polynomial equations in more than one variable.

Example 23.115. Suppose that L is a field, and we have the system

$$\begin{cases} f(x,y) = 0 \\ g(x,y) = 0 \end{cases} \tag{23.16}$$

where $f, g \in L[x, y] - (0)$. To apply the MEA, we may view f and g as elements of $(L[y])[x] \leq (L(y))[x]$, where as usual $L(y)$ is the field of rational functions in the variable y with coefficients from L. Thus we will take $F = L(y)$, $R = L[y]$, and apply the MEA using the given f and g. Now (by Task 23.12.2) the MEA is guaranteed to output a value $r \in L[y]$ such that r is in the ideal of $L[x, y]$ generated by f and g. But if $(\alpha, \beta) \in L^2$ is in the solution set of System 23.16, then since $r = tf_0 + ug_0$ for some $t, u \in L[x, y]$, we must have $r(\alpha, \beta) = 0$. Note that r only involves the variable y, and not x; thus the Modified Euclidean Algorithm has *eliminated* a variable for us, creating a single-variable condition that is easier (we hope) to solve than the original system!

Task 23.12.3. Suppose that L is a field, $a, b, c \in L[y]$, and $f = x - a$, $g = x^2 + bx + c \in L[x, y]$. Let $F = L(y)$. Verify that the MEA applied to these inputs gives the output $r := a^2 + ab + c$ (provided that this quantity is not 0). It is clear from looking at this particular system that if $f = 0$ and $g = 0$ then we must have $r = 0$ too, but the point here is that the MEA produces that result automatically, and can produce similar results for more complicated systems whose solution is not obvious at first sight.

We have seen that, like the original Euclidean Algorithm, the MEA ends by giving us a generator of the gcd of the original two input polynomials, working over the field F. We also expect that the bigger this gcd is (measured by its degree with respect to x), the sooner the MEA procedure will end; the extreme case is when $f = g$, in which case the MEA ends with $i = 1$, giving output f. To gain more insight into the MEA, let us consider *generic* polynomials: that is, polynomials whose coefficients are actually variables, i.e., form an algebraically independent set (over the prime subfield of F, say; we are engaging in a "thought experiment," so we are not being very precise). For simplicity, we will also take our polynomials to be monic for now. We don't expect two generic monic polynomials f and g to have a non-trivial gcd, so the MEA should not end early, but rather should iterate the maximum number of times possible given the degrees of f and g. On the other hand, because the coefficients of f and g are independent variables, we can imagine evaluating them at any values we choose; furthermore, we expect this evaluation to be compatible with the MEA, in the sense that we can perform the evaluation at any stage before, during, or after running the MEA procedure and get the same result—except that, as discussed above, the MEA will end earlier as the degree of the gcd increases. This chain of speculation leads to a concrete conjecture: namely, that the output r of the MEA on two generic polynomials should evaluate to 0 when we substitute values for the coefficients that cause the gcd to be non-trivial. Looking in a splitting field for fg over F, where we can factor $f = \prod_{j=1}^{m}(x - \alpha_j)$ and $g = \prod_{j=1}^{n}(x - \beta_j)$, we expect that if we substitute $\alpha_i = \beta_j$ (for any given i and j) then r will become 0, for then f and g are forced to share a common factor.

Task 23.12.4. Let L be a field, and let $S = L[\alpha_1, \ldots, \alpha_m, \beta_1, \ldots, \beta_n]$ be a polynomial ring over L in $m + n$ variables. Let $r \in S$, and suppose that $\varepsilon(r) = 0$, where $\varepsilon : S \to S$ is the evaluation map over L which fixes all of the variables in S except for mapping β_1 to α_1. Prove that $\beta_1 - \alpha_1$ divides r in S.

The preceding suggests that we should have $\beta_j - \alpha_i$ divides r for all i, j. Unique factorization in the ring S would then force the product of all of these expressions to divide r. With this motivation, we make the following definition.

Definition 23.116. Let F be a field, and let $f, g \in F[x] - (0)$ be monic polynomials. Let K be a splitting field for fg over F, and write $f = \prod_{j=1}^{m}(x - \alpha_j)$ and $g = \prod_{j=1}^{n}(x - \beta_j)$, with $\alpha_j, \beta_j \in K$. The *resultant* of f and g (in the variable x) is

$$\mathrm{res}_x(f, g) := \prod_{i=1}^{m}\prod_{j=1}^{n}(\beta_j - \alpha_i). \tag{23.17}$$

(Recall that an empty product is defined to be 1; so if $f \in F$ or $g \in F$, then $\mathrm{res}_x(f, g) = 1$.)

Task 23.12.5. Let L be a field, and let $S = L[r_1, \ldots, r_{m+n}]$ be a polynomial ring in $m + n$ variables over L. Let $f = \prod_{j=1}^{m}(x - r_j), g = \prod_{j=m+1}^{m+n}(x - r_j) \in S[x]$. Write $f = x^m + \sum_{j=0}^{m-1} a_j x^j$ and $g = x^n + \sum_{j=0}^{n-1} b_j x^j$ with $a_j, b_j \in S$. Let $R = L[a_0, \ldots, a_{m-1}, b_0, \ldots, b_{n-1}] \leq S$, and let F be the field of fractions of R.

(a) Use the definition of the resultant together with Galois theory to prove that we have $\mathrm{res}_x(f, g) \in F$.

(b) Use Exercise 22.20 to prove that $\mathrm{res}_x(f, g) \in R$.

The next task presents equivalent formulas for the resultant of two monic polynomials, which will be used shortly as we investigate non-monic polynomials.

Task 23.12.6. Under the assumptions and notation of Definition 23.116, prove that we have

$$\mathrm{res}_x(f, g) = \prod_{j=1}^{n} f(\beta_j) = (-1)^{mn} \prod_{i=1}^{m} g(\alpha_i) = (-1)^{mn}\mathrm{res}_x(g, f). \tag{23.18}$$

We would like to extend the definition of $\mathrm{res}_x(f, g)$ to the non-monic case, and still avoid denominators, so that the resultant can be written as a polynomial in the coefficients of f and g, as in the monic case. The following task provides motivation for the way we will do this.

Task 23.12.7. Let $f, g \in F[x] - (0)$, where F is a field. Let K be a splitting field for fg over F, and write $f = a \cdot \prod_{j=1}^{m}(x - \alpha_j)$ and $g = b \cdot \prod_{j=1}^{n}(x - \beta_j)$ with $a, b \in F^\times$ and $\alpha_j, \beta_j \in K$. Prove that we have $\prod_{j=1}^{n} f(\beta_j) = a^n \cdot \prod_{i=1}^{m} \prod_{j=1}^{n}(\beta_j - \alpha_i)$ and $(-1)^{mn} \prod_{i=1}^{m} g(\alpha_i) = b^m \cdot \prod_{i=1}^{m} \prod_{j=1}^{n}(\beta_j - \alpha_i)$.

In order to get a formula for a general resultant which avoids denominators, Task 23.12.7 suggests that we should multiply the original resultant formula by $a^n b^m$.

Definition 23.117 (General Resultants). Let F be a field, and let $f, g \in F[x] - (0)$ be polynomials, not necessarily monic. Let K be a splitting field of fg over F, and write $f = a \cdot \prod_{j=1}^{m}(x - \alpha_j)$ and $g = b \cdot \prod_{j=1}^{n}(x - \beta_j)$, with $\alpha_j, \beta_j \in K$. The *resultant* of f and g (in the variable x) is

$$\operatorname{res}_x(f, g) := a^n b^m \cdot \prod_{i=1}^{m} \prod_{j=1}^{n} (\beta_j - \alpha_i). \tag{23.19}$$

Task 23.12.8. With the notation of Definition 23.117, prove that we have

$$\operatorname{res}_x(f, g) = b^m \cdot \prod_{j=1}^{n} f(\beta_j) = (-1)^{mn} \cdot a^n \cdot \prod_{i=1}^{m} g(\alpha_i) = (-1)^{mn} \operatorname{res}_x(g, f).$$

Task 23.12.9. Let F be a field, and let $f, g \in F[x] - (0)$ be polynomials. Write $f = \sum_{j=0}^{m} a_j x^j$ and $g = \sum_{j=0}^{n} b_j x^j$ with $a_j, b_j \in F^\times$. Let R be the image of $\mathbf{Z}$ under the characteristic map $\chi : \mathbf{Z} \to F$. Prove that $\operatorname{res}_x(f, g)$ is an element of the ring $S := R[a_0, \ldots, a_m, b_0, \ldots, b_n]$. Suggestion: Consider $\operatorname{res}_x(f/a_m, g/b_n)$; use Task 23.12.5.

Remark 23.118. Let $\mu(f, g)$ denote the output of the MEA with input f and g. The exact relationship between $\mu(f, g)$ and $\operatorname{res}_x(f, g)$ deserves to be explored, but we leave this to the reader. We note that it can be shown that if $f, g \subset R[x]$ for some domain $R \leq F$, then $\operatorname{res}_x(f, g)$ is in the ideal of $R[x]$ generated by f and g, just as was shown for $\mu(f, g)$ in Task 23.12.2.

23.13 Perfect Numbers and Lucas's Test

Prerequisites. It is recommended to have completed through Chapter 21 before starting this project.

In this project we explore a type of ordinary integer known as a "perfect number," which has held people's interest since ancient times. Then we present a remarkable test from the nineteenth century which allows us to determine enormously large even perfect numbers extremely quickly. Finally, we work out a proof that this test does always give the correct answer.

The idea of a perfect number is that it equals the sum of its "parts," if we interpret *parts* appropriately. We want to say that a "part" of a whole number means a factor of that number; but we must exclude a number from being part of itself, or else the sum will almost always be too big. This leads to the following definitions.

Definition 23.119. Let $n \in \mathbf{Z}^+$. A *proper factor* of n is a positive integer f such that f divides n in $\mathbf{Z}$ and $f \neq n$.

Definition 23.120. Let $n \in \mathbf{Z}$ with $n \geq 2$. Then n is *perfect* if n is equal to the sum of its proper factors.

Example 23.121. The sum of the proper factors of a prime integer is always 1. The sum of the proper factors of 4 is 3, and the sum of the proper factors of 6 is $1 + 2 + 3 = 6$. Therefore, 6 is the smallest perfect number.

Task 23.13.1. Verify that 28 is the second-smallest perfect number.

Even though we excluded n itself when summing the factors of n, it is mathematically more natural to include n; this leads to simpler formulas, as we shall see.

Definition 23.122. Define a function F on the domain $\mathbf{Z}^+$ by letting $F(n)$ be the set of all positive integer factors of n; that is,

$$F(n) = \{f \in \mathbf{Z} \: : \: 1 \leq f \leq n \text{ and } f \text{ divides } n\}.$$

Define a function $\sigma \: : \: \mathbf{Z}^+ \to \mathbf{Z}^+$ by the formula

$$\sigma(n) = \sum_{f \in F(n)} f.$$

The relationship of the function σ to perfect numbers is clear; the reader should verify the following lemma.

Lemma 23.123. *Let n be an integer with $n \geq 2$. Then n is perfect iff $\sigma(n) = 2n$.* $\square$

Next we develop a formula for σ.

Task 23.13.2. Let m and n be relatively prime positive integers. Define a function $\psi \: : \: F(m) \times F(n) \to \mathbf{Z}^+$ by the formula $\psi(f, g) \mapsto f \cdot g$. Prove that the image of ψ is the set $F(mn)$, and that ψ is a bijection from $F(m) \times F(n)$ to $F(mn)$. Use this result to prove Lemma 23.124 below.

Lemma 23.124. *If m and n are relatively prime positive integers, then we have $\sigma(mn) = \sigma(m)\sigma(n)$.* $\square$

Definition 23.125. Let $f \: : \: \mathbf{Z}^+ \to \mathbf{Z}^+$. Then f is called *multiplicative* if we have $f(a \cdot b) = f(a) \cdot f(b)$ for all a and b which are relatively prime.

Task 23.13.3. Define a function $\rho \: : \: \mathbf{Z}^+ \to \mathbf{Z}^+$ by the formula $\rho(n) = \sigma(n)/n$. Prove that ρ is multiplicative.

Task 23.13.4. Use the result of Lemma 23.124 inductively to prove that if $n = \prod_{i=1}^{r} p_i^{e_i}$ where the p_i are distinct positive primes and $e_i \in \mathbf{Z}^+$, then we have $\sigma(n) = \prod_{i=1}^{r} \sigma(p_i^{e_i})$.

Task 23.13.5. Establish that if p is a positive prime integer and $e \in \mathbf{Z}^+$, then we have $\sigma(p^e) = (p^{e+1} - 1)/(p - 1)$. (Hint: Recall the formula for the sum of a finite geometric series.) Use this together with the previous results to prove Lemma 23.126 below.

Lemma 23.126. *Let $n \in \mathbf{Z}$ with $n \geq 2$, and let $n = \prod_{i=1}^{r} p_i^{e_i}$ be the prime factorization of n into powers of distinct primes. Then we have*

$$\sigma(n) = \prod_{i=1}^{r} \frac{p_i^{e_i+1} - 1}{p_i - 1}.$$

From now on, we focus on even perfect numbers. As of this writing, it is still unknown whether there are any odd perfect numbers. On the other hand, even perfect numbers can be shown to have a very special form, of which several dozen examples are known. To reach this form, we first establish some properties of the function ρ.

Task 23.13.6. Prove that if $n \in \mathbf{Z}^+$ and p is a positive prime integer which divides n, then we have $\rho(n) \geq 1 + 1/p$, with equality if and only if $n = p$.

Task 23.13.7. Show that if $e \in \mathbf{Z}^+$, then we have $2/\rho(2^e) = 1 + 1/(2^{e+1} - 1)$.

Task 23.13.8. Suppose that n is an even perfect number. Write $n = 2^e \cdot c$ where c is odd. Use the above results to prove that we have $p \geq 2^{e+1} - 1$ for any prime p which divides c. On the other hand, prove that we also have $2^{e+1} - 1$ divides c. Conclude that $2^{e+1} - 1$ must be prime.

Definition 23.127. A *Mersenne prime* is a prime integer of the form $2^j - 1$, where $j \in \mathbf{Z}^+$.

Task 23.13.9. Prove that if $j \in \mathbf{Z}^+$ and $2^j - 1$ is prime, then j must be prime. Suggestion: Prove the contrapositive, using the fact that $x - 1$ divides $x^a - 1$ in $\mathbf{Z}[x]$ for any $a \in \mathbf{Z}^+$.

With this terminology in place, we now know that every even perfect number n is divisible by a Mersenne prime; and in fact, this Mersenne prime must be equal to $2^j - 1$ where 2^{j-1} exactly divides n. It is natural to investigate the simplest such case, when these two numbers form the entire prime factorization of n; it turns out that we have already struck gold:

Task 23.13.10. Let $j \in \mathbf{Z}^+$, and suppose that $2^j - 1$ is a Mersenne prime. Prove that $2^{j-1} \cdot (2^j - 1)$ is perfect.

Task 23.13.11. Use the previous results to prove Proposition 23.128 below.

Proposition 23.128. *Let p be a positive prime integer such that $2^p - 1$ is prime. Then $2^{p-1} \cdot (2^p - 1)$ is perfect. Furthermore, every even perfect number is of this form.* □

We next present the remarkably simple and fast test that forms the second part of this project's title. Instead of our usual approach of motivating results before stating them, we will start by stating the result, and then try to "reverse-engineer" it to see why it is correct. The test takes an odd prime p as its input, and is supposed to determine whether the number $2^p - 1$ is prime.

Algorithm 23.129 (Lucas's Test). Let p be a positive odd prime integer. Let $q = 2^p - 1$, and let $r = 4$. Repeat the following step $p - 2$ times:
Replace r by $r^2 - 2$ (modulo q).

   If the final value of r is 0 modulo q, then output *true* (q is prime); otherwise, output *false* (q is not prime).

Example 23.130. Let $p = 3$, so $q = 2^3 - 1 = 7$. Then we only need to perform the main step of Lucas's Test once; we compute $4^2 - 2 = 14 \equiv 0 \pmod{7}$, and output *true*. Indeed, 7 is prime, so the test works in this case.

Task 23.13.12. Use Lucas's Test to determine (by hand) whether $2^5 - 1$ is prime. If you know how, then write a computer program to determine all of the Mersenne primes up to $2^{97} - 1$ with Lucas's Test (you may want to use an arbitrary-precision integer type for these computations); compare the running time to that of "trial division," where we simply loop through all possible factors to determine primality.

   Until now, this project has only used elementary number theory; but to prove the correctness of Lucas's Test, we shall need to use some abstract algebra, and also learn a bit more about the properties of numbers.

   There is only one "independent variable" in Lucas's Test, namely the prime p. Sometimes in order to understand a very specific process, it is best to consider a more general situation. In this case, there are not many things to generalize, and we choose to generalize the number 4 into a variable, x. With this change, it's not hard to see that Lucas's Test creates a *polynomial*: the first iteration turns x into $x^2 - 2$; next we get $(x^2 - 2)^2 - 2$; etc. But where should we say that the coefficients of these polynomials live? Since we are told to compute modulo q, it would be natural to make $\mathbf{Z}/q\mathbf{Z}$ our coefficient ring; in the case of most interest to us, namely when q is prime, this ring is a field. But we must somehow also deal with the case when q is not prime, in which case this ring is not even a domain—much worse! After considering the problem for some time, and with a bit of experience, we choose to define these polynomials over a general field:

Definition 23.131. Let K be a field. Let $f_0 = x \in K[x]$, and for $j \in \mathbf{Z}^+$, define f_j recursively by the formula $f_j = f_{j-1}^2 - 2 \in K[x]$.

Task 23.13.13. Let $n \in \mathbf{Z}^+$. Prove by induction that we have

$$f_n = f_1 \circ (f_1 \circ (\cdots \circ f_1) \cdots), \qquad (23.20)$$

where f_1 appears n times in this composition; see Exercise 14.15 for this notation and some of its properties. Conclude using Lemma 3.32 that we have $f_n = f_j \circ f_k$ for any positive integers j, k with $j + k = n$.

Task 23.13.14. Let p be a positive prime integer, and $q = 2^p - 1$. Suppose that q is prime, and let $f = f_{p-2} \in K[x]$ where $K = \mathbf{Z}/q\mathbf{Z}$. Prove that Lucas's Test ouputs "true" if and only if 4 is a root of f in K (where we interpret 4 as $4 \cdot 1_K = 4 + q\mathbf{Z}$).

From Task 23.13.14, we realize that we should study the roots of the polynomials f_j. Since we are more familiar with the field $\mathbf{R}$ of real numbers than with most other fields, let's first take $K = \mathbf{R}$.

Task 23.13.15. Verify that with $K = \mathbf{R}$, the roots of f_1 are $\pm\sqrt{2}$, the roots of f_2 are $\pm\sqrt{2 \pm \sqrt{2}}$, and in general, the roots of f_n are

$$\pm\sqrt{2 \pm \sqrt{2 \pm \sqrt{\cdots \pm \sqrt{2}}}}, \tag{23.21}$$

where there are n nested square root signs, and we can choose all of the $+$ and $-$ signs independently.

To make further progress, we appeal to our knowledge of trigonometric formulas. Specifically, if the reader has ever taken a 45°-angle and bisected it repeatedly, then the expression for the roots of f_n should look somewhat familiar.

Task 23.13.16. Let $K = \mathbf{R}$ and let $n \in \mathbf{Z}^+$. Prove by induction that the roots of f_n are the real numbers $2\cos(\ell\pi/2^{n+1})$ where $\ell \in \mathbf{Z}$, $1 \le \ell \le 2^{n+2}$, and $\ell \equiv 1 \pmod 4$. Suggestion: Use the double-angle formula $\cos(2\alpha) = 2\cos^2(\alpha) - 1$, and Task 23.13.13.

It is fair to ask at this point whether trigonometric formulas can possibly help us to understand the roots of f_n over a finite field, since that is our main interest. The answer is Yes, if we remember Equations 21.2 and 21.4, which allow us to write a cosine in terms of roots of 1.

Task 23.13.17. Let $n, \ell \in \mathbf{Z}^+$. Show that we have

$$2\cos(\ell\pi/2^{n+1}) = \zeta^\ell + \zeta^{-\ell}$$

where $\zeta = \exp(2\pi/2^{n+2})$, a root of unity in $\mathbf{C}^\times$ of order 2^{n+2}.

The point is that roots of 1 are algebraic, so we may be able to translate these ideas over to finite fields by looking for elements which satisfy the same relations. Specifically, to get roots of f_n, we need 2^{n+2}-th roots of 1; that is, elements ζ such that $\zeta^{2^{n+2}} = 1$. To make sure that these roots are distinct, we exclude the case $\mathrm{char}(K) = 2$.

Task 23.13.18. Let K be a field with $\mathrm{char}(K) \neq 2$, let $n \in \mathbf{Z}^+$, and suppose that the polynomial $g_n := x^{2^{n+2}} - 1$ splits completely in $K[x]$.
   (a) Prove that g_n has 2^{n+2} distinct roots in K.
   (b) Prove by induction on n that we have $g_n = (x-1)(x+1)(x^2+1)(x^4+1)(x^8+1)\cdots(x^{2^{n+1}}+1)$.

(c) For an integer j with $0 \leq j \leq n+1$, prove that the roots of $x^{2^j} + 1$ in K are precisely the roots of g_n which have order 2^{j+1} in $K^\times$, and that there are exactly 2^j such roots.

(d) Let ζ be a root of unity of order 2^{n+2} in $K^\times$. Prove by induction on n that $\zeta + \zeta^{-1}$ is root of f_n in K.

(e) Consider the function $\omega : K^\times \to K$ given by the formula $a \mapsto a + a^{-1}$. Prove that for any $b \in K$, we have $|\omega^{-1}(b)| \leq 2$. Conclude that f_n splits completely in K with the 2^n distinct roots

$$\zeta + \zeta^{-1} : \zeta \text{ has order } 2^{n+2} \text{ in } K^\times.$$

Let us focus next on the case when $q = 2^p - 1$ is a Mersenne prime. The previous results suggest that we should look for an extension field K of $F := \mathbf{Z}/q\mathbf{Z}$ which contains roots of unity of order 2^p. Recalling the results we know about finite fields, we look for an extension field of degree n over F such that $q^n - 1$ is divisible by 2^p; because of the form of q, we don't have to look far:

Task 23.13.19. Let p be an odd prime, and suppose that $q := 2^p - 1$ is prime. Let $F := \mathbf{Z}/q\mathbf{Z}$.

(a) Prove that the polynomial $h := x^2 + 1$ is irreducible in $F[x]$.

(b) Let K be a splitting field for h over F, and let i be a root of h in K. Prove that every element of K can be written uniquely in the form $a + bi$ with $a, b \in F$.

(c) Prove that the polynomial $x^{2^p} - 1$ splits completely over K.

Task 23.13.20. Assume the hypotheses and notation of Task 23.13.19. Prove that the following are equivalent:

(i) Lucas's Test returns *true* with input p.

(ii) 4 is a root of f_{p-2} in F.

(iii) There is a solution to the equation $\zeta + \zeta^{-1} = 4$ where $\zeta \in K$ and ζ has order 2^p in $K^\times$.

Now we would like a way to tell whether an element $a + bi$ of $K^\times$ has order 2^p. We know from our earlier results that this is equivalent to saying that $(a + bi)^{2^{p-1}} = -1$. But certainly a necessary condition for this to happen is that $(a + bi)^{2^p} = 1$.

Task 23.13.21. Assume the hypotheses and notation of Task 23.13.19. Justify the following statements: K is Galois over F of degree 2, so $G := \mathrm{Gal}(K/F)$ has order 2. The map $a + bi \mapsto a - bi$ (with $a, b \in F$) is a non-trivial element of G, but so is the map $w \mapsto w^q$. Therefore, these two functions are equal, and we have $(a + bi)^q = a - bi$ for all $a, b \in F$. (We could also deduce this from the Binomial Theorem in characteristic q.) So $a + bi$ (where a, b are in F and are not both 0) has order dividing 2^p iff $(a + bi)^{q+1} = 1$ iff $(a + bi)(a + bi)^q = 1$ iff $(a + bi)(a - bi) = 1$ iff $a^2 + b^2 = 1$ iff $(a + bi)^{-1} = a - bi$.

Task 23.13.22. Assume the hypotheses and notation of Task 23.13.19. Prove that the following are equivalent:

(i) There is a solution to the equation $\zeta + \zeta^{-1} = 4$ where $\zeta \in K$ and ζ has order dividing 2^p in $K^\times$.

(ii) $-3 = b^2$ for some $b \in \mathbf{Z}/q\mathbf{Z}$.

We would now like to answer a particular question in elementary number theory, namely, for which primes q is -3 a perfect square modulo q? We adopt the following notation:

Notation 23.132. Let q be a prime integer and let $F = \mathbf{Z}/q\mathbf{Z}$. Then S_q denotes the set of perfect squares in $F^\times$. That is, $S_q := \{a^2 \; : \; a \in F^\times\}$.

Task 23.13.23. For each prime integer q starting with $q = 5$, perform computations to decide whether -3 is a perfect square modulo q. Repeat until you have a conjecture that describes such primes in a way that is easy to test. A partial answer is given in the following proposition, so try not to look ahead in the text before finishing this task.

Proposition 23.133. *Let q be a prime integer with $q > 3$. If -3 is a perfect square modulo q, then $q \equiv 1 \pmod 3$.*

Task 23.13.24. Prove Proposition 23.133 by induction on q using the following steps. First establish the base case, $q = 5$. Inductively suppose the result is true for all primes strictly between 3 and q. Let $F = \mathbf{Z}/q\mathbf{Z}$. Suppose that $-3 \in S_q$, and show that $a^2 + 3 = cq$ for some $a, c \in \mathbf{Z}$ where c is odd and $1 \le c < q$. Then break into two sub-cases according to whether or not 3 divides c, and use the inductive hypothesis to get the desired conclusion.

Task 23.13.25. Prove Proposition 23.134 below. (You can use the same basic technique that worked to prove Proposition 23.133.)

Proposition 23.134. *Let q be a prime integer with $q > 3$. If 3 is a perfect square modulo q, then $q \equiv 1 \pmod{12}$ or $q \equiv -1 \pmod{12}$.*

Task 23.13.26. Let q be a positive odd prime integer. Prove that we have $-1 \in S_q$ iff $q \equiv 1 \pmod 4$. Then prove that the index of S_q in $(\mathbf{Z}/q\mathbf{Z})^\times$ is 2. Use these results to prove that if $q \equiv 3 \pmod 4$ then exactly one of 3 or -3 is in S_q. Conclude by proving Proposition 23.135 below.

Proposition 23.135. *Let q be a prime integer with $q > 3$ and $q \equiv 3 \pmod 4$. Then -3 is a perfect square modulo q iff $q \equiv 1 \pmod 3$.*

Remark 23.136. Proposition 23.134 is a special case of a result from the late eighteenth century first proved by Gauss and known as the Law of Quadratic Reciprocity (LQR). For distinct positive prime integers p and q, define

$$\left(\frac{q}{p}\right) = \begin{cases} 1, & \text{if } q \in S_p; \\ -1, & \text{otherwise.} \end{cases} \tag{23.22}$$

In full, LQR states that for odd primes p and q we have

$$\left(\frac{q}{p}\right) = (-1)^{\left(\frac{p-1}{2}\right)\left(\frac{q-1}{2}\right)} \left(\frac{p}{q}\right), \tag{23.23}$$

and in the case $q = 2$,

$$\left(\frac{2}{p}\right) = 1 \text{ iff } p \equiv \pm 1 \pmod{8}. \tag{23.24}$$

The fraction symbols enclosed in parentheses are called *Legendre symbols*, except for the fractions in the exponent, which are ordinary fractions(!).

We now know that if q is a Mersenne prime, then the r value in Lucas's Test will be 0 at the end of the jth iteration, where the order of $2 + bi$ is equal to 2^{j+2} and b is either of the two elements of $\mathbf{Z}/q\mathbf{Z}$ such that $b^2 = -3$. We know that $j + 2 \le p$, and we would like to show that equality holds so that Lucas's Test gives the correct output. We would really like to say that $2 + bi = 2 + \sqrt{-3}i = 2 + \sqrt{3}$ in some sense, to simplify our notation. We work on this next.

Task 23.13.27. Assume the hypotheses and notation of Task 23.13.19. Prove that $3 \notin S_q$, and deduce that the polynomial $t := x^2 - 3$ is irreducible in $F[x]$. Show that K is a splitting field for t over F, and let ρ be a root of t in K. Show that every element of K can be written uniquely in the form $a + b\rho$ with $a, b \in F$. Prove that $\omega := 2 + \rho$ has order dividing 2^p in $K^\times$. Use the properties of the cyclic group $K^\times$ to show that ω has order exactly 2^p iff ω is a perfect square but not a perfect fourth power in $K^\times$. Show that the two elements $\pm(a + a\rho)$, where $a = 2^{(p-1)/2}$, are the square roots of ω in K. But prove that neither of these two elements is a perfect square in K (you can reduce this to the assertion that $-1 \notin F$). Conclude that the order of ω in $K^\times$ is exactly 2^p.

We now know that Lucas's Test is correct when given p such that $q := 2^p - 1$ is prime. Let's consider the contrary case, when q is not prime. First we must reconsider our choices of the fields F and K. Since we still need fields, and we want Lucas's Test to have some relevance to these fields, it is natural to look at the field of order ℓ where ℓ is a prime factor of q; for we can recover an integer modulo ℓ from knowing that integer modulo q.

Task 23.13.28. Let p be a positive odd prime integer, and let $q = 2^p - 1$. Suppose that q is not prime, and let ℓ be the smallest prime factor of q in $\mathbf{N}$. Let $F = \mathbf{Z}/\ell\mathbf{Z}$. Let K be a splitting field for the polynomial $x^{2^p} - 1$ over F. Assume for a contradiction that Lucas's Test ouputs *true* with input p.

(a) Show that 4 is a root of f_n in K (actually even in F!).

(b) Show that there is an element $\zeta \in K^\times$ of order 2^p such that $\zeta + \zeta^{-1} = 4$. (See Task 23.13.18.)

(c) Prove that $\zeta \in \tilde{K}$ for some field $\tilde{K}$ such that $F \le \tilde{K} \le K$ and $[\tilde{K} : F] = 2$.

(d) Produce a contradiction using the results above. Hint: We must have $\ell < 2^{p/2}$ (why?).

This concludes the proof of the correctness of Lucas's Test. For such a simple test, there is a lot of math behind the scenes!

23.14   Modules

Prerequisites. Having completed through Chapter 18 is recommended. Having completed Project 23.4 is also recommended.

In this project we introduce a new algebraic category, possibly the most important one that we have not studied yet. To discover this category, let us reflect on our earlier work and look for similarities among some of the structures we studied. Recall that a *vector space* is an abelian group V together with an operation that lets us multiply a field element by an element of V, subject to a few natural-seeming axioms; in the definition, we isolated the field F in question, and said that V was a *vector space over F*. Thus we formed the category of vector spaces over a given field. Meanwhile, within the category of rings, we defined an *ideal* of a ring R to be an abelian group I inside $(R, +)$ which allows us to multiply an element of R by an element of I. The common theme is that in both cases we have an abelian group together with a ring (remember that a field is a very special type of ring!) and an appropriate "multiplication" operation involving an element of the ring and an element of the group. We can capture both of the above cases by generalizing the definition of a vector space over a field, replacing the field with an arbitrary ring. Although it is possible to use general rings here, we will restrict ourselves to commutative rings with 1.

Definition 23.137. Let R be a commutative ring with 1. Then a *module over R*, or *R-module*, is a triple $(M, +, \cdot)$, where $(M, +)$ is an abelian group and $\cdot : R \times M \to M$ is a function, satisfying the following axioms for all $a, b \in R$ and all $m, n \in M$:

(M1) :	$(a \cdot b) \cdot m$	$= \quad a \cdot (b \cdot m)$	[Associativity]
(M2(a)) :	$a \cdot (m + n)$	$= \quad a \cdot m + a \cdot n$	[Left Distributivity]
(M2(b)) :	$(a + b) \cdot m$	$= \quad a \cdot m + b \cdot m$	[Right Distributivity]
(M3) :	$1_R \cdot m$	$= \quad m$	[Unitary Law]

Task 23.14.1. Verify that if F is a field, then an F-module is the same thing as an F-vector space.

Task 23.14.2. Let $(R, +, \cdot)$ be a commutative ring with 1. Let $S \subseteq R$. Prove that $(S, +|_{S \times S}, \cdot|_{R \times S})$ is an R-module if and only if S is an ideal of R. In particular, conclude that R itself is naturally an R-module.

We assume for the remainder of this project that R always denotes a commutative ring with 1. Next, we would like to explore the natural questions we have learned to ask for any category: What are the sub-objects, quotient objects, and morphisms in the category of R-modules?

Definition 23.138. Let $(M, +_M, \cdot_M)$ be an R-module. A *submodule* of M is a triple $(N, +_N, \cdot_N)$ such that $N \subseteq M$, $+_N = +_M | N \times N$, and $\cdot_N = \cdot_M |_{R \times N}$.

We write $N \leq M$ to indicate that N is a submodule of M, and we write $N < M$ if N is a proper submodule of M.

We note that, as in other categories we have studied, there is only one possible choice for $+_N$ and $\cdot_N$ that could make N into a submodule of M.

Task 23.14.3. Let M be an R-module, and let $N \subseteq M$. Prove that $N \leq M$ iff N is non-empty and is closed under addition, negation, and multiplication by elements of R.

Task 23.14.4. Show that an ideal of R is the same thing as a submodule of R.

In the categories of groups and of rings, we needed to have special types of sub-objects in order to form quotients (namely, normal subgroups and ideals, respectively). The situation is simpler in the category of R-modules, as you will prove next: *any* submodule can serve as a "denominator."

Task 23.14.5. Let M be an R-module and suppose $N \leq M$. Let M/N denote the set of all left cosets of N in M under addition. Prove that M/N is an R-module under the operations

$$(m_1 + N) + (m_2 + N) = (m_1 + m_2) + N$$

and

$$r \cdot (m + N) = (rm) + N$$

for $m_1, m_2, m \in M$ and $r \in R$.

The definition of a homomorphism in the category of R-modules may not come as a surprise: we just need a function that commutes with both operations.

Definition 23.139. Let M and N be R-modules. An *R-module homomorphism* from M to N is a function $\sigma : M \to N$ such that for all $m_1, m_2, m \in M$ and all $r \in R$ we have

(**MH1**) $\sigma(m_1 + m_2) = \sigma(m_1) + \sigma(m_2)$ and

(**MH2**) $\sigma(r \cdot m) = r \cdot \sigma(m)$.

Definition 23.140. Let $\sigma : M \to N$ be an R-module homomorphism. The *kernel* of σ is its kernel as a homomorphism of additive groups; that is, $\ker(\sigma) := \sigma^{-1}(\{0_N\}) = \{m \in M : \sigma(m) = 0_N\}$.

Isomorphisms of R-modules fall into the familiar pattern:

Task 23.14.6. Prove that if $\sigma : M \to N$ is a bijective R-module homomorphism, then so is $\sigma^{-1} : N \to M$. Use this to help prove that a bijective R-module homomorphism satisfies the general definition of *isomorphism* given in Definition 23.34.

We have a good analog of the Fundamental Theorem of Ring Homomorphisms that works for modules:

Theorem 23.141 (Fundamental Theorem of Module Homomorphisms). *(i)*
Let M and N be R-modules with $N \leq M$. Then the natural map

$$\sigma \; : \; M \to M/N$$

given by

$$\sigma(m) = m + N$$

is a surjective R-module homomorphism, and $\ker(\sigma) = N$.
(ii) If $\tau \; : \; M \to L$ is any R-module homomorphism, then $\tau(M) \leq L$,
$\ker(\tau) \leq M$, *and we have $\tau(M) \cong M/\ker(\tau)$.*

Task 23.14.7. Prove Theorem 23.141.

As we continue to build the theory of modules over a commutative ring
with 1, the next step is to understand the submodule generated by a subset of
a module. (Even though a vector space is a special case of a module, we follow
tradition by using the term "generate" instead of "span.") It may not surprise
the reader that the result looks very similar to the corresponding results for
ideals and for subspaces.

Notation 23.142. Let M be an R-module, and let $S \subseteq M$. Then the set

$$\langle S \rangle := \{ a_1 s_1 + \cdots + a_k s_k \; : \; k \in \mathbf{N}, a_i \in R, s_i \in S \}$$

is called the *R-module generated by S*.

Lemma 23.143. *Let M be an R-module, and let $S \subseteq M$. Then $\langle S \rangle < M$,*
and $\langle S \rangle$ is the smallest submodule of M which contains S, in the sense that
if $N \leq M$ and $S \subseteq N$, then $\langle S \rangle \subseteq N$.

Task 23.14.8. Prove Lemma 23.143.

We often care about the number of generators needed to get a given module. Accordingly, we make the following definitions.

Definition 23.144. A module M is *cyclic* if there exists a singleton set which
generates M.

Definition 23.145. A module M is *finitely generated* if there exists a finite
set which generates M.

Next, we will study the idea of "freeness" in the category of R-modules.
Careful attention to our previous results about free objects, as highlighted in
Task 23.5.2, leads us to think of freeness as a universal property akin to that
of a direct product or coproduct. We will define a free R-module using this
idea:

Definition 23.146. Let M be an R-module, and let $S \subseteq M$. Then we say
that M is *free* on the set S if for every R-module N and every function
$f : S \to N$, there is a unique extension of f to an R-module homomorphism
$\sigma : M \to N$.

Definition 23.147. An R-module M is called *free* if there exists a non-empty set $S \subseteq M$ such that M is free on S.

In the following task, we will see that free modules are very special objects.

Task 23.14.9. Use Definition 23.146 to prove that a free R-module on a non-empty set S is uniquely determined up to isomorphism by the cardinality of S. Do not assume that S is finite.

Task 23.14.10. Let F be a field. Prove that an F-vector space V is finitely generated as an F-module if and only if V is finite-dimensional over F. Use this result to help prove that every finitely generated F-vector space is free.

It is accepted wisdom among algebraists that if you want to understand a ring, then you should study its modules. In order to see more deeply, we should study these modules *up to isomorphism*. We have seen in Exercise 13.12 that every n-dimensional vector space over a field F is isomorphic to F^n under componentwise operations. Combining this result with that of Task 23.14.10, we see that every finitely-generated module over a field is isomorphic to a finite number of copies of that field. We can interpret this as confirming that fields are indeed the nicest type of ring, in that their modules are particularly easy to describe, with only one parameter, the dimension.

At this point, we make another observation to connect module theory with our work in previous chapters. Recall from Lemma 17.16 and Exercise 17.12 that we were able to construct a coproduct in the category of abelian groups by taking Cartesian products and componentwise operations. We can use the same type of construction with modules.

Proposition 23.148. *Let* $\mathcal{C} = (M_i)_{i \in \mathcal{I}}$ *be an indexed collection of R-modules. Let* $S = \oplus_{i \in \mathcal{I}} M_i$ *denote the subset of* $\prod_{i \in \mathcal{I}} M_i$ *where all but finitely many of the components are 0. Define addition in S componentwise. Also define* $r \cdot (m_i)_{i \in \mathcal{I}}$ *to be* $(r \cdot m_i)_{i \in \mathcal{I}}$ *for* $r \in R$. *Then S is a coproduct of $\mathcal{C}$ in the category of R-modules, where we take the maps* $\tau_i \;:\; M_i \to S$ *to be the natural embeddings.*

Task 23.14.11. Prove Proposition 23.148.

Remark 23.149. Just as with abelian groups, we refer to the coproduct defined in Proposition 23.148 as a *direct sum*.

We know from Task 23.14.9 that free R-modules on a given set are essentially unique. But we still do not officially know whether free R-modules always exist. In fact they do, and the situation for vector spaces turns out to generalize.

Task 23.14.12. Prove that R is free (as an R-module) on the set $\{1\}$.

Task 23.14.13. Prove that every cyclic R-module is isomorphic to R/I for some ideal I of R.

Notice that we have a ring structure on R/I in addition to an R-module structure. These structures are compatible in the sense that for all $a, b \in R$

we have $a \cdot (b + I) = (a + I) \cdot (b + I)$, where the left-hand side of this equation is calculated in the R-module and the right-hand side in the ring structure; both calculations give the coset $(ab) + I$ of I in R.

Task 23.14.14. Let I be an ideal of R, and let $a \in R$. Suppose that $M \cong R/I$ as R-modules. Show that we have $a \in I$ iff for all $m \in M$, $a \cdot m = 0_M$. Thus, we can recover I from M.

Definition 23.150. Let M be an R-module. The *annihilator* of M in R is the set
$$\mathrm{Ann}_R(M) := \{a \in R \ : \ a \cdot m = 0 \text{ for all } m \in M\}.$$

Task 23.14.15. Let M be an R-module. Prove that $\mathrm{Ann}_R(M)$ is an ideal of R.

Task 23.14.16. Let $N \leq M$ be R-modules. Prove that $\mathrm{Ann}_R(M) \subseteq \mathrm{Ann}_R(N)$.

Task 23.14.17. Let M be a cyclic R-module. Prove that R is free iff $\mathrm{Ann}_R(M) = (0)$.

In the case of vector spaces, we have the notion of a basis, which can serve as a coordinate system; the reader may have suspected that for a module which is free on a set S, the set S plays the role of a basis. The next task confirms this.

Task 23.14.18. Let M be an R-module which is free on a set $S \subseteq R$. Prove that every element of M can be written uniquely as a finite R-linear combination of elements of S: that is, prove that for every $m \in M$, we can write $m = \sum_{a \in S} r_a \cdot a$ where $r_a \in R$, all but finitely many of the r_a are 0_R, and the r_a are uniquely determined by m.

Another place we have seen a similar unique sum decomposition is in the study of direct sums of abelian groups, with Proposition 17.20. The next task asks you to prove the corresponding result for modules.

Task 23.14.19. Let M be an R-module, and let $\mathfrak{C} = (M_i)_{i \in \mathcal{I}}$ be an indexed collection of submodules of M.

(a) Prove that the function $\sigma \ : \ \oplus_{i \in \mathcal{I}} M_i \to M$ given by the formula $(m_i)_{i \in \mathcal{I}} \mapsto \sum_{i \in \mathcal{I}} m_i$ is an R-module homomorphism.

(b) Let $N = \mathrm{Im}(\sigma)$. Prove that σ is injective iff every element of N can be written uniquely as a finite R-linear combination of elements of the modules M_i. Note: we naturally use the notation $N = \sum_{i \in \mathcal{I}} M_i$ in case $\mathcal{I}$ is finite.

In light of the preceding results, we expect there to be a connection between direct sums and freeness:

Task 23.14.20. Let $\mathfrak{C} = (M_i)_{i \in \mathcal{I}}$ be an indexed collection of free R-modules. Prove that $\oplus_{i \in \mathcal{I}} M_i$ is free. Conclude that, in particular, $\oplus_{i=1}^{n} R = R^n$ is free (on a set of size n) for any positive integer n.

Combining Tasks 23.14.20 and 23.14.9, we see that every free R-module on a set of size n is isomorphic to R^n. Now that we have a fair understanding of free R-modules, as well as of finitely generated modules over a field (which all turn out to be free), let us explore modules over $\mathbf{Z}$.

To say that M is a $\mathbf{Z}$-module means that M is an abelian group under an operation $+$, and that there is a way to multiply an integer n by an element of M which satisfies the module axioms. But we already know a way to compute $n \cdot m$ for any $n \in \mathbf{Z}$ and $m \in M$, by interpreting this product as exponential notation in the abelian group $(M, +)$.

Task 23.14.21. Let $(M, +)$ be an abelian group. Define $\cdot : \mathbf{Z} \times M \to M$ to be the exponential function of Definition 3.38 in additive notation. Verify that Theorem 3.41 and Lemma 3.44 imply that $(M, +, \cdot)$ is a $\mathbf{Z}$-module.

Thus, given any abelian group, there is a natural way to make it into a $\mathbf{Z}$-module. In the next task, you will show that this is the *only* way.

Task 23.14.22. Let R be a commutative ring with 1, and let M be an R-module. Let $\chi : \mathbf{Z} \to R$ be the characteristic map of Definition 16.21. Prove that for all $n \in \mathbf{Z}$ and all $m \in M$ we have $\chi(n) \cdot m = n \cdot m$, where the left-hand side is multiplication in the R-module M, and the right-hand side is the exponential function in the group $(M, +)$. Conclude that in the case $R = \mathbf{Z}$, there is at most one multiplication function which makes $(M, +)$ into an R-module.

The preceding results suggest that a $\mathbf{Z}$-module is essentially the same thing as an abelian group. Earlier in this project, we said that an F-module is the same thing as an F-vector space (where F is any field); and this is strictly true, because both types of structure are ordered triples which satisfy exactly the same axioms. But it is not strictly true that a $\mathbf{Z}$-module is the same thing as an abelian group, because while a $\mathbf{Z}$-module is an ordered triple $(M, +, \cdot)$, an abelian group is an ordered pair $(M, +)$—there is no way they can be equal! Yet we feel that there should be a nice way to bridge this difference. To do so, we will go a bit further into category theory by introducing the concept of a *functor*, which plays roughly the same role between two categories as a morphism does between two objects within the same category: a functor expresses a relationship between two entire categories.

Definition 23.151. Let $\mathcal{C}$ and $\mathcal{D}$ be two categories. A *functor* F from $\mathcal{C}$ to $\mathcal{D}$ is an assignment for each object A of $\mathcal{C}$ of an object FA of $\mathcal{D}$ and for each morphism $\sigma : A \to B$ in $\mathcal{C}$ of a morphism $F\sigma : FA \to FB$ in $\mathcal{D}$, such that:
    (F1): $F(\mathrm{id}_A) = \mathrm{id}_{F(A)}$ for every object A of $\mathcal{C}$, and
    (F2): $F(\tau \circ \sigma) = F(\tau) \circ F(\sigma)$ for any two morphisms σ and τ of $\mathcal{C}$ such that $\tau \circ \sigma$ is defined.

In case $\mathcal{C} = \mathcal{D}$, there is always at least one functor available, namely the identity functor:

Definition 23.152. The *identity functor* from $\mathcal{C}$ to $\mathcal{C}$ is the functor $\mathrm{id}_{\mathcal{C}}$ defined by $\mathrm{id}_{\mathcal{C}} A = A$ and $\mathrm{id}_{\mathcal{C}} \sigma = \sigma$ for all objects A of $\mathcal{C}$ and all morphisms σ in $\mathcal{C}$.

Task 23.14.23. Prove that functors preserve isomorphisms. More precisely, suppose that F is a functor from $\mathcal{C}$ to $\mathcal{D}$ and that $\sigma : A \to B$ is an isomorphism in $\mathcal{C}$. Prove that $F\sigma : FA \to FB$ is an isomorphism in $\mathcal{D}$.

The notion corresponding to isomorphism at the level of categories is called *equivalence*:

Definition 23.153. Let $\mathcal{C}$ and $\mathcal{D}$ be two categories. We say that these categories are *equivalent* if there is a functor F from $\mathcal{C}$ to $\mathcal{D}$ and a functor G from $\mathcal{D}$ to $\mathcal{C}$ with the following properties:

(E1a): for every object A of $\mathcal{C}$, there is an isomorphism $\delta_A : FGA \to A$;
(E1b): for every object X of $\mathcal{D}$, there is an isomorphism $\varepsilon_X : GFX \to X$;
(E2a): for every morphism $\sigma : A \to B$ in $\mathcal{C}$, the diagram

$$
\begin{array}{ccc}
FGA & \xrightarrow{FG\sigma} & FGB \\
\downarrow{\scriptstyle \delta_A} & & \downarrow{\scriptstyle \delta_B} \\
A & \xrightarrow{\quad \sigma \quad} & B
\end{array}
\tag{23.25}
$$

commutes; and

(E2b): for every morphism $\tau : X \to Y$ in $\mathcal{D}$, the diagram

$$
\begin{array}{ccc}
GFX & \xrightarrow{GF\tau} & GFY \\
\downarrow{\scriptstyle \varepsilon_X} & & \downarrow{\scriptstyle \varepsilon_Y} \\
X & \xrightarrow{\quad \tau \quad} & Y
\end{array}
\tag{23.26}
$$

commutes.

Task 23.14.24. Prove that the category of **Z**-modules is equivalent to the category of abelian groups.

We saw in Task 23.14.17 that the annihilator of a module is an obstacle to freeness. While the annihilator of M looks for solutions $r \in R$ to the equation $r \cdot m = 0$, we next look at solutions $m \in M$:

Definition 23.154. Let M be an R-module. Then $\mathrm{Tor}_R(M) = \{m \in M : r \cdot m = 0 \text{ for some } r \in R - (0)\}$. An element of $\mathrm{Tor}_R(M)$ is called a *torsion element* of M. We say that M is a *torsion module* if $\mathrm{Tor}_R(M) = M$, and that M is *torsion-free* if $\mathrm{Tor}_R(M) = \{0_M\}$.

Task 23.14.25. Find an example of a commutative ring R with $1 \neq 0$ and an R-module M such that $\mathrm{Tor}_R(M)$ is not a submodule of M. But prove that if R is a domain, then we have $\mathrm{Tor}_R(M) \leq M$.

We defined torsion as a way to investigate freeness; and Task 23.14.25 suggests that this idea will be most useful for modules over domains. The next task strengthens the relationship between torsion and freeness for domains:

Task 23.14.26. Let R be a domain. Prove that every free R-module is torsion-free.

Naturally, we want to know whether the converse is true.

Task 23.14.27. Let k be a field, and let $R = k[x, y]$. (Note that R is a UFD.) Consider the R-module $M = (x, y)$, the ideal of R generated by x and y. Prove that M is torsion-free but not free.

Instead of abandoning our question, let us work with an even nicer family of rings. We know that everything works fine over fields, but Task 23.14.27 shows that UFDs are not good enough. Therefore, we turn to PIDs.

Task 23.14.28. Let R be a PID, and let M be a non-trivial finitely-generated torsion-free R-module. Prove that M is free on S for any generating set $S \subseteq M$ of minimum size. Suggestion: Induct on the size of S, and use Exercise 20.22 to help.

Next, we will see that the finitely-generated condition really makes a difference for modules over a PID. (Compare this to modules over a field, where the Axiom of Choice enables us to show that every non-trivial vector space is free.)

Task 23.14.29. Let $M = (\mathbf{Q}, +)$, considered as a $\mathbf{Z}$-module. Prove that M is torsion-free but not free.

We have studied torsion modules and free modules, but these are two extreme cases; Task 23.14.27 shows that they are not exhaustive. A natural question is: If we "remove" the torsion part of a module, are we left with a torsion-free module? This question makes the most sense over a domain, where torsion modules are submodules:

Task 23.14.30. Let R be a domain, let M be an R-module, and let $T = \text{Tor}_R(M)$. Prove that M/T is torsion-free.

Task 23.14.31. Let M be a finitely-generated R-module and let $N \leq M$. Prove that M/N is also finitely-generated. Specifically, if $M = \langle S \rangle$, prove that $M/N = \langle \{a + N : a \in S\} \rangle$.

Over a PID, it turns out that every finitely generated module splits into its torsion part and a free part:

Task 23.14.32. Let R be a PID, and let M be a finitely-generated R-module. Let $T = \text{Tor}_R(M)$, and let $F = M/T$. Prove that F is free on some finite set $S \subseteq F$. Further, prove that M is a direct sum of T and F.

Task 23.14.32 gives an important structure theorem for finitely-generated modules over PIDs. Since $\mathbf{Z}$ is a PID, we will try to find guidance in the Fundamental Theorem of Finite Abelian Groups, Theorem 18.31. We know that a finitely-generated free $\mathbf{Z}$-module is isomorphic to $\mathbf{Z}^n$ for some $n \in \mathbf{Z}^+$, and hence is infinite as a set. Therefore, a finite abelian group must have a trivial free part, so must be pure torsion. Of course, a finite abelian group is also finitely-generated.

Task 23.14.33. Let $(M, +)$ be an abelian group. Prove that M is finite iff M is both finitely-generated and a torsion $\mathbf{Z}$-module.

Before we can apply Theorem 18.31 to the study of finitely-generated abelian groups, we need to know that the torsion part of a finitely-generated abelian group is also finitely-generated. The next task takes care of this.

Task 23.14.34. Let M be an R-module, and suppose that M is an internal direct sum of two submodules M_1 and M_2. Prove that M is finitely-generated iff both M_1 and M_2 are finitely-generated.

We are now ready to generalize Theorem 18.31.

Theorem 23.155 (Fundamental Theorem of Finitely-Generated Abelian Groups). *Every finitely-generated abelian group is a direct sum of cyclic groups.*

Task 23.14.35. Prove Theorem 23.155.

In order to generalize Theorem 23.155 even further, to finitely generated modules over any PID, we must reconsider the proof techniques used in Chapter 18. There we looked at the cardinality of an abelian group and its prime factorization. In the more general case, we cannot do this:

Task 23.14.36. Let $R = \mathbf{Q}[x]$ and $M = R/(x^2 + 1)$. Prove that R is a PID, M is a finitely-generated torsion R-module, and M is infinite.

The key to generalizing the results of Chapter 18 from $\mathbf{Z}$ to any PID lies in replacing cardinalities with annihilators.

Task 23.14.37. Let R be a PID, and let M be a non-zero finitely-generated torsion R-module. Prove that we have $\mathrm{Ann}_R(M) = (a)$ for some $a \in R - \{0\}$.

Next we define the generalization of a p-group for modules over a PID.

Definition 23.156. Let R be a PID, and let p be a non-zero prime ideal of R. A finitely generated R-module M is called a *p-torsion* module if $\mathrm{Ann}_R(M) = p^e$ for some $e \in \mathbf{Z}^+$.

We also want to define the product of an ideal with a module, in order to mimic the construction of kG in Notation 18.14.

Definition 23.157. Let M be an R-module and let A be an ideal of R. Then $A \cdot M$ denotes the submodule $\langle \{a \cdot m \ : \ a \in A \text{ and } m \in M\} \rangle$.

With the above definition in hand, we are ready to generalize Proposition 18.12:

Task 23.14.38. Let M be a non-zero finitely-generated torsion module over a PID R, and write $\mathrm{Ann}_R(M) = A = \prod_{i=1}^{r} p_i^{e_i}$ where p_i are distinct prime ideals of R and $e_i \in \mathbf{Z}^+$. For each i, let $A_i = \prod_{j \neq i} p_j^{e_j}$ and $M_i = A_i \cdot M$. Prove that M_i is a p_i-torsion module, and that $M \cong \oplus_{i=1}^{r} M_i$.

It remains to study the structure of p-torsion modules in order to generalize Proposition 18.23. The following task is a nice substitute for a consequence of Lagrange's Theorem in this setting.

Task 23.14.39. Prove that a submodule of a p-torsion module is also a p-torsion module.

Note that the ideal p^e gets larger as the exponent e gets smaller; this accounts for the use of the word *maximal* in the following proposition.

Proposition 23.158. *Let M be a non-zero finitely-generated p-torsion module over a PID R, where p is a non-zero prime ideal of R. Let $m_1, \ldots, m_n$ be generators of M as an R-module such that $\prod_{i=1}^{n} Ann_R(m_i)$ is maximal. Then the natural map $\sigma : \oplus_{i=1}^{n} \langle m_i \rangle \to M$ is an isomorphism.*
In particular, M is a direct sum of cyclic modules.

Task 23.14.40. Prove Proposition 23.158.

Putting our previous results together, we get:

Theorem 23.159. *Every finitely-generated module over a PID is a direct sum of cyclic modules.*

Task 23.14.41. Prove Theorem 23.159.

Bibliography

[1] Tom M. Apostol. *Introduction to Analytic Number Theory*. Springer, 1976.

[2] C. Bergstrom. Eigenfactor: measuring the value and prestige of scholarly journals. *College and Research Libraries News*, 68(5):314–316, 2007.

[3] David Eisenbud. *Commutative Algebra with a View Toward Algebraic Geometry*. Springer, 1995.

[4] Herbert B. Enderton. *Elements of Set Theory*. Academic Press, 1977.

[5] Joseph A. Gallian. *Contemporary Abstract Algebra*. Houghton Mifflin, sixth edition, 2006.

[6] Roe Goodman. Alice through Looking Glass after Looking Glass: The Mathematics of Mirrors and Kaleidoscopes. *The American Mathematical Monthly*, 111(4):281–298, 2004.

[7] L. J. Gray and D. G. Wilson. Nonnegative Factorization of Positive Semidefinite Nonnegative Matrices. *Linear Algebra and Its Applications*, 31:119–127, 1980.

[8] Richard Hammack. *Book of Proof*. Richard Hammack, third edition, 2018.

[9] Hubert Kiechle. *Theory of K-Loops*. Springer, 2002.

[10] Serge Lang. *Algebra*. Springer, third edition, 2002.

[11] Saunders Mac Lane. *Categories for the Working Mathematician*. Springer, second edition, 1998.

[12] Gregory H. Moore. *Zermelo's Axiom of Choice: Its Origins, Development, and Influence*. Dover, 2013.

[13] Joseph J. Rotman. *An Introduction to Homological Algebra*. Springer, 2009.

[14] L. V. Sabinin. On the gyrogroups of Ungar. *Communications of the Moscow Mathematical Society*, 50:1095–1096, 1995.

[15] Abraham A. Ungar. The holomorphic automorphism group of the complex disk. *Aequationes Mathematicae*, 47:240–254, 1994.

[16] Leonard M. Wapner. *The Pea and the Sun: A Mathematical Paradox.* CRC Press, 2005.

[17] Charles A. Weibel. *An Introduction to Homological Algebra.* Cambridge University Press, 1994.

Index

p-group, 200
p-Sylow subgroup, 217

abelian group, 24
abelianization of a group, 94
adjoining elements to a field, 160
algebraic
    closure, 299
    element, 155
    extension field, 155
algebraically closed field, 219, 265
algebraically independent set, 152
alternating group, 282
annihilator, 359
array notation for a permutation, 27
ascending chain condition, 230
associates (in a ring), 220
automorphism
    of groups, 86
    inner, 88
    of rings, 103
    as a symmetry, 88
automorphism group of a field
        extension, 170
Axiom of Choice, 10, 76, 135, 298

base field, 126
basis
    ordered, 304
    standard, 315
basis of a vector space, 130
    is a minimal spanning set, 132
    size invariance, 133
bijective function, 7
binary operation, 22
    associative, 22
    commutative, 23

identity element for, 23
    inverse of an element with
        respect to, 23
    as "multiplication", 22
Binomial Theorem, 184
block-diagonal, 325

cancellation, 33
    left, 288
Cartesian product, 4
category, 300
Cauchy's Theorem, 216
Cayley table, 26
Cayley's Theorem, 95
center
    of a group, 88
    of a ring, 111
centralizer of a group element, 45
chain of ideals, 122
change-of-basis formula, 305, 306
characteristic
    of a field (classification), 183
    polynomial of a square matrix,
        316
    of a ring, 175
    subgroup, 205
Chinese Remainder Theorem, 189
closure
    under conjugation, 89
    under a group operation, 39
    under inverses, 39
    under ordinary addition and
        multiplication, 19
    under scalar multiplication, 128
commutative
    binary operation, 23
    diagram, 59, 186, 190

367

Printed in the United States
by Baker & Taylor Publisher Services